PLACEBO AND PAIN

PLACEBO AND PAIN

FROM BENCH TO BEDSIDE

Edited by

LUANA COLLOCA
National Institutes of Health, Bethesda, MD, USA

MAGNE ARVE FLATEN
Department of Psychology, University of Tromsø, Trondheim, Norway

KARIN MEISSNER
Institute of Medical Psychology, Ludwig-Maximilians-University, Munich, Germany

AMSTERDAM • BOSTON • HEIDELBERG • LONDON
NEW YORK • OXFORD • PARIS • SAN DIEGO
SAN FRANCISCO • SINGAPORE • SYDNEY • TOKYO

Academic Press is an imprint of Elsevier

Academic Press is an imprint of Elsevier
32 Jamestown Road, London NW1 7BY, UK
225 Wyman Street, Waltham, MA 02451, USA
525 B Street, Suite 1800, San Diego, CA 92101-4495, USA

Copyright © 2013 Elsevier Inc. All rights reserved.
Portions of this book were prepared by U.S. government employees in connection with their official duties, and therefore copyright protection is not available in the United States for such portions of the book pursuant to 17 U.S.C. Section 105. (Preface, Chapter 15, Chapter 22)

No part of this publication may be reproduced, stored in a retrieval system or transmitted in any form or by any means electronic, mechanical, photocopying, recording or otherwise without the prior written permission of the publisher Permissions may be sought directly from Elsevier's Science & Technology Rights Department in Oxford, UK: phone (+44) (0) 1865 843830; fax (+44) (0) 1865 853333; email: permissions@elsevier.com. Alternatively, visit the Science and Technology Books website at www.elsevierdirect.com/rights for further information

Notice
No responsibility is assumed by the publisher for any injury and/or damage to persons or property as a matter of products liability, negligence or otherwise, or from any use or operation of any methods, products, instructions or ideas contained in the material herein. Because of rapid advances in the medical sciences, in particular, independent verification of diagnoses and drug dosages should be made

British Library Cataloguing-in-Publication Data
A catalogue record for this book is available from the British Library

Library of Congress Cataloging-in-Publication Data
A catalog record for this book is available from the Library of Congress

ISBN: 978-0-12-397928-5

For information on all Academic Press publications
visit our website at elsevierdirect.com

Typeset by TNQ Books and Journals Pvt Ltd
www.tnq.co.in

Contents

Preface ix
Contributors xi

1 Historical Aspects of Placebo Analgesia
DAMIEN FINNISS

Introduction 1
Definitions and Conceptualization 1
Placebos as Controls 2
Placebos as a Treatment 2
Placebo in the Early 20th Century 3
Placebo as more than Just an Experimental Control 4
The Emergence of the Study of Placebo Mechanisms 5
Using History to Further Explore Placebo Analgesia 7
References 7

2 Neurochemistry of Placebo Analgesia: Opioids, Cannabinoids and Cholecystokinin
FABRIZIO BENEDETTI, ELISA FRISALDI

Introduction 9
Some Types of Placebo Analgesia are Mediated by Endogenous Opioids 9
Endocannabinoids are Involved in Some Types of Placebo Analgesia 11
Nocebo Hyperalgesia is Mediated by Cholecystokinin 12
Conclusions 13
References 13

3 Placebo Analgesia in Rodents
JIAN-YOU GUO, JIN-YAN WANG, FEI LUO

Introduction 15
The Placebo Effect in Animals 15
Studying Placebo Analgesia in Animal Models 17
Dissection of Placebo Analgesia in Mice 17
Placebo Analgesia Affects the Behavioral Despair Tests in Mice 18
The Opioid Receptors Involved in the Placebo Response in Rats 20
The Pros and Cons of Studying Placebo in the Animal Model 20
Conclusion and Future Directions 22
Acknowledgments 22
References 22

4 Molecular Mechanisms of Placebo Responses in Humans
MARTA PECIÑA, JON-KAR ZUBIETA

Introduction 25
Placebo-Induced Activation of Regional Endogenous Opioid Neurotransmission 28
Dopaminergic Mechanisms in the Formation of Placebo Analgesic Effects 30
Theories of Placebo Analgesia and Placebo-Induced Activation of Regional Endogenous Opioid Neurotransmission 31
Personality Predictors of Placebo-Induced Activation of Regional Endogenous Opioid Neurotransmission 33
Conclusions 34
Acknowledgments 35
References 35

5 How does EEG Contribute to Our Understanding of the Placebo Response?
ANTHONY JONES, CHRISTOPHER BROWN, WAEL EL-DEREDY

Theoretical Models of Placebo Analgesia 37
EEG Measures of Pain and its Anticipation 38
Pain Anticipation and its Role in Pain Perception 38
EEG Studies of Placebo Analgesia 40
Conclusion 42
References 42

6 Spinal Mechanisms of Placebo Analgesia and Nocebo Hyperalgesia: Descending Inhibitory and Facilitatory Influences
SERGE MARCHAND

Introduction 45
Pain and Placebo have Dynamic Interactions 45
Facilitatory Mechanisms 46
Inhibitory Mechanisms 47
Conclusion 50
References 51

7 Spinal and Supraspinal Mechanisms of Placebo Analgesia
FALK EIPPERT, CHRISTIAN BÜCHEL

Introduction 53
The Anatomy of Descending Pain Control 53
Descending Control in Placebo Analgesia 56
Placebo Analgesia and the Spinal Cord 60
Conclusions and Open Questions 64
Acknowledgments 65
References 65

8 Positive and Negative Emotions and Placebo Analgesia
MAGNE ARVE FLATEN, PER M. ASLAKSEN, PETER S. LYBY

Emotion and Motivation 73
Reduction in Negative Emotions: Methodologic Issues and Empirical Studies 74
Individual Differences in Negative Emotions and the Effectiveness of Placebo Interventions on Pain 75
Negative Emotions Reduce the Effectiveness of Opioids 78
Placebo Analgesia, Emotions, and Opioid Activity 78
The Nocebo Response: Negative Placebo Effect or Separate Process? 78
Clinical Implications 79
Conclusion and Future Perspectives 79
References 80

9 Placing Placebo in Normal Brain Function with Neuroimaging
MARTIN INGVAR, PREDRAG PETROVIC, KARIN JENSEN

Acknowledgments 87
References 88

10 Brain Predictors of Individual Differences in Placebo Responding
LEONIE KOBAN, LUKA RUZIC, TOR D. WAGER

Brain Predictors of Individual Differences in Placebo Responding 89
Personality and Brain Predictors of Placebo Analgesia 89
Limitations of Studies on Individual Differences in PA 97
Solutions 99
How can Brain Imaging Studies Find Brain Predictors of PA? Recommendations and Conclusions 99
Acknowledgments 100
References 101

11 Placebo Responses, Antagonistic Responses, and Homeostasis
MAGNE ARVE FLATEN

Placebo Responses and Homeostasis 103
Theoretical Background 103
Classical Conditioning and Pain 104
Conditioning with Administration of Painkillers to Pain-Free Subjects 104
Conditioning with the Administration of Painkillers as the Unconditioned Stimulus to Individuals in Pain 105
Conditioning with Reduction in, or Absence of, Pain as the US 105
Conditioning with an Increase in Pain as the US 106
Active Placebo 111
Compensatory Responses and the Nocebo Effect 111
Summary and Conclusions 112
References 112

12 Placebo Analgesia, Nocebo Hyperalgesia, and Acupuncture
JIAN KONG, RANDY L. GOLLUB

Is Acupuncture a Form of Placebo Treatment? 116
Challenges and Issues in Placebo/Sham Acupuncture Studies 118
Subjective and Objective Measurements in Acupuncture and Placebo Studies 118
Contribution of Neuroimaging to Acupuncture and Placebo/Nocebo Response 119
Summary and Future Directions 123
References 124

13 The Relevance of Placebo and Nocebo Mechanisms for Analgesic Treatments
ULRIKE BINGEL

Placebo and Nocebo in Pain Treatments: Behavioral Evidence 127
Understanding the Neural Mechanisms Underlying the Effects of Expectation and Learning on Drug Efficacy 129
Modulating Expectations to Optimize Analgesic Outcome 132
Exploiting Learning Mechanisms to Optimize Analgesic Outcome 133
Future Aims and Challenges 133
Conclusion 134
References 135

14 How Placebo Responses are Formed: From Bench to Bedside
LUANA COLLOCA

Introduction 137
Instructional Learning 137
Associative Learning 139
Social Learning 143
Expectations 144
Evolutionary Principles Behind Placebo Analgesia 145
Conclusion 146
Conflicts of Interest 146
Acknowledgments 146
References 146

15 Methodologic Aspects of Placebo Research
MAGNE ARVE FLATEN, KARIN MEISSNER, LUANA COLLOCA

Methodology of Studies Investigating Placebo Analgesia and Nocebo Hyperalgesia 149
Induced Pain and Clinical Pain 150
Quantification of Pain 150
Response Bias 151
Design 151
Within-Subjects versus Between-Subjects Designs 152
The Pre-Test 153
Researchers' and Subjects' Perception of the Treatment Allocation 153
Single-Blind versus Double-Blind Designs 154

Induction of Placebo Analgesia by Classic Conditioning:
  Methodological Issues 154
Measurement of Expectations 155
Conclusion 155
Acknowledgment 155
References 156

16 Balanced Placebo Design, Active Placebos, and Other Design Features for Identifying, Minimizing and Characterizing the Placebo Response

PAUL ENCK, KATJA WEIMER, SIBYLLE KLOSTERHALFEN

Introduction 159
Minimize versus Maximize 159
The 'Additive Model' Assumptions 161
The Balanced Placebo Design 161
The Balanced Cross-Over Design 162
The 'Delayed Response' Test 163
Active Placebos 164
Effective Blinding 165
No-Treatment and Waiting-List Controls 166
The Free-Choice Paradigm 167
Ethics of Placebo Research 169
Summary 170
Acknowledgment 170
References 170

17 Psychological Processes that can Bias Responses to Placebo Treatment for Pain

BEN COLAGIURI, PETER F. LOVIBOND

Theoretical Model 175
Demand Characteristics 177
The Hawthorne Effect 178
Response Shift 179
Returning to the Theoretical Model 181
Importance of Objective Outcomes 181
Conclusions and Future Directions 181
References 182

18 Against 'Placebo.' The Case for Changing our Language, and for the Meaning Response

DANIEL E. MOERMAN

A Summary of the Argument 183
A Brief Review of the Data 184
Conclusions 187
References 187

19 Placebo Effects in Complementary and Alternative Medicine: The Self-Healing Response

HARALD WALACH

Background 189
Is Cam 'All Placebo'? A Note on Specificity and the Efficacy
  Paradox 191
Jerome D Frank's Model of General Healing Effects or Common
  Factors in Therapy 194
The Common Myth 194
The Ritual 195
Relationship and the Alleviation of Anxiety 196
Insignia of Power 198
Empowering Patients and Mobilizing Resources 198
Summing Up: The Specificity of Nonspecific Effects and the
  Elegance of Reducing Side-Effects by 'Placebo' 199
References 199

20 Conceptualizations and Magnitudes of Placebo Analgesia Effects Across Meta-Analyses and Experimental Studies

LENE VASE, GITTE LAUE PETERSEN

Introduction 203
Developments in the Conceptualizations and Definitions
  of Placebo Effects 203
Meta-Analyses of Placebo Analgesia Effects 205
Experimental Studies of Factors Influencing the Magnitude of
  Placebo Analgesia Effects 207
Current Status of Meta-Analyses of the Magnitude of Placebo
  Analgesia Effects 210
Acknowledgment 211
References 211

21 The Contribution of Desire, Expectation, and Reduced Negative Emotions to Placebo Anti-Hyperalgesia in Irritable Bowel Syndrome

DONALD D. PRICE, LENE VASE

Introduction 215
Evidence for Visceral and Somatic Hyperalgesia in IBS Patients 215
Visceral and Somatic Hyperalgesia is Dynamically Maintained by
  Tonic Peripheral Impulse Input 216
Animal Models of Hyperalgesia in IBS 218
Psychologic Contributions to Hyperalgesia and Anti-Hyperalgesia
  in IBS 218
Central Nervous System Modulation of Pain in IBS 222
Neurochemical Basis of Anti-Hyperalgesia in Placebo Anti-
  Hyperalgesic Mechanisms 223
A Synergistic Interaction between Peripheral Impulse Input and
  Central Facilitation? 224
Acknowledgment 224
References 225

22 The Wound that Heals: Placebo, Pain and Surgery

WAYNE B. JONAS, CINDY CRAWFORD, KARIN MEISSNER, LUANA COLLOCA

Background 227
Placebo and Brain Stimulation for the Treatment of Pain 230
Conclusions 231
References 232

23 What are the Best Placebo Interventions for the Treatment of Pain?
KARIN MEISSNER, KLAUS LINDE

Introduction 235
The Efficacy Paradox 235
Hypotheses from the Literature 236
Evidence from Direct Comparisons 237
Evidence from Indirect Comparisons 238
Discussion 239
Implications for Clinical Trial Methodology and Decision-Making 240
Conclusions and Future Directions 241
References 241

24 How Communication between Clinicians and Patients may Impact Pain Perception
ARNSTEIN FINSET

Introduction 243
The Impact of Expectancy in Clinical Studies 244
The Impact of Emotional Communication 249
Promoting Patient Involvement and Common Ground: The Patient-Centered Interview 251
Psychosocial Interventions in Pain Management 253
Discussion and Conclusion; Suggestions for Future Research 253
References 254

25 Nocebos in Daily Clinical Practice
BETTINA K. DOERING, WINFRIED RIEF

Introduction 257
Beliefs About Illnesses and Medications 258
Communicating a Diagnosis and Test Results 259
Initiating a Treatment 260
Treatment Implementation 262
The Role of Treatment Experience 263
Conclusions 264
References 265

26 The Potential of the Analgesic Placebo Effect in Clinical Practice – Recommendations for Pain Management
REGINE KLINGER, HERTA FLOR

Introduction 267
Placebo Responses in Patients 267
Comparison of Placebo Effects in Healthy Controls and Patients 269
Use of Placebo Effects in Clinical Practice 270
Placebo Analgesia: Interactions with Attitudes Towards Medication and Prior Experience 274
Summary 274
Acknowledgment 274
References 275

27 Placebo and Nocebo: Ethical Challenges and Solutions
LUANA COLLOCA, FRANKLIN G. MILLER

Introduction 277
Towards Placebos in Clinical Practice 278
Clinicians' Attitudes Towards Placebos 278
Patients' Attitudes Towards Placebos 279
Placebos and the Declaration of Helsinki 280
The Dilemma of Deception 280
The Impact of the Clinician–Patient Relationship 282
The Nocebo and its Implications for how Doctors Consult with their Patients 283
What Translational Research is Being Done, or Should be Done? 284
Acknowledgments 284
References 284

Index 287

Preface

It is with great enthusiasm that we present *Pain and Placebo*, a book outlining key aspects of research on placebo-induced modulation of experimental and clinical pain, explaining the most important mechanistic advances, while discussing the impact of these findings for clinical researchers and health practitioners.

This is the first book that offers a comprehensive view of the themes of pain, placebo and nocebo. Although parts of these topics have been presented in a few books, there was a compelling need for a book specifically devoted to pain, placebo and nocebo research encompassing the realms of both basic science and clinical viewpoints. For this reason, we present this compendium that explains the bases for placebo-induced modulation of pain, with an emphasis on clinical perspectives, and their implications for clinical trials and practices.

The first part of the book is devoted to the description of the mechanisms underlying placebo-induced mediation and modulation of pain. Our expectation is that these sections will be of particular interest to not only scientists in the fields of medicine, psychology, and other health sciences, but also scientists in basic disciplines such as neuroscience, physiology, pharmacology, and biochemistry. Conversely, the second part of the book has been structured to be a helpful tool for senior-level healthcare providers including, but not limited to, anesthesiologists, internists, neurologists, neurosurgeons, neuroscientists, psychologists, nurses, and palliative care providers.

Our vision for this prodigious undertaking could certainly not be accomplished in isolation, and hence, we are honored and privileged to have engaged scientists whose valuable contributions have advanced and elucidated these areas of research. These researchers have paved the way to several of the most relevant neurobiologic discoveries, and have led the discourse about challenges, controversies and potentials of these advances, while providing stimulating and innovative perspectives. Undoubtedly, we could not cover all topics or involve all researchers in the field in this volume. This field of research is growing at such a rapid rate as to make a complete treatise of placebo and pain impossible.

The Editors wish to thank all the authors for their contributions and acknowledge the support they have received from their funders. Luana Colloca would like to thank the Intramural Research Program of the National Center for Complementary and Alternative Medicine (NCCAM), the National Institute of Mental Health (NIMH), and the Department of Bioethics, Clinical Center at the National Institutes of Health for their valuable source of support for her activities. Magne Arve Flaten is grateful to the Bial Foundation and the Research Council of Norway for continuous support, and Karin Meissner thankfully acknowledges the German Ministry of Education and Research and the Schweizer-Arau Foundation for their support.

We trust that this book provides a valuable contribution in promoting future discoveries, providing material for critical discussions, and educating readers about this fascinating and promising field. Our ultimate hope is that by improving the general understanding of the mechanisms of placebo and pain, this book offers a step towards a path of relief for patients in pain.

Luana Colloca
Magne Arve Flaten
Karin Meissner

Contributors

Per M. Aslaksen Department of Psychology, University of Tromsø, Tromsø, Norway

Fabrizio Benedetti Department of Neuroscience, University of Turin Medical School, and National Institute of Neuroscience, Turin, Italy

Ulrike Bingel NeuroImage Nord, Department of Neurology, University Medical Center Hamburg-Eppendorf, Hamburg, Germany

Christopher Brown Human Pain Research Group, University of Manchester, Manchester, UK

Christian Büchel Department of Systems Neuroscience, University Medical Center Hamburg-Eppendorf, Hamburg, Germany

Ben Colagiuri School of Psychology, University of Sydney, Sydney, Australia

Luana Colloca National Center for Complementary and Alternative Medicine (NCCAM), National Institutes of Health, Bethesda, MD, USA, National Institute of Mental Health, National Institutes of Health, Bethesda, MD, USA, Department of Bioethics, Clinical Center, National Institutes of Health, Bethesda, MD, USA

Cindy Crawford Samueli Institute, Alexandria, VA, USA

Bettina K. Doering Department of Clinical Psychology & Psychotherapy, Philipps University Marburg, 35032 Marburg, Germany

Falk Eippert Centre for Functional Magnetic Resonance Imaging of the Brain (FMRIB), University of Oxford, Oxford, UK

Wael El-Deredy Department of Psychological Sciences, University of Manchester, Manchester, UK

Paul Enck Department of Psychosomatic Medicine, University Hospital Tübingen, Tübingen, Germany

Damien Finniss Pain Management Research Institute, University of Sydney Royal North Shore Hospital, Sydney, Australia; School of Rehabilitation Sciencess, Griffith University, Queensland, Australia

Arnstein Finset Department of Behavioral Sciences in Medicine, Institute of Basic Medical Sciences, Faculty of Medicine, University of Oslo, Oslo, Norway

Magne Arve Flaten Department of Psychology, Norwegian University of Science and Technology, Trondheim, Norway

Herta Flor Department of Cognitive and Clinical Neuroscience, Central Institute of Mental Health/Medical Faculty Mannheim, Heidelberg University, Mannheim, Germany

Elisa Frisaldi Department of Neuroscience, University of Turin Medical School, and National Institute of Neuroscience, Turin, Italy

Randy L. Gollub Psychiatry Department, Massachusetts General Hospital and Harvard Medical School, MA, USA

Jian-You Guo Key Laboratory of Mental Health, Institute of Psychology, Chinese Academy of Sciences, Beijing, P. R. China

Martin Ingvar Cognitive Neurophysiology Research Group, Stockholm Brain Institute, Osher Center for Integrative Medicine, Karolinska Institutet, Stockholm, Sweden

Karin Jensen Department of Psychiatry, Massachusetts General Hospital/Harvard Medical School, Charlestown, MA, USA

Wayne B. Jonas Samueli Institute, Alexandria, VA, USA

Anthony Jones Human Pain Research Group, University of Manchester, Manchester, UK

Regine Klinger Outpatient Clinic of Behavior Therapy, Department of Psychology, University of Hamburg, Hamburg, Germany

Sibylle Klosterhalfen Department of Psychosomatic Medicine, University Hospital Tübingen, Tübingen, Germany

Leonie Koban University of Colorado, Boulder CO, USA

Jian Kong Psychiatry Department, Massachusetts General Hospital and Harvard Medical School, MA, USA

Klaus Linde Institute of General Practice, Klinikum rechts der Isar, Technische Universität München, Munich, Germany

Peter F. Lovibond School of Psychology, University of New South Wales, Australia

Fei Luo Key Laboratory of Mental Health, Institute of Psychology, Chinese Academy of Sciences, Beijing, P. R. China

Peter S. Lyby Department of Psychology, University of Tromsø, Tromsø, Norway

Serge Marchand Université de Sherbrooke, Faculté de médecine Centre de recherche clinique Étienne-Le Bel du CHUS Sherbrooke, Québec, Canada

Karin Meissner Institute of Medical Psychology, Ludwig-Maximilians-University Munich, Munich, Germany

Franklin G. Miller Department of Bioethics, Clinical Center, National Institutes of Health, Bethesda, MD, USA

Daniel E. Moerman William E. Stirton Professor Emeritus of Anthropology, University of Michigan-Dearborn, MI, USA

Marta Peciña Department of Psychiatry and Molecular and Behavioral Neuroscience Institute, University of Michigan, Ann Arbor, MI, USA

Gitte Laue Petersen Danish Pain Research Center, Aarhus University Hospital, Aarhus, Denmark

Predrag Petrovic Cognitive Neurophysiology Research Group, Stockholm Brain Institute, Osher Center for Integrative Medicine, Karolinska Institutet, Stockholm, Sweden

Donald D. Price Division of Neuroscience, Department of Oral and Maxillofacial Surgery, University of Florida, Florida, USA

Winfried Rief Department of Clinical Psychology & Psychotherapy, Philipps University Marburg, 35032 Marburg, Germany

Luka Ruzic University of Colorado, Boulder CO, USA

Lene Vase Department of Psychology and Behavioural Sciences, School of Business and Social Sciences, Aarhus University, Aarhus, Denmark

Tor D. Wager University of Colorado, Boulder CO, USA

Harald Walach Institute of Transcultural Health Studies, European University Viadrina, Frankfurt (Oder), Germany

Jin-Yan Wang Key Laboratory of Mental Health, Institute of Psychology, Chinese Academy of Sciences, Beijing, P. R. China

Katja Weimer Department of Psychosomatic Medicine, University Hospital Tübingen, Tübingen, Germany

Jon-Kar Zubieta Department of Psychiatry and Molecular and Behavioral Neuroscience Institute, and Department of Radiology, University of Michigan, Ann Arbor, MI, USA

CHAPTER
1

Historical Aspects of Placebo Analgesia

Damien Finniss

Pain Management Research Institute, University of Sydney & Royal North Shore Hospital, Sydney, Australia; School of Rehabilitation Sciences, Griffith University, Queensland, Australia

INTRODUCTION

Placebos and placebo effects have been an interesting component of medicine and health care for hundreds of years. The topic area of 'placebo' is one that has been appealing to professionals from a wide range of backgrounds, ranging from clinical and laboratory fields to broader anthropologic and sociocultural disciplines. Inherent in this topic area lay a multitude of perspectives and approaches to understanding the meaning of placebo and how it might relate to health care and broader humanity. On this basis, it could be argued that the study of placebo is an important pillar in the practice of medicine.

Recently, significant attention has been given to a modern conceptualization of placebos and placebo effects in an attempt to progress understanding of the area. This has been particularly important in moving the field forward and advancing the study of placebo. However, it is important to appreciate a historical perspective as this provides an understanding of the origin of the word and the initial framework in which it was used. Much of the modern view of placebo is still shaped by its lengthy history.

DEFINITIONS AND CONCEPTUALIZATION

The word 'placebo' seems to have first appeared in a religious context around the time of the 13th century, in the setting of an early Latin translation of the Hebrew Bible.[1,2] It is believed that the word placebo (a Latin word) was actually the result of a mistranslation from the ancient Hebrew word 'ethalech,' which means 'I shall walk.'[3] In the opening phrase of the Vespers for the Dead (Psalm 116, 9th verse), the biblical phrase reads 'Placebo Domino in regione vivorum,' which translated means 'I shall walk before the Lord.'[4,5] However, this phrase seems to have been mistranslated to 'I will please the Lord,' with the word 'placebo' therefore meaning 'to please.'[6] This slight mistranslation was the foundation for the meaning and application of the word for many centuries.

By the late 1300s, the word placebo had taken on a disparaging meaning, where hired mourners were said to 'sing placebos' of false praise to flatter the dead.[2] Similarly, around the same time, in one of Geoffrey Chaucer's Canterbury tales (The Merchant's Tale), a character 'placebo' was depicted as a sycophant.[7] In another of Chaucher's tales (The Parson's Tale), flatterers were described as the Devil's chaplains, always singing placebo.[8] It seems clear that even in these early times, there were some subtle variations in the use and meaning of the word. Although the literal translation of the word was 'to please,' its actual use seemed to be more negative, and these early applications of placebo may well have shaped the direction of its use for many years to come.

The use of the word 'placebo' in a medical context seems to have been due to the work of a British physician, William Cullen.[9] In the year 1772, Cullen delivered a series of lectures in which he used the word placebo in the medical context. He conceptualized giving a placebo as an attempt to comfort or please a particular patient who had incurable disease, and wrote that '... *I did not trust much to it, but I gave it because it is necessary to give a medicine, and as what I call a placebo.*'[2] This is a particularly interesting paper in the history of placebo as it presented placebo in a somewhat positive way, identifying it as a useful tool for both the physician and for the patient. In fact, the notion that placebo may reduce a patient's symptoms (by pleasing them) during the course of an incurable disease introduces the idea that placebo effects may be meaningful in certain therapeutic contexts, particularly when symptomatic improvement is an important goal. This insightful paper may have been the impetus for the first documentation of the term 'placebo' in a medical dictionary in 1785 (Motherby's New Medical Dictionary).[8] In this dictionary, placebo was defined as 'a common placebo method or medicine'; this is not entirely in keeping with the original translation of the word, or even Cullen's initial use, but it nonetheless acknowledged 'placebo' as an entity in the medical setting. However, it was not much later (1811) that a clearer definition was provided in Hooper's Medical Dictionary, which defined a placebo as 'any medicine adapted more to please than benefit the patient,' and therefore this definition was more representative of the origin of the word.[8] Despite many minor modifications over the years, this has arguably remained the foundation of both the definition and meaning of placebo that has persisted to recent times.

PLACEBOS AS CONTROLS

At a similar time to the first use of the word in the medical literature, the notion of using a placebo as a control in the medical setting was presented. In 1784, what is believed to be the first placebo-controlled experiment was conducted by Benjamin Franklin and Antione Lavoisier.[10] The trial was conducted as a result of Louis XVI appointing a Royal Commission to investigate the work of Franz Anton Mesmer, who had claimed to have discovered a new curative technique. This technique was called 'mesmerism,' and was founded on the belief that humans had certain internal channels of fluid, a property Mesmer called 'animal magnetism,' and that targeting these fluid channels with therapy could alleviate many bodily symptoms. Initial results from the elaborate rituals of mesmerism suggested profound effects, and it was on this basis that the controlled trial was established.[10] In this trial, patients were exposed to genuine 'mesmerized' objects (as described by Mesmer) and objects which were reportedly mesmerized, but which had been secretly swapped and were simply a 'dummy.' As a significant number of patients would respond to both objects, the commission concluded that there was no scientific evidence of Mesmer's theories, and that any effects were due to a patient's 'imagination.'[10]

It was not long after that another placebo-controlled trial was conducted, this time in the setting of surgery. In 1799, a British physician by the name of John Haygarth decided to perform a trial on a treatment which involved surgical implantation of metallic rods ('Perkins Tractors').[11] It was believed that the metallic properties of the tractors were able to alleviate the symptoms of disease.[12] In this small trial of five patients, Haygarth implanted the rods in a blinded manner. He first implanted 'imitation' rods (made from wood), and four of the five patients gained relief. The procedure was then repeated with the genuine Perkins Tractors, with the same result. Haygarth quotes *'an important lesson in the physic is here to be learnt, the wonderful and powerful influence of the passions of the mind upon the state and disorder of the body.'*[7] Together, these trials were the first in the medical field to use a placebo in the setting of a control for the purposes of assessing the validity of a treatment. Furthermore, the conclusions of these trials would shape interpretation of placebo-controlled trials to the current day. Frankin, Lavoisier and Haygarth all came to similar conclusions as to a 'response' to the placebo treatment. First, this demonstrated a lack of evidence for the intervention in question. Second, response to placebo was a construct of the 'imagination,' and therefore was quite separate to an effect that existed in the body. Haygarth, however, did make an important link between mind and body, which seems to have been an advanced appreciation for the time.

PLACEBOS AS A TREATMENT

The assessment of placebo use as a treatment is somewhat complex as it depends on whether one assesses the validity of a given treatment, using a modern scientific framework, or attempts to understand the intention of the clinician in prescribing treatments. For the latter, one is particularly reliant on the literature, although it is entirely possible that, despite frequent use of placebos in routine clinical practice, this was not presented in the medical literature due to the negative connotations surrounding placebo use. Regardless of the approach to assess placebo use clinically, this has been a difficult task even for experienced anthropologists and historians.[13-15]

The analysis of 'prescientific' medicine using a current scientific framework has led some authors to propose the idea that the history of prescientific medicine may actually be the history of the placebo effect.[15] Such a view is constructed on the assessment of the many different treatments presented in the literature and in a variety of pharmacopias. For example, some remedies sighted in the London Pharmacopoeia include 'Usnea' (the moss from the skull of a victim of a violent death) and Gascoyne's powder (bezoar, amber, pearls, crabs' eyes, claws and coral).[14,16] Other common remedies included unicorn horn (usually from an elephant) and bezoar stones (allegedly formed from the tears of a deer bitten by a snake, but actually animal gallstones).[17] In fact, it has been estimated that there were at least 5000 ancient remedies with over 16 000 different prescriptions.[14]

Another way of assessing the use of placebo as a treatment has come with the study of Native Americans, who used an incredible number of medicinal plants in different prescriptions for many ailments. It has been assessed that over 219 different cultures of Native Americans used more than 2800 species of medicinal plants in over 25 000 ways.[15] However, this line of sociocultural research places less focus on assessment of scientific validity of these remedies (the actual content) than on their cultural meaning. To this extent, placebo was conceptualized more by the use of remedies as part of a symbolic healing ritual, one which modern medicine would identify as a placebo (by the so-called inert nature of the remedy).[18] However, it is important to keep in mind that so-called 'healing rituals' have not been isolated to specific cultures, with procedures such as 'Royal Touch' dating back many hundreds of years. This relatively common procedure involved laying one's hands on a patient as a treatment for illness, and was one of the most persistent methods of healing that extended into contemporary times.[17] Both qualitative assessment of the content of treatments, and assessment of the broader ritualistic nature of healing using a modern framework, suggest that, by modern definition, these practices resembled the use of placebo as a means of alleviating symptoms of disease.

The assessment of specific placebo use is also difficult and open to interpretation, particularly from a definitional perspective. Many of the above treatments may have been believed to be specific for the cure of disease, rather than simply a treatment to 'please' the patient (using the very original definition of a placebo). In other words, the intention of the clinician may not have been to prescribe a placebo, as he or she may have believed that the treatment was a cure for a disease, even if a placebo was actually given. Although the word placebo is not often used, there are cases where one may interpret the intention of the physician as being 'pleasing' rather than focused on curing the disease. For example, in 1807, Thomas Jefferson was recorded as stating what he termed 'pious fraud' when a successful physician reported that he used more bread pills, drops of coloured water and powders of hickory ashes than all other medicines combined.[7] Similarly, others have noted that part of the practice of medicine was to learn to give placebo, bread pills, subcutaneous water and other devices.[19] Although there is not a great amount of literature using such direct language about placebo use, it does suggest that the practice of using placebos with the specific intention of bringing comfort (and not fixing pathology) may have been quite widespread, although, as mentioned previously, the negative connotations surrounding placebo use may have resulted in a lack of desire to publish on the area.

PLACEBO IN THE EARLY 20TH CENTURY

Placebos continued to be used in the first half of the 20th century, primarily as controls in experiments. This was essentially a progression from the previously described trials some 100 years before. However, there had possibly been a shift in the interpretation of placebo effects in the setting of trials. This may have been shaped by work some years before, such as a small trial conducted by an American physician, Austin Flint, in 1863. In this study, a placebo was given (a diluted remedy) for the management of articular rheumatism. The conclusion was that the placebo would not actually alter the natural course of the disease, rather it would provide symptomatic relief as the disease progressed through its natural history.[7,20] In this instance, response to a placebo was not deemed to be in one's 'imagination,' or not a genuine response, rather it was seen as an observed response to the administration of a placebo treatment. Therefore, one could assess the symptomatic response to a placebo and compare it to the response to the intervention in question (or index intervention). The difference between symptomatic relief to placebo and to that of the index intervention represented the additional effects of the index treatment, which were presumably more specific to the disease process. It is therefore possible that such shifts in thinking may have shaped the introduction of blinded trials using placebos, several of which were conducted in the early 1900s.[20]

The use of placebos in the clinical trial setting increased in the early 20th century, with the first clinical blinded study conducted in 1913.[21] Soon after, several trials using what was called 'the blind method'[22] were conducted, some with numbers approaching 1000 subjects, such as the trial conducted by Adolf Bingel in 1918.[14] In this particular trial, researchers tested an 'antitoxin' and a placebo for the treatment of diphtheria. Bingel made a specific point of using colleagues who were blinded to the allocation as assessors of the treatment, underscoring his beliefs about the importance of controls and blinding in evaluation of treatments. After several years of limited use, controlled blinded trials were again reported, firstly in the setting of assessing different types of ether preparation[23] and then in the assessment of aminophylline for angina pectoris.[24] In fact, over this period of time, Gold and his colleagues introduced the 'placebo-controlled double-blind trial' to modern medicine, which is still in use today.[14] This method was made famous when it was used to study a treatment (Khellin) for angina pectoris in 1950.[25] It was felt that controlling the placebo effect in the study of a treatment was important for understanding the effect of the drug above the response to placebo administration. This paradigm was slowly adopted as standard for both research-funding bodies, and later extended to the approval for new drugs in several countries.

PLACEBO AS MORE THAN JUST AN EXPERIMENTAL CONTROL

Throughout the first half of the 20th century, there were limited publications on the clinical use of placebo. Nonetheless, there were papers such as the one written by Houston in 1938 which was titled 'The Doctor Himself as a Therapeutic Agent.'[26] Although not strictly reporting placebo use, this paper described the effect of the therapeutic relationship or environment on the patient, implying that factors other than the prescribed therapy were also powerful in patient outcomes. Papers such as this one started to evolve thinking about the various components of the therapeutic encounter and how this might result in improvements in patients' symptoms. It was not long after that Wolf and colleagues published a paper advocating the use of placebos in the clinical setting.[27] In this paper, the authors suggested that if a patient was 'pleased' by receiving a placebo, then this was in fact a positive response to treatment. This paper seems to have been the first paper that specifically used the word placebo in the title and advocated use in the routine clinical setting.

Further exploration of placebo in its own right was seen in a pioneering paper by Lasagna and colleagues in 1964. This paper investigated placebo analgesia in the setting of post-operative pain, with the goal of understanding placebo responses and placebo responders. Researchers studied 93 patients with post-operative pain and administered alternating doses of placebo and opioid analgesia (morphine). A clinically meaningful effect was defined as 50% reduction in pain, and multiple alternating doses were able to be given to reach this target. The repeated administration protocol permitted an evaluation of initial and consistent responses to placebo. Researchers found that only 14% of patients responded consistently to placebo, with 31% of patients consistently not responding. A large percentage (55%) of patients were defined as inconsistent responders. An additional qualitative analysis was performed, suggesting that placebo responders may have had more somatic symptoms, higher levels of anxiety, and a more positive view of the hospital compared to non-responders. This led the investigators to make some conclusions as to the different psychologic and behavioral traits seen between responders and non-responders. To this extent, this was a pioneering paper in investigating placebo analgesic responses in a clinical setting.

In the same year (1964), another landmark paper was published.[28] This paper assessed whether 'active placebo action' could be used to improve management of post-operative pain, and therefore represented one of the first attempts to use placebo to improve clinical outcomes. In this study of 97 patients, researchers investigated pain scores and total analgesic medication use in two groups of patients post-operatively. The first group was treated in a standard manner. The second group, or 'special care' group, received additional attention (both in content and time) and advice from the physician regarding management of post-operative pain. At this time, such an intervention was deemed to be 'a placebo' as any benefit was not attributable to a specific pharmacologic action of a medicine. Researchers found that the total analgesic drug dose was significantly lower in the 'special care' group than in the routine group, concluding that the physician interaction was an important component of a therapeutic outcome (Fig. 1.1). This was an important initial paper in linking the doctor–patient interaction with placebo effects and the ability to improve clinical practice.

Perhaps the most significant paper published in many years came in 1955, in that it drew attention to the results of the placebo group in placebo control trials, and, in doing so, estimated the power of 'the placebo effect.'[29]

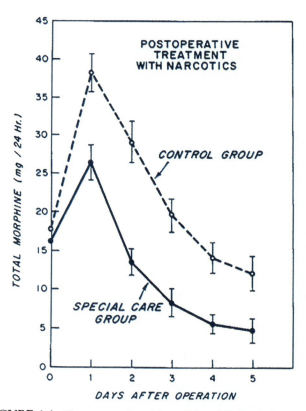

FIGURE 1.1 Figure reproduced from Egbert LD, Battit GE, Welch CE, Bartlett MK. Reduction of postoperative pain by encouragement and instruction of patients – a study of doctor–patient rapport. *N Engl J Med* 1964;270(16):825–827. Note the significant differences between total postoperative morphine use between the control and 'special care' groups.

In this seminal paper, titled 'The Powerful Placebo,' Beecher analysed the results of 15 controlled trials and pooled the results together to estimate the power of the placebo effect. The 'effect' was estimated at an average of 35.2% (range of 21% to 58%).[29,30] After so many years, this paper finally took an empirical approach to quantifying just how important the placebo effect might be in different medical settings. Although there are many papers which question the interpretation of this trial, it marked an important step in research on placebo, as people had become interested in what happened in the placebo group rather than just in the treatment group. Although efforts to study placebo responses in clinical trials were already underway, e.g. in the setting of post-operative pain by Lasagna and colleagues, the paper by Beecher was pivotal in raising the profile of placebo in the medical community.[31]

With further use of placebo as a control in medical and surgical treatments, more interest was generated in challenging the current practice of medicine and also appreciating the power of the placebo effect. Examples of the power of placebo effects were seen in a series of trials in surgery for angina pectoris, conducted not long after the paper by Beecher.

Internal mammary artery ligation, a surgical treatment for angina pectoris, was performed as it was believed to increase blood flow to the myocardium. In turn, patients reported reduced symptoms.[13,15] A series of placebo or 'sham' controlled trials (where half of the patients received the surgery and half were given only a skin incision) demonstrated significant responses to both the placebo (or sham) procedure and the real procedure.[32,33] As in the past, such a response to the placebo (or sham) operation questioned the validity of the actual surgery, suggesting that the symptom relief was merely due to a placebo effect and not to the technical nature of the procedure. Such demonstrations of response to placebo surgical procedures have continued over the course of the last 50 years, with similar powerful placebo responses seen in very recent trials of surgery, such as for arthroscopy for knee pain[34] and vertebroplasty for spinal pain.[35] Even in recent times, interpretation of such powerful placebo responses has involved challenging the validity of different surgical procedures, and although accepted as the 'gold standard' in empirical evaluation of a treatment, placebo responses in this context have still been somewhat negative, although for a different reason, as this can challenge contemporary medical practice.

Continued use of placebo in both medical and surgical trials resulted in further estimates of the power of the placebo effect. In the last 10 years, several papers have been published which have attempted to analyze the power of placebo effects in different settings.[36–38] One of the key findings was that placebo effects were larger in experimental trials (which aim to increase placebo effects) than in clinical trials (which are strictly controls), and that there are significant methodological issues in treating the placebo effect as a single effect and assessing it across many different clinical settings. However, in the bigger picture, these papers again brought the topic of placebo to light and gained an extraordinary amount of attention in medical literature and in the printed media.

THE EMERGENCE OF THE STUDY OF PLACEBO MECHANISMS

Although the recent history of placebo has been dominated by assessment of placebo responses in clinical trials, concurrent interest in the mechanisms of placebo started in the 1960s with a series of elegant experiments studying psychologic conditioning in animals.[39,40] This work was soon expanded to humans,[41] and together demonstrated that placebo analgesic effects may be mediated by conditioning (or learning) processes. Despite uncovering the potential understanding of how placebo effects may be mediated, research into placebo mechanisms only blossomed from this time, representing the start of a new era of research into placebo.

In 1978 a landmark trial was conducted by Levine and colleagues, who tested the hypothesis that placebo analgesia may be mediated by endogenous opioids. In this double-blinded study in post-operative pain, patients received a placebo 2 hours post-surgery and were then categorized into 'responders' and 'non-responders.' When a second drug administration was performed (the opioid antagonist naloxone) in each of these groups, there was no change in pain in the 'non-responder' group; however, there was an increase in pain in the 'responder' group. This was an elegant demonstration that placebo analgesia could be significantly reduced by naloxone, supporting the hypothesis that endogenous opioids mediate placebo analgesia.[42]

In 1983, the role of endogenous opioids was extended by Grevert and colleagues in another important study of the mechanisms of placebo analgesia.[43] In this study, experimentally induced arm pain was created in 30 subjects. Each subject received a placebo injection and was subsequently divided into two groups. Forty minutes after the first placebo injection, subjects in group one received a hidden administration of naloxone, and those in group two received a hidden injection of saline. This 'open–hidden' paradigm represented a novel way to assess the pharmacology of the drug without the psychosocial context (hidden administration) or the pharmacology of the drug and the effects of the psychosocial context (now conceptualized as placebo effects). In this study, hidden naloxone reduced the initial placebo analgesia whereas hidden saline did not, further supporting the role of endogenous opioids in placebo analgesia.

This new paradigm improved experimental validity and was used again a year later (1984) by Levine & Gordon.[44] The hypotheses and results of this experiment were in fact similar to those of Grevert et al in 1983; however, this study was particularly important in that it reproduced the above-mentioned findings with the inclusion of a natural history (no treatment) group. The presence of a natural history group allows for a more valid estimation of a response to a placebo injection as it controls for the natural history of the experimental pain and statistical phenomena such as regression to the mean.[45] Each of these studies conducted between 1978 and 1985 provided critical groundwork for exploration into the role of the endogenous opioid system and placebo analgesia. Furthermore, by virtue of the fact that placebo analgesia was not completely reversed by naloxone, these studies raised the possibility that other physiologic systems may also be involved in placebo analgesia. Together with further research on conditioning in humans, e.g.,[46] and the psychologic construct of expectancy, e.g.,[47] there was a discrete field of work aimed at understanding the mechanisms of placebo effects.

One particularly important study took place at around the same time as the described work into the mechanisms of placebo analgesia. This study, conducted by Gracely and colleagues in 1985, was seemingly designed in a similar manner to previous studies investigating placebo analgesia and the endogenous opioid system.[48] However, the hypotheses were in fact different, and this study aimed to assess placebo analgesia and the role of the therapeutic encounter. In this double-blinded administration of placebo in post-operative dental pain, patients were divided into two groups. Patients in both groups were told that they could receive an opioid analgesic (fentanyl), a placebo, or an opioid antagonist (naloxone) and that this could either improve, worsen or cause no change to their pain. However, the clinicians were told that there would be no active analgesic available for administration in group one, but that there would be a chance of delivering an opioid analgesic in group two. Strict double-blinded conditions remained in place. When analysing the response of patients in both groups to placebo, researchers found that a significant placebo effect existed in group 2 (when the clinician believed that a real analgesic could be given) compared with group 1 (when the clinician believed that the patient could only receive placebo or naloxone). In fact, the average effect in group 1 was negative, and patients reported increases in pain (Fig. 1.2). Whilst this experiment seemed to be designed to extend previous work on

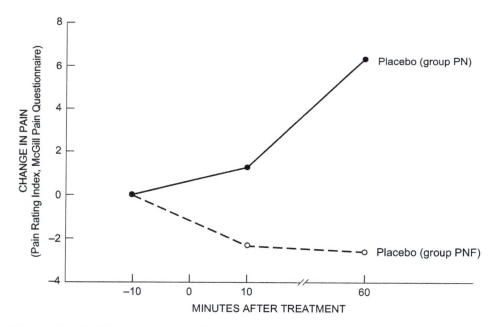

Change in pain rating index between baseline (10 min before injection) and 10 and 60 min after administration of placebo.

PN = group that could have either received placebo or naloxone.
PNF = group that could have received placebo, naloxone, or fentanyl (PNF).

FIGURE 1.2 Figure reproduced from Gracely RH, Dubner R, Deeter WD, Wolskee PJ. Clinicians' expectations influence placebo analgesia. *Lancet* 1985;5:43. Note that in the group PN, where there was no possibility of delivering active analgesia, subjects actually reported increased pain compared with the group PNF, where clinicians believed that there was a possibility of delivering an active analgesic. A significant placebo effect is seen in the PNF group.

opioid mechanisms, it was actually constructed to assess the interaction between the clinician and the patient. The authors concluded that the information given to the clinicians may have resulted in subtle changes in the doctor–patient relationship (as double blinding prohibited exchange of any information) and that the clinician's behaviors and gestures may have the ability to modulate placebo analgesia.

In more recent times, research in placebo mechanisms has expanded exponentially. In 1995, Fabrizio Benedetti published a paper which demonstrated that placebo analgesic effects were not only related to one system (the opioid system), but to another chemical (cholecystokinin).[49] This research group in Italy then drove the research agenda on placebo, discovering that there were multiple psychologic and biologic mechanisms for placebo analgesic effects, e.g.[50–52] Others supported this work (which was primarily in the field of pain and analgesia) and expanded it to include novel imaging of the brain, e.g.,[53,54] and studies of placebo in other conditions such as Parkinson's disease.[55,56] Of great importance was that the mechanism literature prompted re-evaluation of the concept of placebo, which still has strong links to both its biblical origins and early use in the clinical trial setting.

USING HISTORY TO FURTHER EXPLORE PLACEBO ANALGESIA

Only recently has there been a concerted attempt to re-evaluate placebo in the context of clinical trials and clinical practice. Such examples include reviews of placebo mechanisms and their implications for clinical trials and practice,[57] and concerted attempts to reconceptualize the topic area, e.g., Miller and Kaptchuk.[58] Taken together, an analysis of the history of placebo coupled with new experimental data has permitted new avenues for understanding and further exploring placebo effects, particularly in the setting of pain and analgesia.

Recent experiments have demonstrated that there is not one placebo effect, rather there are many, including multiple placebo analgesic effects, e.g.[45,59,60] This has very significant implications for understanding placebo effects in different settings, as different effects are likely to operate in different contexts. Furthermore, it adds to the complexity of attempting to find 'the power of the placebo effect,' and highlights the need for advancement in studies, such as Beecher in 1955 and Hrobjartsson in 2004, which treat placebo as a single entity or effect. This appraisal of older work coupled with new science is important in progressing the field.

Despite some of the unsavory connotations from the origin of the word, there is a strong argument to support the notion that placebo effects exist as part of every medical treatment.[60] This is supported by very interesting research demonstrating differences in drugs when they are given in a normal clinical manner or in a hidden manner (which removes the treatment context or ritual).[61] In fact, when one gives a placebo, it is studying the effect of the psychosocial context (which includes concepts such as meaning and the treatment ritual) on the patient without giving the actual treatment itself.[45] Logic then decrees that any health-care encounter (as there is a psychosocial context) has the ability to activate placebo mechanisms, and that these mechanisms may be very different according to the clinical situation. This reconceptualization draws on the previously mentioned historical literature, some of which places less focus on 'the placebo' and more focus on the therapeutic ritual or context.

Taken together, there is a large transition from the use of placebo to test whether a treatment is real or 'imagined,' or to see placebo in a negative light, to a modern appreciation of a dynamic group of effects related to the therapeutic ritual or context. This appreciation presents the notion that one should not deceptively give placebos but rather understand how placebo mechanisms can be activated when giving valid therapies to patients. It is now understood that it is not the content of the placebo that 'pleases' the patient, rather it is the psychosocial context and ritual of receiving the placebo therapy which changes the mind–brain–body interaction. After some hundreds of years, a new era in placebo research is emerging: one that aims to better understand the mind–brain interaction through understanding placebo effects. Only then will one be able to look back and see how something with such a negative origin has led to very positive steps forward in health care.

References

1. Shapiro A. A historic and heuristic definition of the placebo. *Psychiatry*. 1964;27:52-58.
2. Kerr CE, Milne I, Kaptchuk TJ. William Cullen and a missing mind–body link in the early history of placebos. *J R Soc Med*. 2008; 101(2):89-92.
3. Macedo A, Farre M, Banos JE. Placebo effect and placebos: what are we talking about? Some conceptual and historical considerations. *Eur J Clin Pharmacol*. 2003;59(4):337-342.
4. Jacobs B. Biblical origins of placebo. *J R Soc Med*. 2000;93(4):213-214.
5. Hart C. The mysterious placebo effect. *Modern Drug Discover*. 1999; 2(4):30-40.
6. Lasagna LC. The placebo effect. *J Allergy Clin Immunol*. 1986;78: 161-165.
7. de Craen AJ, Kaptchuk TJ, Tijssen JG, Kleijnen J. Placebos and placebo effects in medicine: historical overview. *J R Soc Med*. 1999; 92(10):511-515.
8. Aronson J. Please, please me. *BMJ*. 1999;318:716.
9. Scott R. Health and virtue: or, how to keep out of harm's way. Lectures on pathology and therapeutics by William Cullen c 1770. *Med Hist*. 1987;31:123-142.
10. Kaptchuk TJ, Kerr CE, Zanger A. Placebo controls, exorcisms and the devil. *Lancet*. 2009;374:1234-1235.

11. Booth C. The rod of Aesculapios: John Haygarth (1740–1827) and Perkins' metallic tractors. *J Med Biogr.* 2005;3(3):155-161.
12. Kaptchuk T. Intentional ignorance: a history of blind assessment and placebo controls in medicine. *Bull Hist Med.* 1998;72:389-433.
13. Moerman D. *Meaning, Medicine and the 'Placebo Effect'*. Cambridge: Cambridge University Press; 2002.
14. Shapiro A, Shapiro E. The placebo: is it much ado about nothing. In: Harrington A, ed. *The Placebo Effect – an Interdisciplinary Exploration.* Cambridge: Harvard University Press; 1997.
15. Moerman D. Physiology and symbols: the anthropological implications of the placebo effect. In: Romanucci-Ross L, Moerman DE, Tancredi LR, eds. *The Anthropology of Medicine: from Culture to Method.* 3rd ed Westport, CT: Bergin & Garvey; 1997.
16. Garrison FH. *An Introduction to the History of Medicine.* 3rd ed. Philadelphia: Saunders; 1921.
17. Shapiro AK. The placebo effect in the history of medical treatment: implications for psychiatry. *Am J Psychiatr.* 1959;116:298-304.
18. Harrington A. Introduction. In: Harrington A, ed. *The Placebo Effect – an Interdisciplinary Exploration.* Cambridge: Harvard University Press; 1997.
19. Cabot RC. The use of truth and falsehood in medicine: an experimental study. *Am Med.* 1903;5:344-349.
20. Kaptchuk TJ. Powerful placebo: the dark side of the randomised controlled trial. *Lancet.* 1998;351(9117):1722-1725.
21. Hewlett AW. Clinical effects of 'natural' and 'synthetic' sodium salicylate. *JAMA.* 1913;61:319-321.
22. Sollmann T. The crucial test of therapeutic evidence. *JAMA.* 1917;7:1439-1442.
23. Hediger EM, Gold H. U.S.P ether from large drums and ether from small cans labeled 'for anesthesia'. *JAMA.* 1935;104:2244-2248.
24. Gold H, Kwit NT, Otto H. The xanthines (theobromine and aminophylline) in the treatment of cardiac pain. *JAMA.* 1937;108:2173-2179.
25. Greiner TH, Gold H, Cattell M, et al. A method for the evaluation of effects of drugs on cardiac pain in patients with angina of effort: a study of Khellin (Visammin). *Am J Med.* 1950;9:143-155.
26. Houston WR. The doctor himself as a therapeutic agent. *Ann Intgern Med.* 1938;11(8):1416-1425.
27. Wolf HG, Duboid EF, Gold H. Use of placebos in therapy. *NY J Med.* 1946;46:1718-1727.
28. Egbert LD, Battit GE, Welch CE, Bartlett MK. Reduction of postoperative pain by encouragement and instruction of patients – A study of doctor–patient rapport. *N Engl J Med.* 1964;270(16):825-827.
29. Beecher HK. The powerful placebo. *JAMA.* 1955;159:1602-1606.
30. Beecher HK. *Measurement of Subjective Responses: Quantitative Effects of Drugs.* New York: Oxford University Press; 1959.
31. Lasagna LC, Mosteller F, Von Felsinger JM, Beecher HK. The study of the placebo response. *Am J Med.* 1954;16(6):770-779.
32. Cobb LA, Thomas GI, Dillard DH, et al. An evaluation of internal mannary artery ligation by a double blind technic. *N Engl J Med.* 1959;260:1115-1118.
33. Dimond EG, Kittle CF, Crockett JE. Comparison of internal mammary artery ligation and sham operation for angina pectoris. *Am J Cardiol.* 1960;5:483-486.
34. Moseley BJ, O'Malley K, Petersen NJ, et al. A controlled trial of arthroscopic surgery for osteoarthritis of the knee. *N Engl J Med.* 2002;347:81-88.
35. Buchbinder R, Osborne RH, Ebeling PR, et al. A randomized trial of vertebroplasty for painful osteoporotic vertebral fractures. *N Engl J Med.* 2009;361:557-568.
36. Hrobjartsson A, Gotzsche PC. Is the placebo powerless? An analysis of clinical trials comparing placebo with no treatment. *N Engl J Med.* 2001;344(21):1594-1602.
37. Hrobjartsson A, Gotzsche PC. Is the placebo effect powerless? Update of a systematic review with 52 new randomized trials comparing placebo with no treatment. *J Int Med.* 2004;256:91-100.
38. Vase L, Riley 3rd JL, Price DD. A comparison of placebo effects in clinical analgesic trials versus studies of placebo analgesia. *Pain.* 2002;99(3):443-452.
39. Herrnstein R. Placebo effect in the rat. *Science.* 1962;138:677-678.
40. Ader R, Cohen N. Behaviourally conditioned immunosuppression. *Psychosomat Med.* 1975;37:333-340.
41. Laska E, Sunshine A. Anticipation of analgesia: a placebo effect. *Headache.* 1973;1:1-11.
42. Levine JD, Gordon NC, Fields HL. The mechanism of placebo analgesia. *Lancet.* 1978;2(8091):654-657.
43. Grevert P, Albert LH, Goldstein A. Partial antagonism of placebo analgesia by naloxone. *Pain.* 1983;16:129-143.
44. Levine JD, Gordon NC. Influence of the method of drug administration on analgesic response. *Nature.* 1984;312(5996):755-756.
45. Price DD, Finniss DG, Benedetti F. A comprehensive review of the placebo effect: recent advances and current thought. *Annu Rev Psychol.* 2008;59:565-590.
46. Voudouris NJ, Peck CL, Coleman G. Conditioned placebo responses. *J Pers Spc Psychol.* 1985;48:47-53.
47. Voudouris NJ, Peck CL, Coleman G. The role of conditioning and verbal expectancy in the placebo response. *Pain.* 1990;43:121-128.
48. Gracely RH, Dubner R, Deeter WD, Wolskee PJ. Clinicians' expectations influence placebo analgesia. *Lancet.* 1985;5:43.
49. Benedetti F, Amanzio M, Maggi G. Potentiation of placebo analgesia by proglumide. *Lancet.* 1995;346(8984):1231.
50. Benedetti F, Arduino C, Amanzio M. Somatotopic activation of opioid systems by target-directed expectations of analgesia. *J Neurosci.* 1999;19(9):3639-3648.
51. Benedetti F, Amanzio M, Baldi S, et al. Inducing placebo respiratory depressant responses in humans via opioid receptors. *Eur J Neurosci.* 1999;11:625-631.
52. Benedetti F, Pollo A, Lopiano L, et al. Conscious expectation and unconscious conditioning in analgesic, motor, and hormonal placebo/nocebo responses. *J Neurosci.* 2003;23(10):4315-4323.
53. Petrovic P, Kalso E, Petersson KM, Ingvar M. Placebo and opioid analgesia – imaging a shared neuronal network. *Science.* 2002;295(5560):1737-1740.
54. Wager TD, Rilling JK, Smith EE, et al. Placebo-induced changes in fMRI in the anticipation and experience of pain. *Science.* 2004;303:1162-1166.
55. Colloca L, Lopiano L, Lanotte M, Benedetti F. Overt versus covert treatment for pain, anxiety, and Parkinson's disease. *Lancet Neurol.* 2004;3:679-684.
56. de la Fuente-Fernandez R, Ruth TJ, Sossi V, et al. Expectation and dopamine release: mechanism of the placebo effect in Parkinson's disease. *Science.* 2001;293(5532):1164-1166.
57. Finniss DG, Benedetti F. Mechanisms of the placebo response and their impact on clinical trials and clinical practice. *Pain.* 2005;114:3-6.
58. Miller FG, Kaptchuk TJ. The power of context: reconceptualizing the placebo effect. *J Roy Soc Med.* 2008;101(5):222-225.
59. Colloca L, Benedetti F. Placebos and painkillers: is mind as real as matter? *Nature Rev Neurosci.* 2005;6(7):545-552.
60. Finniss DG, Kaptchuk TJ, Miller F, Benedetti F. Biological, clinical, and ethical advances of placebo effects. *Lancet.* 2010;375(9715):686-695.
61. Benedetti F, Maggi G, Lopiano L, et al. Open versus hidden medical treatments: the patient's knowledge about a therapy affects the therapy outcome. *Prevent Treat.* 2003;6:article 1.

CHAPTER 2

Neurochemistry of Placebo Analgesia: Opioids, Cannabinoids and Cholecystokinin

Fabrizio Benedetti, Elisa Frisaldi

Department of Neuroscience, University of Turin Medical School, and National Institute of Neuroscience, Turin, Italy

INTRODUCTION

Historically known for their analgesic effects, the endogenous opioid peptides are represented mainly by beta-endorphin, enkephalins and dynorphins. They are the natural ligands of the opioid receptors and, like their relative agonists (e.g. morphine) and antagonists (e.g. naloxone), they are not entirely specific for any one of the receptor subtypes. In fact, different types of opioid receptors exist, all belonging to the G protein-coupled receptor superfamily. The most studied and understood are μ (or MOR, from mu opioid receptor), δ (DOR) and κ (KOR). The opioid receptors are found in a variety of brain regions in different proportions, from the spinal cord to the cerebral cortex. Whereas MOR and DOR receptors probably mediate both spinal and supraspinal analgesia, KOR receptors are involved mainly in spinal analgesia.[1]

The endogenous cannabinoids are closely related to the active ingredients present in *Cannabis sativa* extracts. The endocannabinoids belong to the class of lipid mediators, a series of arachidonic acid derivatives that play pivotal roles in immune regulation, in self-defense, and in the maintenance of homeostasis in living systems. Ligands and receptors of this system are present at various levels in the pain pathways, from peripheral sensory nerve endings to spinal cord and supraspinal centers, in a way that is parallel to, but distinct from, that involving opioid receptors.[2] Like other lipid mediators (e.g. prostaglandins), they appear to be synthesized and released locally on demand. The endocannabinoids in mammalian tissues are mainly anandamide, 2-arachidonylglycerol and 2-arachidonylglyceryl ether.[2] They bind to a couple of distinct types of G-protein-coupled receptor, CB1 and CB2. The former is expressed mainly in the brain, while the latter is expressed in peripheral tissues, such as the immune system.[2-4]

Cholecystokinin (CCK) is an octapeptide which binds to a couple of receptor subtypes: CCK-1 and CCK-2. The distribution of CCK in the brain matches that of the opioid peptides at the spinal and supraspinal level,[5] suggesting a close interaction between the two systems.[6] From a functional point of view, CCK inhibits the analgesic effects of morphine[7] and β-endorphin,[8] so that it is considered as an antagonist of the opioid system.[6]

These three neurotransmitters have been found to powerfully modulate both placebo analgesia and nocebo hyperalgesia. Therefore, today they are considered to have a pivotal role in placebo and nocebo responses (Fig. 2.1). In the following sections we describe how this modulation takes place.

SOME TYPES OF PLACEBO ANALGESIA ARE MEDIATED BY ENDOGENOUS OPIOIDS

Placebo-induced analgesia has been found to be related to the activation of the endogenous opioid systems in some circumstances. The first study that tried to understand the biologic mechanisms of placebo analgesia was aimed at blocking opioid receptors with the antagonist naloxone.[9] This study was performed in the clinical setting in patients who had undergone extraction of the third molar. The investigators found a disruption of placebo analgesia after naloxone administration, which indicates the involvement of endogenous opioid systems in the placebo analgesic effect. One of the major criticisms was the lack of adequate control groups, such as a natural history group, as well as the possibility that naloxone *per se* might have a hyperalgesic effect.[10] From this perspective, the higher pain intensity following naloxone administration would not be due to the blockade

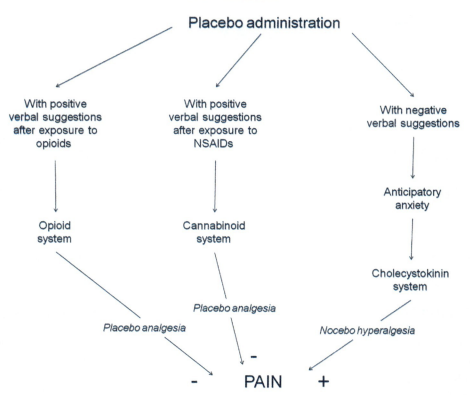

FIGURE 2.1 General schema of the involvement of opioid, cannabinoid and cholecystokinin systems in placebo analgesia and nocebo hyperalgesia.

of placebo-induced activation of endogenous opioids, but rather to the hyperalgesic properties of naloxone. This is a crucial point because naloxone must not have any hyperalgesic effect in order to be used in the study of the placebo effect. Despite these limitations, the study by Levine, Gordon and Fields[9] represented the first attempt to give scientific credibility to the placebo phenomenon by unraveling the underlying biologic mechanisms. Thus, this study represents the passage from the psychologic to the biologic investigation of the placebo effect.

The years that followed the publication of the study by Levine et al[9] were characterized by some attempts to verify and to reproduce these findings.[11,12] However, in 1983, Gracely et al[13] demonstrated that naloxone may indeed have hyperalgesic effects in postoperative pain, thus posing some doubts on the opioid hypothesis of placebo analgesia. Research in this field had a long pause from 1984 to 1995, with the exception of a few isolated studies. For example, Lipman and collaborators[14] studied chronic-pain patients and found that those patients who responded to a placebo administration showed higher concentrations of endorphins in the cerebrospinal fluid compared with those patients who did not respond to the placebo.

From 1995 until 1999, a long series of experiments with rigorous experimental designs were performed by Benedetti and collaborators. During these 5 years, many unanswered questions were clarified, and the role of endogenous opioids in placebo analgesia was explained. By using the model of experimental ischemic arm pain (tourniquet technique), it was definitely clarified that naloxone, which was tailored to individual weights, does not affect this kind of pain, so that any effect following naloxone administration could be attributed to the blockade of placebo-induced opioid activation.[15] Concurrently, the same authors tested the effects of proglumide, a non-specific antagonist for CCK-1 and CCK-2 receptors, on placebo analgesia. It was hypothesized that, on the basis of the anti-opioid action of CCK, the blockade of CCK receptors could enhance the opioids released by the placebo. Indeed, it was found that the CCK-blocker enhanced placebo analgesia, which represents a novel and indirect way to test the opioid hypothesis.[15,16] Therefore, the model we have today to explain the analgesic effect following the administration of a placebo involves two opposing neurotransmitter systems, opioids and CCK.

In 1984, Fields and Levine[17] had hypothesized that the placebo response may be subdivided into opioid and nonopioid components. In particular, Fields and Levine's suggestion was that different physical, psychologic and environmental situations could affect the endogenous opioid systems differently. This concept was further supported by the finding that the placebo analgesic effect is not always mediated by endogenous opioids.[13] This issue was partially addressed by Amanzio and Benedetti,[18] who showed that both expectation and a conditioning procedure can result in placebo analgesia. The former is capable of activating opioid systems whereas the latter activates specific subsystems. In fact, if the placebo response is induced by means of strong expectation cues, it can be blocked by the opioid antagonist naloxone. Similarly, if a placebo is given after repeated administrations of morphine (preconditioning procedure), the placebo response can be blocked by naloxone. Conversely, if the placebo response is induced by means of prior conditioning with a nonopioid drug, it is naloxone-insensitive.[18]

Specific placebo analgesic responses can be obtained in different parts of the body,[19,20] and these responses are naloxone-reversible.[21] For example, if four noxious stimuli are applied to the hands and feet and a placebo cream is applied to one hand only, pain is reduced only on the hand where the placebo cream had been applied. This highly specific effect is blocked by naloxone, suggesting that the placebo-activated endogenous opioid systems have a precise and somatotopic organization, probably at the central level.[21]

A common observation in all these studies is that the naloxone dose necessary to block placebo analgesia is as large as 10 mg, which suggests the involvement of different classes of opioid receptors, like the MOR, DOR and KOR. In fact, the binding affinity of naloxone for the DOR and KOR receptors is about 10–15 times lower than for the MOR receptors, thus large doses are supposed to involve DOR and KOR receptors as well.

In 2002, Petrovic et al[22] found that some brain regions in the cerebral cortex and in the brainstem are affected by both a placebo and the rapidly acting opioid agonist remifentanil, thus indicating a related mechanism of placebo-induced and opioid-induced analgesia. In particular, the administration of a placebo induced the activation of the rostral anterior cingulate cortex and the orbitofrontal cortex. Moreover, there was a significant covariation in activity between the rostral anterior cingulate cortex and the lower pons/medulla, and a subsignificant covariation between the rostral anterior cingulate cortex and the periaqueductal gray, thus suggesting that the descending rostral anterior cingulate/periaqueductal gray/rostral ventromedial medulla pain-modulating circuit is involved in placebo analgesia, as previously hypothesized by Fields and Price.[23]

In 2005, the first direct evidence of opioid-mediated placebo analgesia was published.[24] By using *in vivo* receptor-binding techniques with the radiotracer carfentanil, a MOR agonist, it was shown that a placebo procedure activates MOR neurotransmission in the dorsolateral prefrontal cortex, the anterior cingulate cortex, the insula, and the nucleus accumbens. A more detailed account of MOR neurotransmission after placebo administration was carried out in another study which used noxious thermal stimulation.[25] In these studies, placebos affected opioid activity in a number of predicted opioid-rich regions that play central roles in pain, including periaqueductal gray, dorsal raphe and nucleus cuneiformis, amygdala, orbitofrontal cortex, insula, rostral anterior cingulate, and lateral prefrontal cortex. Connectivity analyses revealed that placebo treatment increased connectivity between the periaqueductal gray and the rostral anterior cingulate cortex.[26]

The pharmacologic approach with opioid antagonists, such as naloxone,[9,18,26,27] and with CCK antagonists, such as proglumide,[15,16,28,29] has been crucial over the past years to understand the neurobiology of placebo analgesia. With a similar approach, Benedetti and collaborators used a CCK-2 receptor agonist, pentagastrin, in order to investigate how CCK hyperactivity affects the placebo analgesic response in the experimental model of ischemic arm pain.[30] In this study, placebo analgesia was tested after morphine preconditioning, which is known to be mediated by endogenous opioids.[18,27] As hypothesized, the activation of CCK-2 receptors by means of pentagastrin completely abolished placebo analgesia.

Therefore, a change in the balance between opioids and CCK, which seems to be crucial in several conditions,[31,32] may influence placebo analgesia in opposite directions, depending on the activity of these two neurotransmitters: when CCK activity outweighs opioid activity, placebo analgesia is reduced, while the opposite situation leads to increased placebo analgesic responses.

These effects have also been studied in mice by Guo et al.[33] On the basis of Amanzio and Benedetti's experiments,[18] these researchers used the hot-plate test in an attempt to measure the reaction time of mice to nociceptive stimulus (hot plate) after different types of pharmacologic conditioning. This was performed by the combination of the conditioned cue stimulus with the unconditioned drug stimulus, either the opioid morphine or the nonopioid aspirin. If mice were conditioned with morphine, placebo analgesia was completely antagonized by naloxone, whereas if mice were conditioned with aspirin, placebo analgesia was naloxone-insensitive. These findings show that, also in mice, the mechanisms underlying placebo analgesia include both opioid and nonopioid components and may depend on the previous exposure to different pharmacologic agents.

ENDOCANNABINOIDS ARE INVOLVED IN SOME TYPES OF PLACEBO ANALGESIA

From the previous studies it is clear that the endogenous opioid systems are not the only mechanisms involved in placebo analgesia. When a nonopioid drug is administered for a few days in a row, and then replaced with a placebo, the placebo analgesic response is not reversed by naloxone, which suggests that specific pharmacologic mechanisms are involved in a learned placebo response, depending on the prior exposure to opioid or nonopioid substances.[18,33] Another placebo analgesic effect that is not mediated by opioids has been described in irritable bowel syndrome patients, indicating that in different medical conditions, and in different circumstances, other systems may be recruited.[34]

There is today accumulating evidence that nonsteroidal anti-inflammatory drugs (NSAIDs), such as ketorolac and aspirin, have effects that go well beyond the inhibition of cyclooxygenase and prostaglandin synthesis. In fact, NSAIDs have been found to interact with endocannabinoids, a class of lipid mediators, both *in vivo* and *in vitro*,[35,36] and cyclooxygenase-2 has been shown to utilize endocannabinoids as substrates.[37] Therefore, the endocannabinoid system may play a pivotal role in both the therapeutic and adverse effects of NSAIDs[38] as well as in NSAID-induced placebo responses.[18,33]

On the basis of all these considerations, Benedetti and collaborators[39] induced opioid or nonopioid placebo analgesic responses and assessed the effects of the CB1 receptor antagonist rimonabant. Unlike naloxone, rimonabant had no effect on opioid-induced placebo analgesia following morphine preconditioning, whereas it completely blocked placebo analgesia following nonopioid preconditioning with ketorolac. These findings indicate that those placebo analgesic responses that are elicited by conditioning with NSAIDs are mediated by CB1 cannabinoid receptors.

Although the site where the CB1 receptors are activated cannot be established, recent *in vivo* studies in baboons[40] and humans[41] indicate that CB1 receptors are abundant in the basal ganglia, for example in the striatum, which has been found to have a key role in the placebo response. In fact, the nucleus accumbens, a part of the ventral striatum which belongs to the dopaminergic reward system, is activated after placebo administration.[42,43] Therefore, a key question for future research will be to understand where in the brain the endocannabinoids are activated during placebo analgesia and how they possibly interact with other systems such as the opioid network.

NOCEBO HYPERALGESIA IS MEDIATED BY CHOLECYSTOKININ

Compared with placebo analgesia, much less is known about nocebo hyperalgesia, mainly due to ethical constraints. In fact, whereas the induction of placebo responses is acceptable in many circumstances,[44] the induction of nocebo responses represents a stressful and anxiogenic procedure, for an inert treatment has to be given along with negative verbal suggestions of pain increase.

In 1997, a trial in postoperative patients was run with the CCK antagonist proglumide.[28] This consisted of a post-surgical manipulation that induced expectations of pain worsening. It was found that proglumide prevented nocebo hyperalgesia in a dose-dependent manner, even though it is not a specific painkiller, thus suggesting that the nocebo hyperalgesic effect is mediated by CCK. This effect was not antagonized by naloxone. A dose of proglumide as low as 0.05 mg was totally ineffective, whereas a dose increase to 0.5 and 5 mg proved to be effective. As CCK is also involved in anxiety mechanisms, it was hypothesized that proglumide affects anticipatory anxiety.[1,28] However, due to ethical limitations in these patients, these effects were not investigated further.

In order to better understand the mechanisms underlying nocebo hyperalgesia and to overcome the ethical constraints that are inherent in the clinical approach, a similar procedure was used in healthy volunteers by inducing experimental pain.[29] The oral administration of an inert substance, along with verbal suggestions of hyperalgesia, was found to induce both hyperalgesia and hyperactivity of the hypothalamic–pituitary–adrenal axis, as assessed by means of adrenocorticotropic hormone (ACTH) and cortisol plasma concentrations. Both nocebo-induced hyperalgesia and hypothalamic–pituitary–adrenal hyperactivity were blocked by the benzodiazepine diazepam, which suggests the involvement of anxiety mechanisms. By contrast, administration of the mixed CCK-1/2 receptor antagonist proglumide blocked nocebo hyperalgesia completely, but had no effect on the hypothalamic–pituitary–adrenal hyperactivity, thus suggesting a specific involvement of CCK in the hyperalgesic but not in the anxiety component of the nocebo effect. Interestingly, both diazepam and proglumide did not show analgesic properties on baseline pain, as they acted only on the nocebo-induced pain increase. These data suggest that a close relationship between anxiety and nocebo hyperalgesia exists, but they also indicate that proglumide does not act by blocking anticipatory anxiety, as previously hypothesized,[1,28] but rather it interrupts a CCKergic link between anxiety and pain. Therefore, unlike the anxiolytic action of diazepam, proglumide blocks a CCKergic pro-nociceptive system which is activated by anxiety and is responsible for anxiety-induced hyperalgesia. Support for this view comes from a study that used a social-defeat model of anxiety in rats, in which CI-988, a selective CCK-2 receptor antagonist, prevented anxiety-induced hyperalgesia.[45]

Nocebo hyperalgesia is thus an interesting model to better understand when and how the endogenous pro-nociceptive systems are activated. The pro-nociceptive and anti-opioid action of CCK has been documented in many brain regions,[46–49] and CCK has been shown to reverse opioid analgesia by acting at the level of the rostral ventromedial medulla[50,51] and to activate pain-facilitating neurons within the same region.[52] The similarity of the pain-facilitating action of CCK, both on brainstem neurons in animals and on nocebo mechanisms in humans, is an interesting point that deserves further investigation.

It is worth pointing out that the discrepancy between anxiety-induced hyperalgesia and stress-induced analgesia may be only apparent. In fact, stress is known to induce analgesia in a variety of situations, in both animals and humans. Indeed, when we are under stress, the threshold of pain is increased. However, the nature of the stressor is likely to play a central role. In fact, whereas hyperalgesia may occur when the anticipatory anxiety is about the pain itself,[29,53–55] analgesia may occur when anxiety is about a stressor that shifts the attention from the pain.[56–58] We should therefore use these two definitions, being aware that they may be different aspects of the same process.[59] In the case of anxiety-induced hyperalgesia, we are talking about anticipation of pain, in which attention is focused on the impending pain. We have seen that the biochemical link between this anticipatory anxiety and the pain increase is represented by the CCKergic systems. Conversely, we should refer to stress-induced analgesia whenever a general state of arousal stems from a stressful situation in the environment, so that attention is now focused on the environmental stressor. In this case, there is experimental evidence that analgesia results from the activation of the endogenous opioid systems.[56,57]

CONCLUSIONS

Considerable progress has been made in our understanding of the neurobiologic mechanisms of the placebo effect, and most of our knowledge originates from the field of pain and analgesia. Today, it is well established that different endogenous neuronal networks are responsible for the modulation of pain by placebos and nocebos. These include opioid, cannabinoid and CCK systems. By using new experimental designs and techniques, such as *in vivo* receptor binding, recording from neurons in awake humans, and a combination of imaging and electrophysiologic techniques, a future challenge will be to understand the interaction between a complex mental activity, like expectation of a future event, and all these neurochemical systems, and this approach will allow us to better describe the intricate connections between mind, brain and body.

However, many questions still remain unanswered. For example, we need to know where, when, and how placebos work across different diseases and therapeutic interventions, and we also need to test the effects of pharmacologic conditioning not only for painkillers but also for other classes of drug as well, such as immunosuppressive and hormone-stimulating agents. Another issue that requires further clarification is why some subjects respond to placebos whereas other subjects do not, a critical point that is likely to be clarified by pursuing further research into both learning and the genetics of neurotransmitters such as opioids, cannabinoids and CCK.

References

1. Benedetti F, Amanzio M. The neurobiology of placebo analgesia: from endogenous opioids to cholecystokinin. *Prog Neurobiol.* 1997;52:109-125.
2. Iversen L. Cannabis and the brain. *Brain.* 2003;126:1252-1270.
3. Felder CC, Glass M. Cannabinoid receptors and their endogenous agonists. *Annu Rev Pharmacol Toxicol.* 1998;3(8):179-200.
4. Munro S, Thomas KL, Abu-Shaar M. Molecular characterization of a peripheral receptor for cannabinoids. *Nature.* 1993;365:61-65.
5. Gall C, Lauterborn J, Burks D, Seroogy K. Co-localization of enkephalin and cholecystokinin in discrete areas of rat brain. *Brain Res.* 1987;403:403-408.
6. Baber NS, Dourish CT, Hill DR. The role of CCK, caerulein, and CCK antagonists in nociception. *Pain.* 1989;39:307-328.
7. Faris PL, Komisaruk BR, Watkins LR, Mayer DJ. Evidence for the neuropeptide cholecystokinin as an antagonist of opiate analgesia. *Science.* 1983;219:310-312.
8. Itoh S, Katsuura G, Maeda Y. Caerulein and cholecystokinin suppress beta-endorphin-induced analgesia in the rat. *Eur J Pharmacol.* 1982;80:421-425.
9. Levine JD, Gordon NC, Fields HL. The mechanisms of placebo analgesia. *Lancet.* 1978;2:654-657.
10. Skrabanek P. Naloxone and placebo. *Lancet.* 1978;2:791.
11. Grevert P, Albert LH, Goldstein A. Partial antagonism of placebo analgesia by naloxone. *Pain.* 1983;16:129-143.
12. Levine JD, Gordon NC. Influence of the method of drug administration on analgesic response. *Nature.* 1984;312:755-756.
13. Gracely RH, Dubner R, Wolskee PJ, Deeter WR. Placebo and naloxone can alter postsurgical pain by separate mechanisms. *Nature.* 1983;306:264-265.
14. Lipman JJ, Miller BE, Mays KS, et al. Peak B endorphin concentration in cerebrospinal fluid: reduced in chronic pain patients and increased during the placebo response. *Psychopharmacol.* 1990;102:112-116.
15. Benedetti F. The opposite effects of the opiate antagonist naloxone and the cholecystokinin antagonist proglumide on placebo analgesia. *Pain.* 1996;64:535-543.
16. Benedetti F, Amanzio M, Maggi G. Potentiation of placebo analgesia by proglumide. *Lancet.* 1995;346:1231.
17. Fields HL, Levine JD. Placebo analgesia-a role for endorphins? *Trends Neurosci.* 1984;7:271-273.
18. Amanzio M, Benedetti F. Neuropharmacological dissection of placebo analgesia: expectation-activated opioid systems versus conditioning-activated specific subsystems. *J Neurosci.* 1999;19:484-494.
19. Montgomery GH, Kirsch I. Mechanisms of placebo pain reduction: an empirical investigation. *Psychol Sci.* 1996;7:174-176.
20. Price DD, Milling LS, Kirsch I, et al. An analysis of factors that contribute to the magnitude of placebo analgesia in an experimental paradigm. *Pain.* 1999;83:147-156.

21. Benedetti F, Arduino C, Amanzio M. Somatotopic activation of opioid systems by target-expectations of analgesia. *J Neurosci*. 1999;9:3639-3648.
22. Petrovic P, Kalso E, Petersson KM, Ingvar M. Placebo and opioid analgesia – imaging a shared neuronal network. *Science*. 2002;295:1737-1740.
23. Fields HL, Price DD. Toward a neurobiology of placebo analgesia. In: Harrington A, ed. *The Placebo Effect: an Interdisciplinary Exploration*. Cambridge, MA: Harvard University Press; 1997:93-116.
24. Zubieta JK, Bueller JA, Jackson LR, et al. Placebo effects mediated by endogenous opioid activity on μ-opioid receptors. *J Neurosci*. 2005;25:7754-7762.
25. Wager TD, Scott DJ, Zubieta JK. Placebo effects on human (micro)-opioid activity during pain. *Proc Natl Acad Sci USA*. 2007;104:11056-11061.
26. Eippert F, Bingel U, Schoell ED, et al. Activation of the opioidergic descending pain control system underlies placebo analgesia. *Neuron*. 2009;63:533-543.
27. Benedetti F, Pollo A, Colloca L. Opioid-mediated placebo responses boost pain endurance and physical performance: is it doping in sport competitions? *J Neurosci*. 2007;27:11934-11939.
28. Benedetti F, Amanzio M, Casadio C, et al. Blockade of nocebo hyperalgesia by the cholecystokinin antagonist proglumide. *Pain*. 1997;71:135-140.
29. Benedetti F, Amanzio M, Vighetti S, Asteggiano G. The biochemical and neuroendocrine bases of the hyperalgesic nocebo effect. *J Neurosci*. 2006;26:12014-12022.
30. Benedetti F, Amanzio M, Thoen W. Disruption of opioid-induced placebo responses by activation of cholecystokinin type-2 receptors. *Psychopharmacol*. 2011;213:791-797.
31. Benedetti F. Mechanisms of placebo and placebo-related effects across diseases and treatments. *Annu Rev Pharmacol Toxicol*. 2008;48:33-60.
32. Enck P, Benedetti F, Schedlowski M. New insights into the placebo and nocebo responses. *Neuron*. 2008;59:195-206.
33. Guo JY, Wang JY, Luo F. Dissection of placebo analgesia in mice: the conditions for activation of opioid and non-opioid systems. *J Psychopharmacol*. 2010;24:1561-1567.
34. Vase L, Robinson ME, Verne GN, Price DD. Increased placebo analgesia over time in irritable bowel syndrome (IBS) patients is associated with desire and expectation but not endogenous opioid mechanisms. *Pain*. 2005;115:338-347.
35. Fowler CJ. The contribution of cyclooxygenase-2 to endocannabinoid metabolism and action. *Br J Pharmacol*. 2007;152:594-601.
36. Shimizu T. Lipid mediators in health and disease: enzymes and receptors as therapeutic targets for the regulation of immunity and inflammation. *Annu Rev Pharmacol Toxicol*. 2009;49:123-150.
37. Rouzer CA, Marnett LJ. Non-redundant functions of cyclooxygenases: oxygenation of endocannabinoids. *J Biol Chem*. 2008;283:8065-8069.
38. Hamza M, Dionne RA. Mechanisms of non-opioid analgesics beyond cycloxygenase enzyme inhibition. *Curr Mol Pharmacol*. 2009;2:1-14.
39. Benedetti F, Amanzio M, Rosato R, Blanchard C. Nonopioid placebo analgesia is mediated by CB1 cannabinoid receptors. *Nat Med*. 2011;17:1228-1230.
40. Horti AG, Fan H, Kuwabara H, et al. 11C-JHU75528: a radiotracer for PET imaging of CB1 cannabinoid receptors. *J Nucl Med*. 2006;47:1689-1696.
41. Wong DF, Kuwabara H, Horti AG, et al. Quantification of cerebral cannabinoid receptors subtype 1 (CB1) in healthy subjects and schizophrenia by the novel PET radioligand [11C]OMAR. *Neuroimage*. 2010;52:1505-1513.
42. de la Fuente-Fernández R, Ruth TJ, Sossi V, et al. Expectation and dopamine release: mechanism of the placebo effect in Parkinson's disease. *Science*. 2001;293:1164-1166.
43. Scott DJ, Stohler CS, Egnatuk CM, et al. Individual differences in reward responding explain placebo-induced expectations and effects. *Neuron*. 2007;55:325-336.
44. Benedetti F, Colloca L. Placebo-induced analgesia: methodology, neurobiology, clinical use, and ethics. *Rev Analgesia*. 2004;7:129-143.
45. Andre J, Zeau B, Pohl M, et al. Involvement of cholecystokininergic systems in anxiety-induced hyperalgesia in male rats: behavioral and biochemical studies. *J Neurosci*. 2005;25:7896-7904.
46. Benedetti F. Cholecystokinin type-A and type-B receptors and their modulation of opioid analgesia. *News Physiol Sci*. 1997;12:263-268.
47. Benedetti F, Lanotte M, Lopiano L, Colloca L. When words are painful – Unraveling the mechanisms of the nocebo effect. *Neuroscience*. 2007;147:260-271.
48. Hebb ALO, Poulin J-F, Roach SP, et al. Cholecystokinin and endogenous opioid peptides: interactive influence on pain, cognition, and emotion. *Prog Neuro-Psychopharmacol Biol Psychiatr*. 2005;29:1225-1238.
49. Lovick TA. Pro-nociceptive action of cholecystokinin in the periaqueductal grey: a role in neuropathic and anxiety-induced hyperalgesic states. *Neurosci Biobehav Rev*. 2008;32:852-862.
50. Mitchell JM, Lowe D, Fields HL. The contribution of the rostral ventromedial medulla to the antinociceptive effects of systemic morphine in restrained and unrestrained rats. *Neuroscience*. 1998;87:123-133.
51. Heinricher MM, McGaraughty S, Tortorici V. Circuitry underlying antiopioid actions of cholecystokinin within the rostral ventromedial medulla. *J Neurophysiol*. 2001;85:280-286.
52. Heinricher MM, Neubert MJ. Neural basis for the hyperalgesic action of cholecystokinin in the rostral ventromedial medulla. *J Neurophysiol*. 2004;92:1982-1989.
53. Sawamoto N, Honda M, Okada T, et al. Expectation of pain enhances responses to nonpainful somatosensory stimulation in the anterior cingulated cortex and parietal operculum/posterior insula: an event-related functional magnetic resonance imaging study. *J Neurosci*. 2000;20:7438-7445.
54. Koyama T, McHaffie JG, Laurienti PJ, Coghill RC. The subjective experience of pain: where expectations become reality. *Proc Natl Acad Sci USA*. 2005;102:12950-12955.
55. Keltner JR, Furst A, Fan C, et al. Isolating the modulatory effect of expectation on pain transmission: a functional magnetic imaging study. *J Neurosci*. 2006;26:4437-4443.
56. Willer JC, Albe-Fessard D. Electrophysiological evidence for a release of endogenous opiates in stress-induced 'analgesia' in man. *Brain Res*. 1980;198:419-426.
57. Terman GW, Morgan MJ, Liebeskind JC. Opioid and non-opioid stress analgesia from cold water swim: importance of stress severity. *Brain Res*. 1986;372:167-171.
58. Flor H, Grusser SM. Conditioned stress-induced analgesia in humans. *Eur J Pain*. 1999;3:317-324.
59. Colloca L, Benedetti F. Nocebo hyperalgesia: how anxiety is turned into pain. *Curr Opin Anaesthesiol*. 2007;20:435-439.

CHAPTER 3

Placebo Analgesia in Rodents

Jian-You Guo, Jin-Yan Wang, Fei Luo

Key Laboratory of Mental Health, Institute of Psychology, Chinese Academy of Sciences, Beijing, P. R. China

INTRODUCTION

Behavioral context can modulate neuronal activity in nociceptive and non-nociceptive somatosensory pathways.[1,2] Placebo analgesia is one of the most striking examples of the cognitive modulation of pain perception. It represents a situation where the administration of an ineffective substance produces an analgesic effect when the subject is convinced that the substance is a potent painkiller. It is not surprising then that a considerable effort has been committed over the last three decades to the study of mechanisms that may account for placebo-induced analgesic effects. The neural basis of placebo analgesia was first established by Levine et al,[3] who discovered that the placebo response could be blocked by the opioid receptor antagonist naloxone. This indicates the involvement of the endogenous opioid system. Following this finding, complex experimental designs have elucidated several components underlying the placebo analgesic response, and other studies have subsequently confirmed this exciting and provocative hypothesis.[4-7] Significant placebo-induced activation of μ-opioid receptor-mediated neurotransmission has been directly observed in higher-order and subcortical brain regions.[8]

Fields and Levine[9] were the first to hypothesize that placebo response may be subdivided into opioid and nonopioid components. In particular, they suggested that different physical, physiologic, and environmental situations could affect the endogenous opioid system differently. This concept was further supported by the finding that the placebo effect was not always mediated by endogenous opioids.[10] Thus, the conditions necessary for the activation of opioid systems needed to be identified. This problem was addressed by Amanzio and Benedetti,[11] who showed that expectation or a conditioning procedure is capable of activating different types of placebo analgesia.

A likely candidate for the mediation of placebo-induced analgesia is the opioid-related neuronal network in the brain.[12] This hypothesis was supported by a recent brain-imaging study in which the authors found that the very same brain regions in the cerebral cortex and brainstem could be affected by either a placebo or the rapidly acting opioid agonist remifentanil, thus indicating a related mechanism in placebo- and opioid-induced analgesia.[13] The direct demonstration of placebo-induced release of endogenous opioids was obtained using *in vivo* receptor binding with positron emission tomography by Wager et al[14] and Scott et al.[15] Although neurochemical mechanisms have not yet been identified in nonopioid-meditated placebo, the possible involvement of some neurotransmitters has been shown in some conditions. For instance, a specific CB1 cannabinoid receptor antagonist could block nonopioid placebo analgesic responses but has no effect on opioid placebo responses. These findings suggest that the endocannabinoid system has a pivotal role in placebo analgesia in some circumstances when the opioid system is not involved.[16,17]

THE PLACEBO EFFECT IN ANIMALS

The placebo effect has been a topic of interest in scientific and clinical communities for many years, and our knowledge of the mechanisms of the placebo effect has been advanced considerably by human studies. However, animal models have been given less attention, and they have provided less information on this subject. For the nonverbal animal, associative learning is considered to play a key role in the placebo effect. Originally demonstrated by Pavlov[18] with animals, associative learning was considered essentially as a pairing of two stimuli. One, initially neutral in that, by itself, it elicits no response, is called the conditioned stimulus (CS);

the other, which consistently elicits a response, is called the unconditioned stimulus (US). The response elicited by the pairing of the CS and the US is called the conditioned response (CR). The formation of CRs may be due to repeated association of the CSs with active medication, the USs. Pairing placebos with effective medication, followed by administering placebos without the medication, can produce a CR that is similar to the response to medication.[19]

Pavlov and others reported the conditioning of some of the effects of morphine on animals.[18,20] Dogs were given subcutaneous doses of morphine sulfate daily. A small amount of morphine produces salivation in the dog and usually produces emesis and sleep. After seven or eight daily administrations of morphine, these dogs salivated profusely, and a few even vomited or slept while the experimenter was preparing the dogs for an injection. Presumably, events associated with the administration of morphine become the conditioned stimuli for some of the reactions that are characteristically induced by the drug. The parallel to the placebo effect is clear, but research has been restricted to morphine, and there have not been enough controls to exclude other interpretations. Herrnstein[21] showed that saline injections mimic the effect of scopolamine hydrobromide (disrupting the learned behavior) in rats. In their schedule, a rat was placed daily in a chamber that was insulated for light and sound and contained a lever and a feeding device. The rat was hungry and was trained to depress the lever by reinforcement with sweetened condensed milk. After 4 months training, the schedule of reinforcement had established a characteristic pattern of responding, one whose primary feature is that, at the beginning of each cycle, there is little or no pressing of the lever, whereas, as the time for reinforcement approaches, the rate of lever-pressing increases continuously but quite gradually. The effect of scopolamine on this behavior is to depress the overall frequency of response and to abolish the orderly progression of the rat. At this time, there was no detectable effect on behavior of the saline administrations. The results showed that physiologic saline could mimic the effect of scopolamine hydrobromide to some extent when the two substances were alternately administered in a series of injections. This might be the first study to mention the placebo effect in animals using the Pavlovian conditioning procedure.

The research of Sherman and Schnitzer[22] further supported a placebo effect in the rat. In their study, rats were assigned randomly to a drug-run (DR), drug-not-run (DNR), or untreated-run (UR) group. Both the DR and DNR groups were given d-amphetamine sulfate in the first week of testing. The DR group was placed in the apparatus and activity was measured for 45 minutes immediately after drug injection. The DNR group was returned to the home cages after drug injection. The UR group was subjected to 45 minutes activity testing after a saline injection. One week later all groups received the untreated-run treatment. The DR group yielded significantly more locomotion than did the UR group, while there was no difference between the DNR group and the UR group. These results confirm the findings that a single administration of a drug may have consequences for later behavior over and above any direct actions of the drug; this apparent placebo effect might represent a conditioned response.

Pihl and Altman[23] investigated whether the number of pairings between the active substance and introduction into an experimental chamber affects the placebo effect in rats. The experiment was run in three phases: a baseline phase, a drug phase, and a post-drug phase. In the baseline phase, all animals were placed in the experimental chamber after administration of 0.1 mL saline (i.p.); these animals received 10 1-hour trials in this baseline phase. In the drug phase, the animals were randomly divided into four groups. Groups differed according to the number of trials they received (put into the chamber after injection of d-amphetamine sulfate solution). The three placebo groups received 3, 9 and 15 trials, respectively. A control group received 15 trials (administered with saline). All animals were put into the chamber after drug or saline injection. In the post-drug phase, all groups received saline injection prior to 15 trials. The results of the post-drug phase showed that the 15-trials group was significantly different from the 3-trials group and from the control group, while the 9-trials group was significantly different from the 3-trials group. These data suggested that the intensity of the placebo response is related to the number of pairings between the active drug and the circumstances under which the drug is presented. Meanwhile, a recent study also demonstrated that the number of trials increases the magnitude of placebo responses in humans.[24]

Regarding immunomodulatory placebo effects, it has been reported that peripheral antigenic stimulation can work as a US, and thus can be associated with a specific external stimulus (CS = placebo). Metalnikov and Chorine are generally credited with having conducted the first studies on behaviorally conditioned immune effects.[25] However, the famous study on conditioned immunosuppression was published by Ader and Cohe.[26] In their experiment, an illness-induced taste aversion was conditioned in rats by pairing saccharin with cyclophosphamide, an immunosuppressive agent. Three days after conditioning, all animals were injected with sheep erythrocytes. Hemagglutinating antibody titers measured 6 days after antigen administration were high in placebo-treated rats. High titers were also observed in nonconditioned animals and in conditioned animals that were not subsequently exposed to saccharin. No agglutinating antibody was detected in conditioned animals treated with cyclophosphamide at the time of antigen administration. Conditioned animals exposed to saccharin at the time of, or following, the injection of antigen were significantly immunosuppressed. An illness-induced taste aversion was also conditioned using lithium chloride (LiCl), a nonimmunosuppressive agent. In this instance, however, there was no attenuation of hemagglutinating antibody titers in response to injection with antigen. This was formally the beginning of psychoneuroimmunology as a modern discipline. To date, a number of innate and adaptive immune responses have been shown to be modulated by behavioral conditioning protocols, in which conditioned immunomodulating responses could be conceptualized as placebo effects.[27-30]

Many CRs can be learned with a single trial, as in fear conditioning and taste-aversion learning, which is another kind of placebo effect (nocebo effect) such as Pavlovian fear conditioning. Pavlovian fear conditioning has become part of the standard arsenal of behavioral tasks used to interrogate the mnemonic capacities of rats and mice.[31] In fear conditioning, neutral stimuli (conditional stimuli or CSs) such as tones, lights, or places (contexts) are arranged to predict aversive outcomes such as footshock (an unconditional stimulus or US). After conditioning, CSs come to evoke learned fear responses (conditional responses or CRs) such as conditioned suppression, freezing and tachycardia.[32] Rodent models of aversive conditioning would become one of the most ubiquitous behavioral paradigms to explore the neural substrates of learning and memory.[33] This placebo effect is beyond the focus of the present topic, therefore we just mention it briefly.

STUDYING PLACEBO ANALGESIA IN ANIMAL MODELS

Study of the placebo effect has yielded its most fruitful results in the field of pain research. An opioid neuronal network in the cerebral cortex and the brainstem was found to mediate placebo-induced analgesia.[34,35] This opioid network belongs to a descending pain-modulating pathway that, either directly or indirectly, connects the cerebral cortex to the brainstem.[36] However, to date, there have been few studies of placebo analgesia in animal models. Guo et al[37] first evoked opioid and nonopioid placebo responses in mice that were either naloxone-reversible or naloxone-insensitive, depending on the drug used in the conditioning procedure. This procedure in mice may serve as a model for further understanding of the opioid and nonopioid mechanisms underlying placebo responses. Furthermore, the established placebo analgesia was considered to be transferable from pain to depression and could produce a significant antidepressant effect in a test on depression in mice. With this animal model of the placebo response after morphine preconditioning, the opioid placebo analgesia was found to be mediated exclusively through a μ-opioid receptor in the rat.

DISSECTION OF PLACEBO ANALGESIA IN MICE

Amanzio and Benedetti[11] have investigated the mechanisms underlying the activation of endogenous opioids in placebo analgesia. Their findings show that the cognitive factors and conditioning are balanced in different ways in placebo analgesia, and this balance is crucial for the activation of opioid or nonopioid systems. Expectation triggers endogenous opioids, whereas conditioning activates specific subsystems. In fact, if conditioning is performed with opioids, placebo analgesia is mediated via opioid receptors; if conditioning is performed with nonopioid drugs, other nonopioid mechanisms are found to be involved. On the basis of Amanzio and Benedetti's experiments[11] and previous animal work on placebo responses, the animal model of placebo analgesia was established in these studies. This animal model might also help to discover whether placebo analgesia is divided into opioid and nonopioid components in mice, in an attempt to clarify the mechanism of activation of opioid and nonopioid responses.[37]

Figure 3.1 shows the animal training procedure of Guo et al[37] that was performed with 4 days of drug conditioning. Female imprinting control region (ICR) mice weighing 18–22 g at the start of the experiment were used. The hot-plate test was used to measure response latencies according to the method described by Hargraves and Hentall.[38] The reason for using the hot plate test was that the supraspinal mechanism was considered to be involved in this pain model.[39] By contrast, some nociceptive tests, such as the tail-flick test, are spinal reflexes, as they persist after section or cold block of upper parts of the spinal cord.[40,41]

As animals received saline treatment and a 30-minute exposure to the cue compartment after morphine conditioning, pain tolerance was significantly elevated compared with the control level, indicating that the previous morphine conditioning was sufficient to evoke a placebo effect. However, if naloxone was administered after morphine conditioning, paw latency was not increased. The same procedures described above were repeated with the nonopioid aspirin conditioning. Interestingly, similar placebo responses were acquired, except that pretreatment with naloxone cannot block the conditioned analgesic response established by prior conditioning with the nonopioid aspirin.

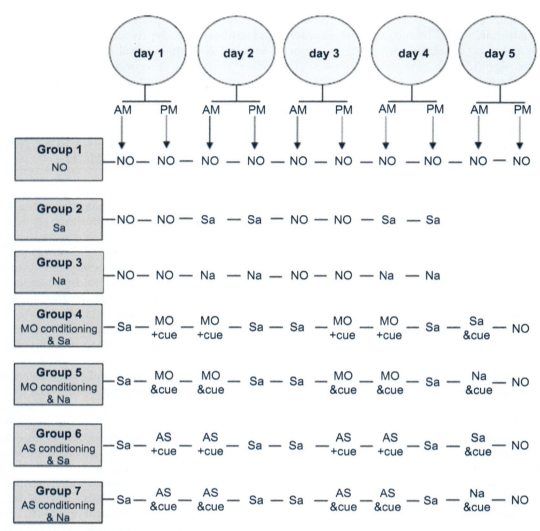

FIGURE 3.1 Experimental paradigm used in the study to identify the opioid and nonopioid components of placebo analgesia in mice. Below each group the experimental condition is specified. Abbreviations: NO, no treatments; Sa, saline; Na, naloxone; AS, aspirin; MO, morphine. Modified from Guo et al.[37]

This research suggested that mice can learn to associate the context cue with elevated pain tolerance via a set of procedures. After mice were given 4 days of drug conditioning with the conditioned cue stimulus (i.e. the chamber) and the unconditioned drug stimulus (morphine or aspirin), saline injection with the contextual cue could produce placebo analgesia at day 5. Moreover, as a placebo response can be subdivided into opioid and nonopioid components in humans, it is also divided into opioid and nonopioid components in mice, depending on the analgesics used during the training procedure; morphine conditioning produced a placebo response that was completely antagonized by naloxone. By contrast, aspirin conditioning elicited a placebo effect that was not blocked by naloxone. This indicates that placebo analgesia can also be dissected into opioid and nonopioid components in mice.

PLACEBO ANALGESIA AFFECTS THE BEHAVIORAL DESPAIR TESTS IN MICE

An abundance of neuroimaging studies has revealed the brain basis of placebo effects on pain and emotion regulation.[42–44] However, in these studies the investigators focused only on the placebo effect obtained within a single domain. That is, they either studied the analgesic effect of a pain-alleviating expectation[45–47] or the ataractic effect of an anxiety-reducing expectation.[42] For a given placebo effect such as the analgesic effect, however, its effective scope might be more general than pain-alleviating. Recent studies in Luo's laboratory[48,49] showed that the placebo effect in humans can be transferred from one domain to the other, namely from pain to emotion. A significant transferable placebo effect that alleviated negative feelings was observed. EEG recordings

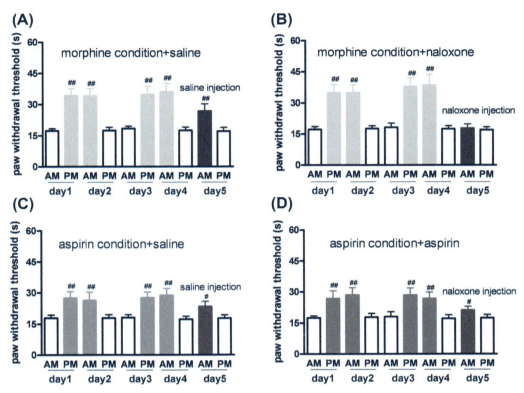

FIGURE 3.2 Morphine- or aspirin-induced placebo effect and its modulation with naloxone. (A) After the procedure of morphine conditioning on days 1–4, mice were injected with saline and put into the conditioned cue box for 30 minutes on day 5 at 8 AM. Paw withdrawal latency was significantly elevated, which mimics the morphine analgesic response. (B) When an injection of naloxone was delivered before exposure to the cue environment after morphine conditioning, the morphine-mimicking effect was completely abolished. (C) After the procedure of aspirin conditioning on days 1–4, mice received a saline injection and stayed in the cue chamber on day 5 at 8 AM for 30 minutes. Paw withdrawal latency was significantly increased, which mimics the aspirin analgesic response. (D) When an injection of naloxone was employed, instead of saline, the placebo effect was not affected. ** $p < 0.01$, * $p < 0.05$, compared with day 1 AM (control condition). Modified from Guo et al.[37]

showed that the transferable placebo treatment which was induced decreased P2 amplitude and increased N2 amplitude, with source location near the posterior cingulate, and fMRI results indicated that this transferable placebo treatment, relative to the control condition, was associated with reduced activity in the amygdala and insula and increased activity in the subgenual anterior cingulate cortex known to be important in emotional regulation. Therefore, this study was performed to test whether placebo effect could be transferred from pain (placebo analgesia) to the other domain (e.g. depressive-like behaviors) in placebo-analgesic mice.[50]

The tail suspension test (TST) and the forced swimming test (FST) are two widely accepted behavioral models for assessing pharmacologic antidepressant activity,[51–53] and it is well known that antidepressant drugs are able to reduce the immobility time in rodents.[54] Thus the antidepressant effect of placebo analgesia was tested with the TST and the FST. The animal model of placebo response after morphine preconditioning was established as described in Figure 3.1. To compare the antidepressant-like effect of placebo analgesia, this study included a group of animals that received treatment with clomipramine hydrochloride (1, 5 or 50 mg.kg^{-1}). Placebo analgesia induced a decrease in immobility (i.e. antidepressant-like effect) that was significantly different from that shown by the control group. Moreover, the placebo analgesia produced a more pronounced antidepressant-like effect than that achieved with 1 mg.kg^{-1} clomipramine, but comparable to that produced by the 5 mg.kg^{-1} dose of clomipramine. Similar results were also obtained in the FST; placebo analgesia significantly decreased the FST-induced immobility time in mice. Comparison of the antidepressant-like actions of placebo analgesia with this effect induced by clomipramine also showed that placebo analgesia produced a more pronounced antidepressant-like effect than that achieved with 1 mg.kg^{-1} clomipramine, but comparable to that produced by the 5 mg.kg^{-1} dose of clomipramine.

The hypothalamus–pituitary–adrenal axis (HPA) axis is a three-gland component of the endocrine system that modulates biologic responses to acute and chronic stress;[55,56] adrenocorticotropic hormone (ACTH) and corticosterone are considered as markers of stress.[57]

Plasma concentrations of corticosterone and ACTH were also measured in this study. The positive control clomipramine significantly reduced the TST-induced increase in the plasma concentrations of corticosterone and ACTH in a dose-dependent manner. Placebo analgesia also markedly decreased the TST-induced plasma concentrations of corticosterone and ACTH compared with control group. Similar results were also obtained in the FST study.

This study confirmed previous observations that mice can learn to associate the context cue with elevated pain tolerance via a set of procedures. The placebo analgesia, which was established by a set of procedures in mice, was transferable and could produce a significant antidepressant effect in a test on depression. Plasma levels of corticosterone and ACTH further proved that the placebo analgesia that was established from pain-reducing training not only induced a significant placebo effect on pain, but also significantly decreased the HPA response to stress and produced a stress-alleviating effect. The immobility time of placebo-analgesia mice in the TST and FST was comparable to that produced by the 5 mg.kg^{-1} dose of clomipramine hydrochloride.

THE OPIOID RECEPTORS INVOLVED IN THE PLACEBO RESPONSE IN RATS

Recent neuroimaging data point towards the rACC as a crucial cortical region involved in placebo analgesia, and the rACC yielded increased activity during both placebo and opioid analgesia.[13] Wager et al[43] found changed activity in pain-sensitive regions such as rACC when comparing the response to noxious stimuli applied to control and placebo cream-treated areas of the skin. In a recent functional magnetic resonance imaging (fMRI) study investigating the activation of the opioidergic descending pain control system in placebo analgesia, Eippert et al[58] have also found that the rACC was an important pain-modulatory cortical structure in behavioral and neural placebo effects as well as in placebo-induced responses.

With a μ-opioid-receptor-selective radiotracer, Zubieta et al[8] showed that significant placebo-induced activation of μ-opioid receptor-mediated neurotransmission was observed in both higher-order and subcortical brain regions. Regional activations were paralleled by lower ratings of pain intensity and reductions in its sensory and affective qualities. Wager et al[14] have demonstrated that the administration of a placebo with implied analgesic properties regionally activates a pain- and stress-inhibitory neurotransmitter system, the endogenous opioid system, through direct effects on the μ-opioid receptors. Besides, the activation of the μ-opioid receptor system is associated with reductions in the sensory and affective ratings of the pain experience. In addition, Scott et al[15] reported that regional μ-opioid activity was associated with the anticipated and subjectively perceived effectiveness of the placebo and reductions in continuous pain ratings. However, to date, it remains unclear whether delta- or kappa-opioid receptors are involved in the placebo analgesia. Very recently, Zhang et al[59] conducted an experiment to investigate whether delta- or kappa-opioid receptors are involved and whether rACC is the key brain structure in placebo analgesia.

The animal training procedure was performed as described above. To test whether rACC is the key brain region involved in placebo analgesia, placebo rats were given intra-rACC injections with naloxone (three doses, 1, 3, 10 μg per rat) 30 minutes before the hot-plate tests. Then, to test whether a μ-, δ- or κ-opioid receptor is involved in the placebo analgesia, these rats were given (before the hot-plate tests) an intra-rACC injection with D-Phe-Cys-Tyr-D-Trp-Orn-Thr-Pen-Thr-NH2 (CTOP), a selective μ-opioid receptor antagonist, or naltrindole (NTI), a highly selective δ-opioid receptor antagonist, or nor-binaltorphimine (nor-BNI), a highly selective κ-opioid receptor antagonist.

After the rats had undergone the placebo analgesia training for 4 days, saline treatment on day 5, and a 30-minute exposure to the cue compartment, pain tolerance was significantly elevated compared with day 1, indicating that the previous morphine conditioning was sufficient to evoke a placebo effect in rats. Microinjection of naloxone produced a dose-related inhibition on the paw withdraw latency. All three doses of naloxone reduced the pain threshold. Moreover, drug injection outside rACC did not alter the pain latency. Furthermore, CTOP dose-dependently inhibited the placebo analgesia. However, neither nor-NBI nor NTI affected the placebo analgesia in rats.

Consistent with previous research in humans, this study also showed that rACC plays a key role in opioid placebo analgesia. Moreover, the opioid placebo analgesia was blocked by microinjection of CTOP into rACC, a selective μ-opioid receptor antagonist, but not by the δ- or κ-opioid receptor antagonists, NTI and nor-BNI, respectively, indicating that the opioid placebo analgesia is mediated exclusively through μ-opioid receptors in the rat.

THE PROS AND CONS OF STUDYING PLACEBO IN THE ANIMAL MODEL

The placebo effect has been a topic of interest in scientific and clinical communities for many years, and our knowledge of the mechanisms of the placebo effect has advanced considerably within the past decade. A significant proportion of the research has occurred in the fields of pain and analgesia, and the placebo analgesic response appears to be the best-understood model of placebo mechanisms. The placebo animal model refers to induction of a placebo response in a nonhuman animal; the psychologic process might be similar to that in humans. A placebo analgesic effect can be fostered by exposure to environmental cues prior to an effective analgesic treatment and contextual cues associated with such treatment. This evidence for the role of associative learning in placebo analgesia suggests the potential use of animal models for studying this phenomenon. Here, we discuss the pros and cons of studying placebo in the animal model.

The Pros of Studying Placebo in the Animal Model

As it is difficult to conduct invasive placebo experiments in human subjects, the biologic mechanisms of opioid and nonopioid placebo responses remain largely unknown. The use of a placebo animal model allows researchers to investigate processes in ways that would be inadmissible in a human subject, performing procedures on the nonhuman animal that imply a level of harm that would not be considered ethical to inflict on a human. Therefore, the use of animal models will help us to better understand the placebo effect at the level of neural mechanisms, as well as the genetic bases. For example, the μ-opioid system has been found to be implicated in the regulation of placebo analgesia in humans.[14,60] However, so far, whether other opioid receptors are involved in the placebo analgesia remains unclear. After rats had undergone the placebo analgesia training for 4 days, they were microinjected into the rACC with mu-, delta- or kappa-opioid receptor antagonists before the hot-plate tests. We first showed that opioid placebo analgesia is mediated exclusively through μ-opioid receptors in the rat.

The most important factor in the direct causes of the placebo effect is that the stimulus features of the placebo 'object' are totally arbitrary and depend on the idiosyncrasies of past experience, cultural meaning, and the social context in which they occur. For example, if a patient learns from experience that intravenous analgesics given by a man in a blue scrub suit are generally more powerful than over-the-counter oral analgesics, then an intravenous placebo would likely become more effective than an oral placebo for that specific patient. For another individual in another therapeutic setting in another culture, a woman in feathers applying a leaf poultice might be more effective than a western-trained physician giving intravenous saline. Furthermore, most adults have had previous exposures to clinical experiences such as taking oral analgesics (opioid or non-opioid drugs). Therefore, although Amanzio and Benedetti[11] showed that expectation triggers endogenous opioids, conditioning activates specific subsystems. If conditioning is performed with opioids, placebo analgesia is mediated via opioid receptors; if conditioning is performed with nonopioid drugs, other nonopioid mechanisms might be involved. However, they cannot preclude a previous conditioning in their experimental subjects. Thus, clear separation of conditioning from other aspects of the placebo response in human experiments is difficult. Use of the animal model for studying the placebo response can overcome the individual experience, culture and social context. As the mice had never been exposed to either opioids or nonopioids in their previous experience, the results should be more convincing for a placebo response.[37]

The equipment required for placebo animal training can be readily scaled to condition several animals simultaneously. The placebo analgesia model requires nothing more than standard rodent conditioning chambers with grid floors and light. The animal training procedures for determining the strength of the response can be manipulated experimentally. These simple procedures can produce robust learning, as the contextual cue produces significant analgesia. The behavior (e.g. pain) data and serum could be easily collected by ordinary researchers. Patently, the procedures in this model are simpler to explain and carry out compared to cognitive personality views, a fact which increases their applicability. In contrast, it must be used very carefully to perform placebo studies in humans. Sophisticated doctors and nurses are needed when drugs are to be administered, or serum collected, in human subjects.

The Cons of Studying Placebo in the Animal Model

Historically, there have been two primary perspectives for approaching the mechanism underlying the placebo effect: expectancy theory[61–65] and classic conditioning.[66–69] Expectancy theory suggests that the placebo effect is achieved through an instruction that initiates a positive expectation toward the preparation. An alternative explanation is the conditioning model, which states that the repeated association of a neutral stimulus with the pharmacologic effect of the agent leads to a conditioned reaction that is similar to the original response to the pharmacologic agent and is now triggered by the placebo. A previous study showed that cognitive expectation cues, drug conditioning, or a combination of both, could evoke different types of placebo analgesic responses in humans.[11] For the nonverbal animal, associative learning is considered to play the key role in the placebo effect. Experience with active treatments may create conditioned associations between treatment context (e.g. an injection) and endogenous neurophysiologic responses. Such conditioned responses may be unconscious and involuntary, engaging separate neural mechanisms from those involved in expectancy. However, conditioning procedures also create expectations that, in turn, may play a key role in the conditioned response.[68] There is considerable overlap between expectancy and conditioning, because learning is one of the major ways that expectancies are formed. Thus, drug-conditioning procedures may also create expectations; whether the placebo animal model is related to expectations or conditioning is unknown. The relative contributions of conditioning and expectancy to placebos in the animal are difficult to disentangle. Therefore, it is impossible to study different types of placebo effect (induced by conditioning or expectancy) in the animal model.

In order to be useful, an animal model must be similar to the human equivalent in the mechanisms of cause and function. It goes without saying that one of the most important factors that triggers placebo effect is represented by verbal suggestions. Thus, animal models are useful models for some components of placebo effects but are intrinsically limited as placebo-effect models because there are no verbally mediated placebo changes. In general, for a placebo response to occur, it would seem necessary that the patient being treated recognizes that there is an intentional effort to treat. Animals appear to lack the ability to comprehend such intentions (although they may not like a particular intervention). As such, animals would not be able to participate in placebo-generating experiences. So, for example, one couldn't rationally suggest to a rat that a particular therapy might help it to get better, or that it was beneficial because it was 'natural'; one presumably wouldn't wax eloquent to a rat that a particular therapy might give it a window of hope for recovery. They just wouldn't understand.

The increase in knowledge of the genomes of nonhuman primates, and other mammals that are genetically close to humans, is allowing the production of genetically engineered animal tissues, organs and even animal species which express human diseases, providing a more robust model of human diseases in an animal model. However, the placebo effect is a very complicated psychobiologic phenomenon that can be attributable to complex brain networks. For example, many brain regions are involved in placebo analgesia in human; these regions include the rACC, hypothalamus (HYPO), periaqueductal gray (PAG), rostroventromedial medulla (RVM), and spinal cord. The dopaminergic reward system, in which dopaminergic neurons in the ventral tegmental area (VTA) project to the nucleus accumbens (NAcc), is also involved.[70] In contrast, the nervous systems of nonmammalian species are much simpler. Furthermore, the rodent species' phylogenetic root is much different from that of humans. These factors might limit the use of animals as a placebo model.

CONCLUSION AND FUTURE DIRECTIONS

Research on the neurobiology of placebo effects is becoming an active and productive area of science with the final aim of understanding their healing potential, as well as their limitations, and of delineating correct and ethical use. The study of placebo effects has yielded its most fruitful results in the field of pain and analgesia, and the placebo analgesic response appears to be the best-understood model of placebo mechanisms.

A placebo analgesic effect can be fostered by exposure to environmental cues prior to an effective analgesic treatment and contextual cues associated with such treatment. This evidence for the role of associative learning in placebo analgesia strongly supports the potential use of animal models for studying this phenomenon. The investigation of placebo responses in animals will pave the way for new research to be done on brain mechanisms in pain modulation as well as into some of the unsolved questions that arise in clinical trials, such as pharmacologic conditioning in crossover designs. These experiments will also provide the foundation for methodologies that would be extremely difficult in human studies, such as recording single-neuron activity and exploring genetic contributions by using knock-out animals.

Acknowledgments

This work was supported by projects from Key Laboratory of Mental Health, Institute of Psychology, Chinese Academy of Sciences; National Natural Science Foundation of China (30800301, 31170992) and the Knowledge Innovation Program of the Chinese Academy of Sciences (KSCX2-YW-R-254, KSCX2-EW-Q-18 and KSCX2-EW-J-8).

References

1. Dubner R, Ren K. Endogenous mechanisms of sensory modulation. *Pain*. 1999;(suppl 6):S45-S53.
2. Bingel U, Lorenz J, Schoell E, et al. Mechanisms of placebo analgesia: rACC recruitment of a subcortical antinociceptive network. *Pain*. 2006;120:8-15.
3. Levine JD, Gordon NC, Fields HL. The mechanism of placebo analgesia. *Lancet*. 1978;2:654-657.
4. Levine JD, Gordon NC. Influence of the method of drug administration on analgesic response. *Nature*. 1984;312:755-756.
5. Grevert P, Albert LH, Goldstein A. Partial antagonism of placebo analgesia by naloxone. *Pain*. 1983;16:129-143.
6. Benedetti F. The opposite effects of the opiate antagonist naloxone and the cholecystokinin antagonist proglumide on placebo analgesia. *Pain*. 1996;64:535-543.
7. Hoehn-Saric R, Masek BJ. Effects of naloxone on normals and chronically anxious patients. *Biol Psychiatry*. 1981;16:1041-1050.
8. Zubieta JK, Bueller JA, Jackson LR, et al. Placebo effects mediated by endogenous opioid activity on mu-opioid receptors. *J Neurosci*. 2005;25:7754-7762.
9. Fields HL, Levine JD. Pain – mechanics and management. *West J Med*. 1984;141:347-357.

10. Gracely RH, Dubner R, Wolskee PJ, Deeter WR. Placebo and naloxone can alter post-surgical pain by separate mechanisms. *Nature*. 1983;306:264-265.
11. Amanzio M, Benedetti F. Neuropharmacological dissection of placebo analgesia: expectation-activated opioid systems versus conditioning-activated specific subsystems. *J Neurosci*. 1999;19:484-494.
12. Fields HL, Price DD. Toward a neurobiology of placebo analgesia. In: Harrington A, ed. *The Placebo Effect: an Interdisciplinary Exploration*. Cambridge: Harvard Univ Press; 1997.
13. Petrovic P, Kalso E, Petersson KM, Ingvar M. Placebo and opioid analgesia – imaging a shared neuronal network. *Science*. 2002;295:1737-1740.
14. Wager TD, Scott DJ, Zubieta JK. Placebo effects on human mu-opioid activity during pain. *Proc Natl Acad Sci USA*. 2007;104:11056-11061.
15. Scott DJ, Stohler CS, Egnatuk CM, et al. Placebo and nocebo effects are defined by opposite opioid and dopaminergic responses. *Arch Gen Psychiatry*. 2008;65:220-231.
16. Benedetti F, Pollo A, Lopiano L, et al. Conscious expectation and unconscious conditioning in analgesic, motor, and hormonal placebo/nocebo responses. *J Neurosci*. 2003;23:4315-4323.
17. Benedetti F, Amanzio M, Rosato R, Blanchard C. Nonopioid placebo analgesia is mediated by CB1 cannabinoid receptors. *Nat Med*. 2011;17:1228-1230.
18. Pavlov IP. *Conditioned Reflexes*. London: Oxford University Press; 1927.
19. Wickramasekera I. A conditioned response model of the placebo effect predictions from the model. *Biofeedb Self Regul*. 1980;5:5-18.
20. Collins KH, Tatum AL. A conditioned salivary reflex established by chronic morphine poisoning. *Am J Physiol*. 1925;74:14-15.
21. Herrnstein RJ. Placebo effect in the rat. *Science*. 1962;138:677-678.
22. Sherman R, Schnitzer SB. Further support for a placebo effect in the rat. *Psychologic Rep*. 1963;13:461-462.
23. Pihl RO, Altman J. An experimental analysis of the placebo effect. *J Clin Pharmacol*. 1972;11:91-95.
24. Colloca L, Petrovic P, Wager TD, et al. How the number of learning trials affects placebo and nocebo responses. *Pain*. 2010;151:430-439.
25. Metal'nikov S, Chorine V. Rôle des réflèxes conditionnnels dans l'immunité. *Annals L'Institute Pasteur*. 1926;40:839-900.
26. Ader R, Cohen N. Behaviorally conditioned immunosuppression. *Psychosom Med*. 1975;37:333-340.
27. Ader R, Cohen N. Conditioning of the immune response. *Neth J Med*. 1991;39:263-273.
28. Pezzone MA, Lee WS, Hoffman GE, et al. Activation of brainstem catecholaminergic neurons by conditioned and unconditioned aversive stimuli as revealed by c-Fos immunoreactivity. *Brain Res*. 1993;608:310-318.
29. Buske-Kirschbaum A, Grota L, Kirschbaum C, et al. Conditioned increase in peripheral blood mononuclear cell (PBMC) number and corticosterone secretion in the rat. *Pharmacol Biochem Behav*. 1996;55:27-32.
30. Bovbjerg D, Ader R, Cohen N. Acquisition and extinction of conditioned suppression of a graft-vs-host response in the rat. *J Immunol*. 1984;132:111-113.
31. Maren S. Neurobiology of Pavlovian fear conditioning. *Annu Rev Neurosci*. 2001;24:897-931.
32. Overmier JB, Leaf RC. Effects of discriminative Pavlovian fear conditioning upon previously or subsequently acquired avoidance responding. *J Comp Physiol Psychol*. 1965;60:213-217.
33. Maren S. Pavlovian fear conditioning as a behavioral assay for hippocampus and amygdala function: cautions and caveats. *Eur J Neurosci*. 2008;28:1661-1666.
34. Benedetti F, Mayberg HS, Wager TD, et al. Neurobiological mechanisms of the placebo effect. *J Neurosci*. 2005;25:10390-10402.
35. Price DD, Finniss DG, Benedetti F. A comprehensive review of the placebo effect: recent advances and current thought. *Annu Rev Psychol*. 2008;59:565-590.
36. Eippert F, Bingel U, Schoell ED, et al. Activation of the opioidergic descending pain control system underlies placebo analgesia. *Neuron*. 2009;63:533-543.
37. Guo JY, Wang JY, Luo F. Dissection of placebo analgesia in mice: the conditions for activation of opioid and non-opioid systems. *J Psychopharmacol*. 2010;24:1561-1567.
38. Hargraves WA, Hentall ID. Analgesic effects of dietary caloric restriction in adult mice. *Pain*. 2005;114:455-461.
39. Le Bars D, Gozariu M, Cadden SW. Animal models of nociception. *Pharmacol Rev*. 2001;53:597-652.
40. Sinclair JG, Main CD, Lo GF. Spinal vs. supraspinal actions of morphine on the rat tail-flick reflex. *Pain*. 1988;33:357-362.
41. Irwin S, Houde RW, Bennett DR, et al. The effects of morphine methadone and meperidine on some reflex responses of spinal animals to nociceptive stimulation. *J Pharmacol Exp Ther*. 1951;101:132-143.
42. Petrovic P, Dietrich T, Fransson P, et al. Placebo in emotional processing – induced expectations of anxiety relief activate a generalized modulatory network. *Neuron*. 2005;46:957-969.
43. Wager TD, Rilling JK, Smith EE, et al. Placebo-induced changes in FMRI in the anticipation and experience of pain. *Science*. 2004;303:1162-1167.
44. Wager TD, Atlas LY, Leotti LA, Rilling JK. Predicting individual differences in placebo analgesia: contributions of brain activity during anticipation and pain experience. *J Neurosci*. 2011;31:439-452.
45. Kong J, Gollub RL, Rosman IS, et al. Brain activity associated with expectancy – enhanced placebo analgesia as measured by functional magnetic resonance imaging. *J Neurosci*. 2006;26:381-388.
46. Wager TD, Matre D, Casey KL. Placebo effects in laser-evoked pain potentials. *Brain Behav Immun*. 2006;20:219-230.
47. Matre D, Casey KL, Knardahl S. Placebo-induced changes in spinal cord pain processing. *J Neurosci*. 2006;26:559-563.
48. Zhang W, Qin S, Guo J, Luo J. A follow-up fMRI study of a transferable placebo anxiolytic effect. *Psychophysiology*. 2011;48:1119-1128.
49. Zhang W, Luo J. The transferable placebo effect from pain to emotion: changes in behavior and EEG activity. *Psychophysiology*. 2009;46:626-634.
50. Guo JY, Yuan XY, Sui F, et al. Placebo analgesia affects the behavioral despair tests and hormonal secretions in mice. *Psychopharmacology (Berl)*. 2011;217:83-90.
51. Rago L, Saano V, Auvinen T, et al. The effect of chronic treatment with peripheral benzodiazepine receptor ligands on behavior and GABAA/benzodiazepine receptors in rat. *Naunyn Schmiedebergs Arch Pharmacol*. 1992;346:432-436.
52. Wada T, Fukuda N. Discriminative stimulus properties of a new anxiolytic, DN-2327, in rats. *Psychopharmacology (Berl)*. 1993;110:280-286.
53. Bourin M, Chenu F, Ripoll N, David DJ. A proposal of decision tree to screen putative antidepressants using forced swim and tail suspension tests. *Behav Brain Res*. 2005;164:266-269.
54. Gardner CR, Ward RA, Deacon RM, et al. Effects of RU33368, a low affinity ligand for neuronal benzodiazepine receptors, on rodent behaviours and GABA-mediated synaptic transmission in rat cerebellar slices. *Gen Pharmacol*. 1992;23:1193-1198.
55. Belzung C, Vogel E, Misslin R. Benzodiazepine antagonist RO 15-1788 partly reverses some anxiolytic effects of ethanol in the mouse. *Psychopharmacology (Berl)*. 1988;95:516-519.
56. Pellow S, File SE. Anxiolytic and anxiogenic drug effects on exploratory activity in an elevated plus-maze: a novel test of anxiety in the rat. *Pharmacol Biochem Behav*. 1986;24:525-529.
57. Treit D. Ro 15-1788, CGS 8216, picrotoxin, and pentylenetetrazol: do they antagonize anxiolytic drug effects through an anxiogenic action? *Brain Res Bull*. 1987;19:401-405.
58. Eippert F, Bingel U, Schoell E, et al. Blockade of endogenous opioid neurotransmission enhances acquisition of conditioned fear in humans. *J Neurosci*. 2008;28:5465-5472.

59. Zhang RR, Zhang WC, Wang JY, Guo JY. The opioid placebo analgesia is mediated exclusively through μ-opioid receptor in rat. *Int J Neuropyschopharmacol.* 2013;16:849-856.
60. Zubieta JK, Ketter TA, Bueller JA, et al. Regulation of human affective responses by anterior cingulate and limbic mu-opioid neurotransmission. *Arch Gen Psychiatry.* 2003;60:1145-1153.
61. Montgomery GH, Kirsch I. Classical conditioning and the placebo effect. *Pain.* 1997;72:107-113.
62. Price DD, Milling LS, Kirsch I, et al. An analysis of factors that contribute to the magnitude of placebo analgesia in an experimental paradigm. *Pain.* 1999;83:147-156.
63. Vase L, Robinson ME, Verne GN, Price DD. The contributions of suggestion, desire, and expectation to placebo effects in irritable bowel syndrome patients. An empirical investigation. *Pain.* 2003;105:17-25.
64. Benedetti F, Arduino C, Amanzio M. Somatotopic activation of opioid systems by target-directed expectations of analgesia. *J Neurosci.* 1999;19:3639-3648.
65. Benedetti F, Amanzio M, Baldi S, et al. Inducing placebo respiratory depressant responses in humans via opioid receptors. *Eur J Neurosci.* 1999;11:625-631.
66. Voudouris NJ, Peck CL, Coleman G. Conditioned response models of placebo phenomena: further support. *Pain.* 1989;38:109-116.
67. Voudouris NJ, Peck CL, Coleman G. Conditioned placebo responses. *J Pers Soc Psychol.* 1985;48:47-53.
68. Wager TD, Nitschke JB. Placebo effects in the brain: linking mental and physiological processes. *Brain Behav Immun.* 2005;19: 281-282.
69. Rescorla RA. Behavioral studies of Pavlovian conditioning. *Annu Rev Neurosci.* 1988;11:329-352.
70. Benedetti F, Carlino E, Pollo A. How placebos change the patient's brain. *Neuropsychopharmacol.* 2011;36:339-354.

CHAPTER 4

Molecular Mechanisms of Placebo Responses in Humans

Marta Peciña[1], Jon-Kar Zubieta[2]

[1]Department of Psychiatry and Molecular and Behavioral Neuroscience Institute, University of Michigan, Ann Arbor, MI, USA, [2]Department of Psychiatry and Molecular and Behavioral Neuroscience Institute, and Department of Radiology, University of Michigan, Ann Arbor, MI, USA

INTRODUCTION

Historically, placebo effects have been reported consistently since the emergence of placebo-controlled trials in the 18th century, when putatively active treatments were for the first time compared against 'sham' controls. It was already recognized that 'the passions of the mind [had a wonderful and powerful influence] upon the state and disorder of the body.'[1] In the widely quoted Beecher report of clinical trials of analgesic drugs,[2] it was noted that placebos exerted significant clinical responses in approximately 30% of patients enrolled in inactive treatment groups (Ch 1). Clinically significant placebo-associated improvements can occur in as few as 5% or as many as 65% of individuals in randomized, controlled trials (RCTs), depending on the disease process under consideration and the particular study sample. Other elements frequently unaccounted for have included the effects of the natural history of the disease (no-treatment), which can spontaneously remit or change in severity in the course of the disease without intervention. The presence of other cognitive–emotional biases, such as the 'halo' effect, related to the individual response, to the characteristics of the experimenting or treatment team or individual, or those induced by the fact that subjects know that they are being studied, termed the Hawthorne effect, have to be additionally considered in the interpretation of placebo-related responses, particularly so when subjective or simple behavioral measures (e.g. improvement in performance) are the primary outcomes.

Considerable effort has been committed to the study of mechanisms that may account for placebo-induced analgesic effects over the last three decades.[3,4] An emerging literature has examined the neurobiology of placebo effects across a variety of domains, such as mood and affective regulation[5,6] as well as motor control in Parkinson's disease.[7,8] However, the neurobiology of the placebo effect was born in 1978, when it was shown that placebo analgesia could be blocked by the opioid receptor antagonist naloxone. This indicated an involvement of the endogenous opioid system in the production of placebo-induced analgesic effects.[9] In patients who had undergone oral surgery 2 hours prior, naloxone, placebo or morphine were administered with the expectation of either pain relief or pain worsening. Naloxone was associated with hyperalgesia, showing that the stress and/or pain associated with the surgical procedure had, by itself, induced the release of endogenous opioids. The administration of placebo induced a significant reduction in pain ratings in 39% of the subjects, which was fully antagonized by naloxone. In subsequent studies by the same group, in which hidden and machine-driven infusions of placebo and naloxone were introduced,[10] the effect of naloxone on placebo analgesia was confirmed, and estimated to approximate that of 8 mg of morphine in that particular experimental setting. Subsequent studies by Gracely et al[11] and Grevert et al[12] using similar opioid receptor-blocking pharmacologic challenges confirmed the existence of opioid-mediated placebo analgesia but also described a time-dependent, nonopioid component that is not reversible by naloxone.

In what has become a classic study of components related to the development of placebo analgesic effects,

Amanzio and Benedetti[13] explored the contribution of verbally-induced expectations and conditioning to the development of placebo analgesic effects. Utilizing ischemic arm pain as an experimental model, it was demonstrated that contextual cues promoting a credible expectation of analgesia during placebo administration induced analgesic effects that were completely blocked by naloxone (i.e. expectation effects were entirely mediated by the activation of opioid mechanisms). Expectation cues that followed a course of morphine (morphine pre-conditioning group) also produced analgesic responses that were also fully antagonized by naloxone. Naloxone reversibility was also achieved in the absence of cues promoting expectation as long as morphine had been pre-administered (i.e. the volunteers were receiving an inactive agent when morphine would normally have been administered). However, conditioning with the nonsteroidal anti-inflammatory drug, ketorolac, paired with additional expectation cues, induced a placebo anti-nociceptive response that was only partially blocked by naloxone, while ketorolac conditioning alone produced analgesia that proved to be naloxone insensitive. Overall, these results showed that while purely cognitive factors were associated with the activation of endogenous opioid systems, conditioning was capable of recruiting other mechanisms in support of analgesia depending on the conditioning agent. More recently, work by the same group showed that conditioning with ketorolac was mediated by CB1 receptors, suggesting that the endocannabinoid system has a pivotal role in placebo analgesia in a conditioning paradigm that is not necessarily related to endogenous opioid system function.[14]

These and other observations have led to the proposition of a number of theoretical constructs to explain the formation of placebo effects, again most typically studied in the context of analgesic responses to pain. All these constructs hinge upon elements of higher order processing involving cognitive and emotional circuits, known to modulate the experience of pain. (A) Verbally-induced expectations of relief, whereby cognitive assessments and beliefs of analgesia trigger the placebo effects.[15,16] (B) Anxiety relief, where placebo administration elicits analgesia through reductions in the anxiety experienced by the subjects.[17,18] (C) The conditioning hypothesis emphasizes the engagement of learned responses through previous exposures to active treatments.[19–22] (D) The so-called response appropriate sensations hypothesis further states that pain and analgesia are experienced after a complex, preconscious assessment of sensory and internal stimuli. Pain experience or pain suppression are then engaged as a process of adaptation to environmental circumstances.[23]

A number of studies have now shown the involvement of distinct brain structures in responses to cognitive manipulation. Hypnotic suggestions have been used to selectively reduce or increase sensory (intensity) and affective (unpleasantness) qualities of pain, with the effects being associated with changes in the metabolic activity of the somatosensory and anterior cingulate cortex, respectively.[24–26] Hypnotic suggestions, however, seem to differ from, and could not account for, typical placebo analgesic responses,[27] albeit some similarities as to the networks involved have emerged.[28] In fact, data show that certain CNS circuits, known to be involved in the perception and integration of the pain experience, are susceptible to various manipulations. The perception of pain can be either diminished or enhanced, depending on the additional presence of cognitive distractors, or the suggestion of pain enhancement or reduction.[29] Theories regarding the placebo analgesic effect uniformly acknowledge the interplay between environmental information and their perception and integration by the individual's organism to induce a positive (placebo) or negative (nocebo) response. The presence of these interactions implies the involvement of higher order, CNS associative processes in the production of analgesic placebo effects. This assertion has been elegantly demonstrated by work in which analgesic agents were administered covertly (subjects were not aware of the actual timing of the administration). Substantially lower and even insignificant effects were obtained from even well-recognized analgesic treatments when the context of drug administration was removed from the treatment.[30–32] These findings call for the elucidation of mechanisms underlying 'mind–body' interactions.

In an initial report, the effects of the short-acting μ-opioid receptor agonist remifentanil on regional cerebral blood flow (rCBF, as measured with positron emission tomography [PET], thought to reflect metabolic demands), were found to overlap with the effects of a placebo under conditions of expectation of analgesia in the rostral anterior cingulate cortex (rACC). Placebo administration increased the correlation between the activity of this region and that of the midbrain periaqueductal gray (PAG), a region known to exert modulatory effects on the experience of pain. Individuals with high placebo analgesic responses further demonstrated greater rCBF responses to remifentanil, suggesting that individual differences in placebo analgesia may involve differences in the concentration or function of μ-opioid receptors.[33] One step further, the question of whether placebo analgesia or opioid analgesia is of an additive nature has been recently investigated.[34] In this study, it was shown that although both remifentanil and expectancy reduced pain, drug effects on pain reports and functional MRI activity did not interact with expectancy. In this study, regions associated with pain processing showed drug-induced modulation during both Open and Hidden conditions, with no differences in drug effects as a function of expectation. Instead, expectancy modulated activity in frontal cortex, with a separable time course from drug effects. These findings suggest that opiates and placebo treatments both influence clinically relevant outcomes and operate without mutual interference, with placebo administration engaging cognitive networks, a so-called 'top-down' regulation.[34]

Substantial work has utilized functional magnetic resonance imaging (fMRI and the blood-oxygenation-level-dependent[BOLD] signal) as well as a covert manipulation to increase individual verbally-induced expectations of relief (Ch 10). This has consisted of a reduction in the heat intensity of the probe used to induce pain during the administration of a placebo. In response to the manipulation, placebo-associated reductions in the activity of the rACC, insular cortex and thalamus were observed, correlating with the subjectively rated pain relief afforded by the placebo administration.[35] Using a similar experimental approach, the opposite effect, activation of the rACC and increased connectivity between this region, the amygdala and PAG during placebo administration, were described.[36] Other work found increases in the activity of the rACC, prefrontal, insular cortex, supramarginal gyrus and inferior parietal cortex, employing sham acupuncture as a form of placebo intervention.[37] While these differences in the directionality of findings may seem difficult to reconcile, particularly when using similar placebo enhancement procedures, several methodologic differences between the studies have been noted.[37] Among them is the selection criteria for the subjects entered in the neuroimaging protocols. In one of them (showing placebo-associated reductions in BOLD responses during placebo administration), only subjects demonstrating substantial placebo analgesia in preceding 'training' trials were studied.[35] In contrast, the reminder of the studies (showing placebo-associated increases in regional BOLD activity) did not eliminate non-responder subjects for imaging.[36,37] Raz et al[28] reported that only highly hypnotizable subjects responded with reductions in rACC BOLD responses during post-hypnotic suggestions in a cognitive conflict resolution task (as opposed to poorly hypnotizable subjects or volunteers in whom no suggestions were used). This may suggest that differences in subject preselection procedures (e.g. the elimination of non-placebo responders) would have contributed to the apparent differences in response directionality between studies. Later work by Wager and colleagues has studied the predictive value of brain activity during placebo administration.[38] Using a cross-validated regression procedure on previous data (n = 47) they showed that increased anticipatory activity in a frontoparietal network, and decreases in a posterior insular/temporal network, predicted a moderate amount of variance in the placebo response. They also showed that the most predictive regions were those associated with emotional appraisal, rather than cognitive control or pain processing (Ch 8). During pain, decreases in limbic and paralimbic regions most strongly predicted placebo analgesia.

In our laboratory, we have primarily focused on the examination of *in vivo* molecular mechanisms and related circuits involved in the formation of placebo effects. For that purpose, we employ positron emission tomography (PET) and validated models to quantify μ-opioid and DA D2/3 receptors while administering a model of sustained experimental pain. Using these types of functional molecular assays, reductions in the *in vivo* availability (binding potential, BP) of the respective receptor population reflect placebo-induced activation of either the opioid or DA neurotransmission, respectively. Subjects were studied under baseline conditions (no stimulus), pain expectation (pain intensity is rated, expected but not actually endured) and actual pain. The latter two were performed with and without the administration of a placebo, consisting of isotonic saline infused intravenously, 1 mL every 4 minutes and with the subject receiving verbal and visual cues at the time of application. The study sample consisted of young healthy males and females, ages 20–30 years. Women were studied in the follicular phase of the menstrual cycle, ascertained by menstrual diaries, timing of menses and plasma levels of estradiol and progesterone prior to scanning. The sustained pain model employed elicits psychophysical responses similar to those of clinical pain states in terms of pain intensity and pain affect.[39] The resulting steady-state of deep muscle pain was maintained for 20 minutes by a computer-controlled closed-loop system through individually titrated infusion of medication-grade hypertonic saline (5%) into the masseter muscle, aiming for a target pain intensity of 40 visual analog scale (VAS) units.[39,40] Volunteers rated pain intensity every 15 seconds using an electronic version of a 10-cm VAS, placed in front of the scanner gantry. For trials where subjects expected to receive pain but a non-painful stimulus was applied, the same procedure was followed, except that isotonic instead of hypertonic saline was administered.

In order to study the molecular mechanisms underlying the placebo effect, our model of sustained experimental pain was used in either one of two modes of operation, producing different experimental conditions: (a) the placebo effect was assessed by measuring the subject-specific infusion volume required to maintain pain at the preset target level for 20 minutes, with or without the administration of the placebo, and (b) by using the subject-specific, pre-established infusion profile with and without administration of the placebo. In the first condition, the placebo effect is perceptually not transparent to the subject as pain intensity is kept at the preset target level for both the 'no placebo' and 'placebo' conditions and with the effect of the placebo being expressed by the difference of the rate of infusion required between the two conditions. For the second scenario, the subject is able to recognize the effect of the administered placebo by experiencing either a lessening or worsening of the pain intensity over the course of the trial as a consequence of the placebo administration.

In addition to the momentary assessments of pain intensity acquired every 15 seconds, subjects completed the McGill Pain Questionnaire with its sensory and affective subscales,[41] 0–100 VAS scores of pain intensity and unpleasantness, the Positive and Negative Affectivity Scale (PANAS) measuring internal affective state,[42] and the Profile of Mood States inventory (POMS), which provides a total mood disturbance score (TMD).[43] These rating scales were completed at the end of the challenges for both conditions, with and without placebo administration.

We were interested in the understanding of individual variations in placebo responses, and all eligible subjects were included in the studies without any consideration given to their potential placebo responsivity. Furthermore, we utilized instructions that were similar to those of typical clinical trials: 'We are testing an agent that has been shown to reduce pain in some subjects. It is thought that it does this through the activation of anti-pain mechanisms in our bodies. You will receive both active and inactive agents during the trial'. In the first series of experiments described below, an additional statement was added to deal with the fact that the placebo effect was not transparent to subjects due to the choice of pain model used: 'You may not be able to tell whether the agent is working, but the investigators will be able to tell with their equipment'.

PLACEBO-INDUCED ACTIVATION OF REGIONAL ENDOGENOUS OPIOID NEUROTRANSMISSION

In an initial investigation involving 14 healthy males, we determined the regional activation of endogenous opioid neurotransmission on μ-opioid receptors with PET and the selective μ-opioid radiotracer [^{11}C]carfentanil.[44] In this experiment, the pain model was operated so that the infusion was individually titrated to the preset level of pain intensity, irrespective of whether placebo was administered or not, potentially preventing subjects from experiencing a difference between conditions. It was observed that the administration of the placebo, with expectation that it represented an analgesic agent, was associated with significant activation of μ-opioid receptor mediated neurotransmission in both higher order and subcortical brain regions. These included the pre- and subgenual rACC, the dorsolateral prefrontal cortex (DLPFC), anterior insular cortex (aINS) and the nucleus accumbens (NAC) (Fig. 4.1). These regional activations were correlated with lower ratings of pain intensity (rACC, aINS, NAC), pain unpleasantness (rACC), reductions in MPQ sensory (rACC, aINS), affective (NAC) and total (rACC, aINS) scores, as well as in the negative emotional state of the volunteers as measured with the POMS (NAC). The magnitude of μ-opioid system activity in the rACC also correlated positively with the increases in pain tolerance (the increase in

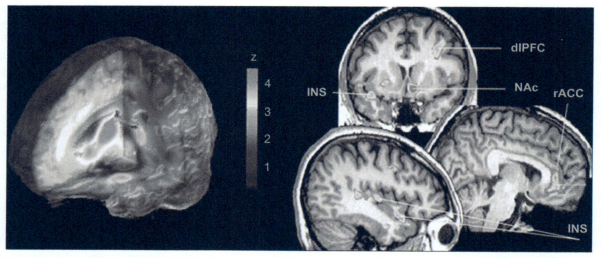

FIGURE 4.1 Placebo-induced activation of regional μ-opioid receptor-mediated neurotransmission. Left: distribution of μ-opioid receptors in the human brain, in a 3D rendering. Right: some of the areas in which significant activation of μ-opioid neurotransmission during sustained pain were observed after the introduction of a placebo with expectation of analgesia. INS: insula; dlPFC: dorsolateral prefrontal cortex; NAC: nucleus accumbens; rACC: rostral anterior cingulate cortex. *This figure is reproduced in color in the color section.*

algesic volume requirements to maintain pain at the target intensity, r = 0.96). This dataset was the first direct evidence that the administration of a placebo with implied analgesic properties was associated with the activation of a pain and stress inhibitory neurotransmitter system, the endogenous opioid system and μ-opioid receptors, involving a number of brain regions. Furthermore, this activation was associated with quantifiable reductions in the physical and emotional attributes of the stressor, a sustained pain challenge. The regions implicated in this phenomenon included some involved in cognitive and emotional integration, including responses to placebo (rACC); the representation and modulation of internal states, both physical and emotional (INS), and reward and saliency assessments (NAC). The DLPFC was not found to be related to changes in the psychophysical properties of the pain challenge, but instead to the expected analgesic effect of the placebo, as rated by the volunteers prior to its administration. This is consistent with the hypothesized function of this brain region in the cognitive adjustments to environmental information for the control of behavior.[45]

A follow-up analysis, conducted in a larger sample (n = 20),[46] examined the variance in endogenous opioid activity as a function of verbally-induced expectations of relief, and psychophysical characteristics of pain. Perhaps counter-intuitively, the largest proportion of the variance in regional endogenous opioid activity (40–68%, depending on the region) was accounted for by a multiple regression model that included the affective (but not sensory) quality of the pain, the PANAS-positive and -negative affect ratings, and a measure of individual pain sensitivity (the volume of algesic substance that had to be infused to maintain pain at target intensity level). This indicated that the individual affective experience during pain, whether pain-specific (MPQ pain affect subscale) or not (PANAS ratings of positive and negative internal affective state), were important predictors of the subsequent development of a placebo response, as was the measure of individual pain sensitivity. This concept seems to be consistent with that advanced by observations that placebo analgesia is achieved proportionally to the relief of anxiety afforded by the placebo.[17,18] It is also in line with the assertion that placebo effects result from the organism's assessment of its internal needs,[23] as pain sensitivity was also found to be a predictor of the formation of placebo responses as reflected by endogenous opioid activation.

A second experimental series was conducted with the same radiotracer, labeling μ-opioid receptors, but this time the infusion profile to achieve target pain levels was determined in advance and repeated in the studies with and without placebo.[47] Pain intensity ratings, acquired every 15 seconds, would be expected to be lower with placebo administration than without, this being the primary evidence of a formation of the placebo effect at a psychophysical level. This series also included PET studies with the dopaminergic (DA) tracer [^{11}C]raclopride, labeling DA D2 receptors in the basal ganglia and D2 and D3 receptors in the NAC.[48] The data acquired with this radiotracer is described in the following section.

In these studies, the expected analgesic effects were rated at 48 ± 23 (range 0–95). After the experiments were conducted, the perceived effectiveness of the placebo was rated at 42 ± 29. Significant endogenous opioid activation was observed in the pre- and subgenual rACC, orbitofrontal cortex (OFC), anterior insula (aINS) and pINS, medial thalamus (mTHA), NAC, amygdala (AMY) and periaqueductal gray (PAG). There was a notable lack of involvement of the DLPFC in these results, while activation in the OFC was observed instead. Regional magnitudes of activation correlated with the subjects' expected analgesia (NAC, PAG), the update of these verbally-induced expectations by the subjectively perceived efficacy of the placebo (the ratio between observed and expected efficacy) (NAC, AMY), as well as with placebo-induced changes in pain intensity (rACC, NAC, OFC). In view of the previous results, where affective state explained a substantial proportion of the variance in placebo responses, we also examined whether increases in positive affect during the placebo condition were related to the opioid response. Positive correlations were obtained between the increases in PANAS-positive affect and the magnitude of placebo-induced endogenous opioid system activity in the NAC.

When individuals were classified as high and low placebo responders, using the median reduction in pain intensity during placebo as the split point, it was opioid activity in the NAC that was significantly different between the two groups. A small group of subjects (n = 5) showed higher ratings of pain (hyperalgesia) during placebo (consistent with a nocebo effect). When compared to high placebo responders, the placebo and nocebo groups demonstrated changes in the opposite direction: regional opioid system activation was observed in high responders, while deactivations were present in the nocebo group.

Besides demonstrating a dynamic modulation of placebo and nocebo responses by the endogenous opioid system, the involvement of NAC opioid neurotransmission in differentiating high and low placebo responders was documented for the first time. This brain region presents high levels of DA innervation arising from the ventral tegmental area (mesolimbic DA circuit) and is known to be involved in responding to rewards and salient stimuli (both rewarding and aversive).[49] It is also thought to respond to updates in verbally-induced expectations that depend on the emotional response to changing environmental information (so-called counterfactual comparisons) through its connections with the OFC and the AMY.[50,51]

DOPAMINERGIC MECHANISMS IN THE FORMATION OF PLACEBO ANALGESIC EFFECTS

The previous results point to a distributed network of regions participating in placebo effects, mediated by the endogenous opioid system. The NAC emerged as a prominent part of it, believed to be responding to the saliency or the reward value of the placebo stimulus. Here, endogenous opioid activation was associated with verbally-induced expectations of analgesia, the update of those expectations over time, and placebo-induced analgesic effects.

The NAC lies at the interface of sensorimotor and limbic systems, and through its connections with the OFC, ventral pallidum and the amygdala, forms part of a circuit involved in the integration of cognitive, affective and motor responses in animal models.[52,53] This circuit and additional interconnected regions (e.g. insular and medial prefrontal cortex, medial thalamus) are heavily modulated by the endogenous opioid system and μ-opioid receptors. It has also been proposed as a primary site of interaction between the effects of DA-releasing drugs, novelty and stressors,[54–56] typically studied in the context of the administration of reinforcing drugs. A possible role of NAC DA in placebo responding was initially postulated following observations that basal ganglia DA release took place in the placebo arm of a RCT in patients with Parkinson's disease[8,57] and major depression.[58] This work, then, suggests an involvement of the ventral basal ganglia in either responding to individual verbally-induced expectations of analgesia or the novelty of a placebo administration.

To further study these processes in the context of placebo analgesia, the same subjects (n = 20) that underwent μ-opioid receptor scanning (2nd experiment above), underwent studies with the DA D2/D3 receptor radiotracer [^{11}C]raclopride.[47] Opioid and DA scans were randomized in order. As in previous studies, scans included a pain anticipation period (pain was expected but not received) where subjects were administered intramuscular non-painful isotonic saline and rated pain intensity in the same manner as the actual pain scans. During the actual pain scans, the same infusion profile was used for studies with and without placebo.[47]

Placebo administration was associated with the activation of DA D2/D3 neurotransmission that was exclusively localized in mesolimbic dopaminergic terminal fields, ventral caudate, ventral putamen and NAC. The magnitude of DA activation in the NAC was positively correlated with the individual verbally-induced expectations of analgesia, the update of those expectations during the study period (the ratio of subjectively rated analgesic efficacy over the initial expectations), and the magnitude of analgesia (the change in pain intensity ratings over the 20-minute study period). As was the case with the opioid system, DA activation in the NAC was also positively correlated with the increase in PANAS-positive affect ratings during placebo. When both regional opioid and DA responses to placebo were examined as to their contribution to placebo analgesia, DA release in the NAC emerged as the most predictive region and neurotransmitter, accounting for 25% of the variance in the formation of placebo analgesic effects. Consistent with the hypothesis that NAC DA responses to placebo constitute a 'trigger' that, responding to the saliency and reward value of the placebo, would allow for the activation of down-stream adaptive (e.g. opioid) responses, placebo-induced NAC DA release was positively correlated with the magnitude of endogenous opioid release in the NAC, ventral putamen, AMY, aINS, pINS and rACC. Similarly to the opioid system, NAC DA release also differentiated volunteers that were above and below the mean in their analgesic responses (high and low placebo responders) in these trials. For the comparison between high placebo and nocebo responders, nocebo responders demonstrated a deactivation of DA neurotransmission during placebo in the NAC and ventral putamen, an effect opposite in direction to that of high placebo responders.

Partly overlapping with the above sample, we then examined the hypothesis that individual variations in placebo responses may be related to differences in the processing of reward expectation.[59] For this purpose, healthy males and females (n = 30 total) were studied with a combination of molecular PET with [^{11}C] raclopride and functional magnetic resonance imaging (fMRI). In this case, and to avoid motivational mechanisms that may be related to individual differences in pain sensitivity, placebo-induced DA release was examined during the pain expectation state. Subjects also underwent an fMRI–BOLD study using a variation of the Monetary Incentive Delay (MID) task. This task is known to activate NAC synaptic activity during anticipation of a monetary reward.[60] Individual variations in placebo-induced NAC DA release were then compared to the synaptic activity of the same region during anticipation of a monetary reward. Both of these measures were also examined as a function of the anticipated analgesic effects of the placebo, deviations from those verbally-induced expectations and the magnitude of placebo analgesic effects in pain challenges. It was hypothesized that in healthy subjects, in the absence of underlying pathology or previous conditioning, individual variations in placebo-induced NAC DA activity, and in the synaptic activity of this region during reward anticipation, would be related to each other and to the variability in placebo effects obtained in the studies.

In a manner similar to what was observed in actual pain studies, the introduction of the placebo during a pain anticipation state was associated with the activation of DA neurotransmission and D2/D3 receptors in the NAC, bilaterally, in a manner proportional to the anticipated analgesic effects as rated by the volunteers, as well as with the difference between anticipated and subjectively perceived effectiveness of the placebo (i.e. the update of expectations over time).

We then examined whether individual variations in the synaptic activity of the NAC during the MID task would be predictive of the magnitude of placebo effects. It was observed that individuals that activated NAC synaptic function to a greater extent during monetary reward anticipation also showed more profound placebo responses. These included greater positive affect scores during pain expectation periods and greater levels of analgesia in pain trials. The NAC BOLD signal during monetary reward anticipation was further correlated with placebo-induced DA activity as measured with PET. In a regression model, NAC synaptic activation during anticipation of the (high, $5) monetary reward accounted for approximately one-third of the variance in the development of placebo-induced analgesia in the pain trials. In a manner similar to the results obtained with NAC DA responses to placebo, the activation of NAC synaptic activity during reward expectation was further correlated to the difference between the anticipated and subjectively perceived analgesic effects of the placebo. It should be noted that the fMRI studies were conducted separately from the pain expectation and pain studies and that the subjects were not aware of any link between the two sets of experiments. Given this situation, these results are believed to reflect intrinsic differences in the response of the NAC during reward anticipation, further defining individual variations in placebo responding.

THEORIES OF PLACEBO ANALGESIA AND PLACEBO-INDUCED ACTIVATION OF REGIONAL ENDOGENOUS OPIOID NEUROTRANSMISSION

Throughout the history of placebo research, an important debate has focused on whether placebo effects depend on conscious expectancy[27,61] or learning mechanisms.[62,63] Though this has typically been framed in terms of expectancy versus conditioning, these alternatives might not be mutually exclusive (Ch 4). Newer learning theories suggest that conditioning is actually a process that is related to both expectancies and association-based plasticity. Rescorla and Wagner formalized a model of classical conditioning in which learning does not depend on simple contiguity between conditioned and unconditioned stimuli. Instead, conditioning depends on prediction error, which signals a discrepancy between expected and observed outcome.[64] Taken together, this line of work implies that expectancies underlie most forms of learning.[65] Therefore, prediction error signal theories, as defined in this work, might provide a mechanism through which classical theories of placebo analgesia, verbally-induced expectations of clinical improvement and conditioning are reconciled, and placebo responses would emerge as a consequence of expectation and outcomes associations. If this were to be the case, the formation and maintenance of placebo effects would represent an instance of expectation and outcome comparisons, and an extension of the mechanisms involved in motivated behavior.

In recent work, we aimed to determine the potential impact of learning and memory in placebo analgesic response. First, we examined the possibility that the recall of placebo responses would be associated with greater opioid neurotransmission during placebo administration.[66] For this purpose, subjects (n = 37) were asked to recall their pain experience by completing the McGill Pain Questionnaire (MPQ) in a phone interview 24 hours after completion using the same scanning protocol used in previous studies.[50] Subjects were considered placebo recall responders when MPQ scores 24 hours after the pain + placebo scans were lower than during the scans when pain alone was introduced (n = 18); and placebo recall non-responders when MPQ scores 24 hours after the pain + placebo studies were the same or greater than during the studies where pain alone was introduced (n = 19).

Consistent with animal models showing an effect of the enkephalinergic system and μ-opioid receptors in learning and memory,[67] our data showed that, in addition to its immediate placebo-analgesic effects, the μ-opioid receptor system is involved in the subsequent recall of placebo responses. Specifically, μ-opioid system activation during placebo administration in the VTA and the Papez circuit (hypothalamus-mammillary bodies, mamillo-thalamic tract, anterior thalamic nuclei, cingulate cortex and hippo/parahippocampus), implicated in reward-motivated learning and memory processing,[68] were associated with attenuated recall of the pain experience 24 hours after the challenge. This report extended previous findings on the role of the μ-opioid neurotransmitter system in acute placebo responses,[47] and highlighted a novel role of this system in the formation of memories of placebo analgesic effects in distinct neural circuits. These provided a framework to understand stimulus learning and therapeutic effect associations, of importance for the sustainability of placebo effects.

In a second series of analyses, we examined whether expectation and outcome comparisons (prediction error signal) predicted placebo responses.[69] This hypothesis was supported by newer learning theories, where conditioning depends on prediction error, which signals a discrepancy between expected and observed outcome.[64] Early classical theories proposed that reward-directed learning depends on the temporal contiguity between stimuli and reward.[70] By contrast, in most modern learning theories,[64] a discrepancy between actual and predicted reward (reward-prediction error) plays an important role for learning stimulus–reward associations. The Rescorla and Wagner model (1972) and its real-time extensions (temporal difference models)[71] postulate that learning is directly influenced by prediction errors that decrease gradually until the predictions match the outcome.

To test this hypothesis we examined the effect of expectation and outcome comparison groups on placebo responses using positron emission tomography (PET) and the μ-opioid receptor selective radiotracer [^{11}C]carfentanil during the same sustained pain challenge with and without placebo administration and the same scanning protocol as previously described.[47] In order to create a measure of expectations and counterfactual comparisons, subjects were assigned to Low (≤50) or High (>50) Expectations or Effectiveness groups based on their answers to the questions: from 0 to 100, how effective do you think the medication will be? (expectations, prior to scan), and from 0 to 100, was the medication effective? (subjective assessment of effectiveness, after scan); 19 subjects were classified as having High Expectations and 29 as having Low Expectations. When the two variables were combined, these resulted in four groups: a High Expectation/Low Effectiveness group (n = 7), Low Expectations/Low Effectiveness group (n = 19), High Expectations/High Effectiveness group (n = 13) and Low Expectations/High Effectiveness group (n = 9).

We observed a lack of significant relationships between the level of initial subjective expectations of pain relief and placebo-associated reductions on pain ratings. Using objective the neurochemical measures acquired with PET, individuals with high expectations showed greater μ-opioid system activation in the DLPFC that were not associated with placebo analgesic effects. These findings presented an apparent discrepancy with classical theories where the formation of placebo responses is dependent on the development of positive expectations. Conversely, a learning mechanism defined by the discrepancy between expectations and subjectively perceived effectiveness was associated with placebo analgesic responses (Fig. 4.2), and with the activation of regional μ-opioid neurotransmission in a substantial number of regions implicated in opioid-mediated antinociception[72] (ACC, OFC, AMYG, THA, INS). The largest placebo responses were observed in those with low expectations and high subjective effectiveness (positive prediction error signal), whereas 'nocebo', hyperalgesic responses were observed in those reporting high expectations and low reported effectiveness (negative prediction error signal). The magnitude of μ-opioid system activation in regions relevant to error detection was further associated with placebo-induced analgesia, as measured by the changes in momentary pain intensity VAS ratings acquired every 15 seconds throughout the study, and the more integrative MPQ ratings acquired at the completion of the pain challenges. Moreover, our data confirmed that the effect of expectation and outcome associations on behavioral placebo responses is mediated by endogenous opioid release and activation of

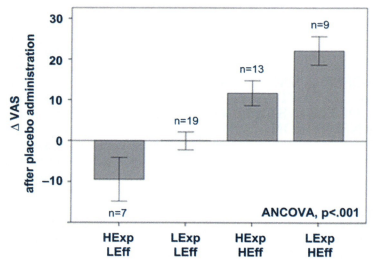

FIGURE 4.2 Effect of expectation–effectiveness comparisons on placebo-induced changes in average VAS intensity ratings acquired over 20 minutes. Greater placebo effects were observed in those with a positive prediction error signal (Low Expectations and High Effectiveness), whereas lower placebo effects were observed in those with a negative prediction error signal (High Expectations and Low Effectiveness).

the μ-opioid receptor system in the dACC. These results provided a mechanism through which classical theories of placebo analgesia (expectation vs. conditioning) could be reconciled, and shed light on individual differences in reward learning and decision-making processes. Expectations and outcomes comparisons then emerge as a cognitive mechanism that, beyond reward-associations, is likely to facilitate the formation and sustainability of placebo responses over time.

PERSONALITY PREDICTORS OF PLACEBO-INDUCED ACTIVATION OF REGIONAL ENDOGENOUS OPIOID NEUROTRANSMISSION

Substantial interest exists in the development of simple measures, such as personality traits, that may provide predictive value for the formation of placebo effects, as they might substantially influence the therapeutic context[73] and therefore placebo effects. Recent work has studied the effect of the patient–practitioner relationship on placebo acupuncture responses.[73] In this study, participants were randomized to a waitlist (observation), limited (placebo acupuncture + neutral practitioner) or augmented placebo acupuncture treatment (placebo acupuncture + empathic practitioner). It was shown that gender (female) and personality traits (extraversion, agreeableness and openness to experience) influenced placebo response, but only in the warm, empathic, augmented group. These results suggested that to the degree a placebo effect is evoked by the patient–practitioner relationship, personality characteristics of the patient will be associated with placebo response and highlight the complexity and variability of placebo effects in the context of a therapeutic relationship.

In line with this work, trait optimism and trait anxiety were found to be positive and negative, reproducible predictors.[74,75] Personality traits thought to be related to dopaminergic function (novelty seeking, harm avoidance, behavioral drive, fun seeking and reward responsiveness) have also been associated with both placebo analgesic effects and gray matter density in the basal ganglia and prefrontal cortex.[76]

We examined the possibility that personality traits would be associated with analgesic placebo responses in a sample of 50 healthy volunteers using the sustained pain paradigm described above.[77] Since our primary hypothesis stated that personality predictors of placebo responses would be associated with stress resiliency, we examined the predictive value of scales assessing emotional, psychologic, and social well-being, dispositional optimism, satisfaction with life and ego-resiliency. Personality traits related to dopaminergic function were also evaluated using the behavioral inhibition/activation scale. Low anticipatory responses to pain have been associated with the subsequent formation of placebo analgesia, therefore trait anxiety was included in the model as a potential negative predictor. Finally, an overall evaluation of personality traits was included using the scores of the five dimensions of the NEO Personality Inventory Revised.

Using the change in average VAS score of pain intensity as a dependent variable, the most predictive traits of placebo analgesia in a univariate model were Ego-Resiliency, NEO Agreeableness, and NEO Neuroticism, which respectively explained 16%, 14%, and 12% of the variance in placebo response (Fig. 4.3). The former two were positive predictors, while the latter was a negative predictor. A multivariate model with Ego-Resiliency, and the NEO facets Altruism, Straightforwardness and Angry Hostility (negative predictor) accounted for 25% of the variation in placebo analgesic responses and had a predictive ability of 18%.

Molecular imaging showed that subjects scoring above the median in a composite of those trait measures also presented greater placebo-induced activation of μ-opioid neurotransmission in the subgenual and dorsal ACC, OFC, INS, NAC, AMYG and at a lower threshold, the PAG. These regions largely overlap with those identified in previous reports as responsive to placebo administration and involved in the regulation of the pain experience.[44,47,78] As previously observed, activation of μ-opioid receptor mediated neurotransmission in some of these regions was associated with reductions in individual pain report (Fig. 4.3). Additionally, we found significant reductions in cortisol plasma levels during placebo administration, which were correlated with reductions in subjective pain report and μ-opioid system activation in the dorsal ACC and PAG. While no significant relationships were observed between the composite scores and reductions in cortisol, these were negatively correlated with NEO Angry Hostility scale scores at trend levels.

Previous studies have suggested that dispositional optimism, as defined by the Life Orientation Test-Revised scale, might be a predictor of placebo analgesia using experimental models of phasic heat and a preconditioning procedure to increase expectations of analgesia,[75] or cold pain (cold pressor test) without preconditioning, where an interaction between Life Orientation Test-Revised scores and experimental condition was reported.[74,75] Similarly, placebo effects have been associated with dopaminergic function, such as novelty seeking, harm avoidance, behavioral drive, fun seeking and reward responsiveness, during intramuscular infusion of hypertonic saline paired with a cream using a conditioning-like procedure.[76] However, these findings were not replicated in the present study using a sustained pain model and no preconditioning procedures.

These results suggested that stable personality traits related to stress resiliency and interpersonal function

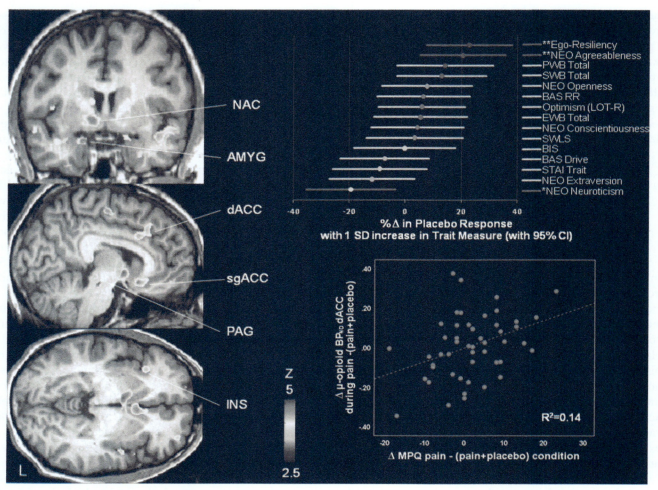

FIGURE 4.3 Personality traits effect on placebo-induced activation of regional μ-opioid receptor mediated neurotransmission. Left: Regions of greater μ-opioid system activation during placebo administration in subjects with high levels of Ego Resilience, Straightforwardness and Altruism and low levels of Angry Hostility. Upper right: Simple linear regression representing percent change in Placebo Response associated with 1 SD increase in Trait Measure (with 95% Confidence Intervals). (** indicates $p < 0.01$; * indicates $p < 0.05$). Lower right: Correlations between Δ μ-opioid BP_{ND} in the dACC during pain compared to (pain + placebo) and Δ in pain ratings (MPQ) during placebo administration. INS: insula; NAC: nucleus accumbens; r/sgACC: rostral and subgenual anterior cingulate cortex; AMYG: amygdala; PAG: periaqueductal gray. *This figure is reproduced in color in the color section.*

have a substantial impact on the capacity to develop placebo effects. If replicated in clinical samples, these trait measures could be employed to reduce variability in treatment trials for conditions, such as persistent pain, mood and movement disorders, where placebo effects can be particularly prominent and obscure the effects of potentially active treatments.

CONCLUSIONS

An emerging literature is demonstrating that cognitive and emotional processes that are engaged during the administration of an otherwise inactive agent, a placebo, are capable of activating internal mechanisms that modify physiology. These processes are of importance to understand the inter-individual variability that leads to recovery from any illness. A network of regions, including the rostral anterior cingulate, dorsolateral prefrontal and orbitofrontal cortices, insula, nucleus accumbens, amygdala, medial thalamus and periaqueductal gray appear to be involved. Opioid and dopamine neurotransmission in these areas modulates various elements of the placebo effect, which appear to include the representation of its subjective value, updates of expectations over time, the recall of pain and placebo experiences and changes in affective state and in pain ratings. The circuitry involved in placebo analgesic effects also has the potential to modulate a number of functions beyond pain, as the brain regions involved have been implicated in the regulation of stress responses, neuroendocrine and autonomic functions, mood, reward and integrative cognitive processes, such as decision-making. Besides the perspective that placebo effects confound RCTs, the information so far acquired points to neurobiologic systems that, when activated by positive expectations or conditioning, are capable of inducing physiologic change. They should therefore be considered as resiliency mechanisms with the potential to aid in the recovery from challenges to the organism.

Acknowledgments

Support was provided by the National Center for Complementary and Alternative Medicine grant R01 AT 001415 and R01 DA 016423 to J.K.Z.

References

1. Haygarth J. Of the Imagination, as a Cause and as a Cure of Disorders of the Body, Exemplified by Ficticious Tractors, and Epidemical Convulsions. *Bath: Crutwell*. 1801.
2. Beecher H. The powerful placebo. *JAMA*. 1955;159:1602-1606.
3. Benedetti F, Mayberg HS, Wager TD, et al. Neurobiological mechanisms of the placebo effect. *J Neurosci*. 2005;25(45):10390-10402.
4. Price DD, Finniss DG, Benedetti F. A comprehensive review of the placebo effect: recent advances and current thought. *Annu Rev Psychol*. 2008;59:565-590.
5. Mayberg HS. Limbic-cortical dysregulation: a proposed model of depression. *J Neuropsychiatr Clin Neurosci*. 1997;9(3):471-481.
6. Petrovic P, Dietrich T, Fransson P, et al. Placebo in emotional processing – induced expectations of anxiety relief activate a generalized modulatory network. *Neuron*. 2005;46(6):957-969.
7. Benedetti F, Colloca L, Torre E, et al. Placebo-responsive Parkinson patients show decreased activity in single neurons of subthalamic nucleus. *Nat Neurosci*. 2004;7(6):587-588.
8. de la Fuente-Fernandez R, Ruth TJ, Sossi V, et al. Expectation and dopamine release: mechanism of the placebo effect in Parkinson's disease. *Science*. 2001;293(5532):1164-1166.
9. Levine J, Gordon N, Fields H. The mechanism of placebo analgesia. *Lancet*. 1978;2(8091):654-657.
10. Levine JD, Gordon NC. Influence of the method of drug administration on analgesic response. *Nature*. 1984/1985;312(5996):755-756.
11. Gracely RH, Dubner R, Wolskee PJ, Deeter WR. Placebo and naloxone can alter post-surgical pain by separate mechanisms. *Nature*. 1983;306(5940):264-265.
12. Grevert P, Albert L, Goldstein A. Partial antagonism of placebo analgesia by naloxone. *Pain*. 1983;16(2):129-143.
13. Amanzio M, Benedetti F. Neuropharmacological dissection of placebo analgesia: expectation-activated opioid systems versus conditioning-activated specific subsystems. *J Neurosci*. 1999;19(1):484-494.
14. Benedetti F, Amanzio M, Rosato R, Blanchard C. Nonopioid placebo analgesia is mediated by CB1 cannabinoid receptors. *Nat Med*. 2011;17(10):1228-1230.
15. Montgomery G, Kirsch I. Classical conditioning and the placebo effect. *Pain*. 1997;72:107-113.
16. Price D, Milling L, Kirsch I, et al. An analysis of factors that contribute to the magnitude of placebo analgesia in an experimental paradigm. *Pain*. 1999;83:147-156.
17. McGlashan TH, Evans FJ, Orne MT. The nature of hypnotic analgesia and placebo response to experimental pain. *Psychosom Med*. 1969;31(3):227-246.
18. Evans FJ. Expectancy, therapeutic instructions, and the placebo response. In: White L, Tursky B, Schwartz G, eds. *Placebo: Theory, Research and Mechanisms*. New York: Guilford; 1985:215-228.
19. Gleidman L, Gantt W, Teitelbaum H. Some implications of conditioned reflex studies for placebo research. *Am J Psychiatry*. 1957;113:1103-1107.
20. Voudouris NJ, Peck CL, Coleman G. The role of conditioning and verbal expectancy in the placebo response. *Pain*. 1990;43(1):121-128.
21. Siegel S. Drug anticipatory responses in animals. In: White L, Tursky B, Schwartz G, eds. *Placebo: Theory, Research and Mechanisms*. New York: Guilford; 1985:288-305.
22. Ader R. The role of conditioiining in pharmacotherapy. In: Harrington A, ed. *The Placebo Effect: an Interdisciplinary Exploration*. Cambridge, MA: Harvard University Press; 1997:138-165.
23. Wall P. *Pain and the Placebo Response*. New York: Wiley; 1993.
24. Rainville P, Duncan G, Price D, et al. Pain affect encoded in human anterior cingulate but not somatosensory cortex. *Science*. 1997;277(5328):968-971.
25. Hofbauer RK, Rainville P, Duncan GH, Bushnell MC. Cortical representation of the sensory dimension of pain. *J Neurophysiol*. 2001;86(1):402-411.
26. Willoch F, Rosen G, Tolle T, et al. Phantom limb pain in the human brain: unraveling neural circuitries of phantom limb sensations using positron emission tomography. *Ann Neurol*. 2000;48(6):842-849.
27. Price DD, Barrell JJ. Mechanisms of analgesia produced by hypnosis and placebo suggestions. *Prog Brain Res*. 2000;122:255-271.
28. Raz A, Fan J, Posner MI. Hypnotic suggestion reduces conflict in the human brain. *Proc Natl Acad Sci USA*. 2005;102(28):9978-9983.
29. Petrovic P, Ingvar M. Imaging cognitive modulation of pain processing. *Pain*. 2002;95(1-2):1-5.
30. Levine JD, Gordon NC, Smith R, Fields HL. Analgesic responses to morphine and placebo in individuals with postoperative pain. *Pain*. 1981;10(3):379-389.
31. Amanzio M, Pollo A, Maggi G, Benedetti F. Response variability to analgesics: a role for non-specific activation of endogenous opioids. *Pain*. 2001;90(3):205-215.
32. Benedetti F, Pollo A, Lopiano L, et al. Conscious expectation and unconscious conditioning in analgesic, motor, and hormonal placebo/nocebo responses. *J Neurosci*. 2003;23(10):4315-4323.
33. Petrovic P, Kalso E, Petersson KM, Ingvar M. Placebo and opioid analgesia – imaging a shared neuronal network. *Science*. 2002;295(5560):1737-1740.
34. Atlas LY, Whittington RA, Lindquist MA, et al. Dissociable influences of opiates and expectations on pain. *J Neurosci*. 2012;32(23):8053-8064.
35. Wager TD, Rilling JK, Smith EE, et al. Placebo-induced changes in FMRI in the anticipation and experience of pain. *Science*. 2004;303(5661):1162-1167.
36. Bingel U, Lorenz J, Schoell E, et al. Mechanisms of placebo analgesia: rACC recruitment of a subcortical antinociceptive network. *Pain*. 2006;120(1-2):8-15.
37. Kong J, Gollub RL, Rosman IS, et al. Brain activity associated with expectancy-enhanced placebo analgesia as measured by functional magnetic resonance imaging. *J Neurosci*. 2006;26(2):381-388.
38. Wager TD, Atlas LY, Leotti LA, Rilling JK. Predicting individual differences in placebo analgesia: contributions of brain activity during anticipation and pain experience. *J Neurosci*. 2011;31(2):439-452.
39. Stohler C, Kowalski C. Spatial and temporal summation of sensory and affective dimensions of deep somatic pain. *Pain*. 1999;79(2-3):165-173.
40. Zhang X, Ashton-Miller JA, Stohler CS. A closed-loop system for maintaining constant experimental muscle pain in man. *IEEE Trans Biomed Eng*. 1993;40(4):344-352.
41. Melzack R, Katz J. The McGill pain questionnaire: appraisal and current status. In: Turk D, Melzack R, eds. *Handbook of Pain Assessment*. New York: Guilford Press; 2000:152-168.
42. Watson D, Clark LA, Tellegen A. Development and validation of brief measures of positive and negative affect: the PANAS scales. *J Personal Soc Psychol*. 1988;54:1063-1070.

43. EdITS Manual for the Profile of Mood States. San Diego, CA: Educational and Industrial Testing Service (EdITS); 1992.
44. Zubieta JK, Bueller JA, Jackson LR, et al. Placebo effects mediated by endogenous opioid activity on mu-opioid receptors. *J Neurosci.* 2005;25(34):7754-7762.
45. Fuster JM. Executive frontal functions. *Exp Brain Res.* 2000;133(1):66-70.
46. Zubieta JK, Yau WY, Scott DJ, Stohler CS. Belief or need? Accounting for individual variations in the neurochemistry of the placebo effect. *Brain Behav Immun.* 2006;20(1):15-26.
47. Scott DJ, Stohler CS, Egnatuk CM, et al. Placebo and nocebo effects are defined by opposite opioid and dopaminergic responses. *Arch Gen Psychiatr.* 2008;65(2):220-231.
48. Seeman P, Wilson A, Gmeiner P, Kapur S. Dopamine D2 and D3 receptors in human putamen, caudate nucleus, and globus pallidus. *Synapse.* 2006;60(3):205-211.
49. Horvitz J. Mesolimbic and nigrostriatal dopamine responses to salient non-rewarding stimuli. *Neuroscience.* 2000;96:651-656.
50. Schultz W. Behavioral theories and the neurophysiology of reward. *Annu Rev Psychol.* 2006;57:87-115.
51. Tobler PN, Fiorillo CD, Schultz W. Adaptive coding of reward value by dopamine neurons. *Science.* 2005;307(5715):1642-1645.
52. Kalivas PW, Churchill L, Romanides A. Involvement of the pallidal-thalamocortical circuit in adaptive behavior. *Ann N Y Acad Sci.* 1999;877:64-70.
53. Mogenson GJ, Yang CR. The contribution of basal forebrain to limbic-motor integration and the mediation of motivation to action. *Adv Exp Med Biol.* 1991;295:267-290.
54. Badiani A, Oates MM, Day HE, et al. Environmental modulation of amphetamine-induced c-fos expression in D1 versus D2 striatal neurons. *Behav Brain Res.* 1999;103(2):203-209.
55. Day HE, Badiani A, Uslaner JM, et al. Environmental novelty differentially affects c-fos mRNA expression induced by amphetamine or cocaine in subregions of the bed nucleus of the stria terminalis and amygdala. *J Neurosci.* 2001;21(2):732-740.
56. Uslaner J, Badiani A, Day HE, et al. Environmental context modulates the ability of cocaine and amphetamine to induce c-fos mRNA expression in the neocortex, caudate nucleus, and nucleus accumbens. *Brain Res.* 2001;920(1–2):106-116.
57. de la Fuente-Fernandez R, Phillips AG, Zamburlini M, et al. Dopamine release in human ventral striatum and expectation of reward. *Behav Brain Res.* 2002;136(2):359-363.
58. Mayberg HS, Silva JA, Brannan SK, et al. The functional neuroanatomy of the placebo effect. *Am J Psychiatr.* 2002;159(5):728-737.
59. Scott DJ, Stohler CS, Egnatuk CM, et al. Individual differences in reward responding explain placebo-induced expectations and effects. *Neuron.* 2007;55(2):325-336.
60. Knutson B, Bjork JM, Fong GW, et al. Amphetamine modulates human incentive processing. *Neuron.* 2004;43(2):261-269.
61. Kirsch I. Response expectancy as a determinant of experience and behavior. *Am Psychologist.* 1985;40(11):1185-1202.
62. Ader R, Cohen N. Behaviorally conditioned immunosuppression. *Psychosom Med.* 1975;37(4):333-340.
63. Voudouris NJ, Peck CL, Coleman G. Conditioned response models of placebo phenomena: further support. *Pain.* 1989;38(1):109-116.
64. Rescorla RA, Wagner AR. A theory of Pavlovian conditioning: variations in the effectiveness of reinforcement and nonreinforcemecent. In: Black AH, Prokasy WF, eds. *Classical Conditioning, II, Current Research and Theory.* New York: Appleton Century Crofts; 1972:64-69.
65. Reiss S. Pavlovian conditioning and human fear: an expectancy model. *Behav Ther.* 1980;11(3):380-396.
66. Pecina M, Stohler CS, Zubieta JK. Role of mu-opioid system in the formation of memory of placebo responses. *Mol Psychiatr.* 2013;18(2):135-137.
67. Rigter H. Attenuation of amnesia in rats by systemically administered enkephalins. *Science.* 1978;200(4337):83-85.
68. Adcock RA, Thangavel A, Whitfield-Gabrieli S, et al. Reward-motivated learning: mesolimbic activation precedes memory formation. *Neuron.* 2006;50(3):507-517.
69. Pecina M, Stohler CS, Zubieta JK. Neurobiology of placebo effects: expectations or learning? *Soc Cogn Affect Neurosci.* 2013. [Epub ahead of print.]
70. Pavlov IP. *Conditional Reflexes.* London: Oxford University Press; 1927.
71. Sutton RS, Barto AG. Toward a modern theory of adaptive networks: expectation and prediction. *Psychol Rev.* 1981;88(2):135-170.
72. Zubieta JK, Smith YR, Bueller JA, et al. Regional mu opioid receptor regulation of sensory and affective dimensions of pain. *Science.* 2001;293(5528):311-315.
73. Kelley JM, Lembo AJ, Ablon JS, et al. Patient and practitioner influences on the placebo effect in irritable bowel syndrome. *Psychosom Med.* 2009;71(7):789-797.
74. Geers AL, Wellman JA, Fowler SL, et al. Dispositional optimism predicts placebo analgesia. *J Pain.* 2010;11(11):1165-1171.
75. Morton DL, Watson A, El-Deredy W, Jones AK. Reproducibility of placebo analgesia: Effect of dispositional optimism. *Pain.* 2009;146(1–2):194-198.
76. Schweinhardt P, Seminowicz DA, Jaeger E, et al. The anatomy of the mesolimbic reward system: a link between personality and the placebo analgesic response. *J Neurosci.* 2009;29(15):4882-4887.
77. Pecina M, Azhar H, Love TM, et al. Personality trait predictors of placebo analgesia and neurobiological correlates. *Neuropsychopharmacol.* 2013;38(4):639-646.
78. Wager TD, Scott DJ, Zubieta JK. Placebo effects on human mu-opioid activity during pain. *Proc Natl Acad Sci USA.* 2007;104(26):11056-11061.

CHAPTER 5

How does EEG Contribute to Our Understanding of the Placebo Response?
Insights from the Perspective of Bayesian Inference

Anthony Jones[1], Christopher Brown[1], Wael El-Deredy[2]

[1]Human Pain Research Group, University of Manchester, Manchester, UK, [2]Department of Psychological Sciences, University of Manchester, Manchester, UK

THEORETICAL MODELS OF PLACEBO ANALGESIA

A scientific understanding of the mechanisms of placebo analgesia must conform to a theoretical model that is consistent across multiple domains of knowledge (e.g. neurophysiologic, pharmacologic, psychologic, social). In the psychologic domain, it has been argued that expectations are largely responsible for placebo effects.[1-3] Expectation may be partly related to the information provided about the treatment by healthcare professionals, and partly to psychologic processes particular to the patient. Wager[1] and Morton et al[4] proposed overlapping models of the role of expectancy, positing a role also for a second factor, anxiety reduction, in placebo analgesia. Morton et al[5] argued that dispositional optimism, the tendency to expect positive outcomes, also plays a role in facilitating the reduction in anxiety.

There are also noncognitive approaches to account for placebo responding, which may or may not rely on expectation, most notably classical conditioning.[6] A true drug or treatment is an unconditioned stimulus (US) leading to an unconditioned response (symptom relief). A placebo treatment repeatedly pairs a neutral stimulus (e.g. sugar pill) with the notion of the US and thus become conditioned stimuli that can elicit a response similar to that of the active drug, a conditioned response.[7] Early experiments showed a role for classical conditioning in the placebo effect, with studies by Voudouris et al[6] showing that surreptitious reduction in the intensity of stimulation, when paired with an inert treatment, produce strong placebo responses. However, Montgomery and Kirsch[8] found, using an almost identical procedure, that when subjects were told that the treatment was inert, there was no placebo response, suggesting that cognitive expectations were essential. A comprehensive discussion on the theoretical debate between cognitive expectancy versus noncognitive conditioning models is offered by Stewart-Williams and Podd.[9] For the purposes of this chapter, it suffices to point out that Montgomery and Kirsch[8] argued that conditioning will produce placebo response expectancies rather than actual placebo responses. The idea that it is the expectancies that actually elicit the response bears close resemblance to assimilation theory, which is a framework that predicts the effects of expectations on sensory perception.[10] Grounded in ideas around cognitive dissonance, it assumes that the discrepancy between expected and actual experience is undesirable, and therefore the perception is altered to match the expectation. The idea is consistent with the concept of the placebo response being an erroneous non-detection of pain signals.[11]

The concept of cognitive dissonance is also consistent with more complex probabilistic approaches to understanding pain perception, such as used within the framework of Bayesian theory.[1] In this conceptual approach, which builds on the notion of perception as inference,[12] pain perception is thought of as a product of prior information (informing expectation/anticipation) and current information (nociceptive processing). In applying these Bayesian concepts to sensory neuroscience the balance between the influence of prior expectation over sensory evidence is governed by the uncertainty in those expectations,[13] such that sensory evidence has more influence when expectations are more uncertain, due to processes of attention and learning that become activated during uncertainty. However, it has been suggested also that uncertainty necessarily increases pain reports,[14] a view that appears to conflict with a purely Bayesian model of pain. The evidence we review below favours the Bayesian model, although it is possible that uncertain anticipation may induce anxiety and higher pain ratings under a psychologic context in which uncertainty is associated with greater threat value.[15]

Based on the above theoretical considerations, here we propose an integrated model. The model argues that the placebo response is primarily driven by cognitive processes of expectation and assimilation with expectation, processes that are influenced by a range of other cognitive processes such as confidence in treatment cues, implicit conditioning based on past treatments, and explicit memories of those treatments. Nonspecific psychologic processes such as positive emotional responses arising from an empathic therapeutic relationship have also been shown to be involved in placebo responding,[16] which may then influence treatment-specific expectancy processes by increasing confidence in treatment cues (i.e. increasing the subject's perception that those cues are reliable). We also propose that one of the mediators of the effects of expectations on pain may be a reduction in anxiety and stress.

The remainder of this chapter will review in more detail the evidence for four key components of this model: expectation, conditioning, attention and anxiety reduction, and will provide evidence from EEG studies for the involvement of these processes.

EEG MEASURES OF PAIN AND ITS ANTICIPATION

Electroencephalography (EEG) allows us to record changes in cortical neuronal activity using surface electrodes on the scalp. Two types of information can be analyzed from EEG recordings: (1) the frequency characteristics of the EEG, either at rest or in response (time-locked) to specific events, (2) event-related potentials (ERPs), which are responses that are consistent across multiple trials of data, determined by averaged time and phase-locked responses to stimuli or actions, or in anticipation of these. In this chapter we focus mainly on ERPs to painful or non-painful laser stimuli (laser-evoked potentials, LEPs) and anticipation-evoked potentials known as the stimulus-preceding negativity (SPN) (Fig. 5.1). The LEPs that are most reliably recorded are an early negative component N1, subsequent N2 component and ensuing positive component P2.[18–21] N2/P2 are generally thought to be related to the conscious experience of pain,[19,21] are modulated by attention and mood[22,23] and are correlated with anticipatory processing as measured by the amplitude of the SPN.[24]

PAIN ANTICIPATION AND ITS ROLE IN PAIN PERCEPTION

EEG, because of its high temporal resolution, has the potential to investigate top-down mechanisms such as expectation, attention, and anxiety in terms of anticipatory processes preceding pain. Specifically, EEG provides the means to clearly differentiate between neural processes during anticipation and experience of pain and to begin to unpick the somewhat complex relationship between these two aspects of pain processing in the brain.[24] The SPN generated prior to a pain stimulus can be used as a measure of pain anticipation. In this section we review EEG studies of anticipation,[24,25] which show that anticipatory activity is correlated with the expected intensity of pain and is a predictor of subsequent pain perception and neural correlates of pain such as laser-evoked potentials (LEPs).

Behaviorally, Brown et al,[24] have shown that when high-, medium- and low-intensity laser stimuli are delivered to the back of the arm in a pseudo-randomized order, the perception of those stimuli is influenced by preceding cues that provide either certain or uncertain information about subsequent stimulus intensity. In other words, prior expectation generated from the cues significantly influenced pain ratings. Certain expectations of high pain increased pain perception, whereas certain expectations of low pain decreased pain perception, relative to uncertain conditions in which the same intensity of pain was delivered. The probability distribution of stimulus intensity was roughly centered on the medium intensity stimuli, as high, medium and low stimuli had equal likelihood. The ratings during relative uncertainty were more clustered towards the medium intensity and this would favor a probabilistic or Bayesian model of pain perception, rather than the view that uncertainty increases pain via anxiety.

When EEG data from the above experiment were studied,[24] it was found that under 'certain' conditions, when the future pain intensity was known, anticipatory processing was a predictor of pain processing, but not during 'uncertain' conditions. During uncertainty, anticipation, as measured by the SPN, did not correlate with LEPs, whereas during certain expectation the SPN was

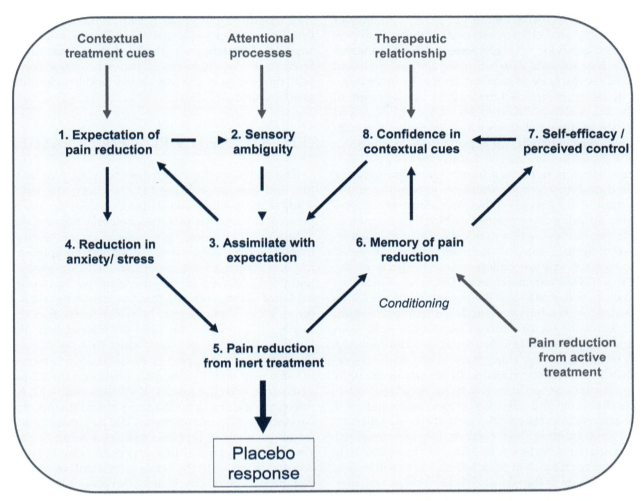

FIGURE 5.1 A model of the core components of the placebo response. An expectation of effect (1) causes ambiguity (2) regarding the intensity of the pain experience. Sensory ambiguity is further compounded by attentional processes such as distraction from the pain. When faced with the ambiguity of the perceived pain signal, subjects can either assimilate (3) or contrast their experience to their prior expectation of treatment. Assimilation results in enhancement of the effects of expectation (1). Expectation then goes on to cause a decrease in anxiety (4), which is known to cause decreases in perceived pain intensity (5). The reduction in pain from placebo and real analgesics will act as a conditioning stimulus that enhances memory of analgesic efficacy (6) and subjective confidence (8) in any cues indicating than an analgesic treatment is taking place. Positive emotional responses arising from an empathic therapeutic relationship may also increase confidence in treatment cues. Confidence acts as a facilitator of assimilation with expectation (3), thus further enhancing placebo effects. Reproduced from Brown et al.[17]

indeed correlated with subsequent LEPs. In other words, not only did certain information about forthcoming pain bias pain perception, it also caused anticipatory processes to better predict subsequent pain processing. To shed light on this, further analysis looked at the brain sources of anticipatory activity during certain and uncertain conditions, showing that they appeared to be processed in different circuitry in the brain.[24] Uncertain anticipation was processed in areas of the brain more concerned with attention, such as dorsolateral prefrontal cortex (DLPF), posterior cingulate cortex (PCC) and inferior parietal cortex (IPC). This is possibly because uncertainty promotes mechanisms of learning by directing attention to current sensory information, as suggested by Bayesian accounts of sensory perception.[13] Conversely, certain conditions (compared to uncertain) were associated with anticipation being processed in areas of the brain more associated with semantic memory, conditioning and emotional/autonomic responses such as anterior prefrontal cortex (aPFC), inferior frontal and temporal cortices and subgenual cingulate cortex.[24] This is consistent with greater reliance on past information in determining pain experience (via expectation) when the intensity of forthcoming pain is known. In sum, these differences in processing provide further evidence in favor of a Bayesian model of pain perception and placebo analgesia. During more uncertain expectation the Bayesian model predicts that more attention will be placed on the incoming sensory information (nociceptive information presented to the brain), with less reliance on any prior information.

According to Bayesian theories of perception, greater reliability of a predictor increases its influence on perception and behavior.[13,26,27] Consistent with this, there is evidence that subjects who are more certain/confident of their likely response to pain will tend to be more influenced by information about the forthcoming intensity of pain. Brown et al[25] studied the effects of prior beliefs and confidence in those beliefs as a way of measuring subjects' perceived reliability of expectations. In an extension of the previously described study by Brown et al,[24] the same subjects rated how much emotional distress they predicted they would experience in response to the pain stimuli. They also rated their confidence in this prediction as an index of the extent to which they might rely on their expectations when making sensory judgments. Analysis of the high-intensity laser stimulations revealed that confidence in prior beliefs was a predictor of the extent to which certain anticipation (relative to uncertain anticipation) was able to bias the perception of pain. This is consistent with a Bayesian model of pain perception. Further evidence for this model came from data from the same study[25] showing that greater confidence, and a greater effect of certain vs. uncertain anticipation cues on pain, was positively correlated with activity in the right anterior insula (a region known to be involved with aversive conditioning and affective responses to pain)[28,29] and negatively correlated with activity in the inferior parietal and midcingulate cortices (regions involved with attentional orientation to pain).[30,31] In individuals with greater confidence in prior beliefs, there will be less need for attention to the sensory–discriminatory components of pain and therefore less resources given to attentional areas of the pain matrix such as inferior parietal and mid-cingulate cortices. On the other hand, greater confidence in prior beliefs may induce processing within areas concerned with conditioned affective responses to pain such as anterior insula.

These EEG studies in normal volunteers have some potentially important clinical implications. The results suggest that patients who attend poorly to the details of sensory pain processing may appraise pain in a way that is weighted towards prior beliefs. This may mean that such patients are also more prone to manipulation of these beliefs, and may therefore be more susceptible to both placebo and nocebo responses. On the other hand, if chronic pain and attentional deficits are associated with significant psychologic comorbidity such as anxiety and depression, these may result in less psychologic flexibility that may interfere with the ability to induce placebo responses. Indeed, recent experiments from our laboratory[32,33] suggest that patients with fibromyalgia (FM), who can often present with such psychologic comorbities, have abnormalities of both their ability to allocate attentional resources (less focus on sensory discrimination and more on affective processing) and the way they anticipate pain (less modulation of anticipatory processing by external cues). Recent work also that suggests that patients with FM demonstrate subtle differences in experimental placebo response consistent with differences in the way they attend to prior information (unpublished data).

EEG STUDIES OF PLACEBO ANALGESIA

The above methodology for studying pain anticipation and perception using EEG has been applied to the study of the mechanisms of placebo analgesia. The model presented above for placebo analgesia requires placebo to be a true physiologic phenomenon rather than due to increased compliance (i.e. the subject complying with implicit wishes of the experimenter to observe a pain reduction). Watson et al[34] provided the first clear EEG evidence of this, showing that placebo analgesia is a real physiologic phenomenon in which cortical responses to pain (P2 peak of the LEP) are diminished. Further work by Wager et al[35] established that the N2 peak of the LEP does not appear to be affected as a part of the placebo response, finding only effects on the P2 peak in agreement with the results of Watson et al.[34] Of further interest to the study by Wager et al was the finding that variance in P2 peak responses did not entirely account for the changes in reported pain as a result of placebo. The authors discussed this as possibly due to later pain processing (post-LEP) being important in modifying pain reports. Similarly, Colloca[36] did not establish a correlational relationship between the neurophysiologic response and pain reports. In a study by Fiorio et al of 'placebo-like' enhancement of non-noxious stimulus perception,[37] reported stimulus intensity was also not correlated with late evoked-potential amplitudes. In this case, the authors discussed the possibility that greater evoked potential amplitudes represented a 'cortical sensory gain' that enhanced stimulus salience, possibly via enhancement of attention, but that this is only one out of many possible processes that may influence perceptual judgements.[37]

Studies have shown that, in healthy populations, both placebo analgesia and physiologic changes as measured with EEG are reproducible across different experimental sessions.[4,5,38] For example, it has been clearly demonstrated that the experimental placebo response is associated with reductions in SPN.[38] However, the way in which this reduction occurred was unexpected. In this study, experimental placebo was induced in a group of subjects by providing a sham treatment (a cream) to the skin, which was paired to reinforced expectation (conditioning) by telling subjects that the cream may have been an analgesic. A control group had the same procedure but were fully informed that the cream did not contain an anaesthetic. Unexpectedly, there was no reduction in SPN in either group in this first session. There was also a lack of correlation between changes in the SPN and changes in pain as a result of the sham. These negative results may indicate a persistence of fear and/or lack in confidence in predictions of reduced pain. However, in both groups, the experiment was repeated with a minimum gap of two weeks (range 2–6 weeks). The SPN was substantially reduced in the repeat sham treatment group both before and after application of cream, although with bigger reductions post-application. This may be because there was greater confidence in the belief that the cream would work on the second session having experienced the beneficial effects of the placebo treatment in the first session.

These reproducibility findings may be related psychologic traits that predispose towards placebo responding. For example, in studies by Morton et al,[4,5] reproducibility of placebo analgesia has been positively correlated with the psychologic trait of optimism with a trend towards a negative association with generalized anxiety. EEG work has gone some way towards supporting the role of cognitive factors and anxiety in placebo analgesia and its reproducibility. Morton et al[38] showed that reductions in state anxiety preceded the large reductions in anticipatory activity occurring when the experimental placebo was repeated in the same individuals. It was suggested that this reduction in anxiety may be one of the main mediators of placebo analgesia, although this was not correlated with changes in anxiety. However, the cognitive trait of dispositional optimism was found to be significantly correlated with changes in anxiety, SPN and LEP (N2/P2) over the two sessions.

There are some caveats to using the SPN to make inferences about changes in cognitive mediators of placebo analgesia. The SPN is also known to be increased by reward,[39,40] and it is possible that placebo analgesia (in particular, expectations of reduced pain) may increase reward processing and contribute to larger SPN amplitudes. In other words, the SPN could represent a mixture of anticipatory processes related to both aversion and reward originating from different regions of the brain, but that cannot be differentiated using a simple analysis of scalp amplitudes. This may explain why a reduction in the SPN was not found in the first session of the above-mentioned reproducibility study.[38] Also, the results of Morton et al are not entirely consistent with recent fMRI experiments[41,42] (reviewed in Chapter 10) that clearly show focal reductions in brain activity during pain anticipation in a single placebo session after changes in reinforced altered expectancy ('conditioning'). This highlights the importance of identifying the different brain sources that give rise to the SPN and interrogating them individually, as done in the studies by Brown et al[24,25] on pain anticipation.

Research has begun to use EEG to differentiate between verbal expectancy and conditioning mechanisms underlying placebo analgesia. Research by Colloca et al[36] has shown that both conditioning and expectation alone lead to reductions in LEP responses, but only conditioning leads to an actual reduction in pain consistent with a placebo analgesic response. Colloca et al assessed the effect of verbal suggestion alone in one group and reinforced suggestion (i.e. conditioning) in a second group, whereby the intensity of the laser stimulus was surreptitiously turned down after bland cream application. Only significant reduction in pain was produced by reinforced suggestion. Interestingly, the LEP (N2/P2) was significantly reduced in both groups but this was only associated with pain reduction in the group who were exposed to reinforced suggestion. This provides evidence that learning in the context of experience of pain reduction is a more powerful mediator of placebo analgesia than learning from verbal suggestion alone.

One crucial question is: does attention influence placebo responses? 'Attention' is a broad concept that includes a number of processes and ways of classifying these processes. A key example is the dual-processing theory of Schneider and Shiffrin[43,44] in which a distinction is made between automatic and controlled processing. Automatic processing is fast, parallel, and not limited by short-term memory, while controlled processing allows little subjective control and requires extensive and consistent training to develop. This is an important distinction in relation to the model of placebo (Fig. 5.2), as it may be the case that only automatic processes of attention interact with pain expectancy. For example, work by Buhle et al[45] has usefully established that controlled processes of attention are additive rather than interactive with placebo analgesia, suggesting that controlled attentional processes are not required for a placebo response. In their paper, Buhle et al describe using a cognitive distraction task (the n-back task) that requires working (short-term) memory, and is therefore an example of controlled attention.

Going back to the study by Colloca,[36] *automatic* processes of attention may have been a factor in the results, in that the suggestion-alone group were less certain and (according to the Bayesian model) would have allocated greater attentional resources to nociceptive processing to resolve this uncertainty, compared to the group exposed to reinforced suggestion. As such, changes in attentional processing may partially explain a degree of variance in the LEPs that was not present under the more certain conditions induced by reinforcement. This hypothesis could be tested using measurement of anticipatory processing (SPN) and source reconstruction to identify whether there were any differences in attentional processing regions, but behavioral experiments that assess automatic attentional processes may also be required.

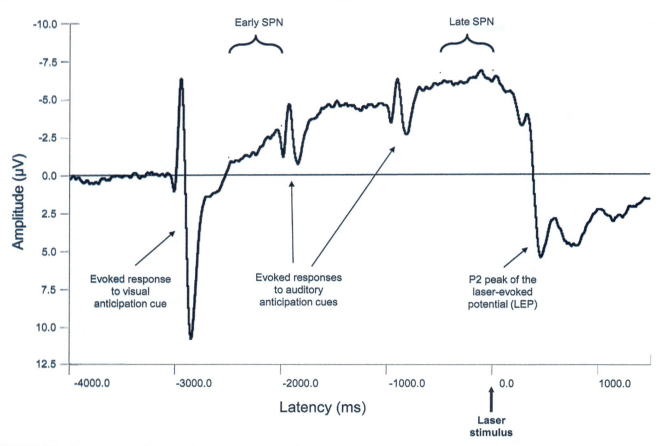

FIGURE 5.2 An example of the evoked-potential responses that can be recorded using EEG during the anticipation and experience of acute laser pain. Early and late phases of the stimulus-preceding negativity (SPN) can be differentiated that may have functionally distinct roles in pain expectancy.

CONCLUSION

EEG provides the means to understand some of the variability in placebo response, and the mechanisms behind this variability. It provides a relatively cost-effective way of monitoring some key components of placebo response and the role of psychologic traits and phenotypes in modifying these components. So far, we have learnt that expectations drive pain perception in a way that is consistent with a Bayesian model. Within this model, more certain expectation (as a result of perceptions of pain reduction cues being reliable) weights pain perception towards prior experience and beliefs, while uncertain expectation promotes learning by increasing attention towards nociceptive processing. Studies have begun to study these mechanisms in relation to placebo analgesia, showing reduction of anticipatory processing (SPN) and related psychologic factors such as anxiety, leading to a reduction in LEP amplitudes consistent with a true physiologic response. Recent work to differentiate the effects of verbal expectancies and conditioning on these responses is a promising example of the use of EEG for interrogating the mechanisms of placebo analgesia.

However, more work is required to further refine our current cognitive model of placebo analgesia. In particular, research should focus on (1) the role of anticipatory reward processing, (2) whether anxiety reduction is a necessary mediator of expectancy effects, (3) whether there are distinct neuronal mechanisms related to conditioning effects or whether these simply act to reinforce expectations of pain relief, and (4) the possible role of attention in explaining physiologic differences between verbal expectancy and conditioning (reinforcement) mechanisms.

References

1. Wager TD. Expectations and anxiety as mediators of placebo effects in pain. *Pain*. 2005;115:225-226.
2. Benedetti F, Pollo A, Lopiano L, et al. Conscious expectation and unconscious conditioning in analgesic, motor, and hormonal placebo/nocebo responses. *J Neurosci*. 2003;23:4315-4323.
3. Ross M, Olson JM. An expectancy-attribution model of the effects of placebos. *Psychol Rev*. 1981;88:408-437.
4. Morton DL, El-Deredy W, Watson A, Jones AKP. Placebo analgesia as a case of a cognitive style driven by prior expectation. *Brain Res*. 2010;1359:137-141.

5. Morton DL, Watson A, El-Deredy W, Jones AKP. Reproducibility of placebo analgesia: Effect of dispositional optimism. *Pain*. 2009;146:194-198.
6. Voudouris NJ, Peck CL, Coleman G. The role of conditioning and verbal expectancy in the placebo response. *Pain*. 1990;43:121-128.
7. Pavlov IP. *Conditional Reflexes*. Oxford: Oxford University Press; 1927.
8. Montgomery GH, Kirsch I. Classical conditioning and the placebo effect. *Pain*. 1997;72:107-113.
9. Stewart-Williams S, Podd J. The placebo effect: dissolving the expectancy versus conditioning debate. *Psychol Bull*. 2004;130:324-340.
10. Deliza R, Macfie HJH. The generation of sensory expectation by external cues and its effect on sensory perception and hedonic ratings: a review. *J Sensory Stud*. 1996;11:103-128.
11. Allan LG, Siegel S. A signal detection theory analysis of the placebo effect. *Eval Health Profess*. 2002;25:410-420.
12. Shams L, Beierholm UR. Causal inference in perception. *Trends Cognit Sci*. 2010;14:425-432.
13. Yu AJ, Dayan P. Uncertainty, neuromodulation, and attention. *Neuron*. 2005;46:681-692.
14. Ploghaus A, Becerra L, Borras C, Borsook D. Neural circuitry underlying pain modulation: expectation, hypnosis, placebo. *Trends Cogn Sci*. 2003;7:197-200.
15. Ploghaus A, Narain C, Beckmann CF, et al. Exacerbation of pain by anxiety is associated with activity in a hippocampal network. *J Neurosci*. 2001;21:9896-9903.
16. Kaptchuk TJ, Kelley JM, Conboy LA, et al. Components of placebo effect: Randomised controlled trial in patients with irritable bowel syndrome. *BMJ*. 2008;336:999-1003.
17. Brown C, Watson A, Morton D, et al. Role of central neurophysiological systems in placebo analgesia and their relationships with cognitive processes mediating placebo responding. *Future Neurol*. 2011;6:389-398.
18. Garcia-Larrea L, Frot M, Valeriani M. Brain generators of laser-evoked potentials: from dipoles to functional significance. *Neurophysiol Clin*. 2003;33:279-292.
19. Garcia-Larrea L, Peyron R, Laurent B, Mauguiere F. Association and dissociation between laser-evoked potentials and pain perception. *Neurorep*. 1997;8:3785-3789.
20. Carmon A, Friedman Y, Coger R, Kenton B. Single trial analysis of evoked potentials to noxious thermal stimulation in man. *Pain*. 1980;8:21-32.
21. Carmon A, Dotan Y, Sarne Y. Correlation of subjective pain experience with cerebral evoked responses to noxious thermal stimulations. *Exp Brain Res*. 1978;33:445-453.
22. Legrain V, Guerit JM, Bruyer R, Plaghki L. Attentional modulation of the nociceptive processing into the human brain: selective spatial attention, probability of stimulus occurrence, and target detection effects on laser evoked potentials. *Pain*. 2002;99:21-39.
23. Beydoun A, Morrow TJ, Shen JF, Casey KI. Variability of laser-evoked potentials: attention, arousal and lateralized differences. *Electroencephalogr Clin Neurophysiol*. 1993;88:173-181.
24. Brown CA, Seymour B, Boyle Y, et al. Modulation of pain perception by expectation and uncertainty: behavioral characteristics and anticipatory neural correlates. *Pain*. 2008;135:240-250.
25. Brown CA, Seymour B, El-Deredy W, Jones AKP. Confidence in beliefs about pain predicts expectancy effects on pain perception and anticipatory processing in right anterior insula. *Pain*. 2008;139:324-332.
26. Friston K. Learning and inference in the brain. *Neural Netw*. 2003;16:1325-1352.
27. Kersten D, Yuille A. Bayesian models of object perception. *Curr Opin Neurobiol*. 2003;13:150-158.
28. Wiech K, Tracey I. The influence of negative emotions on pain: Behavioral effects and neural mechanisms. *Neuroimage*. 2009;47:987-994.
29. Sarinopoulos I, Dixon GE, Short SJ, et al. Brain mechanisms of expectation associated with insula and amygdala response to aversive taste: implications for placebo. *Brain Behav Immun*. 2006;20:120-132.
30. Peyron R, Laurent B, Garcia-Larrea L. Functional imaging of brain responses to pain. A review and meta-analysis. *Neurophysiol Clin*. 2000;30:263-288.
31. Peyron R, Garcia-Larrea L, Gregoire MC, et al. Haemodynamic brain responses to acute pain in humans: sensory and attentional networks. *Brain*. 1999;122(9):1765-1780.
32. Brown CA, Jones AKP. *Anticipatory and pain-evoked neural correlates of pain symptoms, mood and coping in patients with fibromyalgia and osteoarthritis*: The 14th World Congress on Pain; 2012 (abstract).
33. Kulkarni B, Boger EJ, Watson A, et al. *Cortical Mechanisms of Attentional Dysfunction in Fibromyalgia: a PET Study*. Sydney: IASP 11th World Congress on Pain; 2005 (abstract).
34. Watson A, El-Deredy W, Vogt BA, Jones AK. Placebo analgesia is not due to compliance or habituation: EEG and behavioural evidence. *Neurorep*. 2007;18:771-775.
35. Wager TD, Matre D, Casey KL. Placebo effects in laser-evoked pain potentials. *Brain Behav Immun*. 2006;20:219-230.
36. Colloca L, Tinazzi M, Recchia S, et al. Learning potentiates neurophysiological and behavioral placebo analgesic responses. *Pain*. 2008;139:306-314.
37. Fiorio M, Recchia S, Corrá F, et al. Enhancing non-noxious perception: Behavioural and neurophysiological correlates of a placebo-like manipulation. *Neuroscience*. 2012;217:96-104.
38. Morton DL, Brown CA, Watson A, et al. Cognitive changes as a result of a single exposure to placebo. *Neuropsychologia*. 2010;48:1958-1964.
39. Ohgami Y, Kotani Y, Tsukamoto T, et al. Effects of monetary reward and punishment on stimulus-preceding negativity. *Psychophysiology*. 2006;43:227-236.
40. Kotani Y, Kishida S, Hiraku S, et al. Effects of information and reward on stimulus-preceding negativity prior to feedback stimuli. *Psychophysiology*. 2003;40:818-826.
41. Watson A, El-Deredy W, Iannetti GD, et al. Placebo conditioning and placebo analgesia modulate a common brain network during pain anticipation and perception. *Pain*. 2009;145:24-30.
42. Wager TD, Atlas LJ, Leotti LA, Rilling JK. Predicting individual differences in placebo analgesia: contributions of brain activity during anticipation and pain experience. *J Neurosci*. 2011;31:439-452.
43. Shiffrin RM, Schneider W. Controlled and automatic human information processing: II. Perceptual learning, automatic attending and a general theory. *Psychol Rev*. 1977;84:127-190.
44. Schneider W, Shiffrin RM. Controlled and automatic human information processing: I. Detection, search, and attention. *Psychol Rev*. 1977;84:1-66.
45. Buhle JT, Stevens BL, Friedman JJ, Wager TD. Distraction and placebo: two separate routes to pain control. *Psychol Sci*. 2012;23(3):246-253.

CHAPTER 6

Spinal Mechanisms of Placebo Analgesia and Nocebo Hyperalgesia: Descending Inhibitory and Facilitatory Influences

Serge Marchand

Université de Sherbrooke, Faculté de médecine and Centre de recherche clinique Étienne-Le Bel du CHUS Sherbrooke, Québec, Canada

INTRODUCTION

The outcomes of an inactive treatment that mimics the effect of a real treatment are called placebo or nocebo effects. The placebo effect defines the positive response of a subject to a substance or to any procedure that has no therapeutic effect for the treated condition.[1] More specifically, this effect can be defined by the physiologic and/or psychologic changes, such as a reduction of pain associated with the use of an inert medication, a simulated therapeutic procedure or a meeting with therapeutic symbols. The nocebo effect is the appearance of adverse effects accompanying the use of an inert substance. Placebo and nocebo responses can be present without the administration of an inert substance, such as when you have just received a diagnosis or when you are warned of an upcoming particularly painful stimulus.[2]

The least we can say is that these effects are frequently perceived as mysterious or questionable by the caregiver and the patient. The care-giver may interpret these responses to an inactive treatment as the proof of the absence of a real pathology, while the patient may feel that the caregiver has fooled him/her and perceives his/her response as a psychologic weakness. These impressions about the placebo response are probably based on the fact that most people are seeing the placebo effect as a mere reinterpretation of the patient's perception, like the impression that the pain is now reduced or has just disappeared while, in fact, it is exactly the same. Physiologically, this point of view on the placebo mechanisms will be described as a purely cortical effect.

The reality is a little more complex than that. As described in the other chapters of this book, the placebo response is clearly triggering cortical changes that several good laboratories are recording. However, this effect is not restricted to the cortex and will influence the rest of the central nervous system (CNS), from the cortex to the spinal cord.

In this chapter, I concentrate on how the placebo and nocebo mechanisms are closely linked to endogenous pain modulation mechanisms and how placebo analgesia or nocebo hyperalgesia acts through brainstem and spinal facilitatory and inhibitory mechanisms.

PAIN AND PLACEBO HAVE DYNAMIC INTERACTIONS

To better appreciate the mechanisms and the clinical implications of the placebo and nocebo responses in pain, we need to understand the basic neurophysiology of the development and persistence of pain.

Pain is a complex phenomenon playing a major role in protecting us from injury or illness. The pain signal needs to be clear and emotionally salient for an individual to act rapidly and adequately to get away from the nociceptive source and take care of the injury. However, in some conditions, the nociceptive signals have to be temporarily silenced to focus on actions required to reduce further harm and thus, increasing chances of survival.

The pain perceived following a nociceptive stimulus would then be completely different depending on the context and situation. To avoid an injury, the CNS needs to be able to rapidly encode the localization and intensity of a nociceptive stimulus. However, the nervous system also needs to be able to ignore pain in other situations, such as getting out of a car on fire after an accident, even if you have fractures or lacerations. It is most likely for these reasons that the CNS has developed several complex endogenous facilitatory and inhibitory mechanisms that can either emphasize or reduce the perception of pain following a nociceptive stimulus depending on the circumstances.

Nociceptive afferents can be modulated at all levels from periphery to higher CNS centers. Because of the dynamic and plastic characteristics of the nervous system, pain perception is not the sole result of the nociceptive activity, but the endpoint of complex facilitatory and inhibitory endogenous pain modulation mechanisms. This plasticity speaks to the nervous system's ability to adapt and change. It is then not surprising that the etiology of pain of two patients suffering from apparently similar clinical conditions may be related to different mechanisms such as increased facilitatory mechanisms in one case and a reduction of endogenous pain inhibitory mechanisms in the other case. As we will see, the placebo and nocebo responses modulate the nociceptive activity and pain responses by acting through these mechanisms.

To better understand the link between pain and placebo mechanisms, we have to study the nociceptive signal from the periphery to the higher centers of the CNS, but we also need to understand the descending pain modulation controls arising from the higher centers and projecting to the spinal cord.

Based on our knowledge of the neurophysiology of pain, we can conclude that the development, maintenance and recovery from pain depend on several factors. Persistent pain can result from the activity of nociceptive afferents, but it can also be related to a reduction of endogenous inhibition and/or an increase of endogenous facilitatory mechanisms. Central sensitization supports the importance of endogenous pain facilitatory circuitry in the development and maintenance of pain. The facilitatory and inhibitory roles on pain modulation played by different structures of the brainstem have been well documented.[3-5]

In order to better understand the effects of placebo on descending facilitatory and inhibitory mechanisms and on the spinal nociceptive activity, we will shortly review some of these pain modulatory mechanisms.

FACILITATORY MECHANISMS

Spinal Sensitization

Central sensitization refers to a phenomenon whereby the second neuron membrane permeability changes and responds at higher frequency when recruited by nociceptive (hyperalgesia) and non-nociceptive primary input (allodynia). Central sensitization is defined as an increase of excitability and spontaneous discharge of the dorsal horn neurons with an associated increase in the receptive field of these neurons. This phenomenon will principally affect the wide dynamic range (WDR) neurons from the dorsal horn and is dependent on the activity of the N-methyl-D-aspartate (NMDA) receptors.[6,7] These neurophysiologic and neurochemical mechanisms involved in central sensitization are responsible for the modification of the spinal nociceptive circuitry and contribute to the maintenance of pain.

In the spinal cord, secondary hyperalgesia is a phenomenon that refers to sensitization.[8] Repeated recruitment of C fibers following an injury will produce central sensitization by changing the response properties of the membrane of secondary neurons. This will result in an increase of the firing rate, a phenomenon known as windup.[9] The high frequency recruitment of C fibers, either by increased repetitive stimuli or by a tonic stimulation,[10] will then induce an increase in the perceived pain, even if the intensity of the stimulation remains constant. This central sensitization at the spinal level can persist for minutes, but can also be present for hours and even days.[11] The prolonged activation of the NMDA receptors will induce the transcription of rapidly expressed genes (c-*fos*, c-*jun*), resulting in sensitization of nociceptors. This neuronal plasticity of the secondary neuron will result in a reduced threshold and enlargement of their receptive field in the spinal cord and produce hyperalgesic and allodynic responses that may persist even after the injury is healed.

Descending Facilitatory Mechanisms

It is now well documented that several supraspinal facilitatory and inhibitory mechanisms play a major role in pain perception and most probably in certain chronic pain conditions.[12] The work of Fields describing activation of 'ON' cells and inhibition of 'OFF' cells in the brainstem during nociceptive activity has demonstrated the importance of facilitatory mechanisms in amplifying the nociceptive response.[13] Considering that proglumide, an antagonist of cholecystokinin (CCK), blocked nocebo hyperalgesia,[14] and that CCK directly activates 'ON' cells,[15] it is possible that the hyperalgesia reported during the nocebo effect depends on these excitatory bulbospinal circuits that facilitate spinal nociceptive activity.

Recent studies have also demonstrated that certain physiologic conditions, such as nociceptive hyperactivity, may change the usual neuronal response to specific neurotransmitters. A particular example is the hyperalgesic effect that can be observed in some patients using opioid medications.[16] Therefore, drugs with opioid activity could, under certain circumstances, produce a completely opposite effect and enhance pain by producing hyperalgesic responses.[16,17] The same is also true for gamma-aminobutyric acid (GABA) that has been clearly identified as an inhibitory neurotransmitter, but in certain conditions may cause depolarization of neurons.[18] These observations support the concept of pain as a dynamic phenomenon. An understanding of these complex mechanisms can help explain the clinical variability of response to treatments in patients with chronic pain.

Placebo and Nocebo Effects on Facilitatory Mechanisms

As we have just seen, both spinal and supraspinal facilitatory mechanisms are responsible for the increase in pain perception by the recruitment of endogenous pain facilitatory mechanisms. We will now examine how nocebo responses are closely linked to these endogenous pain modulation mechanisms.

As previously described, a CCK antagonist, proglumide, is capable of blocking the nocebo response.[2] Interestingly, the pronociceptive effects of CCK act on descending facilitatory mechanisms by activating ON cells in the brainstem that will send facilitatory nociceptive efferences to the spinal cord.[15] These results suggest that the nocebo response is related to the triggering of a brainstem–spinal cord circuitry rather than only a cortical reappraisal of the nociceptive input. This suggestion is reinforced by a study demonstrating that the pronociceptive effect of CCK could also be related to an inhibitory effect of CCK on the GABA inhibitory interneurons in the spinal cord.[19]

In order to see whether placebo was acting at the spinal level rather than only at the cortical level, Matre and colleague induced mechanical hyperalgesia by applying a 5-minute thermal nociceptive stimulus.[20] They found that the territory of mechanical hyperalgesia and allodynia was significantly reduced during the placebo session as compared to the control session, suggesting that the placebo effect was spinal, because secondary hyperalgesia (outside the stimulation territory) is a spinal sensitization phenomenon.

In a recent study, Peterson and colleagues demonstrated that a placebo manipulation was able to reduce hyperalgesia in patients suffering from neuropathic pain.[21] Nineteen patients who had developed neuropathic pain after thoracotomy received open or hidden lidocaine for their pain. The open administration session reduced significantly more the area of pinprick hyperalgesia than did the hidden session, suggesting that psychologic factors related to the consciousness of receiving the treatment had an effect on central sensitization.

The most direct demonstration of a placebo effect at the spinal level comes from Eippert and colleagues who performed functional magnetic resonance imaging (fMRI) of the brainstem and spinal cord to painful stimuli before and after applying a placebo cream. The subjects had previously been conditioned to believe that the cream was a strong analgesic by reducing the nociceptive stimulus after the application.[22] They found that the pain rating and spinal activity recorded by fMRI were significantly reduced under the placebo condition.

It is well documented that negative emotions, expectation or conditioning that result from a nocebo response trigger activity from the prefrontal cortex (ventrolateral PFC – vlPFC; ventral-medial PFC – vmPFC) that will recruit descending facilitatory mechanisms from the brainstem (rostroventral medulla – RVM).[23–25] Considering that descending facilitatory mechanisms may play an important role in some chronic pain conditions,[2,23,26,27] we can then conclude that context, situations and conditioning that trigger nocebo responses may activate descending facilitatory mechanisms and spinal changes that will play a role in pain chronification.

INHIBITORY MECHANISMS

As we have just seen, descending facilitation can play a major role in hyperalgesia and can be implicated in the development and persistence of chronic pain. However, a reduction in endogenous pain inhibitory mechanisms can also play a role in pain, even in some chronic pain conditions.[28–30] In order to better understand the role of endogenous pain inhibitory mechanisms in the development and treatment of pain, we will introduce three levels of modulation in the CNS (Fig. 6.1): (1) spinal mechanisms producing localized analgesia; (2) descending inhibitory mechanisms from the brainstem producing diffuse inhibition; and (3) higher center effects that will either modulate descending mechanisms or change the perception of pain by reinterpreting the nociceptive signal.

Spinal Mechanisms

Since the proposal of the gate control theory by Melzack and Wall,[31] the modulation of nociceptive afferents at their entry into the spinal cord has been well documented. The gate control theory hypothesizes that, amongst other mechanisms, selective activation of non-nociceptive afferent Aβ fibers will recruit inhibitory interneurons in the *substantia gelatinosa* of the posterior spinal cord, producing a localized analgesia and decreasing pain perception. In contrast, in certain neuropathic pain conditions, the nociceptive secondary projection neurons will be recruited by non-nociceptive

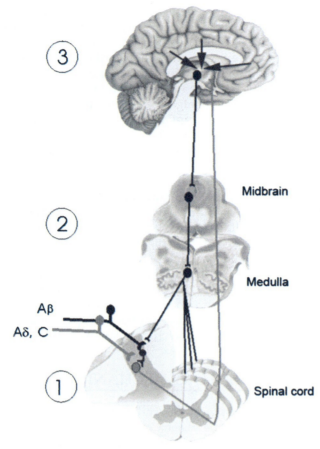

FIGURE 6.1 A schematic representation of the three main levels of endogenous pain modulation: (1) spinal inhibitory mechanisms, (2) inhibitory mechanisms descending from the brainstem, and (3) inhibitory mechanisms descending from higher centers. As described in the text, placebo and nocebo responses act by modulating these mechanisms and changing the spinal cord response to nociceptive activity. *This figure is reproduced in color in the color section.*

afferent Aβ fibers to transmit a pain signal following an innocuous stimulation, a phenomenon known as allodynia. Certain pain conditions may also result from a reduced efficacy of tonic inhibitory controls within the spinal cord.[32,33]

Diffuse Noxious Inhibitory Controls

A few years after the gate control theory was proposed in 1965, Reynolds demonstrated that the stimulation of the periaqueductal gray (PAG) produced a strong inhibition of nociceptive activity.[34] The role of the rostroventral medulla in the modulation of pain has since been well documented.[9,35] Regions such as the PAG and the nucleus raphe magnus (NRM) have been identified as important serotonergic and noradrenergic descending inhibitory pathways. These inhibitory pathways then recruit enkephalinergic interneurons in the spinal cord to produce the analgesic response.

We had to wait until the end of the 1970s before a model known as diffuse noxious inhibitory control (DNIC) was proposed.[4,36] This model is based on the observation that a localized nociceptive stimulation can produce a diffuse analgesic effect over the rest of the body, an analgesic approach known as counter-irritation. In the DNIC model, Le Bars (1979) proposed that a nociceptive stimulus would send input to higher centers, but would also send afferences to the PAG and NRM of the brainstem, recruiting inhibitory output at multiple levels of the spinal cord.

Animal studies demonstrate that a lesion in the dorsolateral funiculus, the main descending inhibitory pathway, will produce hyperalgesia, suggesting the existence of a tonic descending inhibition under normal conditions.[37,38] Certain clinical conditions are related to a deficit of endogenous pain inhibition. For example, the low concentration of serotonin and noradrenaline in the cerebrospinal fluid of patients suffering from fibromyalgia (FM)[39] suggests a deficit of DNIC, with increasing evidence corroborated by other studies.[28,30,40–43]

Documenting the role of descending inhibitory mechanisms will help us to better understand certain chronic pain conditions, such as FM. It will also help towards understanding the mechanism of action of pharmacologic approaches, such as the use of antidepressants in chronic pain conditions by enhancing DNIC efficacy by their serotonergic and noradrenergic activity. Moreover, the deficit of DNIC (conditioned pain modulation – CPM) seems to be a good predictor of the response to these classes of medication.[44]

Control of Higher Centers

There has been an increased appreciation of the role of supraspinal centers in pain and pain modulation in recent years. Several cortical regions receive input from the spinothalamic tract and interact to produce the multidimensional experience of pain perception.[45] The use of brain-imaging techniques has shown robust activation of certain cortical regions, including the primary and secondary somatosensory cortices, related to the sensory aspect of pain, and the anterior cingulate cortex (ACC) and insular cortex (IC) for the affective component of the pain experience.[46]

There is no doubt that cognitive manipulations, such as distraction, hypnosis and expectation influence pain perception.[47] Hypnosis has been demonstrated to change both the sensory and affective component of pain perception. Subjects given the same nociceptive stimulus perceived both intensity and unpleasantness of pain differently depending upon the suggestion given.[48] Using positron emission tomography (PET) to obtain brain activity images, activity of the primary somatosensory cortex was proportional to the perceived intensity of pain,[49] whereas cingulate cortex activity reflected unpleasantness

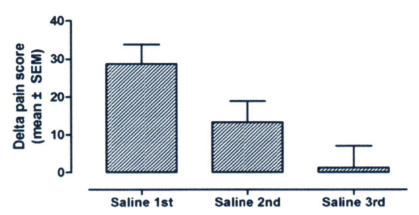

FIGURE 6.2 TENS analgesia experienced during the saline condition (control condition) between participants who received (1) saline at their first session, (2) naloxone at their first session and saline at their second session, and (3) naloxone at their first and second sessions and saline at their third session. As we can see, the analgesic effect of TENS, when it was presented at the first session, showed a larger analgesic effect than the other participants, suggesting a negative (nocebo) conditioning of the previous naloxone sessions.

of pain.[50] These data confirm that simple analgesic or hyperalgesic suggestions can change the activity of specific brain regions related to pain perception.[23,51]

Placebo Effects on Inhibitory Mechanisms

Pain not only arises from an increase in facilitatory mechanisms but also from a decrease in inhibitory mechanisms – just like the balance between the sympathetic and parasympathetic systems gives an accurate control of the autonomic nervous system.

Conditioning and expectation are two important factors in inducing placebo and nocebo responses.[51,52] In the next examples, we will see how being exposed to an ineffective procedure can reduce or totally block the subsequent treatment with a similar procedure. The second example will demonstrate how the efficacy of a treatment can be significantly reduced by the sole expectation of negative outcomes.

Effect of Conditioning on TENS Analgesia: Spinal Mechanisms and Nocebo Response

Transcutaneous electrical nerve stimulation (TENS) is an analgesic procedure that consists of a stimulation of non-nociceptive afferences by high-frequency, low-intensity electrical stimulations. The inhibition of pain through the stimulation of non-nociceptive afferents primarily comes under the gate control theory,[31] according to which the selective recruitment of non-nociceptive afferents (Aβ) inhibits nociceptive afferents (A∂ and C) by the activation of inhibitory interneurons in the *substantia gelatinosa* of the posterior horns of the spinal cord. In order to explore the possible contribution of the opioidergic component in TENS analgesia an opioidergic antagonist with high mu opioid receptor affinity, naloxone, was used. Healthy subjects were randomly assigned in a double-blind, crossover design to one of the following conditions: (1) high dose of naloxone (0.14 mg kg^{-1}), (2) low dose of naloxone (0.02 mg kg^{-1}) and (3) saline (0.9% NaCl). We found that this type of TENS was indeed blocked by the highest dose of naloxone, suggesting that previous studies that supported a non-opioidergic mechanism only blocked mu opioid receptors with their low dose of naloxone.[53]

However, the most interesting part of this project for this chapter was the negative effect that conditioning could do to this type of spinal analgesia. In order to have a strong scientific design, we randomized the three injections: naloxone high-dose, naloxone low-dose and saline. All the subjects knew that they had an equal chance to receive any order of the tested drugs. Interestingly, we found that when the saline condition (TENS without the antagonist naloxone) was presented at the first session, the analgesic effect was the strongest. However, the efficacy of TENS was significantly reduced if it was presented after a first session with naloxone and was almost totally blocked after two sessions of naloxone, suggesting an important order effect.

Even if the subject knew that it was possible that the first two sessions could be the naloxone sessions, the prior exposition to ineffective treatments significantly reduced subsequent treatment efficacy. Since high-frequency, low-intensity TENS produces its analgesia in the spinal cord,[54,55] this suggests that the negative conditioning effect was a spinal nocebo effect (Fig. 6.2).

Effect of Expectation on Endogenous Analgesia

In a recent study, we were able to demonstrate that manipulating the expectation related to an analgesic procedure can completely reverse the analgesic effect of endogenous pain modulation and the related pain experience. By suggesting that a procedure that is normally analgesic would produce more pain, subjects indeed reported more pain. Experimental pain was evoked

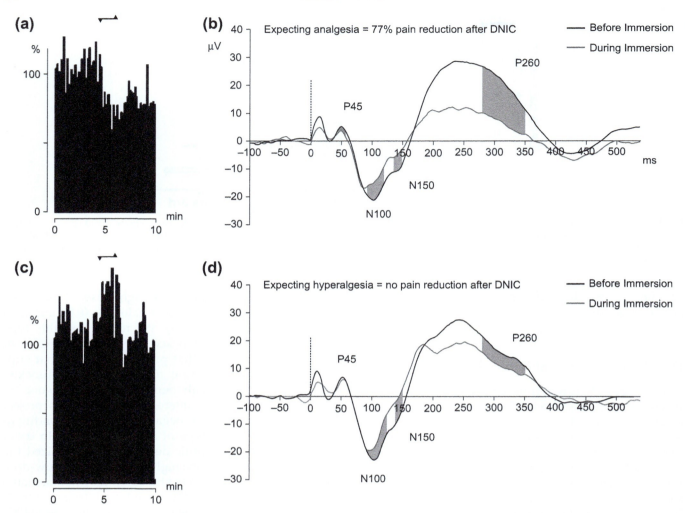

FIGURE 6.3 Immersion-induced change in spinal and cortical activity differs depending on expectations of pain relief. (a) Analgesia-expected group (77% perceived pain relief) and (c) hyperalgesia-expected group (0% perceived pain relief) show the average reflex amplitudes recorded for each sural nerve stimulation and expressed as a percentage value from baseline. Group differences in reflex amplitude show a reduction in reflex amplitude only for the analgesia-expected group. (b) Analgesia-expected group and (d) hyperalgesia-expected group show the average sural nerve-evoked brain potentials recorded prior to and during arm immersion. Larger P260 reductions are observed when analgesia is expected (b), than when hyperalgesia is expected (d).

through intermittent electrical stimulations of the left ankle over the retromalleolar path of the sural nerve. When sufficiently intense, this type of stimulation triggers a nociceptive spinal withdrawal reflex (measured by electromyographic recordings of the knee flexor muscle) and somatosensory evoked potentials (SEP) (by scalp electroencephalographic electrodes), whose amplitude correlates with stimulation intensity. During immersion, there was a significant reduction in perceived sural nerve pain, reflex amplitude and SEP in patients who correctly expected that the immersion would have analgesic properties. On the other hand, participants who expected that the immersion would have pain-enhancing properties showed an increase in perceived sural nerve pain and a complete abolition of the normal reduction in reflex amplitude and SEP. Therefore, suggestion was able to totally block the endogenous analgesia normally recorded with DNIC.[56] Similar results were obtained by another group of investigators, but with a strong exogenous analgesic, morphine. Morphine analgesia was potentiated or inhibited depending on the instruction that was given to the subject.[57]

These results support the idea that cognitive information can modulate the efficacy of endogenous pain modulation and emphasize the importance of the patient's expectations regarding analgesia (Fig. 6.3).

CONCLUSION

Pain is a complex phenomenon. On the one hand, it is critical to maintain our homeostatic state for our survival. On the other hand, some situations need the pain signal to be silenced in order to concentrate and act adequately. Fortunately, the central nervous system offers complementary endogenous mechanisms that permit sensitization or inhibition of the nociceptive signal. By the activation of these endogenous systems we can either be more prone to reacting to a potentially harmful stimulus, or completely ignore it in order to concentrate on a more important aspect at the time. Psychologic factors such as conditioning and expectation are very efficient in recruiting these facilitatory and inhibitory mechanisms. This is not surprising considering that most often the need for these mechanisms is based on our appraisal of a perceived situation. Unfortunately, in chronic pain conditions the signal is amplified and loses its informative properties. In order to better understand the importance of the psychologic factors in pain, and their potential role in pain chronification, we need to better understand their mechanisms and how they interact with pain treatments.

Placebo and nocebo responses are probably the most intriguing psychologic outcomes in pain perception and treatment. Contrary to the popular belief that a placebo or nocebo response is only a reflection of psychologic reappraisal of our perception, we now have strong evidence that such responses are related to measurable changes of both facilitatory and inhibitory endogenous neurophysiologic mechanisms from the higher centers to the spinal cord. Therefore, we must understand that a placebo or nocebo response has the potential to change not only your brain,[51] but also your spinal cord.[20,56] These brain and spinal cord changes clearly play a role in the variability between individuals in response to pain treatments.[58] They also participate in pain chronification.[59]

Placebo and nocebo are intrinsic factors that are present in several pain conditions. They therefore must be studied as important factors in the development and treatment of pain in order to control their undesirable side effects, nocebo responses, and enhance the desirable effects, the placebo responses.

References

1. Benedetti F, Amanzio M. The neurobiology of placebo analgesia: from endogenous opioids to cholecystokinin. *Prog Neurobiol.* 1997;51:109-125.
2. Benedetti F, Lanotte M, Lopiano L, Colloca L. When words are painful: unraveling the mechanisms of the nocebo effect. *Neurosci.* 2007;147(2):260-271.
3. Fields HL, Heinricher M. Anatomy and physiology of a nociceptive modulatory system. *Phil Tran R Soc Lond.* 1985;308(1136):361-374.
4. Le Bars D, Dickenson AH, Besson JM. Diffuse noxious inhibitory controls (DNIC). 1. Effects on dorsal horn convergent neurones in the rat. *Pain.* 1979;6(3):283-304.
5. Basbaum AI, Fields HL. Endogenous pain control mechanisms: review and hypothesis. *Ann Neurol.* 1978;4(5):451-462.
6. Eide PK. Wind-up and the NMDA receptor complex from a clinical perspective. *Eur J Pain.* 2000;4(1):5-15.
7. Woolf CJ, Thompson SW. The induction and maintenance of central sensitization is dependent on N-methyl-D-aspartic acid receptor activation; implications for the treatment of post-injury pain hypersensitivity states. *Pain.* 1991;44(3):293-299.
8. Terman GW, Bonica JJ, Loeser JD. Spinal Mechanisms and their Modulation. *Management of Pain.* 2001; Vol. 3:73-152: New York: Lippincott Williams & Wilkins.
9. Fields HL, Basbaum A, Heinrich RL. Central nervous system mechanisms of pain modulation: vol. 5. In: McMahon SB, Koltzenburg M, eds. *Wall and Melzack's Texbook of Pain.* Philadelphia: Elsevier; 2006:125-142.
10. Granot M, Granovsky Y, Sprecher E, et al. Contact heat-evoked temporal summation: tonic versus repetitive-phasic stimulation. *Pain.* 2006;122(3):295-305.
11. Woolf CJ. Windup and central sensitization are not equivalent. *Pain.* 1996;66(2–3):105-108.
12. Yarnitsky D. Conditioned pain modulation (the diffuse noxious inhibitory control-like effect): its relevance for acute and chronic pain states. *Curr Opin Anaesthesiol.* 2010;23(5):611-615.
13. Fields HL, Malick A, Burstein R. Dorsal horn projection targets of ON and OFF cells in the rostral ventromedial medulla. *J Neurophysiol.* 1995;74(4):1742-1759.
14. Benedetti F, Amanzio M, Vighetti S, Asteggiano G. The biochemical and neuroendocrine bases of the hyperalgesic nocebo effect. *J Neurosci.* 2006;26(46):12014-12022.
15. Heinricher MM, Neubert MJ. Neural basis for the hyperalgesic action of cholecystokinin in the rostral ventromedial medulla. *J Neurophysiol.* 2004;92(4):1982-1989.
16. Simonnet G, Rivat C. Opioid-induced hyperalgesia: abnormal or normal pain? *NeuroRep.* 2003;14(1):1-7.
17. Davis MP, Shaiova LA, Angst MS. When opioids cause pain. *J Clin Oncol.* 2007;25(28):4497-4498.
18. Coull JA, Boudreau D, Bachand K, et al. Trans-synaptic shift in anion gradient in spinal lamina I neurons as a mechanism of neuropathic pain. *Nature.* 2003;424(6951):938-942.
19. Ma KT, Si JQ, Zhang ZQ, et al. Modulatory effect of CCK-8S on GABA-induced depolarization from rat dorsal root ganglion. *Brain Res.* 2006;1121(1):66-75.
20. Matre D, Casey KL, Knardahl S. Placebo-induced changes in spinal cord pain processing. *J Neurosci.* 2006;26(2):559-563.
21. Petersen GL, Finnerup NB, Norskov KN, et al. Placebo manipulations reduce hyperalgesia in neuropathic pain. *Pain.* 2012; 153(6):1292-1300.
22. Eippert F, Finsterbusch J, Bingel U, Buchel C. Direct evidence for spinal cord involvement in placebo analgesia. *Science.* 2009; 326(5951):404.
23. Tracey I. Getting the pain you expect: mechanisms of placebo, nocebo and reappraisal effects in humans. *Nature Med.* 2010;16(11):1277-1283.
24. Wiech K, Tracey I. The influence of negative emotions on pain: behavioral effects and neural mechanisms. *Neuroimage.* 2009;47(3): 987-994.
25. Wiech K, Ploner M, Tracey I. Neurocognitive aspects of pain perception. *Trends Cognit Sci.* 2008;12(8):306-313.
26. Benedetti F, Amanzio M. The placebo response: how words and rituals change the patient's brain. *Patient Educ Couns.* 2011;84(3): 413-419.
27. Porreca F, Ossipov MH, Gebhart GF. Chronic pain and medullary descending facilitation. *Trends Neurosci.* 2002;25(6):319-325.
28. Julien N, Goffaux P, Arsenault P, Marchand S. Widespread pain in fibromyalgia is related to a deficit of endogenous pain inhibition. *Pain.* 2005;114(1–2):295-302.
29. Leonard G, Goffaux P, Mathieu D, et al. Evidence of descending inhibition deficits in atypical but not classical trigeminal neuralgia. *Pain.* 2009;147(1–3):217-223.

30. Yarnitsky D, Crispel Y, Eisenberg E, et al. Prediction of chronic post-operative pain: pre-operative DNIC testing identifies patients at risk. *Pain*. 2008;138(1):22-28.
31. Melzack R, Wall PD. Pain mechanisms: a new theory. *Science*. 1965;150:971-979.
32. Traub RJ. Spinal modulation of the induction of central sensitization. *Brain Res*. 1997;778(1):34-42.
33. Millan MJ. The induction of pain: an integrative review. *Prog Neurobiol*. 1999;57(1):1-164.
34. Reynolds DV. Surgery in the rat during electrical analgesia. *Science*. 1969;164(878):444-445.
35. Ossipov MH, Dussor GO, Porreca F. Central modulation of pain. *J Clin Invest*. 2010;120(11):3779-3787.
36. Le Bars D, Dickenson AH, Besson JM. Diffuse noxious inhibitory controls (DNIC). II.Lack of effect on non-convergent neurones, supraspinal involvement and theoretical implications. *Pain*. 1979;6(3):305-327.
37. Davies JE, Marsden CA, Roberts MH. Hyperalgesia and the reduction of monoamines resulting from lesions of the dorsolateral funiculus. *Brain Res*. 1983;261(1):59-68.
38. Abbott FV, Hong Y, Franklin KB. The effect of lesions of the dorsolateral funiculus on formalin pain and morphine analgesia: a dose–response analysis. *Pain*. 1996;65(1):17-23.
39. Russell IJ. Neurochemical pathogenesis of fibromyalgia. *Z Rheumatol*. 1998;57(suppl 2):63-66.
40. Normand E, Potvin S, Gaumond I, et al. Pain inhibition is deficient in chronic widespread pain but normal in major depressive disorder. *J Clin Psychiatr*. 2011;72(2):219-224.
41. de Souza JB, Potvin S, Goffaux P, et al. The deficit of pain inhibition in fibromyalgia is more pronounced in patients with comorbid depressive symptoms. *Clin J Pain*. 2009;25(2):123-127.
42. Lautenbacher S, Rollman GB. Possible deficiencies of pain modulation in fibromyalgia. *Clin J Pain*. 1997;13(3):189-196.
43. Kosek E, Hansson P. Modulatory influence on somatosensory perception from vibration and heterotopic noxious conditioning stimulation (HNCS) in fibromyalgia patients and healthy subjects. *Pain*. 1997;70(1):41-51.
44. Yarnitsky D, Granot M, Nahman-Averbuch H, et al. Conditioned pain modulation predicts duloxetine efficacy in painful diabetic neuropathy. *Pain*. 2012;153(6):1193-1198.
45. Price DD. Psychological and neural mechanisms of the affective dimension of pain. *Science*. 2000;288(5472):1769-1772.
46. Casey KL. The imaging of pain: Background and rationale. In: Casey KL, Bushnell MC, eds. *Pain Imaging: Progress in Pain Research and Management*. Seattle: IASP Press; 2000:1-29.
47. Apkarian AV, Bushnell MC, Treede RD, Zubieta JK. Human brain mechanisms of pain perception and regulation in health and disease. *Eur J Pain*. 2005;9(4):463-484.
48. Rainville P, Carrier B, Hofbauer RK, et al. Dissociation of sensory and affective dimensions of pain using hypnotic modulation. *Pain*. 1999;82(2):159-171.
49. Hofbauer RK, Rainville P, Duncan GH, Bushnell MC. Cortical representation of the sensory dimension of pain. *J Neurophysiol*. 2001;86(1):402-411.
50. Rainville P, Duncan GH, Price DD, et al. Pain affect encoded in human anterior cingulate but not somatosensory cortex. *Science*. 1997;277:968-971.
51. Benedetti F, Carlino E, Pollo A. How placebos change the patient's brain. *Neuropsychopharmacol*. 2011;36(1):339-354.
52. Goffaux P, Leonard G, Marchand S, Rainville P. Placebo analgesia. In: Beaulieu P, Lussier D, Porreca F, Dickenson A, eds. *Pharmacology of Pain*. Seattle: IASP Press; 2010:451-473.
53. Leonard G, Goffaux P, Marchand S. Deciphering the role of endogenous opioids in high-frequency TENS using low and high doses of naloxone. *Pain*. 2010;151(1):215-219.
54. Handwerker HO, Iggo A, Zimmermann M. Segmental and supraspinal actions on dorsal horn neurons responding to noxious and non-noxious skin stimuli. *Pain*. 1975;1:147-165.
55. Salter MW, Henry JL. Differential responses of nociceptive vs. non-nociceptive spinal dorsal horn neurones to cutaneously applied vibration in the cat. *Pain*. 1990;40:311-322.
56. Goffaux P, Redmond WJ, Rainville P, Marchand S. Descending analgesia – when the spine echoes what the brain expects. *Pain*. 2007;130(1-2):137-143.
57. Bingel U, Wanigasekera V, Wiech K, et al. The effect of treatment expectation on drug efficacy: imaging the analgesic benefit of the opioid remifentanil. *Sci Transl Med*. 2011;3(70):70ra14.
58. Amanzio M, Pollo A, Maggi G, Benedetti F. Response variability to analgesics: a role for non-specific activation of endogenous opioids. *Pain*. 2001;90(3):205-215.
59. Flor H, Turk DC. *Chronic Pain. An Integrated Biobehavioral Approach*. Seattle: IASP Press; 2011.

CHAPTER 7

Spinal and Supraspinal Mechanisms of Placebo Analgesia

Falk Eippert[1], Christian Büchel[2]

[1]Centre for Functional Magnetic Resonance Imaging of the Brain (FMRIB), University of Oxford, Oxford, UK,
[2]Department of Systems Neuroscience, University Medical Center Hamburg-Eppendorf, Hamburg, Germany

INTRODUCTION

Pain is a subjective experience that is not invariably related to the amount of nociceptive input to the central nervous system. While a monotonic relationship between strength of nociceptor stimulation and perceived pain intensity is often observed,[1,2] deviations from such a relation are just as abundant and are probably best appreciated in extreme cases where a traumatic injury does not lead to a strong feeling of pain, as in competition or combat.[3] Such a reduction of pain perception—although certainly to a lesser extent—can also be induced experimentally, as for example via hypnotic suggestions,[4] expectations of reduced pain,[5] or attentional distraction.[6]

Placebo analgesia is thus just one example of the impact that psychological factors can have on pain perception, yet due to the pervasiveness of placebo effects in clinical trials, it is possibly one of the most investigated forms of psychological pain modulation. It has become clear that placebo effects in pain are determined by multiple psychological factors and rely on various different neurobiological mechanisms.[7–14] In this chapter we look at one of the earliest explanations of placebo analgesia at the neurobiological level,[15] namely, that a descending pain control system, which relies heavily on opioidergic neurotransmission and controls nociceptive processing already at the level of the spinal cord, contributes to placebo effects in pain.

In the following, we first give an overview of descending pain control as established in animal studies. We then move on to ask what evidence there is that descending control of pain underlies placebo analgesia; we focus on studies using functional magnetic resonance imaging (fMRI) of the human brain. Finally, we move away from the brain and, instead, look at the role of the spinal cord for placebo effects in the context of pain, again focusing on fMRI data (as well as the challenges that exist in spinal fMRI).

THE ANATOMY OF DESCENDING PAIN CONTROL

The functional neuroanatomy of pain processing is not a one-way road starting in the body periphery and ending in the cerebral cortex; instead, it contains both ascending pathways (i.e. from the body periphery over the spinal cord to higher brain structures) and descending pathways (i.e. from higher brain structures to the spinal cord; Fig. 7.1). Nociceptive information from the body periphery reaches the central nervous system via primary afferents that terminate in the dorsal horn of the spinal cord. The main targets of nociceptive afferents are the superficial laminae I and II, although some also synapse in the deeper laminae of the dorsal horn; these also play a prominent role in nociceptive processing (especially lamina V and its wide-dynamic-range neurons). The dorsal horn contains a large number of inhibitory and excitatory interneurons, which allows for complex processing of nociceptive information. From the dorsal horn, nociceptive information is transmitted to numerous higher regions via several ascending pathways to specific parts of the brainstem, midbrain, thalamus and hypothalamus amongst others, and eventually reaches the cortex. Modulation of spinal nociceptive processing, i.e. a modulation at the earliest station of nociceptive input to the central nervous system, will therefore have a profound effect on subsequently

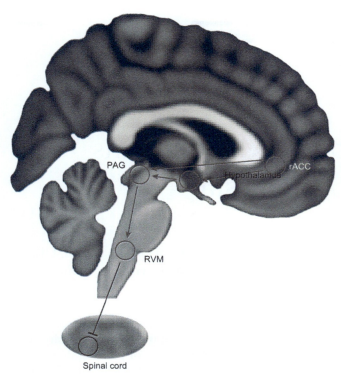

FIGURE 7.1 Neuroanatomy of descending pain control. This simplified diagram shows key regions involved in opioidergic descending pain control, as identified by both animal studies and human imaging studies. The endpoint of this system is the spinal cord, where nociceptive processing is inhibited by projections from the RVM (blue arrow). The RVM in turn receives a substantial input from the PAG, which is innervated by the hypothalamus as well as medial prefrontal regions (red arrows). Note that several connections (such as reciprocal ones) are omitted for the sake of clarity and that several non-midline regions (such as the amygdala) are not depicted. The sagittal T1-weighted brain section stems from the MNI152 brain, whereas the transversal T2*-weighted spinal cord section stems from a recent spinal cord study (Eippert et al, unpublished data). PAG: periaqueductal gray; rACC: rostral anterior cingulate cortex; RVM: rostral ventromedial medulla. *This figure is reproduced in color in the color section.*

occurring supra-spinal processing and the resulting experience of pain.

The concept of a descending pain-modulating system was introduced by Melzack and Wall in their article on the gate control theory of pain,[16] in which a system of supra-spinal origin that is able to control spinal nociceptive processing was proposed. Consistent with this proposal, animal experiments highlighted that several regions, especially in the midbrain and brainstem, are involved in modulating the responses of spinal cord neurons to noxious stimuli and that opioidergic neurotransmission plays a crucial role.[17-24] The core of the descending pain control system consists of the periaqueductal gray (PAG; the gray matter adjacent to the cerebral aqueduct in the midbrain) and the rostral ventromedial medulla (RVM; a functionally defined collection of nuclei that includes the nucleus raphe magnus and adjacent reticular nuclei).

The PAG is reciprocally connected with the RVM, but its efferent projections predominate and constitute a major source of input for the RVM, which, in turn, projects caudally to the dorsal horn of the spinal cord. While most previous research has concentrated on the PAG–RVM system, it is by no means the only system involved in descending pain control.[24] For example, a noradrenergic system involving nuclei in the dorsolateral pontine tegmentum also contributes to modulation of spinal nociceptive processing,[25] as do regions in the caudal medulla, one of which is also involved in the phenomenon of diffuse noxious inhibitory controls.[26] While the focus in this chapter is on the inhibition of pain by descending control systems, it should not be forgotten that these systems have an equally important role in pain facilitation, as for example demonstrated in recent experiments.[27,28]

Initial evidence for the role of the PAG in pain control was obtained in experiments which showed that electrical stimulation of the PAG resulted in analgesia in rats,[29,30] while leaving responses in other sensory modalities intact. This selective inhibitory effect on noxious stimulation was later demonstrated to occur as well with stimulation of the human PAG[31] and it was shown to depend on opioidergic neurotransmission as it could be blocked by administration of the opioid antagonist naloxone. This is especially interesting with regard to a very early study[32] which showed that the PAG is one of the most effective sites for eliciting analgesia through morphine injection, see also Ref.[33] However, stimulation of the PAG leads not only to a behaviorally observable analgesia, but also to selective inhibition of spinal nociceptive neurons.[34,35] This descending control of spinal nociception is mainly realized via the RVM, in line with the PAG's efferent connections, which are sparse to the spinal cord but dense to the RVM, for which the PAG is a major source of input.[36,37]

That the RVM is a critical relay for anti-nociceptive impulses to flow from the PAG to the spinal cord can be demonstrated by lesioning or inactivating the RVM, which prevents the PAG from exerting its anti-nociceptive effects in the dorsal horn.[38,39] Similar to the PAG, there is a powerful analgesic effect of RVM morphine injections, and a stimulation of RVM neurons leads to analgesia as well as an inhibition of neuronal responses in the dorsal horn.[40-42] The RVM projects to the spinal cord via the dorsolateral funiculus,[43] a lesion of which blocks the anti-nociceptive effects of PAG stimulation.[44] A systematic description of the functional properties of RVM neurons with regard to nociceptive processing was first carried out by Fields and colleagues,[45] who identified three classes of cell: off-cells (which stop responding just before a nociceptive reflex), on-cells (which show a burst just before a nociceptive reflex), and neutral-cells (whose behavior is unchanged). Opioid injection into either the

PAG or RVM leads to behavioral anti-nociceptive effects that are paralleld by an activation of off-cells and an inhibition of on-cells,[46] both of which show projections to the spinal cord.[47] Importantly though, it is the activation of OFF-cells that supports the anti-nociceptive effect of opioids[46,48] (although recent experimental evidence suggests a somewhat different perspective[49]). Note that neurons with similar functional properties are also found in other structures of the descending circuit, such as the PAG.[50]

The modulatory effects of RVM projections to the dorsal horn of the spinal cord occur most prominently in layers I, II and V, which is also where the projections of ON- and OFF-cells terminate.[51,52] There are several distinct ways in which nociceptive information transfer can be modulated in the dorsal horn[21]: (a) inhibition of primary afferents at the presynaptic terminals, (b) inhibition of excitatory interneurons that relay information from primary afferents to projection neurons, (c) excitation of inhibitory interneurons that could inhibit both primary afferents and excitatory interneurons, and (d) direct postsynaptic inhibition of ascending projection neurons. Obviously, each of these possibilities will differ in the pharmacological mediators that are involved, and a large number of neurotransmitters have been implicated in the descending modulation of spinal nociception, such as opioids, serotonin, noradrenalin, GABA and glycine.[19,53]

The inputs to the PAG–RVM system are numerous. While both regions receive afferents from the spinal cord,[54] and could thus be engaged by noxious stimuli alone, they are also extensively innervated by more rostral structures, with the strongest input to the PAG originating from the hypothalamus.[55] This input is highly relevant for descending inhibition, as hypothalamic activation leads to an inhibition of spinal nociceptive neurons[56] as well as an anti-nociceptive effect that is mediated by projections from the hypothalamus to the PAG–RVM system.[57] Another important subcortical input to the PAG arises from the amygdala.[58] Similar to the hypothalamus, this projection is functionally relevant, as opioid administration leads to amygdala-mediated anti-nociceptive effects,[59] for which the integrity of the PAG–RVM pathway is paramount.[60] Finally, the PAG receives a significant number of afferents from cortical regions, with medial prefrontal regions (which also project to the hypothalamus) having received the greatest interest.[61–64] With regard to the pharmacology of descending pain control, it is interesting to note that opioid receptors are present in all of the aforementioned regions, and that there are opioid-dependent links between them.[65]

The PAG–RVM system is thus optimally situated to integrate ascending nociceptive information with that provided by more rostral regions that are known to be involved in homeostatic and emotional processes. Along these lines, it has been suggested that the opioidergic PAG–RVM system might be a major anatomical substrate via which emotional and cognitive variables influence nociceptive processing. A large amount of animal research has shown that pain inhibition occurs in a variety of situations, with the most prominent example being stress-induced analgesia (stress being usually induced by footshocks). While stress-induced analgesia can also be mediated by nonopioid mechanisms, a strong opioigergical component is evident and has been demonstrated to rely on the PAG–RVM system.[66] However, analgesia can be induced not only by stimuli that are inherently stressful or aversive, but also by stimuli that are predictive of such stimulation, as demonstrated by the phenomenon of conditioned analgesia.[67] The acquisition of conditioned fear leads to an analgesic state and an inhibition of spinal nociceptive processing,[68] which is mediated by an opioid-dependent engagement of the PAG–RVM system via the amygdala,[69–71] a structure that plays a central role in conditioned fear. Inhibition of nociceptive processing is observed not only during situations that can be classified as aversive, but also during behaviors essential for survival, such as micturition and feeding.[72,73] Going even further, pain is also inhibited in appetitive states, e.g. when animals are presented with sucrose or sucrose-predictive cues.[74,75] One prominent branch of theories that aims to explain these divergent findings rests on the idea of prioritizing conflicting motivations, as suggested by Bolles and Fanselow,[76] Fanselow,[77] and Fields.[20,78] According to these theories, pain can be considered as a motivational state that often occurs concurrently with other motivational states. As an animal has a limited set of behaviors that can be carried out at the same time, a decision has to be made which motivational state is given priority and thus allowed to drive behavior. In some circumstances, it will be clearly beneficial for the animal if pain-related behavior is inhibited. It is in these states that the PAG–RVM system exerts its inhibitory influences on nociceptive processing (note, however, that the reverse is also true: in some circumstances (such as illness), nociceptive processing should be prioritized in order to allow for recuperation). With regard to studies in humans, the results of the above-mentioned animal studies in the aversive domain have been partly replicated (stress-induced analgesia[79,80] and conditioned analgesia[81]). Interestingly, placebo analgesia has many ties to reward processing,[82–84] and a placebo can even be considered as a reward-predicting cue, because it implies subsequent pain relief, which is rewarding in the context of pain.[20] This leads to the question of whether the descending pain control system is also engaged during the pain-modulatory phenomenon of placebo analgesia.

DESCENDING CONTROL IN PLACEBO ANALGESIA

Behavioral Studies

Levine and colleagues provided the first neuropharmacological exploration of the placebo analgesic effect and were also the first to suggest that an opioid-dependent system of descending pain control might be involved in placebo analgesia.[15] Using a model of postoperative pain, they were able to show that the opioid antagonist naloxone blocks placebo analgesia, suggesting that endogenous opioids play a major role in the generation of placebo analgesic effects. While this seminal study has initially been criticized on the grounds of experimental design and data analysis,[85–87] it paved the way for subsequent investigations that showed a significantly decreased placebo analgesic effect under naloxone in both experimental and clinical pain models.[88–90] Note, however, that two other early studies investigating opioidergic involvement in placebo analgesia showed that placebo effects can occur despite a naloxone-induced blockade of the opioid system in both clinical[91] and experimental pain models,[92] which hints at the existence of multiple mechanisms for the generation of placebo analgesia; these early studies are reviewed in detail elsewhere.[93–95] The specific contributions of endogenous opioids to placebo analgesia were investigated in much more detail in an elegant series of studies by Benedetti and colleagues. Of particular interest for the present discussion is a study in which these authors induced somatotopically specific placebo effects, which could be antagonized by systemic naloxone administration,[96] indicating that endogenous opioids can act very selectively. In line with these data, there is evidence (though far from unequivocal[97]) that descending control mechanisms can indeed exert their effects in a crude form of somatotopy.[97,98]

While the above-mentioned behavioral studies provided significant evidence for opioidergic underpinnings of placebo analgesic effects (thereby also showing that placebo analgesia is far more than altered reporting behavior), they did not directly investigate the neurobiological mechanisms (i.e. the involvement and interplay of different brain regions) that underlie placebo analgesia. To this end, and to thus determine whether there is evidence for involvement of the PAG–RVM descending pain control system in placebo analgesia, we now turn to neuroimaging studies.

Neuroimaging Studies

The investigation of placebo analgesia with neuroimaging techniques began with a seminal positron emission tomography (PET) study by Petrovic and colleagues.[99] The authors were able to show that—in an experimental pain setting—placebo and opioid analgesia (the latter being induced by administration of the μ-opioid agonist remifentanil) activated overlapping brain regions, suggestive of a shared underlying neural mechanism. More specifically, both conditions (in comparison to a control condition of pain only) showed a significant activation of the rostral part of the anterior cingulate cortex (rACC)—a region that has a high concentration of opioid receptors[100,101] and is strongly activated by opioid agonists,[102,103]—as well as an enhanced correlation of the response in rACC with responses in the brainstem (pons and PAG). Obviously these data are only of correlative nature, but they hint at the possibility that placebo analgesia involves an activation of the opioid-dependent descending pain control system in the brainstem via the rACC, which is in agreement with the PAG's afferent connectivity.[61,62]

Should descending pain control indeed play a role in placebo analgesia, one would expect subcortical and cortical brain areas involved in processing painful stimuli[2,104] to show reduced responses under placebo analgesia (due to the inhibition of ascending nociceptive traffic that occurs at the level of the spinal cord). While Petrovic and colleagues did not report any such data, later functional magnetic resonance imaging (fMRI) experiments by Wager et al[105] provided the first hints that such a mechanism might indeed be involved in placebo analgesia: they observed that several pain-responsive regions—including the thalamus, the secondary somatosensory cortex (SII), the insula, and a dorsal part of the anterior cingulate cortex—either showed reduced BOLD responses to painful stimulation under placebo or correlated with the reported pain reduction under placebo. Several other studies have replicated this effect of placebo-induced reductions in activation of pain-processing regions,[106–115] adding weight to the suggestion that this might be an important underlying mechanism (see also Figure 2 in Meissner et al[14] and Amanzio et al[116] for an aggregation of results across studies). Nevertheless, it is important to note that the extent of placebo-induced reductions varies strongly across studies and that not all studies observe such a pattern of responses,[117] indicating that inhibition of ascending nociceptive traffic is not the only explanation for placebo effects in pain and that these effects are likely configured via multiple neurobiological mechanisms.

Following up on the findings of Petrovic et al,[99] Wager and colleagues[105] also investigated whether regions of the descending pain control system would show enhanced activation under the placebo condition. Supporting this idea, they observed that a midbrain region in the vicinity of the PAG showed enhanced responses under the placebo condition already during the anticipation of painful stimuli. Responses in this region furthermore correlated with pain-anticipatory responses under placebo in dorsolateral prefrontal cortex (dlPFC), a region that is heavily involved in executive functions and general top-down control influencing behavior and neural processing in multiple modalities.[118,119] Perhaps most importantly, they could show that the anticipatory responses in PAG and dlPFC correlated with the behavioral index of placebo analgesia (reduction in pain ratings), as well as with the later occurring placebo-induced reduction of BOLD responses in several pain-processing regions, such as the thalamus. Following these findings, placebo-enhanced responses in rACC[107,108,110,113,117,120] and dlPFC[108,110,112,113,117,120] have been found consistently across studies both during anticipation of pain and during painful stimulation itself (sometimes correlating with the strength of the placebo effect), although other regions are also frequently found in the same analysis (e.g. anterior insula, ventrolateral prefrontal cortex, as well as orbitofrontal and parietal regions); note, however, that the locations designated as 'rACC' vary considerably across studies,[121] raising the possibility that responses of these regions might actually tap into different mechanisms. A study by Bingel et al[107] is especially interesting as these authors were able to show that, under placebo, there is a stronger coupling between the rACC (subgenual part) and subcortical structures important for antinociception, such as the amygdala and PAG. In contrast to the earlier studies,[99,105] which showed correlations between cingulo-prefrontal regions and PAG across subjects, this study demonstrated such correlations within subjects, i.e. based on the actual neurophysiological timecourses of each region.

While all these results are consistent with an involvement of opioid-dependent descending pain control, they provide only indirect evidence as they have no pharmacological specificity and thus cannot ascertain whether any of the described effects are indeed caused by opioids. In contrast to this, PET imaging studies using the μ-opioid receptor binding tracer [11C]carfentanil[122–124] were able to show enhanced opioid release during placebo analgesia in dlPFC, rACC, amygdala and PAG, thus significantly adding to the fMRI results by means of neurochemical specificity and allowing some mechanistic insight into how opioids contribute to placebo analgesia (which the earlier pharmacological challenge studies[15,88–90] were not able to do).

Pharmacological fMRI and Placebo Analgesia

The two strands of neuroimaging research mentioned so far—fMRI and PET studies showing reduced hemodynamic responses in pain-processing regions as well as increased hemodynamic responses in putative 'pain modulatory' regions, and μ-opioid selective tracer PET studies showing increased opioid release in 'pain modulatory' regions—clearly hint at the involvement of the opioidergic descending pain control system. However, a critical test would actually require (a) that one observes decreased responses in pain-processing regions under placebo analgesia concomitant with increased responses in core regions of the descending pain control system (amygdala, hypothalamus, PAG, RVM), and (b) that one can assign both of these effects to be opioid-dependent by challenging this system with an opioid antagonist, which should not only disrupt behavioral placebo effects but also both of the aforementioned response patterns. Neither tracer PET nor standard fMRI can satisfy both of these conditions, as tracer PET is not informative with regard to decreases of neural (or hemodynamic) responses, and standard fMRI has no neurochemical specificity. Furthermore, evidence for the involvement of the lower opioid system in placebo analgesia was not too prominent, as only the PAG and the amygdala had been implicated in placebo analgesia,[99,105,107,123–125] but not the hypothalamus or the RVM. As the hypothalamus significantly contributes to descending pain control[56] (providing the major input to the PAG[55]) and the RVM directly controls spinal nociception[126] (mediating the anti-nociceptive effects of the PAG,[38] which is its major input[37]), one would expect these structures to be involved in placebo analgesia as well. That hypothalamus and RVM responses had not been reported might be partly due to the rather low spatial resolution of the above-mentioned imaging studies, which is not optimal for investigating small midbrain/brainstem structures.[127] To address all of these issues, we conducted a pharmacological fMRI study[128] which employed (a) a higher spatial resolution than previous studies, thus increasing the sensitivity for detecting responses in small subcortical structures, and (b) a between-subjects pharmacological challenge using the opioid receptor antagonist naloxone, thus allowing us to record placebo-induced decreases in pain-processing regions and increases in core regions of the descending pain control system, as well as assessing their opioid dependence at the same time.

For this study, we recruited 48 participants who were assigned to two groups on a double-blind basis, one receiving the opioid antagonist naloxone and the other receiving saline. The experimental paradigm we used (Fig. 7.2) contained both expectation and conditioning components and had been employed in a similar form by many other groups before.[105,129] The experiment took place on two consecutive days and consisted of three phases: manipulation day 1, manipulation day 2, and test day 2. Before each phase, participants were treated with two identical creams on two separate areas of their left forearm. Participants were told that one cream was a highly effective pain reliever, whereas the other served as a sensory control. During the manipulation phases,

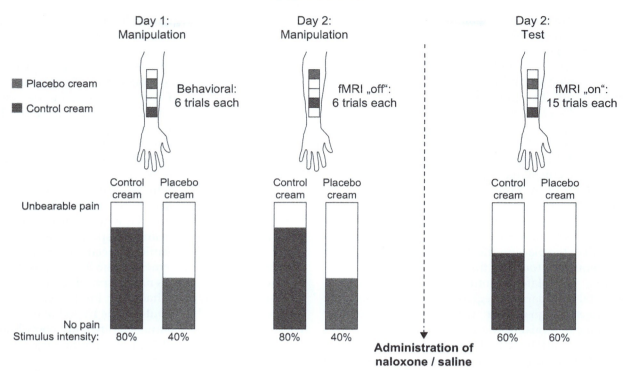

FIGURE 7.2 Experimental paradigm of a pharmacological fMRI study on placebo analgesia. Participants were recruited with the understanding that we were investigating the effects of a peripherally acting analgesic ('lidocaine' cream) on brain responses to noxious heat. The experiment took place on two consecutive days and consisted of three phases: manipulation day 1, manipulation day 2, and test day 2. Before each phase, subjects were treated with two identical creams on their left forearm and were told that one cream was a highly effective pain reliever, whereas the other served as sensory control. During the manipulation phases (which consisted of six trials under placebo cream and control cream, respectively), painful stimulation on the placebo-treated patch was surreptitiously lowered (from 80 [score on a visual analog scale (VAS)] under control to 40 under placebo) to convince the subjects that they had received a potent analgesic cream and to create expectations of future pain relief when treated with this cream. On day 2, the manipulation phase was carried out inside the (resting) MR scanner, to reactivate and strengthen the expectations of pain relief in this context. Before the test phase started, subjects either received an injection of saline or naloxone. fMRI data were collected during the test phase, which consisted of 15 trials under each condition. Importantly, during this phase the strength of painful stimulation was identical on both skin patches (60 on a VAS), in order to test for placebo analgesic effects. Note that in the spinal imaging study (see section 'Spinal fMRI of placebo analgesia'), we omitted the day 1 manipulation session—as subjects had participated in the previous study—and also did not administer any drugs. Reproduced and modified, with permission, from reference 128. *This figure is reproduced in color in the color section.*

painful contact heat stimulation on the placebo-treated patch was surreptitiously lowered (from 80 [score on a visual analogue scale (VAS) from 0 (no pain) to 100 (unbearable pain)] under control to 40 under placebo) to convince the participants that they had received a potent analgesic cream. Before the test phase started, participants either received an intravenous injection of saline or naloxone. fMRI data were collected only during the test phase, during which the strength of painful stimulation was identical on both skin patches (60 on a VAS), in order to test for placebo effects.

We observed that the saline group (i.e. the group with an intact opioid system) showed a large placebo effect (more than 30% reduction in pain ratings), which was significantly smaller in the naloxone group (i.e. the group with a blocked opioid system; about 10% reduction). Importantly, naloxone selectively influenced pain ratings in the placebo condition, but had no effect whatsoever on pain ratings in the control condition. Participants in the saline group also tended to rate the analgesic efficacy of the cream as higher when asked after the experiment. While these results conceptually replicate the early reports of naloxone blocking placebo analgesia,[15,88–90] it is important to remember that self-report measures of placebo analgesia such as pain ratings can be influenced by demand characteristics.[130] We therefore also tested whether a similar pattern of responses (stronger placebo effects in the saline group) could be found on an autonomic measure and indeed observed that skin conductance responses to the painful stimulation paralleled the findings in the subjective ratings: see also Ref.[131]. Next, we asked the critical question of whether the behavioral effect of naloxone (blockade of placebo effects) would be reflected in the responses of pain-processing regions, i.e. whether these regions would show reduced responses to the painful stimulation under placebo in comparison to control, and whether naloxone would abolish this reduction. Such an effect was observed in numerous regions involved in pain processing, such as the thalamus, the primary and secondary somatosensory cortices, the anterior cingulate cortex and the insula, as well as the amygdala and basal ganglia. This not only replicated the results of previous imaging studies[105–107,109] but also extended these findings by demonstrating a causal role of opioidergic neurotransmission in their generation, and furthermore suggested that the disruptive effect of naloxone on placebo analgesia is implemented by naloxone blocking the placebo-induced reduction of responses in pain-processing regions: but see Refs.[132,133] The widespread placebo-induced reductions of responses in pain-processing regions (ranging from thalamus to primary somatosensory cortex) and their reversal by naloxone are somewhat reminiscent of effects observed under exogenous opiate administration[102,134] and suggest that an inhibition of ascending traffic at the level of the spinal cord might be involved.

In a next step, we therefore investigated whether core regions of the descending pain control system, as revealed by animal studies (i.e. amygdala, hypothalamus, PAG, RVM) as well as additional cortical regions revealed by human imaging studies (rACC, dlPFC), would show enhanced activation under placebo. Comparing placebo to control in the saline group, we indeed observed enhanced responses in both dlPFC and rACC, which were significantly weaker under naloxone. This integrates observations of previous fMRI studies, which showed enhanced hemodynamic responses in these regions,[99,105,107,117,120] with tracer PET studies, which showed enhanced opioid release in these regions,[122–124] and also suggests a degree of functional relevance as these responses were blocked when behavioral placebo effects could also not be observed due to naloxone. Such a blockade of placebo analgesia has also been observed when disrupting dlPFC function via repetitive transcranial magnetic stimulation[135] and when assessing Alzheimer patients who had disrupted coupling from PFC to the rest of the brain,[136] indicating the importance of prefrontal regions for placebo effects in pain.

To investigate the responses of the lower opioid system with greater accuracy and sensitivity, we employed a brainstem-dedicated image processing strategy. This allowed us to detect placebo-induced responses in core regions of the descending pain control system, namely the hypothalamus, the PAG and the RVM (but not the amygdala), all of which were significantly reduced by naloxone (Fig. 7.3). Interestingly, the responses in the lower opioid system were highly correlated with the behavioral placebo effect in the saline group, but significantly less in the naloxone group, indicating that the responses in this phylogenetically conserved system of pain control are functionally relevant, as one would expect from previous stimulation studies in animals.[29,57,126] While several studies had previously shown

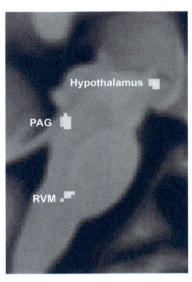

FIGURE 7.3 Placebo-induced midbrain and brainstem responses. This sagittal slice shows hypothalamus, PAG and RVM responses that were significantly stronger under placebo than under control; the response is overlaid on the group-averaged T1 image. Importantly, the responses in these key regions of descending pain control were significantly weaker under naloxone, indicating that these responses are opioid-dependent. All three structures furthermore showed responses that were correlated with the strength of the behavioral placebo effect. Reproduced and modified, with permission, from reference 128. *This figure is reproduced in color in the color section.*

enhanced PAG responses under placebo,[105,107,123–125] we were now able to show that not only the PAG, but also its main subcortical input, the hypothalamus,[55] as well as its main mediator for effects on the spinal cord, the RVM,[39] are actively involved in an opioid-dependent form of placebo analgesia; note that the extent of activation in the pontine region identified by Petrovic and colleagues[99] seems to overlap with the responses in the RVM we observed in this study. The placebo-induced activation of the hypothalamus is especially interesting, not only because of its substantial input to the PAG, but also because its stimulation has inhibitory effects on nociceptive processing in the spinal cord.[56] Intriguingly, the anti-nociceptive effects of hypothalamic stimulation are mediated by the PAG and RVM,[57] all of which were identified as playing a role in placebo analgesia in this study. These responses were furthermore all on the ipsilateral to the side of painful stimulation, which is consistent with observations in animal studies that show predominantly ipsilateral projections from both the hypothalamus and PAG.[18,57] It is currently unclear why we did not observe amygdala responses under placebo analgesia.

Finally, we investigated the integration of the cortical responses and the lower opioid system. Based on anatomical data from animal studies,[61,62] as well as previous reports of increased co-variation between rACC and PAG under placebo[99,107,123] (but see Ref.[124]), we focussed on these two regions as proxies for the cortical and lower opioid system. Consistent with data from Bingel et al,[107] we observed that the coupling of rACC and PAG (as estimated by within-subject time-course correlations) was significantly enhanced under placebo, but on top of that we could also demonstrate that this enhancement was selectively blocked by naloxone. The strength of rACC–PAG coupling under placebo was furthermore functionally highly relevant, as it predicted both the strength of behavioral placebo effects as well as the reduction of responses to painful stimulation in a key pain-processing region—the secondary somatosensory cortex, which has previously been shown to be susceptible to pain modulation[137]—in the saline group; these relationships were significantly weaker in the naloxone group. In analogy to these functional data, Stein et al[138] have recently reported that estimates of structural connectivity (based on diffusion tensor imaging data) between rACC and PAG predicted the strength of placebo analgesia; it remains to be seen how these measures of functional and structural connectivity relate to each other.[139] In a last analysis, we also observed that the strength of rACC–PAG coupling predicted activation of the RVM, which clearly relates this cortico-mesencephalic connectivity to a modulation of spinal processing, as the RVM is the final brainstem station of descending control and is critical for the anti-nociceptive effects of PAG excitation.[38,39]

Taken together, in this study we observed placebo-enhanced responses in the complete hierarchy of the descending pain control system, suggestive of a pathway from cortical areas to the brainstem, the activation of which presumably leads to an inhibition of nociceptive processing in the dorsal horn of the spinal cord. Responses in this system were opioid-dependent and also predictive of placebo-analgesic effects on the behavioral and neural level. These data are not only consistent with previous imaging studies, but also with a rich literature on descending control from animal studies. However, several limitations of the current approach should also be noted. First, the pharmacological challenge of the endogenous opioid system via naloxone is nonspecific with regard to the type of opioid receptors involved, because naloxone binds to all three classes of opioid receptors (μ, δ, and κ; note though that it has a slightly higher affinity for μ-opioid receptors[140]). While one could suspect—and previous tracer-PET and animal studies suggest—that actions at μ-opioid receptors account for the effects we observed, it will remain for future studies to tease apart the role of different opioid receptors once specific antagonists or tracers are developed for use in human participants.[141] Second, it is important to keep in mind that the opioidergic system is only one of several systems mediating pain modulation. For example, studies in experimental[92,142] and clinical pain states[91,143] have shown non-opioidergic placebo effects—some of which are dependent on endocannabinoid signalling[144]—and a reanalysis of the data by Petrovic and colleagues[99] suggests that ventrolateral prefrontal cortex activity represents non-opioidergic contributions to placebo analgesia.[145] Similarly, dopaminergic mechanisms have also been implicated in placebo analgesia,[83,84,124] but it is currently not only unclear whether these findings are related to dopamine involvement in descending control,[146] but also whether they have causal relevance for placebo analgesia, as pharmacological challenge studies of the dopamine system are still lacking. Third, the categorization of areas as 'pain responsive' and 'pain modulatory' is a clear oversimplification: all of the subcortical areas we observed to be activated by the placebo condition (hypothalamus, PAG, RVM) receive direct input from the spinal cord[54] and the prototypical 'top-down control' area—the dlPFC—is also responsive to pain under conditions where no explicit modulation takes place. Most instructive in this regard is probably a finding made by both Wager et al[105] and Watson et al,[110] who noted that very similar parts of anterior cingulate cortex (slightly above and posterior to the genu of the corpus callosum) exhibited placebo-enhanced responses in the anticipation phase, but significant reductions under placebo during painful stimulation, clearly calling into question the dichotomy of 'pain responsive' versus 'pain modulatory'. Thus, while the current state of knowledge is clearly suggestive of an involvement of the descending pain control system in placebo analgesia, there are obviously many unanswered questions.

PLACEBO ANALGESIA AND THE SPINAL CORD

In fMRI studies of placebo analgesia, as the ones mentioned above, responses in brain structures such as the PAG are often taken as evidence that one has observed activation of the descending pain control system. While this is certainly a worthwhile interpretation due to the large animal literature that demonstrated involvement of these structures in descending modulation of pain processing, it is far from certain that responses in these structures can be automatically equated with a modulation of spinal nociceptive processing. First, PAG stimulation has been shown to exert supra-spinal anti-nociceptive effects in addition to its well known spinal contributions,[147,148] in line with ascending projections from the PAG.[149] Second, the PAG is not only involved in pain modulation, but contributes to a variety of behavioral processes, such as reproductive behavior, fear and anxiety, autonomic regulation, etc.[127,150] and changes in some of these likely go along with pain modulation.

Related problems also exist for other structures: for example, the RVM—which is highly relevant for mediating anti-nociceptive effects at the level of the spinal cord via its direct projections—does not solely control nociceptive processes, but also contributes substantially to various forms of autonomic regulation;[151] similarly, the rACC seems to have a more general role in the modulation of sensory and affective experience,[152] as for example shown in gustatory processing[153] and the processing of emotional visual stimuli[82] and its activation in placebo analgesia studies might thus be related to modulation of affective processes instead of nociception. fMRI signals in these regions can thus not be *selectively* associated with descending anti-nociception, and care should be taken when employing such reverse inference[154] (but see Ref.[155]). To establish whether certain pain-modulatory phenomena, such as placebo analgesia, indeed employ descending pain control to modify spinal mechanisms, it is important to investigate the endpoint of the descending pain control system, i.e. the spinal cord itself, and several studies have attempted to do exactly this.

Behavioral Studies

The first study that investigated this issue used the nociceptive R-III reflex to measure modulation of spinal nociceptive processing[156] and aimed to induce placebo effects by suggesting to participants that an intravenous injection prior to painful electrical stimulation contained the analgesic drug fentanyl (when it contained only saline). Using this expectation-based model, the authors did not observe a significant reduction of subjective pain reports under placebo (nor did they find an effect on the R-III reflex) and could thus not clarify the role of the spinal cord in placebo analgesia. In a more recent study, Matre et al[157] used a model of secondary hyperalgesia in order to obtain evidence for a modulation of spinal processing under placebo analgesia. Combining conditioning and expectation components in their paradigm, the authors were first of all able to show a significant placebo analgesic effect on the subjective level, i.e. participants rated the applied heat pain stimuli as significantly less intense under placebo. The authors then went on to assess the extent of secondary hyperalgesia—which develops around the site of the noxious thermal stimulation—in both conditions. Importantly, the zone of secondary hyperalgesia where punctate hyperalgesia and stroking hyperalgesia (allodynia) could be observed was significantly reduced under placebo in comparison to a control condition. As these forms of secondary hyperalgesia are dependent on sensitization of neurons in the dorsal horn of the spinal cord, this study provided evidence for the hypothesis that placebo analgesia involves a modulation of spinal nociceptive processing. Despite this striking behavioral finding (as well as others that demonstrate expectation-induced modulation of spinal processing[158]), the study did not provide *direct* evidence for spinal cord involvement in placebo analgesia. Obtaining *direct* evidence, i.e. placebo-induced changes in recorded spinal cord responses, might however be achieved by non-invasively measuring spinal cord responses via fMRI.

Spinal fMRI

While fMRI of the brain is well established, fMRI of the spinal cord[159–161] is still in a somewhat early stage. The first reports of spinal fMRI appeared in the late 1990s[162,163] and even now there are only a limited number of laboratories employing this technique. This is likely due to the fact that in comparison to fMRI of the brain, performing fMRI studies of the spinal cord is very challenging for several reasons. First, the spinal cord has a very small cross-sectional diameter (at the level of the 6th cervical segment about 12 mm in left–right and 8 mm in anterior–posterior direction[164]), making it impossible to use standard brain-based fMRI protocols (Fig. 7.4). Second, static magnetic field inhomogeneities predominate due to the alternation of vertebrae and connective tissue,[165] leading to a periodic loss of signal along the rostrocaudal axis of the spinal cord.[166] Third, spinal cord BOLD responses are strongly affected by physiological noise arising from various respiratory and cardiovascular sources,[167–169] which is of greater magnitude than in the brain and can thus possibly obscure task-related BOLD signal changes.[169,170] However, numerous improvements in fMRI acquisition and analysis techniques (such as the development of multi-array coils,[171,172] the use of high-resolution imaging in combination with slice-specific shimming,[166] or the careful modeling of physiological noise[168,173,174]) have greatly enhanced the feasibility of spinal fMRI and interesting findings with respect to pain-related responses begin to emerge in both animal and human studies.

Following pioneering studies in rodent spinal fMRI,[175,176] Lilja and colleagues[177] have shown that BOLD responses can be observed in the ipsilateral dorsal horn of rats who received electrical stimulation of the hind limbs and that the strength and extent of these responses mirror the intensity of electrical stimulation. Interestingly, the observed spinal cord BOLD responses could be significantly suppressed by administration of morphine, an effect that was reversed by administration of the opioid antagonist naloxone. In another set of studies, Zhao and colleagues[178,179] used both BOLD and other fMRI contrast mechanisms to demonstrate responses in the ipsilateral dorsal horn of both the lumbar and cervical spinal cord of rats exposed to electrical stimulation of the hindpaw and forepaw, respectively. They were furthermore able to show that local lidocaine administration completely abolishes spinal cord fMRI responses to electrical stimulation,[180] as would be expected by blocking

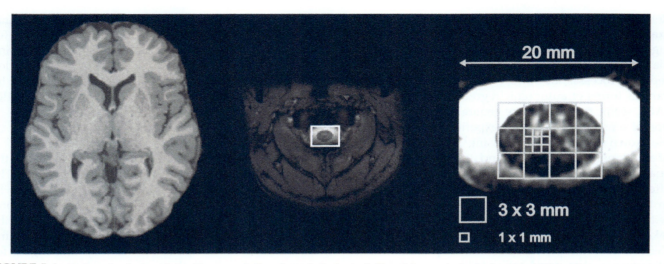

FIGURE 7.4 Brain and spinal cord size. Transversal slices through the brain (left) and the cervical spinal cord (middle) at the same scale show how minuscule the spine is in relation to the brain. The enlarged section (right) indicates that a standard in-plane voxel size of 3 × 3 mm would be much too coarse to image the spinal cord. Therefore, we used a 1 × 1 mm in-plane voxel size, which is more adequate to disentangle white and gray matter within the spinal cord, as well as to dissociate responses in the anterior–posterior and left–right dimensions. Note that due to the imaging sequences used, cerebrospinal fluid is black in the brain section and white in the spinal cord section, whereas gray matter is dark in the brain section and white in the spinal cord section. *This figure is reproduced in color in the color section.*

peripheral nerve transmission. Together with the finding of a morphine-induced depression of spinal cord responses to painful stimuli,[177] this study demonstrates that it is feasible to employ spinal fMRI for investigating the pharmacological modulation of pain processing in the rat spinal cord, which is obviously of high relevance for analgesic drug development. However, these data do not speak to the possibility of measuring the *endogenous* modulation of spinal nociceptive processing with fMRI, as would be required for directly investigating the spinal cord's role in placebo analgesia.

This experimental question might instead be better interrogated in humans, where subjective pain experience can be recorded at the same time as spinal fMRI responses to painful stimuli (note though that opioid-related placebo analgesic effects have recently been shown to exist in mice and rats[181–183]). Following the first reports of responses in the human spinal cord to sensory stimulation,[184,185] a number of fMRI studies by Stroman and colleagues have investigated the feasibility of imaging nociceptive processing in the human spinal cord,[186–189] by employing a T2-weighted fMRI protocol. Of particular relevance is a recent study by Summers et al who observed that spinal cord BOLD responses are significantly stronger for noxious stimuli (laser heat) than for innocuous stimuli (soft brushing).[190] While a lateralization of responses was not found in this study, Brooks and colleagues[191] have very recently provided strong evidence for lateralization of spinal cord BOLD responses: both painful thermal stimuli and non-painful (but nociceptive) punctate stimuli led to significantly greater responses in the ipsilateral part of the spinal cord than in the contralateral one. This is not only in accordance with the known spinal anatomy, but also mirrors autoradiographic and electrophysiological animal data:[192,193] but see Ref.[194]. While neither segmental nor anterior–posterior specificity was demonstrated in these two experiments, they nevertheless suggest that non-invasive imaging of basic components of pain-related responses (intensity and laterality) in the human spinal cord is feasible, and they highlight the possibility of imaging endogenous modulation of nociceptive spinal processing.

Spinal fMRI of Placebo Analgesia

We recently employed spinal fMRI to investigate whether changes in spinal nociceptive processing can indeed be observed under placebo analgesia,[195] thus indicating that descending control of pain is one of the mechanisms underlying placebo analgesic effects. To this end, we invited 15 volunteers who had participated in our previous fMRI study on placebo analgesia[128] and subjected them to a similar paradigm as in the previous study (Fig. 7.2). Pain was induced by contact thermal stimulation of the left radial volar forearm (dermatome C6) and fMRI data were recorded from the 5th cervical to the 1st thoracic spinal segment, using a protocol that was optimized for spinal fMRI. In particular, we positioned the slices perpendicular to the spinal cord, using a slice thickness of 5 mm in order to achieve an adequate signal-to-noise ratio despite the rather high in-plane resolution (1 mm × 1 mm). Such a prescription optimally conforms to the functional neuroanatomy of the spinal cord, because (1) 5 mm thick slices adequately sample along the rostro-caudal axis of the cord since each cervical segment is about 15 mm long,[164] and (2) a 1 mm² in-plane resolution aims at minimizing partial volume effects (i.e. combining gray and white matter signal within a single voxel) that will occur due to the fine-scale organization of spinal gray matter. We furthermore optimized spatial image processing procedures (i.e. motion correction, inter-subject registration, etc.) of the spinal fMRI data and corrected for physiological noise by recording respiratory and cardiac data during the fMRI session and then included these data in the statistical model as confound regressors.[196]

Participants clearly experienced the thermal stimulation as painful, as evidenced by their pain ratings. Regarding spinal fMRI data, we observed significant BOLD responses to the painful stimulus in the ipsilateral dorsal horn at the expected segmental level (top of 5th cervical vertebra, corresponding approximately to segment C6; Fig. 7.5a). This was the strongest response in all the imaged spinal segments, suggesting a good specificity of our method with regard to lateralization and segmental localization. Participants also showed a significant placebo effect, as evidenced by lower pain ratings under the placebo condition as compared to the control condition. Most importantly, spinal cord BOLD responses in the ipsilateral dorsal horn—obtained from the voxel that showed the strongest pain-related activity—mirrored the behavioral effect by exhibiting a strong BOLD response in the control condition, but almost no response in the placebo condition (Fig. 7.5a). These data thus provided direct evidence that placebo analgesia can affect nociceptive processing already at the earliest stage of the central nervous system, namely the dorsal horn of the spinal cord, as was already hypothesized in the first neurobiological study on placebo analgesia.[15] This finding also lends support to the statement that descending pain control is a mechanism that contributes to placebo analgesic effects, as has been stated in numerous imaging studies[99,107,128] and clearly demonstrates that placebo effects are not only mediated by supra-spinal mechanisms, but involve spinal mechanisms as well.

Nevertheless, these results obviously also raise many novel questions. First of all, we do not know via which brainstem mechanisms inhibition of spinal processing is realized. Although we would strongly suspect that opioid-dependent activation of the PAG–RVM axis is involved,[128] this cannot be ascertained as we did not measure brainstem and spinal responses at the same time (thus precluding any inference regarding connectivity between these parts of the central nervous system) and inhibition of spinal processing as observed here can also be realized by other routes, such as the noradrenergic nuclei of the dorsolateral pontine tegmentum.[24,25] Along similar lines, we do not know on which neurons placebo-induced descending control exerts its effects in

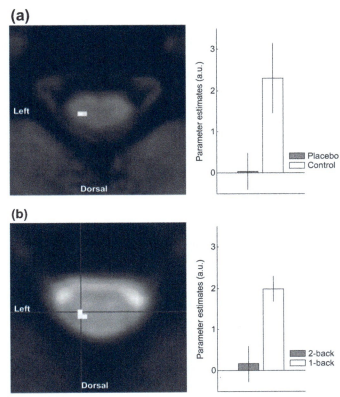

FIGURE 7.5 Modulation of nociceptive processing in the human spinal cord. (a) In the spinal fMRI study on placebo analgesia, we observed significant responses to the painful stimulation in the ipsilateral dorsal spinal cord (where nociceptive afferents terminate), as shown by the transversal section (level C6); the response is overlaid on the group-averaged T1 image. The group-averaged parameter estimates (reflecting the strength of activation) on the right were obtained from the voxel that exhibited the strongest response to pain and clearly show a significant reduction under placebo compared to control. (b) A similar result was observed in the spinal fMRI study on distraction, where pain-related responses in the ipsilateral dorsal spinal cord (at a location nearly identical to the one shown in panel (a)) were significantly reduced when participants where distracted from pain by high working memory load under the 2-back condition (see transversal section and parameter estimates); the response is overlaid on the group-averaged T2* image. Reproduced and modified, with permission, from references 195 and 205. *This figure is reproduced in color in the color section.*

the spinal cord, because the BOLD response is a very coarse and indirect measure of neuronal activation (with each voxel containing a huge number of neurons and activation being inferred via changes in blood oxygenation[197]). Effects might occur at the level of primary afferents, inhibitory interneurons, excitatory interneurons, or projection neurons, but such a differentiation cannot be obtained with current methods. Second, it is unclear whether the observed effect is specific for noxious stimuli, as it could be of a more general nature and might thus equally occur for innocuous stimuli. We would argue against this idea, because (a) previous investigations demonstrated that placebo effects are more evident in painful than in non-painful conditions[123,198] and (b)

PAG-mediated descending control is quite specific, in that it targets nociceptive spinal processing but leaves innocuous spinal processing intact.[35,199] Third, one might ask why the observed BOLD responses occurred only in the deep dorsal horn and were not evident in the superficial dorsal horn (i.e. laminae I and II), which is the main target for primary afferents conveying nociceptive information and which thus strongly contributes to nociceptive processing. A possible explanation relates to the fact that signal dropout due to magnetic field inhomogeneities is especially strong at the edge of the cord,[165,166] which might thus limit the ability to observe BOLD responses in the more superficial parts of the dorsal horn (a further limitation might be imperfect inter-subject alignment). One should also not discount the possibility that differences in the characteristics of the intrinsic neuronal circuitry in superficial and deep dorsal horn[200] might lead to responses in the deep part being more easily translated into a BOLD signal. Finally, it appears to be somewhat paradox that BOLD responses in the dorsal horn were completely suppressed under placebo, yet subjective pain ratings showed that participants still experienced pain under placebo (although to a lesser degree). While there are several mechanisms that might explain this apparent paradox—relating, for example, to (1) differential descending control of wide dynamic range versus nociceptive specific neurons[201,202] and C-fiber versus A-fibre input[23], (2) laminar specificity of descending control[203] and (3) different electrophysiological properties of nociceptive specific versus wide dynamic range neurons[204] or neurons in superficial versus deep lamina[200]—they are mere speculation at the current point and will not be discussed in detail.

Reassuringly, this spinal fMRI study is not the only one demonstrating a direct effect of psychological factors on spinal nociceptive processing. In a recent study[205], we employed an attention-distracting manipulation and observed that high levels of distraction not only led to reduced pain (which had already been demonstrated by many others[6,206]), but importantly also to reduced spinal cord BOLD responses (Fig. 7.5b). In a separate behavioral study, we were furthermore able to show that the observed pain reduction could be attributed to opioidergic processes, as it was strongly reduced when the opioid antagonist naloxone was given. Intriguingly, the location of the spinal cord BOLD responses in this study was nearly identical to the location observed in the study on placebo analgesia, suggesting good reliability in using fMRI of the human spinal cord.

CONCLUSIONS AND OPEN QUESTIONS

In our opinion, the studies reviewed here clearly suggest that descending control of spinal nociceptive processing via the opioid-dependent PAG–RVM system is one of the mechanisms underlying placebo analgesia. This is not to say that it is the only mechanism of descending control involved, because other types of descending control might contribute as well[24,49] and a network analysis carried out by Wager and colleagues[123] suggests heterogeneity even within the μ-opioid system. Furthermore, descending control might not feature equally strong in all kinds of placebo effects,[117] and cortico-cortical or cortico-subcortical contributions likely also play a substantial role.[133,207–210] It will be an important endeavor to tease apart the conditions under which spinal versus supra-spinal mechanisms of pain modulation predominate.

A further interesting question relates to the psychological constraints under which the PAG–RVM system can be brought into play. While a large part of placebo analgesia is probably consciously mediated, there is also evidence for non-conscious placebo responses in the context of pain.[211] It will be interesting to see whether the descending pain control system can be engaged only via conscious expectations of pain relief or also via non-consciously learned predictive relationships between cues and pain relief. We tend to favor the latter possibility, because responses in this system (although notably without dlPFC involvement) were evident in a conditioned analgesia paradigm, where only a small minority of participants was aware of the experimental contingency.[81]

The importance of placebo-induced reductions of activity in pain-processing areas for the experienced pain reduction under placebo is another topic of current debate.[14,133] As alluded to above, these reductions have been observed across a large number of studies (for an overview, see Figure 2 in Meissner et al[14]), but the extent of this effect varies considerably across studies and it is unclear how it relates to experienced pain reduction. While we could show that both a behavioral index of placebo analgesia (pain ratings) and widespread placebo-induced reductions of activity in pain-processing regions were significantly diminished by naloxone, we cannot infer that reductions in neural activity of these regions actually drive reductions in pain perception. A promising step in this direction is the use of mediation analysis, which has shown that activity in several pain-processing regions mediates the effect of predictive cues on pain perception.[132] One might also ask whether it is neurobiologically plausible that placebo analgesia would involve a reduction of activity in all regions associated with pain processing. For example, tracer PET studies[122–124] have shown enhanced opioid signalling under placebo in the insula—a core region in pain processing[104]—which goes along with evidence from animal studies that found a specific insular region mediating opioidergic anti-nociception via descending control mechanisms;[212]

 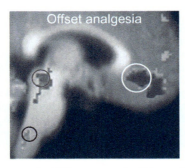 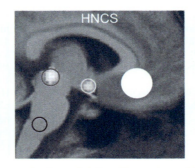 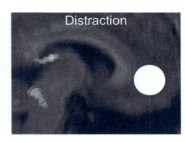

FIGURE 7.6 Brainstem and rACC responses in different forms of pain modulation. Key regions of the descending pain control system show responses in different forms of pain modulation, such as placebo,[128] offset analgesia,[222] heterotopic noxious conditioning stimulation (HNCS)[218] and distraction.[215] There is a general overlap of activated regions—most clearly seen in the PAG—but response locations obviously vary, as do the underlying mechanisms (for example, offset analgesia is mediated by non-opioidergic mechanisms,[223] while the other depicted forms of pain modulation do have an opiodergic component). Black circles indicate the RVM, red circles indicate the PAG, yellow circles indicate the hypothalamus and white circles indicate the rACC (filled circles indicate that these regions were used as seeds in functional connectivity analyses). Reproduced and modified, with permission, from references 128, 215, 218 and 222. Note that this figure has been reproduced with permission of the International Association for the Study of Pain® (IASP). The figure may not be reproduced for any other purpose without permission. *This figure is reproduced in color in the color section.*

together, these findings speak against dichotomizing regions as 'pain responsive' and 'pain modulatory'. Furthermore, based on the selectivity of descending control that is sometimes observed,[23] one could also envision a scenario where descending control might preferentially target neuronal populations that give rise to projections to affective brain regions, such as the recently identified non-peptidergical population in lamina V that directly projects e.g. to the amygdala.[213] Similarly, a modulation of spinal nociceptive processing does not necessarily influence all supra-spinal indices of pain processing.[158]

Finally one might ask how placebo analgesia and its underlying mechanism of descending pain control relate to other forms of pain modulation (e.g. hypnosis, distraction, stress-induced analgesia, offset analgesia, etc.). While each of these phenomena has a distinct repertoire of mechanisms, there are also some interesting overlaps. First of all, decreased activity in pain-processing regions has, for example, also been observed in studies on distraction,[214,215] conditioned analgesia,[81] perceived controllability,[216] hypnosis[217] and heterotopic noxious conditioning stimulation.[218] Similarly, a reduction in spinal nociceptive processing has been demonstrated not only during placebo analgesia,[157,195] but also during distraction,[205,219] affective picture viewing[220] and hypnosis.[4] Enhanced responses in parts of the descending pain control system, such as the rACC and PAG, have also been observed in numerous studies (Fig. 7.6), such as in distraction,[214,215,221] conditioned analgesia,[81] and offset analgesia,[222] as has increased rACC–PAG coupling.[215,218] A core mechanism of descending pain control might thus contribute to various forms of pain modulation and it will be interesting to see whether consistent inter-individual differences that predict the capability at various forms of pain modulation can be identified on the neural or genetic level.

In conclusion, we believe that there is abundant evidence for an involvement of descending pain control in placebo analgesia. This evidence comes from the convergence of data obtained by behavioral studies of spinal nociception, pharmacological challenge studies of the opioid system, tracer PET studies targeting the opioid system, and fMRI studies that not only target the brain, but also the spinal cord. When viewed collectively, these studies suggest that placebo analgesia is configured via multiple mechanisms, but that descending control of nociception, as first described by Melzack and Wall[16] in their gate-control theory, is clearly one of those mechanisms. This also speaks to the power of psychological factors in modifying pain perception, as they can exert their influence already at the earliest stages of the nociceptive processing stream.

Acknowledgments

This work was supported by a Marie Curie Fellowship (grant agreement number 273805, 'Pain modulation') to F.E. and an ERC Advanced Grant (grant agreement number 269661, 'Placebo' ERC-2010-AdG 20100407) to C.B., who also acknowledges support from the SFB 936 (project A6).

References

1. Kenshalo Jr DR, Anton F, Dubner R. The detection and perceived intensity of noxious thermal stimuli in monkey and in human. *J Neurophysiol.* 1989;62(2):429-436.
2. Coghill RC, Sang CN, Maisog JM, Iadarola MJ. Pain intensity processing within the human brain: a bilateral, distributed mechanism. *J Neurophysiol.* 1999;82(4):1934-1943.
3. Wall PD. On the relation of injury to pain. The John J. Bonica lecture. *Pain.* 1979;6(3):253-264.
4. Kiernan BD, Dane JR, Phillips LH, Price DD. Hypnotic analgesia reduces R-III nociceptive reflex: further evidence concerning the multifactorial nature of hypnotic analgesia. *Pain.* 1995;60(1):39-47.

5. Koyama T, McHaffie JG, Laurienti PJ, Coghill RC. The subjective experience of pain: where expectations become reality. *Proc Natl Acad Sci USA*. 2005;102(36):12950-12955.
6. Miron D, Duncan GH, Bushnell MC. Effects of attention on the intensity and unpleasantness of thermal pain. *Pain*. 1989;39(3):345-352.
7. Stewart-Williams S, Podd J. The placebo effect: dissolving the expectancy versus conditioning debate. *Psychol Bull*. 2004;130(2):324-340.
8. Benedetti F, Mayberg HS, Wager TD, et al. Neurobiological mechanisms of the placebo effect. *J Neurosci*. 2005;25(45):10390-10402.
9. Colloca L, Benedetti F. Placebos and painkillers: is mind as real as matter? *Nat Rev Neurosci*. 2005;6(7):545-552.
10. Enck P, Benedetti F, Schedlowski M. New insights into the placebo and nocebo responses. *Neuron*. 2008;59(2):195-206.
11. Price DD, Finniss DG, Benedetti F. A comprehensive review of the placebo effect: recent advances and current thought. *Annu Rev Psychol*. 2008;59:565-590.
12. Zubieta JK, Stohler CS. Neurobiological mechanisms of placebo responses. *Ann N Y Acad Sci*. 2009;1156:198-210.
13. Tracey I. Getting the pain you expect: mechanisms of placebo, nocebo and reappraisal effects in humans. *Nat Med*. 2010;16(11):1277-1283.
14. Meissner K, Bingel U, Colloca L, et al. The placebo effect: advances from different methodological approaches. *J Neurosci*. 2011;31(45):16117-16124.
15. Levine JD, Gordon NC, Fields HL. The mechanism of placebo analgesia. *Lancet*. 1978;2(8091):654-657.
16. Melzack R, Wall PD. Pain mechanisms: a new theory. *Science*. 1965;150(3699):971-979.
17. Basbaum AI, Fields HL. Endogenous pain control systems: brainstem spinal pathways and endorphin circuitry. *Annu Rev Neurosci*. 1984;7:309-338.
18. Jones S. Descending control of nociception. In: Light A, ed. *The Initial Processing of Pain and its Descending Control: Spinal and Trigeminal Systems*. Basel, Switzerland: Karger; 1992:203-295.
19. Millan MJ. Descending control of pain. *Prog Neurobiol*. 2002;66(6):355-474.
20. Fields H. State-dependent opioid control of pain. *Nat Rev Neurosci*. 2004;5(7):565-575.
21. Fields H, Basbaum A, Heinricher M. Central nervous system mechanisms of pain modulation. In: McMahon S, Koltzenburg M, eds. *Wall and Melzack's Textbook of Pain*. London, UK: Elsevier; 2006:125-142.
22. Heinricher M, Ingram S. The brainstem and nociceptive modulation. In: Basbaum A, Bushnell M, eds. *Science of Pain*. Oxford, UK: Academic Press; 2009:593-626.
23. Heinricher MM, Tavares I, Leith JL, Lumb BM. Descending control of nociception: specificity, recruitment and plasticity. *Brain Res Rev*. 2009;60(1):214-225.
24. Ren K, Dubner R. Descending control mechanisms. In: Basbaum A, Bushnell M, eds. *Science of Pain*. Oxford, UK: Academic Press; 2009:723-762.
25. Pertovaara A. Noradrenergic pain modulation. *Prog Neurobiol*. 2006;80(2):53-83.
26. Le Bars D. The whole body receptive field of dorsal horn multireceptive neurones. *Brain Res Rev*. 2002;40(1-3):29-44.
27. King T, Vera-Portocarrero L, Gutierrez T, et al. Unmasking the tonic-aversive state in neuropathic pain. *Nat Neurosci*. 2009;12(11):1364-1366.
28. Marshall TM, Herman DS, Largent-Milnes TM, et al. Activation of descending pain-facilitatory pathways from the rostral ventromedial medulla by cholecystokinin elicits release of prostaglandin-E(2) in the spinal cord. *Pain*. 2012;153(1):86-94.
29. Reynolds DV. Surgery in the rat during electrical analgesia induced by focal brain stimulation. *Science*. 1969;164(3878):444-445.
30. Mayer DJ, Wolfle TL, Akil H, et al. Analgesia from electrical stimulation in the brainstem of the rat. *Science*. 1971;174(4016):1351-1354.
31. Hosobuchi Y, Adams JE, Linchitz R. Pain relief by electrical stimulation of the central gray matter in humans and its reversal by naloxone. *Science*. 1977;197(4299):183-186.
32. Tsou K, Jang CS. Studies on the site of analgesic action of morphine by intracerebral micro-injection. *Sci Sin*. 1964;13:1099-1109.
33. Yaksh TL, Yeung JC, Rudy TA. Systematic examination in the rat of brain sites sensitive to the direct application of morphine: observation of differential effects within the periaqueductal gray. *Brain Res*. 1976;114(1):83-103.
34. Gebhart GF, Sandkühler J, Thalhammer JG, Zimmermann M. Quantitative comparison of inhibition in spinal cord of nociceptive information by stimulation in periaqueductal gray or nucleus raphe magnus of the cat. *J Neurophysiol*. 1983;50(6):1433-1445.
35. Leith JL, Koutsikou S, Lumb BM, Apps R. Spinal processing of noxious and innocuous cold information: differential modulation by the periaqueductal gray. *J Neurosci*. 2010;30(14):4933-4942.
36. Abols IA, Basbaum AI. Afferent connections of the rostral medulla of the cat: a neural substrate for midbrain–medullary interactions in the modulation of pain. *J Comp Neurol*. 1981;201(2):285-297.
37. Beitz AJ. The sites of origin of brain stem neurotensin and serotonin projections to the rodent nucleus raphe magnus. *J Neurosci*. 1982;2(7):829-842.
38. Behbehani MM, Fields HL. Evidence that an excitatory connection between the periaqueductal gray and nucleus raphe magnus mediates stimulation produced analgesia. *Brain Res*. 1979;170(1):85-93.
39. Aimone LD, Gebhart GF. Stimulation-produced spinal inhibition from the midbrain in the rat is mediated by an excitatory amino acid neurotransmitter in the medial medulla. *J Neurosci*. 1986;6(6):1803-1813.
40. Dickenson AH, Oliveras JL, Besson JM. Role of the nucleus raphe magnus in opiate analgesia as studied by the microinjection technique in the rat. *Brain Res*. 1979;170(1):95-111.
41. Du HJ, Kitahata LM, Thalhammer JG, Zimmermann M. Inhibition of nociceptive neuronal responses in the cat's spinal dorsal horn by electrical stimulation and morphine microinjection in nucleus raphe magnus. *Pain*. 1984;19(3):249-257.
42. Jones SL, Gebhart GF. Inhibition of spinal nociceptive transmission from the midbrain, pons and medulla in the rat: activation of descending inhibition by morphine, glutamate and electrical stimulation. *Brain Res*. 1988;460(2):281-296.
43. Basbaum AI, Fields HL. The origin of descending pathways in the dorsolateral funiculus of the spinal cord of the cat and rat: further studies on the anatomy of pain modulation. *J Comp Neurol*. 1979;187(3):513-531.
44. Basbaum AI, Marley NJ, O'Keefe J, Clanton CH. Reversal of morphine and stimulus-produced analgesia by subtotal spinal cord lesions. *Pain*. 1977;3(1):43-56.
45. Fields HL, Bry J, Hentall I, Zorman G. The activity of neurons in the rostral medulla of the rat during withdrawal from noxious heat. *J Neurosci*. 1983;3(12):2545-2552.
46. Heinricher MM, Morgan MM, Tortorici V, Fields HL. Disinhibition of off-cells and antinociception produced by an opioid action within the rostral ventromedial medulla. *Neuroscience*. 1994;63(1):279-288.
47. Vanegas H, Barbaro NM, Fields HL. Tail-flick related activity in medullospinal neurons. *Brain Res*. 1984;321(1):135-141.

48. Heinricher MM, McGaraughty S, Tortorici V. Circuitry underlying antiopioid actions of cholecystokinin within the rostral ventromedial medulla. *J Neurophysiol.* 2001;85(1):280-286.
49. Mason P. Medullary circuits for nociceptive modulation. *Curr Opin Neurobiol.* 2012;22(4):640-645.
50. Heinricher MM, Cheng ZF, Fields HL. Evidence for two classes of nociceptive modulating neurons in the periaqueductal gray. *J Neurosci.* 1987;7(1):271-278.
51. Basbaum AI, Ralston DD, Ralston 3rd HJ. Bulbospinal projections in the primate: a light and electron microscopic study of a pain modulating system. *J Comp Neurol.* 1986;250(3):311-323.
52. Fields HL, Malick A, Burstein R. Dorsal horn projection targets of ON and OFF cells in the rostral ventromedial medulla. *J Neurophysiol.* 1995;74(4):1742-1759.
53. Fields HL, Heinricher MM, Mason P. Neurotransmitters in nociceptive modulatory circuits. *Annu Rev Neurosci.* 1991;14:219-245.
54. Lima D. Ascending pathways: anatomy and physiology. In: Basbaum A, Bushnell M, eds. *Science of Pain.* Oxford, UK: Academic Press; 2009:477-526.
55. Beitz AJ. The organization of afferent projections to the midbrain periaqueductal gray of the rat. *Neuroscience.* 1982;7(1):133-159.
56. Workman BJ, Lumb BM. Inhibitory effects evoked from the anterior hypothalamus are selective for the nociceptive responses of dorsal horn neurons with high- and low-threshold inputs. *J Neurophysiol.* 1997;77(5):2831-2835.
57. Aimone LD, Bauer CA, Gebhart GF. Brain-stem relays mediating stimulation-produced antinociception from the lateral hypothalamus in the rat. *J Neurosci.* 1988;8(7):2652-2663.
58. Rizvi TA, Ennis M, Behbehani MM, Shipley MT. Connections between the central nucleus of the amygdala and the midbrain periaqueductal gray: topography and reciprocity. *J Comp Neurol.* 1991;303(1):121-131.
59. Manning BH, Mayer DJ. The central nucleus of the amygdala contributes to the production of morphine antinociception in the rat tail-flick test. *J Neurosci.* 1995;15(12):8199-8213.
60. Helmstetter FJ, Tershner SA, Poore LH, Bellgowan PS. Antinociception following opioid stimulation of the basolateral amygdala is expressed through the periaqueductal gray and rostral ventromedial medulla. *Brain Res.* 1998;779(1-2):104-118.
61. An X, Bandler R, Öngür D, Price JL. Prefrontal cortical projections to longitudinal columns in the midbrain periaqueductal gray in macaque monkeys. *J Comp Neurol.* 1998;401(4):455-479.
62. Floyd NS, Price JL, Ferry AT, et al. Orbitomedial prefrontal cortical projections to distinct longitudinal columns of the periaqueductal gray in the rat. *J Comp Neurol.* 2000;422(4):556-578.
63. Öngür D, An X, Price JL. Prefrontal cortical projections to the hypothalamus in macaque monkeys. *J Comp Neurol.* 1998;401(4):480-505.
64. Floyd NS, Price JL, Ferry AT, et al. Orbitomedial prefrontal cortical projections to hypothalamus in the rat. *J Comp Neurol.* 2001;432(3):307-328.
65. Tershner SA, Helmstetter FJ. Antinociception produced by mu opioid receptor activation in the amygdala is partly dependent on activation of mu opioid and neurotensin receptors in the ventral periaqueductal gray. *Brain Res.* 2000;865(1):17-26.
66. Terman GW, Shavit Y, Lewis JW, et al. Intrinsic mechanisms of pain inhibition: activation by stress. *Science.* 1984;226(4680):1270-1277.
67. Fanselow MS, Baackes MP. Conditioned fear-induced opiate analgesia on the Formalin test: evidence for two aversive motivational systems. *Learning Motiv.* 1982;13(2):200-221.
68. Harris JA, Westbrook RF, Duffield TQ, Bentivoglio M. Fos expression in the spinal cord is suppressed in rats displaying conditioned hypoalgesia. *Behav Neurosci.* 1995;109(2):320-328.
69. Helmstetter FJ. The amygdala is essential for the expression of conditional hypoalgesia. *Behav Neurosci.* 1992;106(3):518-528.
70. Bellgowan PS, Helmstetter FJ. The role of mu and kappa opioid receptors within the periaqueductal gray in the expression of conditional hypoalgesia. *Brain Res.* 1998;791(1-2):83-89.
71. Foo H, Helmstetter FJ. Hypoalgesia elicited by a conditioned stimulus is blocked by a mu, but not a delta or a kappa, opioid antagonist injected into the rostral ventromedial medulla. *Pain.* 1999;83(3):427-431.
72. Baez MA, Brink TS, Mason P. Roles for pain modulatory cells during micturition and continence. *J Neurosci.* 2005;25(2):384-394.
73. Foo H, Mason P. Sensory suppression during feeding. *Proc Natl Acad Sci USA.* 2005;102(46):16865-16869.
74. Dum J, Herz A. Endorphinergic modulation of neural reward systems indicated by behavioral changes. *Pharmacol Biochem Behav.* 1984;21(2):259-266.
75. Segato FN, Castro-Souza C, Segato EN, Morato S, Coimbra NC. Sucrose ingestion causes opioid analgesia. *Braz J Med Biol Res.* 1997;30(8):981-984.
76. Bolles RC, Fanselow MS. A perceptual-defensive-recuperative model of fear and pain. *Behav Brain Sci.* 1980;3(2):291-301.
77. Fanselow MS. Conditioned fear-induced opiate analgesia: a competing motivational state theory of stress analgesia. *Ann N Y Acad Sci.* 1986;467:40-54.
78. Fields H. Motivation-decision model of pain: the role of opioids. In: Flor H, Kalso E, Dostrovsky J, eds. *Proceedings of the 11th World Congress on Pain.* Seattle, USA: IASP Press; 2006:449-459.
79. Willer JC, Dehen H, Cambier J. Stress-induced analgesia in humans: endogenous opioids and naloxone-reversible depression of pain reflexes. *Science.* 1981;212(4495):689-691.
80. Pitman RK, Van der Kolk BA, Orr SP, Greenberg MS. Naloxone-reversible analgesic response to combat-related stimuli in posttraumatic stress disorder. A pilot study. *Arch Gen Psychiatry.* 1990;47(6):541-544.
81. Eippert F, Bingel U, Schoell E, et al. Blockade of endogenous opioid neurotransmission enhances acquisition of conditioned fear in humans. *J Neurosci.* 2008;28(21):5465-5472.
82. Petrovic P, Dietrich T, Fransson P, et al. Placebo in emotional processing–induced expectations of anxiety relief activate a generalized modulatory network. *Neuron.* 2005;46(6):957-969.
83. Scott DJ, Stohler CS, Egnatuk CM, et al. Individual differences in reward responding explain placebo-induced expectations and effects. *Neuron.* 2007;55(2):325-336.
84. Schweinhardt P, Seminowicz DA, Jaeger E, et al. The anatomy of the mesolimbic reward system: a link between personality and the placebo analgesic response. *J Neurosci.* 2009;29(15):4882-4887.
85. Goldstein A, Grevert P. Placebo analgesia, endorphins, and naloxone. *Lancet.* 1978;2(8104-5):1385.
86. Korczyn AD. Mechanism of placebo analgesia. *Lancet.* 1978;2(8103):1304-1305.
87. Skrabanek P. Naloxone and placebo. *Lancet.* 1978;2(8093):791.
88. Grevert P, Albert LH, Goldstein A. Partial antagonism of placebo analgesia by naloxone. *Pain.* 1983;16(2):129-143.
89. Levine JD, Gordon NC. Influence of the method of drug administration on analgesic response. *Nature.* 1984;312(5996):755-756.
90. Benedetti F. The opposite effects of the opiate antagonist naloxone and the cholecystokinin antagonist proglumide on placebo analgesia. *Pain.* 1996;64(3):535-543.
91. Gracely RH, Dubner R, Wolskee PJ, Deeter WR. Placebo and naloxone can alter post-surgical pain by separate mechanisms. *Nature.* 1983;306(5940):264-265.

92. Posner J, Burke CA. The effects of naloxone on opiate and placebo analgesia in healthy volunteers. *Psychopharmacology (Berl.)*. 1985;87(4):468-472.
93. Fields HL, Levine JD. Placebo analgesia – a role for endorphins? *Trends Neurosci*. 1984;7(8):271-273.
94. Benedetti F, Amanzio M. The neurobiology of placebo analgesia: from endogenous opioids to cholecystokinin. *Prog Neurobiol*. 1997;52(2):109-125.
95. Ter Riet G, De Craen AJ, De Boer A, Kessels AG. Is placebo analgesia mediated by endogenous opioids? A systematic review. *Pain*. 1998;76(3):273-275.
96. Benedetti F, Arduino C, Amanzio M. Somatotopic activation of opioid systems by target-directed expectations of analgesia. *J Neurosci*. 1999;19(9):3639-3648.
97. Levine R, Morgan MM, Cannon JT, Liebeskind JC. Stimulation of the periaqueductal gray matter of the rat produces a preferential ipsilateral antinociception. *Brain Res*. 1991;567(1):140-144.
98. Soper WY, Melzack R. Stimulation-produced analgesia: evidence for somatotopic organization in the midbrain. *Brain Res*. 1982;251(2):301-311.
99. Petrovic P, Kalso E, Petersson KM, Ingvar M. Placebo and opioid analgesia – imaging a shared neuronal network. *Science*. 2002;295(5560):1737-1740.
100. Willoch F, Tölle TR, Wester HJ, et al. Central pain after pontine infarction is associated with changes in opioid receptor binding: a PET study with 11C-diprenorphine. *AJNR Am J Neuroradiol*. 1999;20(4):686-690.
101. Baumgärtner U, Buchholz HG, Bellosevich A, et al. High opiate receptor binding potential in the human lateral pain system. *Neuroimage*. 2006;30(3):692-699.
102. Casey KL, Svensson P, Morrow TJ, et al. Selective opiate modulation of nociceptive processing in the human brain. *J Neurophysiol*. 2000;84(1):525-533.
103. Wagner KJ, Willoch F, Kochs EF, et al. Dose-dependent regional cerebral blood flow changes during remifentanil infusion in humans: a positron emission tomography study. *Anesthesiology*. 2001;94(5):732-739.
104. Apkarian AV, Bushnell MC, Treede R-D, Zubieta J-K. Human brain mechanisms of pain perception and regulation in health and disease. *Eur J Pain*. 2005;9(4):463-484.
105. Wager TD, Rilling JK, Smith EE, et al. Placebo-induced changes in FMRI in the anticipation and experience of pain. *Science*. 2004;303(5661):1162-1167.
106. Lieberman MD, Jarcho JM, Berman S, et al. The neural correlates of placebo effects: a disruption account. *Neuroimage*. 2004;22(1):447-455.
107. Bingel U, Lorenz J, Schoell E, et al. Mechanisms of placebo analgesia: rACC recruitment of a subcortical antinociceptive network. *Pain*. 2006;120(1-2):8-15.
108. Bingel U, Wanigasekera V, Wiech K, et al. The effect of treatment expectation on drug efficacy: imaging the analgesic benefit of the opioid remifentanil. *Sci Transl Med*. 2011;3(70):70ra14.
109. Price DD, Craggs J, Verne GN, et al. Placebo analgesia is accompanied by large reductions in pain-related brain activity in irritable bowel syndrome patients. *Pain*. 2007;127(1-2):63-72.
110. Watson A, El-Deredy W, Iannetti GD, et al. Placebo conditioning and placebo analgesia modulate a common brain network during pain anticipation and perception. *Pain*. 2009;145(1–2):24-30.
111. Lu HC, Hsieh JC, Lu CL, et al. Neuronal correlates in the modulation of placebo analgesia in experimentally-induced esophageal pain: a 3T-fMRI study. *Pain*. 2010;148(1):75-83.
112. Lui F, Colloca L, Duzzi D, et al. Neural bases of conditioned placebo analgesia. *Pain*. 2010;151(3):816-824.
113. Atlas LY, Whittington RA, Lindquist MA, et al. Dissociable influences of opiates and expectations on pain. *J Neurosci.*. 2012;32(23):8053-8064.
114. Elsenbruch S, Kotsis V, Benson S, et al. Neural mechanisms mediating the effects of expectation in visceral placebo analgesia: an fMRI study in healthy placebo responders and nonresponders. *Pain*. 2012;153(2):382-390.
115. Lee H-F, Hsieh J-C, Lu C-L, et al. Enhanced affect/cognition-related brain responses during visceral placebo analgesia in irritable bowel syndrome patients. *Pain*. 2012;153(6):1301-1310.
116. Amanzio M, Benedetti F, Porro CA, et al. Activation likelihood estimation meta-analysis of brain correlates of placebo analgesia in human experimental pain. *Hum Brain Mapp*. 2013;34(3):738-752.
117. Kong J, Gollub RL, Rosman IS, et al. Brain activity associated with expectancy-enhanced placebo analgesia as measured by functional magnetic resonance imaging. *J Neurosci*. 2006;26(2):381-388.
118. Miller EK. The prefrontal cortex and cognitive control. *Nat Rev Neurosci*. 2000;1(1):59-65.
119. Miller EK, Cohen JD. An integrative theory of prefrontal cortex function. *Annu Rev Neurosci*. 2001;24:167-202.
120. Craggs JG, Price DD, Perlstein WM, et al. The dynamic mechanisms of placebo induced analgesia: evidence of sustained and transient regional involvement. *Pain*. 2008;139(3):660-669.
121. Seminowicz DA. Believe in your placebo. *J Neurosci*. 2006;26(17):4453-4454.
122. Zubieta JK, Bueller JA, Jackson LR, et al. Placebo effects mediated by endogenous opioid activity on mu-opioid receptors. *J Neurosci*. 2005;25(34):7754-7762.
123. Wager TD, Scott DJ, Zubieta JK. Placebo effects on human mu-opioid activity during pain. *Proc Natl Acad Sci USA*. 2007;104(26):11056-11061.
124. Scott DJ, Stohler CS, Egnatuk CM, et al. Placebo and nocebo effects are defined by opposite opioid and dopaminergic responses. *Arch Gen Psychiatry*. 2008;65(2):220-231.
125. Pariente J, White P, Frackowiak RSJ, Lewith G. Expectancy and belief modulate the neuronal substrates of pain treated by acupuncture. *Neuroimage*. 2005;25(4):1161-1167.
126. Jensen TS, Yaksh TL. Comparison of the antinociceptive effect of morphine and glutamate at coincidental sites in the periaqueductal gray and medial medulla in rats. *Brain Res*. 1989;476(1):1-9.
127. Linnman C, Moulton EA, Barmettler G, et al. Neuroimaging of the periaqueductal gray: state of the field. *Neuroimage*. 2012;60(1):505-522.
128. Eippert F, Bingel U, Schoell ED, et al. Activation of the opioidergic descending pain control system underlies placebo analgesia. *Neuron*. 2009;63(4):533-543.
129. Price DD, Milling LS, Kirsch I, et al. An analysis of factors that contribute to the magnitude of placebo analgesia in an experimental paradigm. *Pain*. 1999;83(2):147-156.
130. Wager TD. Expectations and anxiety as mediators of placebo effects in pain. *Pain*. 2005;115(3):225-226.
131. Nakamura Y, Donaldson GW, Kuhn R, et al. Investigating dose-dependent effects of placebo analgesia: a psychophysiological approach. *Pain*. 2012;153(1):227-237.
132. Atlas LY, Bolger N, Lindquist MA, Wager TD. Brain mediators of predictive cue effects on perceived pain. *J Neurosci*. 2010;30(39):12964-12977.
133. Wager TD, Atlas LY, Leotti LA, Rilling JK. Predicting individual differences in placebo analgesia: contributions of brain activity during anticipation and pain experience. *J Neurosci*. 2011;31(2):439-452.
134. Wagner KJ, Sprenger T, Kochs EF, et al. Imaging human cerebral pain modulation by dose-dependent opioid analgesia: a positron emission tomography activation study using remifentanil. *Anesthesiology*. 2007;106(3):548-556.

135. Krummenacher P, Candia V, Folkers G, et al. Prefrontal cortex modulates placebo analgesia. *Pain*. 2010;148(3):368-374.
136. Benedetti F, Arduino C, Costa S, et al. Loss of expectation-related mechanisms in Alzheimer's disease makes analgesic therapies less effective. *Pain*. 2006;121(1–2):133-144.
137. Lorenz J, Hauck M, Paur RC, et al. Cortical correlates of false expectations during pain intensity judgments – a possible manifestation of placebo/nocebo cognitions. *Brain Behav Immun*. 2005;19(4):283-295.
138. Stein N, Sprenger C, Scholz J, et al. White matter integrity of the descending pain modulatory system is associated with interindividual differences in placebo analgesia. *Pain*. 2012;153(11):2210-2217.
139. Honey CJ, Sporns O, Cammoun L, et al. Predicting human resting-state functional connectivity from structural connectivity. *Proc Natl Acad Sci USA*. 2009;106(6):2035-2040.
140. Corbett AD, Henderson G, McKnight AT, Paterson SJ. 75 years of opioid research: the exciting but vain quest for the Holy Grail. *Br J Pharmacol*. 2006;147(suppl 1):S153-S162.
141. Henriksen G, Willoch F. Imaging of opioid receptors in the central nervous system. *Brain*. 2008;131(Pt 5):1171-1196.
142. Amanzio M, Benedetti F. Neuropharmacological dissection of placebo analgesia: expectation-activated opioid systems versus conditioning-activated specific subsystems. *J Neurosci*. 1999;19(1):484-494.
143. Vase L, Robinson ME, Verne GN, Price DD. Increased placebo analgesia over time in irritable bowel syndrome (IBS) patients is associated with desire and expectation but not endogenous opioid mechanisms. *Pain*. 2005;115(3):338-347.
144. Benedetti F, Amanzio M, Rosato R, Blanchard C. Nonopioid placebo analgesia is mediated by CB1 cannabinoid receptors. *Nat Med*. 2011;17(10):1228-1230.
145. Petrovic P, Kalso E, Petersson KM, et al. A prefrontal non-opioid mechanism in placebo analgesia. *Pain*. 2010;150(1):59-65.
146. Tamae A, Nakatsuka T, Koga K, et al. Direct inhibition of substantia gelatinosa neurones in the rat spinal cord by activation of dopamine D2-like receptors. *J Physiol (Lond.)*. 2005;568(Pt 1):243-253.
147. Morgan MM, Sohn JH, Liebeskind JC. Stimulation of the periaqueductal gray matter inhibits nociception at the supraspinal as well as spinal level. *Brain Res*. 1989;502(1):61-66.
148. Borszcz GS, Johnson CP, Thorp MV. The differential contribution of spinopetal projections to increases in vocalization and motor reflex thresholds generated by the microinjection of morphine into the periaqueductal gray. *Behav Neurosci*. 1996;110(2):368-388.
149. Mantyh PW. Connections of midbrain periaqueductal gray in the monkey. I. Ascending efferent projections. *J Neurophysiol*. 1983;49(3):567-581.
150. Behbehani MM. Functional characteristics of the midbrain periaqueductal gray. *Prog Neurobiol*. 1995;46(6):575-605.
151. Mason P. Contributions of the medullary raphe and ventromedial reticular region to pain modulation and other homeostatic functions. *Annu Rev Neurosci*. 2001;24:737-777.
152. Etkin A, Egner T, Kalisch R. Emotional processing in anterior cingulate and medial prefrontal cortex. *Trends Cogn Sci*. 2011;15(2):85-93.
153. Sarinopoulos I, Dixon GE, Short SJ, et al. Brain mechanisms of expectation associated with insula and amygdala response to aversive taste: implications for placebo. *Brain Behav Immun*. 2006;20(2):120-132.
154. Poldrack RA. Can cognitive processes be inferred from neuroimaging data? *Trends Cogn Sci*. 2006;10(2):59-63.
155. Yarkoni T, Poldrack RA, Nichols TE, et al. Large-scale automated synthesis of human functional neuroimaging data. *Nat Meth*. 2011;8(8):665-670.
156. Roelofs J, Ter Riet G, Peters ML, et al. Expectations of analgesia do not affect spinal nociceptive R-III reflex activity: an experimental study into the mechanism of placebo-induced analgesia. *Pain*. 2000;89(1):75-80.
157. Matre D, Casey KL, Knardahl S. Placebo-induced changes in spinal cord pain processing. *J Neurosci*. 2006;26(2):559-563.
158. Goffaux P, Redmond WJ, Rainville P, Marchand S. Descending analgesia – when the spine echoes what the brain expects. *Pain*. 2007;130(1–2):137-143.
159. Giove F, Garreffa G, Giulietti G, et al. Issues about the fMRI of the human spinal cord. *Magn Reson Imaging*. 2004;22(10):1505-1516.
160. Stroman PW. Magnetic resonance imaging of neuronal function in the spinal cord: spinal FMRI. *Clin Med Res*. 2005;3(3):146-156.
161. Summers PE, Iannetti GD, Porro CA. Functional exploration of the human spinal cord during voluntary movement and somatosensory stimulation. *Magn Reson Imaging*. 2010;28(8):1216-1224.
162. Yoshizawa T, Nose T, Moore GJ, Sillerud LO. Functional magnetic resonance imaging of motor activation in the human cervical spinal cord. *Neuroimage*. 1996;4(3 Pt 1):174-182.
163. Stroman PW, Nance PW, Ryner LN. BOLD MRI of the human cervical spinal cord at 3 tesla. *Magn Reson Med*. 1999;42(3):571-576.
164. Ko HY, Park JH, Shin YB, Baek SY. Gross quantitative measurements of spinal cord segments in human. *Spinal Cord*. 2004;42(1):35-40.
165. Cooke FJ, Blamire AM, Manners DN, et al. Quantitative proton magnetic resonance spectroscopy of the cervical spinal cord. *Magn Reson Med*. 2004;51(6):1122-1128.
166. Finsterbusch J, Eippert F, Büchel C. Single, slice-specific z-shim gradient pulses improve T2*-weighted imaging of the spinal cord. *Neuroimage*. 2012;59(3):2307-2315.
167. Stroman PW. Discrimination of errors from neuronal activity in functional MRI of the human spinal cord by means of general linear model analysis. *Magn Reson Med*. 2006;56(2):452-456.
168. Brooks JCW, Beckmann CF, Miller KL, et al. Physiological noise modelling for spinal functional magnetic resonance imaging studies. *Neuroimage*. 2008;39(2):680-692.
169. Piché M, Cohen-Adad J, Nejad MK, et al. Characterization of cardiac-related noise in fMRI of the cervical spinal cord. *Magn Reson Imaging*. 2009;27(3):300-310.
170. Cohen-Adad J, Gauthier CJ, Brooks JCW, et al. BOLD signal responses to controlled hypercapnia in human spinal cord. *Neuroimage*. 2010;50(3):1074-1084.
171. Maieron M, Iannetti GD, Bodurka J, et al. Functional responses in the human spinal cord during willed motor actions: evidence for side- and rate-dependent activity. *J Neurosci*. 2007;27(15):4182-4190.
172. Cohen-Adad J, Mareyam A, Keil B, et al. 32-Channel RF coil optimized for brain and cervical spinal cord at 3 T. *Magn Reson Med*. 2011;66(4):1198-1208.
173. Kong Y, Jenkinson M, Andersson J, et al. Assessment of physiological noise modelling methods for functional imaging of the spinal cord. *NeuroImage*. 2012;60(2):1538-1549.
174. Xie G, Piché M, Khoshejad M, et al. Reduction of physiological noise with independent component analysis improves the detection of nociceptive responses with fMRI of the human spinal cord. *NeuroImage*. 2012.
175. Pórszász R, Beckmann N, Bruttel K, et al. Signal changes in the spinal cord of the rat after injection of formalin into the hindpaw: characterization using functional magnetic resonance imaging. *Proc Natl Acad Sci USA*. 1997;94(10):5034-5039.
176. Malisza KL, Stroman PW. Functional imaging of the rat cervical spinal cord. *J Magn Reson Imaging*. 2002;16(5):553-558.
177. Lilja J, Endo T, Hofstetter C, et al. Blood oxygenation level-dependent visualization of synaptic relay stations of sensory pathways along the neuroaxis in response to graded sensory stimulation of a limb. *J Neurosci*. 2006;26(23):6330-6336.

178. Zhao F, Williams M, Meng X, et al. BOLD and blood volume-weighted fMRI of rat lumbar spinal cord during non-noxious and noxious electrical hindpaw stimulation. *Neuroimage*. 2008;40(1):133-147.
179. Zhao F, Williams M, Meng X, et al. Pain fMRI in rat cervical spinal cord: an echo planar imaging evaluation of sensitivity of BOLD and blood volume-weighted fMRI. *Neuroimage*. 2009;44(2):349-362.
180. Zhao F, Williams M, Welsh DC, et al. fMRI investigation of the effect of local and systemic lidocaine on noxious electrical stimulation-induced activation in spinal cord. *Pain*. 2009;145(1-2):110-119.
181. Bryant CD, Roberts KW, Culbertson CS, et al. Pavlovian conditioning of multiple opioid-like responses in mice. *Drug Alcohol Depend*. 2009;103(1-2):74-83.
182. Guo JY, Wang JY, Luo F. Dissection of placebo analgesia in mice: the conditions for activation of opioid and non-opioid systems. *J Psychopharmacol (Oxford)*. 2010;24(10):1561-1567.
183. Nolan TA, Price DD, Caudle RM, et al. Placebo-induced analgesia in an operant pain model in rats. *Pain*. 2012;153(10):2009-2016.
184. Backes WH, Mess WH, Wilmink JT. Functional MR imaging of the cervical spinal cord by use of median nerve stimulation and fist clenching. *AJNR Am J Neuroradiol*. 2001;22(10):1854-1859.
185. Stroman PW, Ryner LN. Functional MRI of motor and sensory activation in the human spinal cord. *Magn Reson Imaging*. 2001;19(1):27-32.
186. Stroman PW, Tomanek B, Krause V, et al. Mapping of neuronal function in the healthy and injured human spinal cord with spinal fMRI. *Neuroimage*. 2002;17(4):1854-1860.
187. Stroman PW, Kornelsen J, Bergman A, et al. Noninvasive assessment of the injured human spinal cord by means of functional magnetic resonance imaging. *Spinal Cord*. 2004;42(2):59-66.
188. Ghazni NF, Cahill CM, Stroman PW. Tactile sensory and pain networks in the human spinal cord and brain stem mapped by means of functional MR imaging. *AJNR Am J Neuroradiol*. 2010;31(4):661-667.
189. Cahill CM, Stroman PW. Mapping of neural activity produced by thermal pain in the healthy human spinal cord and brain stem: a functional magnetic resonance imaging study. *Magn Reson Imaging*. 2011;29(3):342-352.
190. Summers PE, Ferraro D, Duzzi D, et al. A quantitative comparison of BOLD fMRI responses to noxious and innocuous stimuli in the human spinal cord. *Neuroimage*. 2010;50(4):1408-1415.
191. Brooks JCW, Kong Y, Lee MC, et al. Stimulus site and modality dependence of functional activity within the human spinal cord. *J Neurosci*. 2012;32(18):6231-6239.
192. Fitzgerald M. The contralateral input to the dorsal horn of the spinal cord in the decerebrate spinal rat. *Brain Res*. 1982;236(2):275-287.
193. Coghill RC, Price DD, Hayes RL, Mayer DJ. Spatial distribution of nociceptive processing in the rat spinal cord. *J Neurophysiol*. 1991;65(1):133-140.
194. Porro CA, Cavazzuti M, Galetti A, et al. Functional activity mapping of the rat spinal cord during formalin-induced noxious stimulation. *Neuroscience*. 1991;41(2-3):655-665.
195. Eippert F, Finsterbusch J, Bingel U, Büchel C. Direct evidence for spinal cord involvement in placebo analgesia. *Science*. 2009;326(5951):404.
196. Deckers RHR, Van Gelderen P, Ries M, et al. An adaptive filter for suppression of cardiac and respiratory noise in MRI time series data. *Neuroimage*. 2006;33(4):1072-1081.
197. Logothetis NK. What we can do and what we cannot do with fMRI. *Nature*. 2008;453(7197):869-878.
198. Levine JD, Gordon NC, Bornstein JC, Fields HL. Role of pain in placebo analgesia. *Proc Natl Acad Sci USA*. 1979;76(7):3528-3531.
199. Bennett GJ, Mayer DJ. Inhibition of spinal cord interneurons by narcotic microinjection and focal electrical stimulation in the periaqueductal central gray matter. *Brain Res*. 1979;172(2):243-257.
200. Ruscheweyh R, Sandkühler J. Lamina-specific membrane and discharge properties of rat spinal dorsal horn neurones in vitro. *J Physiol (Lond.)*. 2002;541(Pt 1):231-244.
201. Dostrovsky JO, Shah Y, Gray BG. Descending inhibitory influences from periaqueductal gray, nucleus raphe magnus, and adjacent reticular formation. II. Effects on medullary dorsal horn nociceptive and nonnociceptive neurons. *J Neurophysiol*. 1983;49(4):948-960.
202. Viisanen H, Ansah OB, Pertovaara A. The role of the dopamine D2 receptor in descending control of pain induced by motor cortex stimulation in the neuropathic rat. *Brain Res Bull*. 2012;89(3-4):133-143.
203. Koutsikou S, Parry DM, MacMillan FM, Lumb BM. Laminar organization of spinal dorsal horn neurones activated by C- vs. A-heat nociceptors and their descending control from the periaqueductal grey in the rat. *Eur J Neurosci*. 2007;26(4):943-952.
204. Lopez-Garcia JA, King AE. Membrane properties of physiologically classified rat dorsal horn neurons in vitro: correlation with cutaneous sensory afferent input. *Eur J Neurosci*. 1994;6(6):998-1007.
205. Sprenger C, Eippert F, Finsterbusch J, et al. Attention modulates spinal cord responses to pain. *Curr Biol*. 2012;22(11):1019-1022.
206. Buhle J, Wager TD. Performance-dependent inhibition of pain by an executive working memory task. *Pain*. 2010;149(1):19-26.
207. Lorenz J, Minoshima S, Casey KL. Keeping pain out of mind: the role of the dorsolateral prefrontal cortex in pain modulation. *Brain*. 2003;126(Pt 5):1079-1091.
208. Wager TD, Matre D, Casey KL. Placebo effects in laser-evoked pain potentials. *Brain Behav Immun*. 2006;20(3):219-230.
209. Craggs JG, Price DD, Verne GN, et al. Functional brain interactions that serve cognitive-affective processing during pain and placebo analgesia. *Neuroimage*. 2007;38(4):720-729.
210. Goffaux P, De Souza JB, Potvin S, Marchand S. Pain relief through expectation supersedes descending inhibitory deficits in fibromyalgia patients. *Pain*. 2009;145(1-2):18-23.
211. Jensen KB, Kaptchuk TJ, Kirsch I, et al. Nonconscious activation of placebo and nocebo pain responses. *Proc Natl Acad Sci USA*. 2012;109(39):15959-15964.
212. Burkey AR, Carstens E, Wenniger JJ, et al. An opioidergic cortical antinociception triggering site in the agranular insular cortex of the rat that contributes to morphine antinociception. *J Neurosci*. 1996;16(20):6612-6623.
213. Braz JM, Nassar MA, Wood JN, Basbaum AI. Parallel 'pain' pathways arise from subpopulations of primary afferent nociceptor. *Neuron*. 2005;47(6):787-793.
214. Bantick SJ, Wise RG, Ploghaus A, et al. Imaging how attention modulates pain in humans using functional MRI. *Brain*. 2002;125(Pt 2):310-319.
215. Valet M, Sprenger T, Boecker H, et al. Distraction modulates connectivity of the cingulo-frontal cortex and the midbrain during pain – an fMRI analysis. *Pain*. 2004;109(3):399-408.
216. Salomons TV, Johnstone T, Backonja M-M, Davidson RJ. Perceived controllability modulates the neural response to pain. *J Neurosci*. 2004;24(32):7199-7203.
217. Hofbauer RK, Rainville P, Duncan GH, Bushnell MC. Cortical representation of the sensory dimension of pain. *J Neurophysiol*. 2001;86(1):402-411.
218. Sprenger C, Bingel U, Büchel C. Treating pain with pain: supraspinal mechanisms of endogenous analgesia elicited by heterotopic noxious conditioning stimulation. *Pain*. 2011;152(2):428-439.

219. Willer JC, Boureau F, Albe-Fessard D. Supraspinal influences on nociceptive flexion reflex and pain sensation in man. *Brain Res.* 1979;179(1):61-68.
220. Rhudy JL, Williams AE, McCabe KM, et al. Affective modulation of nociception at spinal and supraspinal levels. *Psychophysiology.* 2005;42(5):579-587.
221. Tracey I, Ploghaus A, Gati JS, et al. Imaging attentional modulation of pain in the periaqueductal gray in humans. *J Neurosci.* 2002;22(7):2748-2752.
222. Yelle MD, Oshiro Y, Kraft RA, Coghill RC. Temporal filtering of nociceptive information by dynamic activation of endogenous pain modulatory systems. *J Neurosci.* 2009;29(33):10264-10271.
223. Martucci KT, Eisenach JC, Tong C, Coghill RC. Opioid-independent mechanisms supporting offset analgesia and temporal sharpening of nociceptive information. *Pain.* 2012;153(6):1232-1243.

CHAPTER 8

Positive and Negative Emotions and Placebo Analgesia

Magne Arve Flaten[1], Per M. Aslaksen[2], Peter S. Lyby[2]

[1]Department of Psychology, Norwegian University of Science and Technology, Trondheim, Norway,
[2]Department of Psychology, University of Tromsø, Tromsø, Norway

Several reports have shown that the expectation of reduced pain is a central element in placebo analgesia.[1-3] Thus, granting that expectation is a cognitive concept, placebo analgesia could be understood as a cognitive process. However, it has been proposed that reduction in negative emotions is an important mediator of placebo analgesia,[1-5] meaning that the cognition of expecting less (or more) pain exerts its effect on pain via activation of emotional processes.[5] Several studies have shown that emotions can moderate pain, and especially that negative emotions often increase pain perception,[6-9] although this may depend, for example, on the intensity of emotional arousal. In most studies investigating the impact of emotions on pain perception, emotions have typically been produced by films,[10] affective pictures[11] and odors[12] that were independent of the pain stimulus used to induce pain.[13] Thus, in many studies, emotions have been induced by stimuli not related to pain. Price et al,[1] on the other hand, proposed that pain-related emotions are triggered by the unpleasantness of pain. Thus, the context in which pain is experienced, and the relationship of the emotions to pain, i.e. whether they are elicited by pain or are not pain-related, may play a role.

According to Price et al,[1] pain-related negative emotions should increase concomitantly with the intensity of the pain stimulus, a prediction that has been supported in several studies.[4,14] This shows that pain itself is a potent stressor,[14] and high levels of negative affect and arousal from pain stimulation may activate the endogenous opioid system.[15] Thus, it has been suggested that the relationship between emotional activation and pain can be described as an inverted 'U', where pain is rated highest with moderate negative emotions, and lower with highly positive and highly negative emotions[5,15] (Fig. 8.1). According to Rhudy and Meagher,[15] negative emotions of low to moderate intensity may increase attention towards and amplify pain via neural circuits in the amygdala and periaqueductal gray (PAG). As in many pain studies, attention could possibly also play a role in placebo analgesia. It is possible that administration of placebo induces an affective or motivational state that reduces attention towards the painful stimuli. The motivational state that regulates attention can be partly under conscious control, and the placebo may serve as a safety signal and permit attention to be directed to stimuli other than pain.[16] Consequently, attention may be conceived as a moderator of the effect emotions have on pain.

EMOTION AND MOTIVATION

Studies by Vase et al[2,3] have shown that placebo analgesic responses may be partially mediated by reductions in anxiety levels. Similar results have also been obtained in studies where subjects reporting higher levels of anxiety report higher pain compared to subjects scoring lower on anxiety.[17,18] The reduction in anxiety levels, as observed in the studies of Vase et al,[2,3] was related to expectations, but this association was not as strong as the relationship between pain ratings and expectations of the effect of treatment on pain. As shown above, a reduced negative affect is associated with decreased pain, and it can be predicted that if information about a painkiller reduces stress and negative emotions, then pain should be reduced as well. However, the findings of Vase et al[2,3] suggest that anxiety is just one of several possible emotional states that may modulate the placebo analgesic effect. This view is also supported in pain studies by Rhudy and Meagher[8,15] in which subjects

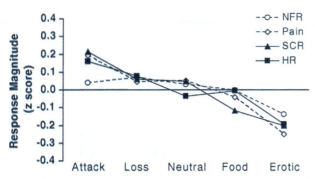

FIGURE 8.1 The effect of emotional picture content on pain ratings (pain), the nociceptive flexion reflex (NFR), skin conductance responses (SCR), and heart rate (HR) after noxious stimulation of the sural nerve. Unpleasant pictures tended to enhance nociceptive reactions, whereas pleasant pictures inhibited nociceptive reactions. The degree of pain modulation was moderated by the emotional intensity (arousal) elicited by the pictures, so only the most arousing pictures (attack, erotic) elicited modulation that was significantly different from the neutral pictures. These findings suggest that valence and arousal both contribute to the modulation of nociception and pain. From Rhudy et al[16] with permission from the publisher.

react with various emotional feelings, ranging from fear to surprise, under noxious stimulation. The latter notion fits well with results from neurobiologic placebo studies that find activation of distinct, but overlapping, networks involved in the processing of sensory and affective aspects of nociceptive stimuli;[19–21] this implies that a broad spectrum of emotions could be important for the placebo analgesic response. Petrovic et al[22] found that subjects with large placebo responses displayed decreases in neuronal activity in the emotional networks of the cortex, suggesting that placebo effects are linked to neuronal emotional modulation.

In those studies in which patients were asked to rate their expected pain levels and their desire for pain relief,[2,23] the results showed that both factors acted independently; the interaction between desire and expectation explained about 40% of the variance in pain intensity. In the studies of Vase et al[2] and Verne et al,[23] anxiety levels were reduced after placebo administration, a finding that was supported in a later study by Vase et al.[3] The findings from the studies of Vase et al[2,3] and Verne et al,[23] that support the desire–expectation model proposed by Price and Barrell,[24,25] suggest a clear connection between placebo analgesia and emotional factors. Petersen et al[26] found that placebo analgesia, observed as a reduced area of hyperalgesia in neuropathic pain, was higher when the negative affect was lower. There was no indication that a positive affect increased during placebo analgesia, unlike the result found in a study by Scott et al,[27] where a positive affect slightly (but non-significantly) increased after administration of a placebo when the subjects were informed that it would reduce pain. However, the samples in the two studies were rather small—19 and 14 participants in the studies of Petersen et al and Scott et al, respectively—and this could explain the differing results. A possible problem in studies like those of Vase et al[2,3] and Verne et al[23] was that, before the experimental procedure was being carried out, participants rated the pain levels and emotional state that they expected after the (placebo) treatment. It is possible that these ratings of expected pain levels may be viewed as a sort of social contract.[28] Thus, the reporting of expectations and emotions prior to the pain-induction procedure may establish a norm and a commitment which shapes the later self-reports during pain.

In contrast to findings that suggest that emotional modulation is important in placebo analgesia, Flaten et al[4] found that placebo analgesia might also be observed without a concomitant reduction in negative emotions, at least at the group level. This could be true especially when initial levels of negative emotions are too low to permit the observation of a decrease after placebo administration.

Vase et al[2,3] suggested that the mechanism of emotional modulation in placebo analgesia involves the sympathetic nervous system. This is plausible because several studies have established that pain sensations increase sympathetic activity as measured by skin conductance[8] and heart-rate variability.[13] Pollo et al[29] tested the hypothesis that placebo analgesia is accompanied by modulations in cardiovascular activity by measuring heart rate variability; the results from these authors showed that the low-frequency cardiac responses, indicating sympathetic input to the heart, were decreased during placebo analgesia, suggesting that reduction in cardiac autonomic arousal is a part of placebo analgesia.

In sum, several studies have pointed to the fact that emotions are important factors for the placebo analgesic response, and these observations support the following straightforward hypothesis: expectation of having received effective treatment should reduce negative emotions like stress and anxiety, and the reduction in negative emotions should then reduce pain.

REDUCTION IN NEGATIVE EMOTIONS: METHODOLOGIC ISSUES AND EMPIRICAL STUDIES

Even if the hypothesis above is simple, experimental studies have struggled to show that administration of a placebo reduces negative emotions. Aslaksen and Flaten[30] performed an experimental study with a within-subjects design to test whether placebo administration reduces the subjective feeling of stress and concomitant activity in the autonomic nervous system measured by heart-rate variability. The volunteers participated in two conditions on two separate days where heat was repeatedly used to produce pain stimuli. In the placebo condition, participants received placebo capsules and were told that the capsules were effective painkillers. In the natural history control condition, the same pain procedure was performed, but without placebo. Half of the volunteers underwent the placebo condition before the natural history condition; the other half participated in the natural history condition before the placebo condition. A placebo—consisting of two capsules containing lactose—was administered after the first of five pain stimulations. Five pain stimuli in one pre-test and four post-tests were fixed at 46°C, with a duration of 240 seconds each. Pain and emotional ratings were obtained by pen-and-paper visual analog scales. In the natural history condition, pain was inflicted five times without administration of placebo. The results showed that pain intensity was significantly reduced in the placebo condition compared with the natural history condition; these results were also obtained when statistically controlling for the order of the conditions. Thus, placebo analgesia was induced at the group level. As expected, stress levels and cardiac activity, related to sympathetic arousal, were also significantly reduced after administration of placebo. Stepwise regression analysis revealed that the reduction in stress was the only significant predictor for the placebo response when controlling for baseline pain, arousal, mood, order of conditions, subject gender, experimenter gender, previous experience with non-prescribed analgesics, and heart rate variability. The results from Aslaksen and Flaten[29] lend support to the stress-reduction hypothesis of placebo analgesia, but the findings could not be regarded as conclusive because stress measures were obtained during pain stimulation and not prior to the painful stimuli. Thus, the results could not answer the question of whether the reduction in emotions reduced the pain, i.e. the placebo analgesic response, or if it was the other way around, that the placebo analgesic response, i.e. the reduction in pain, reduced the negative emotions.

To further explore the stress reduction hypothesis, Aslaksen and colleagues[31] performed a study in which stress measurements were obtained in the absence of experimental pain induction. As in the studies of Aslaksen & Flaten,[30] a within-subjects design with a placebo condition and a natural history condition was employed. The placebo capsules were administered after the pre-test, and the subjects were told that the medication was an effective painkiller with an excellent effect on pain induced by heat. Pulses of heat pain, with a peak temperature of 52°C and a heating/cooling rate of 70°C/40°C per second, were delivered by a PC-controlled thermode to elicit the N2/P2 components; these can be observed in event-related potentials (ERPs). Pain pulses were delivered in blocks of 24 pulses. To be able to answer more precisely whether stress reduction is a key mechanism in placebo analgesic responses, emotional measures were obtained before the administration of placebo, immediately after, and in the intervals between pain blocks. Thus, stress was measured in the absence of pain, so that these measures were not confounded. A significant stress reduction was observed in the placebo condition, and regression analysis showed that the stress reduction recorded in the absence of pain significantly predicted the placebo response on pain. The P2 wave in the ERP was significantly reduced in the placebo condition compared to the natural history condition. The results from Aslaksen et al[31] showed that placebo administration reduces negative emotions, compatible with findings in other studies on the placebo analgesic response. For example, Scott et al[27] and Vase et al[3] measured emotions in the absence of pain, and revealed that negative emotional activation decreased after placebo medication.

INDIVIDUAL DIFFERENCES IN NEGATIVE EMOTIONS AND THE EFFECTIVENESS OF PLACEBO INTERVENTIONS ON PAIN

A placebo intervention may lead to a reduction in pain, often accompanied by a reduction in negative affect. However, there are individuals who do not respond to the placebo intervention; these are called placebo non-responders. A great deal of attention has been devoted to the placebo responders and to the states and traits that are associated with placebo analgesic responses, leaving the second half of the story, that of the placebo non-responder, unexplored. By paying attention to those who do not respond in an experiment, one may see the placebo analgesic response from a different viewpoint. An interesting, important and overlooked question then, is whether new mechanistic insights into placebo analgesia might be gained by attempting to unravel the mechanisms behind placebo non-responding.

Individuals taking part in a pain experiment, or who show up for a hospital appointment, carry with them their personalities and habitual tendencies to appraise, react and respond to pain and discomfort. These individual differences may be related to some of the variability observed in placebo responding. Lyby and colleagues[32–34] systematically investigated the relationship of dispositional fear of pain (FoP) to placebo analgesia in a series of experiments. Fear and anxiety are the two most frequently reported emotions in the context of pain,[35] and a number of studies have shown that these emotions, together with low-to-moderate arousal, exacerbate pain, cause activity in the hypothalamic–pituitary–adrenal axis, and trigger endogenous peptides that have hyperalgesic and anti-opioid effects.[36] Fear and anxiety thus appear to cause an increase in pain, which is opposite to the emotional, homeostatic and neurochemical reactions that have been observed in placebo responders. Accordingly, Lyby et al[34] proposed that fear would reduce or abolish placebo analgesia.

A correlational approach was used in two studies[32,33] in which fear of pain, as assessed by the Fear of Pain Questionnaire III, was used as a predictor of response to placebo analgesia. Both studies used a repeated-measures within-subject design, in which the order of the natural history condition and the placebo condition was counterbalanced across subjects. Thermal pain was induced in the lower forearm. Placebo capsules were administered in the placebo condition, the subjects being told that the capsules would reduce pain. Placebo capsules were not administered in the natural history condition.

In the first study,[32] participants reported pain intensity, pain unpleasantness, and stress while being exposed to the thermal stimuli. A mean placebo effect in reported pain intensity was observed. About one-third of the participants, however, responded with a negative placebo effect or nocebo effect, i.e. increased pain, and not a decrease in pain, and increased FoP was associated with a larger negative placebo effect. FoP was also associated with higher reported stress before the start of the experiment, and higher reported stress during pain in the natural history condition.

In the second study[33] the aim was to replicate and extend the findings from the first study by employing ERPs elicited by painful stimulation, in addition to subjective report. Subjective pain report is based on cognitively construed representations, and many studies report significant placebo effects in reported pain. One advantage of combining subjective report with an objective method measuring cortical responses to pain is that reporting bias may be excluded as an explanation of placebo analgesia.[37] Additionally, ERPs are useful in the investigation of important empirical questions such as localization of the source of the placebo-modulated ERP, and the ERP can be used to test theories of the level of the central nervous system at which placebos modulate pain. Event-related potentials are accepted as a method for answering this question because the N2/P2 components in the ERP, at least partially, reflect early or pre-cognitive nociceptive processing. In line with the hypothesis that FoP might be associated with reduced placebo analgesia, it was hypothesized that FoP should be related to absence of placebo analgesic effects in the ERP.

The results revealed that higher FoP was associated with lower placebo responding in P2 amplitude and reported pain unpleasantness. The results linking FoP to placebo unresponsiveness in P2 amplitude suggest that the effect of FoP on placebo analgesia is partially pre-cognitive and not only confined to cognitively construed representations of pain. This finding might reflect a nociceptive system that is more easily activated due to anticipatory fear in high-FoP subjects, as has been suggested by other studies.[38]

Analyses also revealed that gender predicted placebo responding on P2 amplitude, as the female participants did not respond with a reduction in P2 amplitude, i.e. females did not display a physiologic placebo analgesic response. Moreover, female subjects scored significantly higher on FoP than did male subjects. Hierarchical blockwise regression was applied, and, after removing the linear effect of FoP, no differences between male and female subjects in placebo responding were observed. Thus, the absence of placebo responses in females could be explained by higher fear of pain in female subjects.

Thus, females had significantly lower placebo responses compared to males, both on subjective pain measurements and on the P2 wave in the ERP. The gender difference was also found in the stress data, where females had higher stress levels than males. The interaction of gender × stress was the only significant predictor for the placebo response and explained 23% of the variance in the placebo effect on pain unpleasantness, suggesting that males responded with larger stress reduction after placebo administration compared to females.

Even though the literature on gender and gender differences in pain is large, only a limited number of studies have addressed such differences in placebo analgesia.[39] The few studies on gender differences in placebo analgesia that exist suggest that males respond with stronger pain reduction after placebo administration compared to females.[4,30,31] Furthermore, several studies have suggested that gender differences in the experience of pain are mainly caused by differences in emotional processing.[40–42] A recent functional magnetic resonance imaging (fMRI) study on the placebo analgesic response,[43] revealed a consistent pattern of cerebral activation differences between the sexes during anticipation of pain, in the early and late phases of pain stimulation after injection of a placebo. As in the studies of Aslaksen et al,[31] regression analysis showed that the interaction of gender × emotional activation was the only significant predictor for the behavioral placebo response, suggesting that gender differences in emotional modulation produces gender differences in placebo analgesia. The cerebral gender difference in pain processing after placebo medication included the insular cortex, the left hippocampus and Brodmanns area 10, all cerebral areas involved in emotional modulation.[44,45] Negative emotions and fear of pain are known to decrease placebo analgesic effects,[4,5,32–34] and there is now evidence from imaging studies suggesting that the placebo analgesic response is related to engagement of cerebral emotional processes, rather than early suppression of nociceptive processing.[45,46]

Lyby et al[32,33] thus showed that the magnitude of the placebo analgesic effect in subjective report and ERPs depended on the level of FoP, in which those with a higher FoP were less responsive to the placebo interventions. These results are in line with a few other studies on placebo analgesia that have employed state measures of negative affect and in which higher negative affects have been inversely correlated with placebo analgesic magnitude.[47,48] The problem with these results, however, is that they are correlative in nature, and can thus say nothing about cause and effect, even though one could hypothesize that FoP, reflecting trait qualities, precedes the experimental procedures in time and thus justifies a causal interpretation.

The primary aim of Lyby et al[34] was therefore to investigate the causal effect of fear on placebo analgesia by inducing fear experimentally by informing the subjects that electric shocks would be administrated within a certain time period, a procedure termed instructed fear.[49] In the natural history condition, subjects received painful stimulation across three test sessions, with no intervention. In the placebo condition, the subjects received the same three sessions of painful stimulation as in the natural history condition, but in addition they received capsules containing lactose between the first and second sessions of painful stimulation; they were told that the capsules contained a powerful painkiller. To investigate the effect of fear on placebo analgesia, a condition was introduced that was identical to the placebo condition, but, in addition, fear was induced after the administration of the placebo capsules, i.e. the '2 Condition by Test' design used in Lyby et al[33] was expanded to a '3 Condition (natural history, placebo, placebo + fear) by Test', within-subject design.

Anticipation of electric shock was chosen as the fear-induction procedure because it is a well validated method, consistently inducing fear,[50,51] and also because it is relevant to FoP because electric shocks are expected to be painful. Additionally, the acoustic startle reflex was used as an outcome measure because fear-potentiated startle is a well validated marker of fear.[51] Potentiation of the startle reflex reflects early and pre-cognitive processing of fear,[52-54] and its amplitude is related to amygdala activation and to networks mediating defensive activation and action preparedness.[49,55,56] Thus, for individual difference analyses, fear-potentiated startle and FoP were used as predictors of placebo responding. Self-reported effectiveness of how well the fear-induction procedure produced fear was also used as a predictor. It was hypothesized that measures of fear should predict placebo responding. We also expected that placebo responding in startle should be positively related to corresponding placebo responses in subjective reports.

A placebo effect was observed in the startle data, i.e. startle reflexes were smaller in the placebo condition compared to the natural history condition. In subjects with high FoP, however, startle reflexes were not reduced in the placebo condition (Fig. 8.2). In the pain-intensity data, there was a trend towards a placebo effect that was abolished by induced fear, and again, this was most pronounced in subjects who were highest in FoP. Moreover, the placebo effect on startle, and the disruption of this effect by induced fear, predicted the corresponding effects (i.e. placebo effect and its disruption) in the pain-intensity

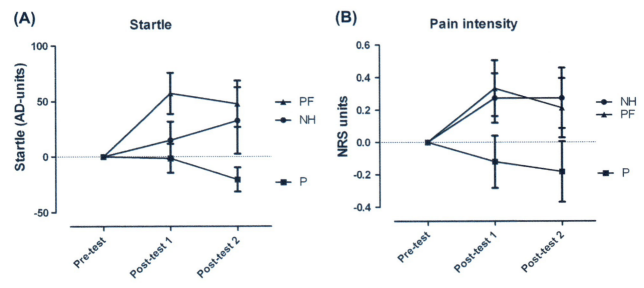

FIGURE 8.2 Interactions of condition by test in the startle reflex data (reported in analog/digital units) (A) and in reported pain intensity (numerical rating scale [NRS]) (B). Negative numbers indicate a reduction in response compared to the pretest, whereas positive numbers indicate an increase in response as compared to pretests. From Lyby et al[34] with permission from the publisher.

data. This suggests shared underlying mechanisms for fear and reduced placebo analgesia, and the regression analyses suggest that the expression of these mechanisms correlate with individual differences in fear.

NEGATIVE EMOTIONS REDUCE THE EFFECTIVENESS OF OPIOIDS

One way in which negative emotions might reduce the effectiveness of placebo interventions on pain is by reducing endogenous opioid neurotransmission. In a recent experiment placebo analgesia was induced by classical conditioning with an opioid drug, and the resultant placebo effect was completely abolished by the injection of the cholecystokinin agonist pentagastrin.[57] Pentagastrin and cholecystokinin are both considered to be neurochemical correlates of anxiety,[58] and administration of cholecystokinin is a validated method of inducing fear and anxiety in research participants. However, subjective levels of fear or anxiety were measured in this study.

In a study by Wang and colleagues[59] a sample of 614 postoperative patients were offered morphine intravenously infused via a patient-controlled analgesia (PCA) technique. The subjects were randomized into four groups differing in the information given about the PCA treatment: a no-information group, a positive-information group, a partially negative information group which received information that the pump had only a limited effect, and a totally negative information group where the patients were told that the pump had no effect. The results showed that negative information increased pain intensity and stress (plasma cortisol) as well as morphine use.

In the studies of Zubieta and colleagues,[60] μ-opioid receptor binding was monitored during sustained neutral and sad mood situations in a sample of healthy volunteers. The results showed a reduction in μ-opioid neurotransmission (i.e. increases in μ-receptor availability) from the neutral to the sadness condition in several limbic brain areas. The reduction in opioid neurotransmission also correlated with self-reported increases in negative affect and decreases in positive affect. This finding was later replicated twice by the same research group.[61,62] These findings suggest that endogenous opioid tone is reduced by negative emotion. Thus, negative emotions, exemplified by fear, anxiety and sadness, seem to increase pain and reduce the effectiveness of both exogenous and endogenous opioids. This finding is supported by reviews on patient-controlled analgesia techniques and on nocebo hyperalgesia, in which the overall conclusion is that the hyperalgesic and anti-opioid effects of negative emotions compromises the analgesic effect of painkillers.[63,64]

PLACEBO ANALGESIA, EMOTIONS, AND OPIOID ACTIVITY

The finding that relaxation training is mediated via opioid mechanisms fits well with the present notion of placebo analgesia as mediated by emotional modulation. Increased opioid activity has been hypothesized to increase positive emotions and/or to decrease negative emotions, as opioids are implicated in sex,[65] food consumption and use of addictive drugs,[66] as well as in pain relief,[67] all of which activate positive emotions. In a positron emission tomography (PET) study, Koepp et al[68] induced positive emotions via film, music and positive statements, and contrasted this with neutral emotions. They found reduced binding of a mu-receptor agonist while the subjects reported positive emotions, indicating that positive emotions were related to increased opioid activity. Moreover, a number of studies have shown that placebo analgesic responding is mediated via opioid release. Levine and Gordon[69] showed that an injection of naloxone partly reversed the placebo analgesic effect, and other studies have shown similar results.[70,71] Brain-imaging studies have shown activation of the descending pain-inhibitory system in the brain and in the spinal cord during placebo analgesia.[72,73] These findings indicate that placebo analgesia is partly due to reduced pain transmission at the level of the spinal cord. Of special interest is a study by McCubbin et al[74] in which one group of males with mildly elevated mean arterial blood pressure were subjected to relaxation training, whereas a control group did not receive such training. The group that received relaxation training displayed decreased blood pressure reactivity to a mental stressor compared with the group that did not receive training. Naltrexone, an opioid antagonist, antagonized the effect of relaxation training, showing that the effect of relaxation training was mediated by opioid mechanisms. Taken together, these findings suggest that opioid mechanisms are involved in the modulation of emotions, as well as in placebo analgesia.

THE NOCEBO RESPONSE: NEGATIVE PLACEBO EFFECT OR SEPARATE PROCESS?

Expectations that a procedure is painful, or if information is provided that a stimulus will be even more painful, has been found to increase pain.[58,75] Hence, the expectation that pain will occur or will increase induces negative emotions like nervousness or fear, which increase pain. Thus, placebo and nocebo responses may be seen as opposite results depending on induction of positive or negative emotions, respectively. The nocebo is dependent on the anxiety-inducing content of the information, which is opposite to that of the placebo effect that is hypothesized to be dependent on the anxiety-reducing content of the information.

Information that a substance or a procedure increases pain, increases stress as measured by cortisol[14] (Fig. 8.3). Induction of fear or anxiety increases pain. Schweiger and Parducci[75] told healthy subjects that an electrical current was being passed through their heads, whereas no stimulation was administered, and found that most subjects responded with feelings of pain.

Benedetti et al[71] showed that administration of proglumide, a cholecystokinin antagonist, abolished the nocebo response of increased pain. Cholecystokinin is a peptide that acts as an opioid antagonist; it induces subjective and physiologic stress, and thus increases pain. Administration of proglumide enhanced the placebo analgesic response. Thus, the data are consistent with the hypothesis that a reduction in stress reduced the level of pain, which is similar to a placebo analgesic effect. An experimental situation where pain is induced is stressful to the participants, and by reducing some of the stress a moderating effect on pain is obtained. It is not clear whether administration of proglumide increased positive feelings, via indirect effects on the endogenous opioid system, or whether proglumide decreased negative feelings, via its effect on cholecystokinin. However, emotional modulation seems to underlie both the placebo and nocebo responses that are correlated with activity at the opposite ends of a continuum of emotional valence. When placebos change emotional valence in a positive direction, placebo effects are observed. When placebos change emotional valence in a negative direction, nocebo effects are the result. However, negative emotions have been associated with stress-induced analgesia, even when arousal levels are not very high.[76] Thus, the effect of negative emotions on pain seems to depend on factors other than emotional valence and arousal. It has been suggested, for example, that fear or stress, on the one hand, and anxiety on the other, may have opposite effects on pain.[34] Others have suggested that negative emotions that are not pain-relevant may decrease pain, whereas pain-relevant emotions may increase pain.[1,77] This issue is still debated.

CLINICAL IMPLICATIONS

A clinical implication of the finding that negative emotions reduce or abolish placebo analgesia becomes evident in light of research that demonstrates the additive impact of placebos on conventional analgesics. Placebo effects on pain, due to expectations of pain relief, conditioning, or both, are an integrated part of analgesic treatment. Hidden administrations of analgesics have the potential to reveal such additive effects. The purpose of hidden treatments is to eliminate the effects of psychologic mechanisms on analgesic outcome, e.g. the patient does not expect pain to be reduced under hidden administration. In contrast, open administration implies that the patient knows that a painkiller is administered, and when it is administered. Levine and Gordon[69] examined the effects of open versus hidden treatments. They found that an open administration of placebo was as effective as about 8 mg of hidden administration of morphine. In this experiment the investigators had to increase the hidden morphine to 12 mg before the effect of morphine became significantly larger than the placebo treatment. Levine and Gordon,[69] and later studies, demonstrated that hidden administration of drugs reduces the total therapeutic impact by preventing the inhibitory effect of expectations.[78,79]

Thus, in terms of the detrimental effects of fear on placebo analgesia, preventing fear and negative affect is one way in which treatment outcomes and patient care can be optimized. Different approaches of how to buffer negative affective states, either by working with them directly (i.e. psychotherapy, relaxation, medication) or indirectly (i.e. providing conditions for positive affect) should be a primary goal for all health-care personnel.

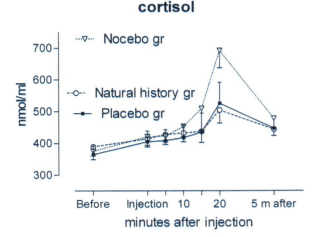

FIGURE 8.3 Mean plasma cortisol in three groups in which pain was induced and the participants received an injection with three types of information: one group of participants received no information about the injection; another group was told that the injection would reduce pain (placebo group); and a third group was told that the injection would increase pain (nocebo group). Information that the injection increased pain led to an increase in cortisol 20 minutes after the injection. Reprinted from Johansen et al[14] with permission from the publisher.

CONCLUSION AND FUTURE PERSPECTIVES

Studies described in this chapter have demonstrated that the inhibitory effect of the placebo interventions on pain is not sufficient to cause a decrease in pain for those high in FoP or negative affect.[2,3,32–34,47] Other studies that have demonstrated that negative affects cause a deactivation in mu-opioid receptors (i.e. decrease in opioid binding) and increased opioid consumption, suggest that negative affects interfere with opioid neurotransmission. Lyby et al[34] and Benedetti et al[57] directly demonstrated that fear, and neural correlates of fear, abolish placebo analgesia.

The experimental designs used by Lyby et al[34] and Benedetti et al[57] represent operationalizations of simultaneous inhibitory (i.e. placebo) and facilitatory (i.e. negative emotions/cholecystokinin) activation. The finding that fear and a cholecystokinin agonist abolished placebo analgesia suggests that the facilitatory effect of fear and cholecystokinin canceled out the inhibitory effect of the placebo intervention. This interpretation is in line with converging evidence from both animal and human studies which demonstrate that the two divisions of the pain-modulatory system can be activated at the same time and to different degrees, independently of one another (i.e. blocking one system will not affect activity in the other).[67,80,81] This means that it is the end result of both (i.e. parallel) inhibitory and facilitatory activation that determines the direction and the degree of modulation. This perspective, primarily drawn from the study of placebo non-responders, may provide a more nuanced way of understanding placebo responding and the role of emotional modulation in placebo responding.

References

1. Price DD. *Psychological Mechanisms of Pain and Analgesia*. Seattle: International Association for the study of Pain Press; 1999.
2. Vase L, Robinson ME, Verne GN, Price DD. The contributions of suggestion, desire, and expectation to placebo effects in irritable bowel syndrome patients – an empirical investigation. *Pain*. 2003;105:17-25.
3. Vase L, Robinson ME, Verne GN, Price DD. Increased placebo analgesia over time in irritable bowel syndrome (IBS) patients is associated with desire and expectation but not endogenous opioid mechanisms. *Pain*. 2005;115:338-347.
4. Flaten MA, Aslaksen PM, Finset A, et al. Cognitive and emotional factors in placebo analgesia. *J Psychosom Res*. 2006;61:81-89.
5. Flaten MA, Aslaksen PM, Lyby PS, Bjorkedal E. The relation of emotions to placebo responses. *Phil Trans Roy Soc B Biol Sci*. 2011;366:1818-1827.
6. Rhudy JL, Meagher MW. Fear and anxiety: divergent effects on human pain thresholds. *Pain*. 2000;84:65-75.
7. Keogh E, Cochrane M. Anxiety sensitivity, cognitive biases, and the experience of pain. *J Pain*. 2002;3:320-329.
8. Rhudy JL, Meagher MW. Individual differences in the emotional reaction to shock determine whether hypoalgesia is observed. *Pain Med*. 2003;4:244-256.
9. Roy M, Piche M, Chen JI, et al. Cerebral and spinal modulation of pain by emotions. *Proc Natl Acad Sci USA*. 2009;106:20900-20905.
10. Zillmann D, deWied M, KingJablonski C, Jenzowsky S. Drama-induced affect and pain sensitivity. *Psychosom Med*. 1996;58:333-341.
11. Meagher MW, Arnau RC, Rhudy JL. Pain and emotion: effects of affective picture modulation. *Psychosom Med*. 2001;63:79-90.
12. Villemure C, Slotnick BM, Bushnell MC. Effects of odors on pain perception: deciphering the roles of emotion and attention. *Pain*. 2003;106:101-108.
13. Rainville P, Bao QVH, Chretien P. Pain-related emotions modulate experimental pain perception and autonomic responses. *Pain*. 2005;118:306-318.
14. Johansen O, Brox J, Flaten MA. Placebo and nocebo responses, cortisol, and circulating beta-endorphin. *Psychosom Med*. 2003;65:786-790.
15. Rhudy JL, Meagher MW. The role of emotion in pain modulation. *Curr Opin Psychiatry*. 2001;14:241-245.
16. Rhudy JL, Williams AE, McCabe KM, et al. Emotional control of nociceptive reactions (ECON): do affective valence and arousal play a role? *Pain*. 2008;127:250-261.
17. McGlashan TH, Evans FJ, Orne MT. Nature of hypnotic analgesia and placebo response to experimental pain. *Psychosom Med*. 1969;31:227-246.
18. Staats PS, Staats A, Hekmat H. The additive impact of anxiety and a placebo on pain. *Pain Med*. 2001;2:267-279.
19. Wager TD, Rilling JK, Smith EE, et al. Placebo-induced changes in fMRI in the anticipation and experience of pain. *Science*. 2004;303:1162-1167.
20. Kong J, Gollub RL, Rosman IS, et al. Brain activity associated with expectancy-enhanced placebo analgesia as measured by functional magnetic resonance Imaging. *J Neurosci*. 2006;26:381-388.
21. Craggs JG, Price DD, Verne GN, et al. Functional brain interactions that serve cognitive-affective processing during pain and placebo analgesia. *Neuroimage*. 2007;38:720-729.
22. Petrovic P, Dietrich T, Fransson P, et al. Placebo in emotional processing – induced expectations of anxiety relief activate a generalized modulatory network. *Neuron*. 2005;46:957-969.
23. Verne GN, Robinson ME, Vase L, Price DD. Reversal of visceral and cutaneous hyperalgesia by local rectal anesthesia in irritable bowel syndrome (IBS) patients. *Pain*. 2003;105:223-230.
24. Price DD. Neuroscience – psychological and neural mechanisms of the affective dimension of pain. *Science*. 2000;288:1769-1772.
25. Price DD, Barrell JE, Barrell JJ. A quantitative-experiential analysis of human emotions. *Motiv Emot*. 1985;9:19-38.
26. Petersen GL, Finnerup NB, Nørskova KN, et al. Placebo manipulations reduce hyperalgesia in neuropathic pain. *Pain*. 2012;153:1292-1300.
27. Scott DJ, Stohler CS, Egnatuk CM, et al. Individual differences in reward responding explain placebo-induced expectations and effects. *Neuron*. 2007;55:325-336.
28. Wager TD. The neural bases of placebo effects in pain. *Curr Dir Psychol Sci*. 2005;14:175-179.
29. Pollo A, Vighetti S, Rainero I, Benedetti F. Placebo analgesia and the heart. *Pain*. 2003;102:125-133.
30. Aslaksen PM, Flaten MA. The roles of physiological and subjective stress in the effectiveness of a placebo on experimentally induced pain. *Psychosom Med*. 2008;70:811-818.
31. Aslaksen PM, Bystad M, Vambheim SM, Flaten MA. Gender differences in placebo analgesia: event-related potentials and emotional modulation. *Psychosom Med*. 2011;73:193-199.
32. Lyby PS, Aslaksen PM, Flaten MA. Is fear of pain related to placebo analgesia? *J Psychosom Res*. 2010;68:369-377.
33. Lyby PS, Aslaksen PM, Flaten MA. Variability in placebo analgesia and the role of fear of pain-an ERP study. *Pain*. 2011;152:2405-2412.
34. Lyby PS, Åsli O, Forsberg JT, Flaten MA. Induced fear reduces the effectiveness of a placebo intervention on pain. *Pain*. 2012;153(5):1114-1121.
35. Craig KD. Emotional aspects of pain. In: Wall PD, Melzack R, eds. *Textbook of Pain*. New York: Churchill Livingstone; 1994:261-274.
36. Benedetti F, Amanzio M, Vighetti S, Asteggiano G. The biochemical and neuroendocrine bases of the hyperalgesic nocebo effect. *J Neurosci*. 2006;26(46):12014-12022.

37. Hrobjartsson A, Gotzsche C. Is the placebo powerless? An analysis of clinical trials comparing placebo with no treatment. *New Eng J Med.* 2001;344(21):1594-1602.
38. Bradley MM, Silakowski T, Lang PJ. Fear of pain and defensive activation. *Pain.* 2008;137(1):156-163.
39. Enck P, Benedetti F, Schedlowski M. New insights into the placebo and nocebo responses. *Neuron.* 2008;59:195-206.
40. Rhudy JL, Williams AE. Gender differences in pain: do emotions play a role? *Gend Med.* 2008;2:208-226.
41. Goffaux P, Michaud K, Gaudreau J, et al. Sex differences in perceived pain are affected by an anxious brain. *Pain.* 2011;152:2065-2073.
42. Linnman C, Beucke JC, Jensen KB, et al. Sex similarities and differences in pain-related periaqueductal gray connectivity. *Pain.* 2012;153:444-454.
43. Aslaksen PM, Vambheim SM, Bystad M, Vangberg TR. Imaging sex differences in placebo analgesia. Submitted.
44. Paulesu E, Sambugaro E, Torti T, et al. Neural correlates of worry in generalized anxiety disorder and in normal controls: a functional MRI study. *Psychol Med.* 2010;40:117-124.
45. Amanzio M, Benedetti F, Porro CA, et al. Activation likelihood estimation meta-analysis of brain correlates of placebo analgesia in human experimental pain. *Hum Brain Mapp.* 2011. doi:10.1002/hbm.21471.
46. Wager TD, Atlas LY, Leotti LA, Rilling JK. Predicting individual differences in placebo analgesia: contributions of brain activity during anticipation and pain experience. *J Neurosci.* 2011;31:439-452.
47. Morton DL, Brown CA, Watson A, et al. Cognitive changes as a result of a single exposure to placebo. *Neuropsychologica.* 2010;48(7):1958-1964.
48. Zubieta JK, Bueller JA, Jackson LR, et al. Placebo effects mediated by endogenous opioid activity on mu-opioid receptors. *J Neurosci.* 2005;25(34):7754-7762.
49. Phelps EA, O'Connor KJ, Gatenby JC, et al. Activation of the left amygdala to a cognitive representation of fear. *Nature Neurosci.* 2001;4(4):437-441.
50. Grillon C, Ameli R. Effects of threat and safety signals on startle during anticipation of aversive shocks, sounds, and airblasts. *J Psychophysiol.* 1998;12(4):329-337.
51. Lang PJ, Bradley MM, Cuthbert BN. Emotion, attention, and the startle reflex. *Psychol Rev.* 1990;97(3):377-395.
52. Åsli O, Kulvedrøsten S, Solbakken L, Flaten MA. Fear potentiated startle at short intervals following conditioned stimulus onset during delay but not trace conditioning. *Psychophysiol.* 2009;46:880-888.
53. Åsli O, Flaten MA. How fast is fear? Automatic and controlled processing after delay and trace fear conditioning. *J Psychophysiol.* 2012;26(1):20-28.
54. Flaten MA. A comparison of electromyographic and photoelectric techniques in the study of classical eyeblink conditioning and startle reflex modification. *J Psychophysiol.* 1993;7:230-237.
55. Funayama ES, Grillon C, Davis M, Phelps EA. A double dissociation in the affective modulation of startle in humans: effects of unilateral temporal lobectomy. *J Cogn Neurosci.* 2004;13(6):721-729.
56. Buchanan TW, Tranel D, Adolphs R. Anteromedial temporal lobe damage blocks startle modulation by fear and disgust. *Behav Neurosci.* 2004;118(2):429-437.
57. Benedetti F, Amanzio M, Thoen W. Disruption of opioid-induced placebo responses by activation of cholecystokinin type-2 receptors. *Psychopharmacol.* 2011;213(4):791-797.
58. Benedetti F, Lanotte M, Lopiano L, Colloca L. When words are painful: unraveling the mechanisms of the nocebo effect. *Neurosci.* 2007;147(2):260-271.
59. Wang F, Shen X, Xu S, et al. Negative words on surgical wards result in therapeutic failure of patient-controlled analgesia and further release of cortisol after abdominal surgeries. *Minerva Anesthesiol.* 2008;74:353-365.
60. Zubieta JK, Ketter TA, Bueller JA, et al. Regulation of human affective responses by anterior cingulate and limbic mu-opioid neurotransmission. *Arch Gen Psychiat.* 2003;60(11):1145-1153.
61. Kennedy SE, Koeppe RA, Young EA, Zubieta JK. Dysregulation of endogenous opioid emotion regulation circuitry in major depression in women. *Arch Gen Psychiat.* 2006;63(11):1199-1208.
62. Prossin AR, Love TM, Koeppe RA, et al. Dysregulation of regional endogenous opioid function in borderline personality disorder. *Am J Psychiat.* 2010;167(8):925-933.
63. Colloca L, Benedetti F. Nocebo hyperalgesia: how anxiety is turned into pain. *Curr Opin Anaesthesiol.* 2007;20(5):435-439.
64. Svedman P, Ingvar M, Gordh T. 'Anxiebo', placebo, and postoperative pain. *BMC Anesthesiol.* 2005;5:9.
65. Agmo A, Berenfeld R. Reinforcing properties of ejaculation in the male-rat: role of opioids and dopamine. *Behav Neurosci.* 1990;104:177-182.
66. Koob GF, Volkow ND. Neurocircuitry of addiction. *Neuropsychopharmacol.* 2010;35:217-238.
67. Fields H. State-dependent opioid control of pain. *Nature Rev Neurosci.* 2004;5(7):565-575.
68. Koepp MJ, Hammers A, Lawrence AD, et al. Evidence for endogenous opioid release in the amygdala during positive emotion. *Neuroimage.* 2009;44:252-256.
69. Levine JD, Gordon NC. Influence of the method of drug administration on analgesic response. *Nature.* 1984;312:755-756.
70. Grevert P, Albert LH, Goldstein A. Partial antagonism of placebo analgesia by naloxone. *Pain.* 1983;16:129-143.
71. Benedetti FB. The opposite effects of the opiate antagonist naloxone and the cholecystokinin antagonist proglumide on placebo analgesia. *Pain.* 1996;64:535-543.
72. Eippert F, Bingel U, Schoell ED, et al. Activation of the opioidergic descending pain control system underlies placebo analgesia. *Neuron.* 2009;63:533-543.
73. Eippert F, Finsterbusch J, Bingel U, Büchel C. Direct evidence for spinal cord involvement in placebo analgesia. *Science.* 2009;326:404.
74. McCubbin JA, Wilson JF, Bruehl S, et al. Relaxation training and opioid inhibition of blood pressure response to stress. *J Consult Clin Psychol.* 1996;64:593-601.
75. Schweiger A, Parducci A. Nocebo: the psychological induction of pain. Pavlovian. *J Biol Sci.* 1981;16:140-143.
76. Flor H, Grüsser SM. Conditioned stress-induced analgesia in humans. *Eur J Pain.* 1999;3(4):317-324.
77. al Absi M, Rokke PD. Can anxiety help us tolerate pain? *Pain.* 1991;46(1):43-51.
78. Pollo A, Amanzio M, Arslanian A, et al. Response expectancies in placebo analgesia and their clinical relevance. *Pain.* 2001;93(1):77-84.
79. Benedetti F, Carlino E, Pollo A. Hidden administration of drugs. *Clin Pharmacol Ther.* 2011;90(5):651-661.
80. Crown ED, Grau JW, Meagher MW. Pain in a balance: noxious events engage opposing processes that concurrently modulate nociceptive reactivity. *Behav Neurosci.* 2004;118(6):1418-1426.
81. Heinricher MM, Neubert MJ. Neural basis for the hyperalgesic action of cholecystokinin in the rostral ventromedial medulla. *J Neurophysiol.* 2004;92(4):1982-1989.

CHAPTER 9

Placing Placebo in Normal Brain Function with Neuroimaging

Martin Ingvar[1], Predrag Petrovic[1], Karin Jensen[2]

[1]Cognitive Neurophysiology Research Group, Stockholm Brain Institute, Osher Center for Integrative Medicine, Karolinska Institutet, Stockholm, Sweden, [2]Department of Psychiatry, Massachusetts General Hospital/Harvard Medical School, Charlestown, MA, USA

The literature on placebo mechanisms is diverse, and many questions remain to be answered before a comprehensive understanding may reach general acceptance. Here, we argue that the neural mechanisms underlying placebo responses must be understood in the context of a model-driven approach to brain function, i.e. using the methods of mainstream cognitive neuroscience.

The early literature on placebo is concentrated on the magic of doctors' actions in medical rituals and the use of inert treatments such as 'placebo pills'. As the power of placebo effects reached a wider understanding, clinical medicine was presented with multiple examples of scientifically unfounded hopes of placebo treatments as powerful tools in clinical practice.[1] This early phase led to a longstanding debate whether a placebo effect is real or not[2] and whether the use of placebo in clinical medicine is ethical or not. Today, there is a widespread use of deceptive placebos in clinical medicine[3] in spite of the ethical problems connected to such practice.[4] The debate regarding deceptive usage of placebos is still active and has presented a hurdle to cross for placebo research.

The second phase of placebo research has contributed to the inclusion of inactive treatment in drug trials so as to handle the nuisance effects of the placebo response. In our early understanding, the placebo response was considered to be a non-specific nuisance effect separated from specific drug effects. In its most simple form, using a generalized model for many placebo effects, a placebo response is represented by a linear additive model, i.e. there is a fixed response constant over time for a certain treatment that may be separated into (1) specific treatment effects and (2) the nonspecific placebo effects. This model is still the basis for randomized controlled trials (RCTs) aimed at drug development and the way authorities and industries generally assess the efficacy of new drugs. The linear additive model may be seen as overly simplistic and fails to recognize the dynamics of the placebo phenomenon. The stubborn use of this model could potentially harm the estimations of real drug effects by decreasing the sensitivity of the testing procedures. The lack of dynamic estimations of placebo–drug interactions may contribute to the failure of many clinical trials and the lack of innovation in pharmacologic treatment of e.g. chronic pain and depression.

The third phase of placebo research has been characterized by groundbreaking discoveries regarding the neural mechanisms of placebo responses, facilitated by the advances in cognitive neuroscience and access to functional neuroimaging. The important first evidence that placebo mechanisms could be the target for systematic studies, and could be biomedically anchored, was provided by the seminal study by Levine and colleagues more than two decades before the advent of functional neuroimaging.[5] Levine and colleagues demonstrated, in an elegant concept-driven design, that placebo analgesia could rationally be tied to the function of endorphine release. Since then, many behavioral studies have validated the original findings by Levine and co-workers, and some have added new information about the neurobiology of placebos. However, the original study by Levine and colleagues provided an essential conceptual framework that has, since then, formed a foundation for current theories of placebo mechanisms.

An important step towards the understanding of placebo mechanisms was taken when some early accounts of a functional anatomy of placebo responses were published.[6–8] Since then, the neuroscientifically based research has grown tremendously and contributed to important knowledge about the brain processes involved in placebo responses.

In summary, three important lines of scientific progress have taken place and moved the placebo phenomenon into an important subject for neuroscience. Imaging has added information on the functional anatomic basis for different behavioral findings in placebo studies. Thereby, the proposed models of placebo mechanisms have been integrated into the framework of cognitive regulation, offering a more meaningful interpretation as to why a placebo phenomenon may occur. The evolutionary drive for adaptability of the individual has been essential for human brain development. In spite of its billions of cells, the constant activity, and its massively parallel construction and self-organizing properties, the brain is energy efficient.[9,10] This computational efficiency, and thereby energy efficiency, is upheld by general mechanisms that may apply directly to our understanding of placebo effects.

A descriptive quote from Steven Pinker[11] nicely illustrates some of the problems we face when trying to connect intuitive models of brain function with placebo theories:

> 'The main lesson of thirty-five years of artificial intelligence (AI) research is that the hard problems are easy and the easy problems are hard. The mental abilities of a four-year-old that we take for granted – recognizing a face, lifting a pencil, walking across a room, answering a question – in fact solve some of the hardest engineering problems ever conceived... As the new generation of intelligent devices appears, it will be the stock analysts and petrochemical engineers and parole board members who are in danger of being replaced by machines. The gardeners, receptionists, and cooks are secure in their jobs for decades to come'.

Pinker referred to Moravec's paradox, stating that it is the high-level reasoning preformed by AI and robotics researchers that, contrary to traditional assumptions, requires very little computation, whereas low-level sensorimotor skills require enormous computational resources. Placebo responses mainly pertain to the latter and thereby require a sophisticated approach to be understood.

The brain is a complex system, i.e. it is composed of interconnected parts that, taken together, exhibit one or more properties not obvious from the properties of its individual parts. At the same time, local modules can hold several functions and this economizes with wiring and thereby with energy and time.[12] The representation of information in the brain network is related to the local network structure, and thus any information fed into the module will be handled in relation to the present network status. Several theories now point in the direction that any internal representations (models) of the external world are performed directly in the systems for perception. Grounded theories on cognition (GTC) are gaining popularity with their ability to house the need for internal representation models of the environment directly bound to functional units. Furthermore, GTC allows for computational sparseness and a decreased need for synchrony in time. GTC also assumes that there is no central module for cognition.[13] According to this view, all cognitions, including amodal cognitions such as reasoning, numeric, and language processing, emerge from a combination of bodily, affective, perceptual, and motor processes. GTC posits dynamic brain–body–environment interactions and perception–action links as the common bases of simple behaviors as well as complex cognitive and social skills, without representational separations between these domains.[14,15] An example of this reasoning can be found in the discussion on mirror neurons.[16] Taken together, these theories of the brain's continuous internal representation and integration of external events create a foundation for a neurocognitive explanatory model of placebo responses. Based on GTC, the model points to the general principle of local discrimination between expected value and input, and any recorded discrepancies would be the output that needs to be accounted for.[17] For example, the mismatch between an expected and perceived painful stimulus creates a discrepancy in the neural representation of that event. According to this model, the brain's internal representation of the painful stimulus can be adjusted and accounted for by altered top-down modulation of the nociceptive input.

The core concept of the perception–action loop as the major mechanism for non-declarative learning (unconscious, implicit skills) has, together with the development of the GTC concept, moved to entail all feedback systems in the brain, including those of emotional regulation. For example, an important addition to our understanding of pain regulation comes from the neuroanatomically based concepts developed by Craig, describing pain as a homeostatic emotion.[18] According to Craig, homeostatic regulation of pain builds its activity on a mix of information based on nociception and expectations of nociceptive input, largely working in a top-down manner.

The concept of homeostatic feedback loops is also compatible with one of the most influential model systems in computational neuroscience, namely the adaptive resonance theory (ART) and its developments.[19] This theoretical framework has the distinct property of allowing decisions based on sparse information, and thereby, to

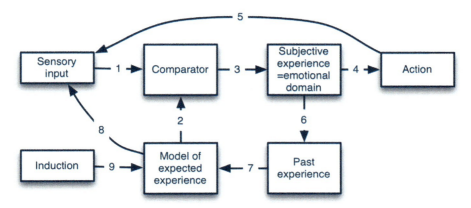

FIGURE 9.1 The sensory input is fed directly to the model comparator (1) that compares the input with the internal representation of the expected information (the expectancy model) (2). The resulting subjective experience (3) then feeds into action (4), its feedback (5) and memory (6) that, in turn, influences the expectancy model (7). This information content has, in certain systems, also a possibility of directly influencing lower order systems via top-down influence (8). The model may also be directly influenced by induced changes (9) of e.g. context or other perturbations, such as a placebo induction.

be energy effective. The latter is of highest significance for any model that claims biologic relevance. A minimal model based on the above is illustrated in Fig. 9.1.

This model-driven understanding of the placebo concept automatically generates a number of testable hypotheses, and in this chapter we exemplify findings from neuroimaging according to the suggested model, referred to with arrow numbers from Fig. 9.1.

In the suggested model, all sensory input is compared to the expectancy of that particular input, directly in the primary input region, and any deviance will incur an activation (arrows 1 and 2). Any sensory activation could potentially invoke model comparisons, not only in the sensory regions, but also in motor regions downstream in the perception–action loop (arrows 3 and 4). Thus, an activation resulting from a placebo manipulation should be particularly expressed in the region where the comparison is made and in functional units downstream. Indeed, the original finding of changes both in the rostral anterior cingulate cortex (ACC) and its connected regions in the lateral orbitofrontal and anterior insular cortices[8] has been re-reported many times and thus seems to be a robust finding. The interpretation is that the activation in the rostral ACC/lateral orbitofrontal cortex is part of the network for mismatch detection and regulation. The finding is compatible with the proposed model of the comparison of sensory input and expectation in Fig. 9.1 (arrows 1, 2 and 3, 4).

Artificial stimuli also have the capacity to incur changes in brain regions as a response to inconsistencies between sensory input and sensory expectation. The thermal grill illusion[20] is a perfect example where an expected pattern of thermal stimuli is violated, which results in an awkward painful experience paired with thalamic hypoactivity.[21,22] The illusion is a result of an unexpected thermal sensation by using a grill where every other grid element is warm and the others cool. It is likely that the unpleasantness alarm is set off due to the inadequacy of our inner model of what to expect.

In order to achieve an agreement between expected and actual input, either one can be adjusted. The expectancy can be changed by feedback from the output (experience) (arrows 3-6-7). Thus, upon repeated sensory stimuli, the primary sensory response should wain quickly as a result of adaptation in all components of the sensory system.[23] In random passive movements such adaptation is limited. For example, if somebody else moves your finger or toe this limits predictions in space and time and therefore upsets the comparison system with increased activations.[24] A placebo perturbation (arrow 9) of the internally expected input might, by corresponding mechanisms, also wain with time irrespective of whether it is an induction of placebo or nocebo (negative placebo).[25] This finding is, in turn, a clear pointer towards the difficulties of accepting some placebo assumptions in drug trials, positing that placebo is additive to a specific treatment effect and more or less constant over time.

Conversely, model preparedness for a change that ultimately does not take place will also evoke an increased load and computational cost. We tested this in a paradigm where a sensory stimulus (tickling) was administered in a random manner. Subjects believed that each stimulus was preceded by a visual cue, but following the induction of the expectation (the cue) the experiment was varied so that the visual cue was not always followed by the tickle. The findings demonstrated how a comprehensive representation of the tickle sensation was in preparedness, even in the absence of the sensory input, illustrating the power of an inner predictive model with strong priors.[26] In an early study we demonstrated how inhibition of motor function could be elucidated with positron emisson tomography (PET). We induced itch, with all its motivational drive to scratch, but gave strict verbal instruction not to do so.[27] The motor program for scratching is presumably well encoded, and hence, a number of brain regions connected to the actual behavior of scratching were all activated.

Our theoretical model posits that the mechanism of top-down regulation should be possible to invoke by instruction or manipulations of context (arrows 9 and 8). In an experimental pain experiment, we manipulated the instructions and told the research subjects that the painful stimulus (cold pressure test) would either last 60 or 120 seconds. Unknowingly to the subjects, all measurements were made during the first minute. Thus, the nociceptive input was constant, but the expectations were different due to the information that some sessions would include a more prolonged pain stimulus. The main finding was a suppressed activity bilaterally in the amygdala, which was interpreted as mirroring the adaptive response and the need to suppress the emotional response in order to endure the experimental situation.[28]

A specific trait of all different forms of adaptation (learning) is the variation of the time span for different forms of learning. While social learning often involves evolutionarily facilitated mechanisms for rapid learning, other forms of learning, such as skill learning, may need robust learning efforts with multiple repetitions to reach the desired level. The understanding of placebo as a modification of expectations necessitates the possibility to quickly update and refresh one's expectations. A number of scientific articles have demonstrated such an update of expectancies, and one of the first studies of sensory expectation[29] elegantly demonstrated this.

The understanding of adaptability is connected to the concept of pattern recognition in space and time. A prediction from our general model, as presented in Fig. 9.1, is that withholding information will incur a computational cost as this invokes uncertainty and a possible mismatch between the expected and actual sensory input. We tested this by demonstrating how a time-locked visual and sensory stimulus quickly allowed for associative learning, and the generation of a robust expectancy model. When the time lock was broken, the model for predictability was incapacitated and the cost was noted both in sensory regions and regions associated with timing and pattern recognition.[30]

Placebo designs for brain imaging should ideally be anchored in cognitive theory or they run the risk of becoming phenomenologic in their interpretations, with poor explanatory power of the biologic mechanisms underlying placebo responses. Given the general concept that the model in Figure 9.1 presents, a placebo response should be possible to invoke in any system characterized by homeostatic regulation, e.g. in all emotional domains. Our first generalization outside the domain of pain regulation was an experiment where we manipulated the expected emotional reactions to images with aversive content (so-called IAPS images).[31] We induced expectations by suggesting that we could lower the level of anxiety in response to the images by injection of a benzodiazepine, and conversely, remove this effect by a blocking antagonist. In line with the instructions, subjects were injected with real benzodiazepine and with the blocker (Lanexate (r)) while rating the invoked unpleasantness of the pictures. As expected, the rating decreased during benzodiazepine injection and increased following the injection of the blocker. On Day 2 the subjects returned and were studied with functional magnetic resonance imaging (fMRI). The same scripted instructions were given as day one, but all the injections were exchanged for saline. The fMRI investigation demonstrated that the same modulatory network, including the rostral ACC and the lateral orbitofrontal cortex, was involved in the anxiolytic placebo effect as in our previous findings in placebo analgesia. However, the top-down target of manipulation was situated, as the theory would predict, in the extrastriatal visual cortices and the amygdala.[32] The relevance of this finding was confirmed in a recently published fMRI study, using a paradigm where we studied top-down regulation of aversive pictures in response to explicit instructions of cognitive control. Subjects were simply given the instruction to reappraise the emotional content to its fullest emotional expression (take in) or to reappraise with suppression (hold back). The latter induced a top-down reappraisal mechanism based in the dorsolateral prefrontal cortex, both for unpleasant and neutral picture content, whereas the lateral orbitofrontal cortex selectively changed its activity in response to reappraisal of pictures with negative emotional content. Notably, this paradigm did not include any perturbation of the expectancy model and therefore no need for regulation between the prior and the sensory input which might explain why no activation was incurred in the rostral ACC.[33]

Another important generalization of the theoretical models of placebo stems from the work by Schedlowski and colleagues. They have presented systematic studies of manipulations of expectancy in regard to the immune reflex, as described in the seminal work by Tracey and others.[34,35] Tracey and co-workers have elegantly demonstrated the nervous system regulatory mechanisms of e.g. macrophage activation in the periphery, and the existence of both afferent and efferent neuronal mechanisms. This mechanism seems possible to manipulate by means of psychologic treatment interventions.[36] Thus, as given in the general model (Fig. 9.1), the inflammatory response, as mirrored in e.g. levels of cytokines, is possible to manipulate by means of cognitive representations of induced expectations.[37] The induction of a response is not static; it is dependent on several factors that together determine the intensity of the induction. A verbal instruction is often not enough, as the activated mechanisms belong to an unaware physiologic domain.[38] The clinical relevance, seen as the possibility of translating these findings into benefits for patients in clinical practice, may thus far not be fully understood. Yet, there are several important clues in the literature to the importance of this mechanism.

The massively parallel structure of the brain has a hierarchical organization along the neural axis. We hypothesize that a placebo response may be instantiated at any level of the neural axis, as all levels carry the same principle of adjusting signalling between an internal comparator and priors. Thus, any level of the neural adaptive system could potentially be a target for a placebo manipulation. By generalizing placebo responses to an induced mismatch between an expected and actual input, it should be possible to get a placebo response also by induction via the unaware domain. A recent study[39] demonstrated the ability not only to use implicit learning for induction of a placebo response, but importantly also to evoke this learnt association via stimuli of which the subject was unaware. The method used in this experiment was that of backward masking of visual stimuli, paired with painful stimuli. Interestingly, the application of a standardized pain intensity varied in subjective experience according to the preceding conditioning that had been made, even if subjects were not aware of the conditioned cues they were presented with. The placebo response can thus be activated by learning mechanisms that operate outside of conscious awareness, possibly through subcortical neural circuits. Conversely, our proposed model posits that if a cognitive instruction preceds a placebo response, this should have an instantiation in the frontal cortex. In a re-analysis[40] of our previous data[8,32] we were able to demonstrate a separation between the neural correlates to placebo vs. specific drug treatment. While the effects in the rostral ACC were activated in both treatment conditions, the orbitofrontal activation was only found during placebo. In a direct challenge of the difference between placebo and drug analgesia, the finding of separable frontal activations related to expectancies was replicated.[41] This points to the great importance of furthering studies where pharmacologic and psychologic treatments are combined in the treatment of chronic pain, possibly elucidating segregated neural mechanisms, or interactions, that have instrumental value for the development of new treatments.

Placing placebo mechanisms into an adaptive general brain function brings on the hypothesis that placebo responses could possible be detected along the effector axis of any appropriate system (arrow 8). As an example, in our first paper, we could demonstrate that the placebo response incurred system-appropriate top-down regulation at the level of the brainstem, including the periaqueductal gray (PAG) and medulla.[8] Several other studies have replicated this,[42] and correspondingly direct imaging of the opiate systems adds confirmation to this conclusion.[43,44]

An important addition to the understanding of system-relevant placebo responses was made when a proper and reproducible method to study fMRI pain activations was developed for the spinal cord.[45] That team has directly demonstrated the top-down effects of a placebo analgesia manipulation in the spinal cord[46] and that this response is modulated the same way as other pain modulatory interventions, such as distraction.[47]

Because a placebo response is possibly instantiated in the same system as the underlying process it is affecting, it is expected that the dynamics of placebo is reflected in the dynamics of that specific response. For example, the better the response to a drug, the better the placebo response. This has been repeatedly reported[8,32,39] and is therefore no surprise. Importantly, variations in the intensity of the induction of the placebo have been reported to also influence the activity in the primary region for coding of the regulation of the mismatch between expectancy and sensory input, represented in the rostral ACC.[48]

When placebo is put into the general framework of cognitive neuroscience it reduces the mystic view, and a picture emerges of placebo research as an area that may contribute to the understanding of brain function in general, and emotional regulation in particular. Another exciting area is the further developments in neuropsychoimmunology, where several tools now allow us to further operationalize the mechanisms of the interaction between bodily systems.

Here, we have established a foundation of placebo in learning theories. Knowing that cognitive behavioral therapy, and other modern psychologic interventions, are anchored in learning theory, it is possible to see that lessons from placebo research may be brought into the understanding of the brain mechanisms of effects of such therapies. Indeed, we have demonstrated the ability to invoke a cortically driven top-down control function of pain following treatment with behavioral therapy in subjects with chronic pain.[49] This was indicated by an increased activity in the lateral frontal cortex upon pain provocation.

In summary, we have proposed an argument that placebo effects are the responses of an inherent brain mechanism that uses expectancy values as a mode to minimize discrepancies and computational load.

Acknowledgments

The authors' work is supported by The Swedish Research Council 2012-1999, 2009-3191, The Osher Center for Integrative Medicine, Karolinska Institutet, Stockholm County Council and the Knut and Alice Wallenberg Foundation, the Swedish Society for Medical Research and the Swedish Council for Working Life and Social Research.

References

1. Harrington A. *The Placebo Effect: an Interdisciplinary Exploration*. Boston, MA: Harvard University; 1997.
2. Shetty N, Friedman JH, Kieburtz K, et al. The placebo response in Parkinson's disease. Parkinson Study Group. *Clin Neuropharmacol*. 1999;22(4):207-212.
3. Tilburt JC, Emanuel EJ, Kaptchuk TJ, et al. Prescribing 'placebo treatments': results of national survey of US internists and rheumatologists. *BMJ*. 2008;337:a1938.
4. Bostick NA, Sade R, Levine MA, Stewart Jr DM. Placebo use in clinical practice: report of the American Medical Association Council on Ethical and Judicial Affairs. *J Clin Ethics*. 2008;19(1):58-61.
5. Levine JD, Gordon NC, Fields HL. The mechanism of placebo analgesia. *Lancet*. 1978;2(8091):654-657.
6. de la Fuente-Fernandez R, Schulzer M, Stoessl AJ. The placebo effect in neurological disorders. *Lancet Neurol*. 2002;1(2):85-91.
7. Mayberg HS, Silva JA, Brannan SK, et al. The functional neuroanatomy of the placebo effect. *Am J Psychiatry*. 2002;159(5):728-737.
8. Petrovic P, Kalso E, Petersson KM, Ingvar M. Placebo and opioid analgesia – imaging a shared neuronal network. *Science*. 2002;295(5560):1737-1740.
9. McClelland JL, Rumelhart DE, the PDP Research Group. *Parallel Distributed Processing: Explorations in the Microstructure of Cognition, vol. 2: Psychological and Biological Models*. Cambridge, MA: MIT Press; 1986.
10. Rumelhart D, McClelland J, the t.P.R. Group. *Parallel Distributed Processing: Explorations in the Microstructure of Cognition, vol. 1: Foundations*. Cambridge, MA: MIT Press; 1986.
11. Pinker S. *The Language Instinct*: New York, Harper; 1994.
12. Bullmore E, Sporns O. The economy of brain network organization. *Nat Rev Neurosci*. 2012;13(5):336-349.
13. Pezzulo G, Barsalou LW, Cangelosi A, et al. Computational grounded cognition: a new alliance between grounded cognition and computational modeling. *Front Psychol*. 2012;3:612.
14. Spivey M. *The Continuity of Mind*. New York, NY: Oxford University Press; 2007.
15. Barsalou LW. Grounded cognition. *Annu Rev Psychol*. 2008;59:617-645.
16. Rizzolatti G, Craighero L. The mirror-neuron system. *Annu Rev Neurosci*. 2004;27:169-692.
17. Friston K. Hierarchical models in the brain. *PLoS Comput Biol*. 2008;4:e1000211.
18. Craig AD. A new view of pain as a homeostatic emotion. *Trends Neurosci*. 2003;26(6):303-307.
19. Grossberg S. Cortical and subcortical predictive dynamics and learning during perception, cognition, emotion and action. *Philos Trans R Soc Lond B Biol Sci*. 2009;64(1521):1223-1234.
20. Craig AD, Bushnell MC. The thermal grill illusion: unmasking the burn of cold pain. *Science*. 1994;265(5169):252-255.
21. Craig AD, Bushnell MC, Zhang ET, Blomqvist A. A thalamic nucleus specific for pain and temperature sensation. *Nature*. 1994;372(6508):770-773.
22. Lindstedt F, Johansson B, Martinsen S, et al. Evidence for thalamic involvement in the thermal grill illusion: an FMRI study. *PLoS One*. 2011;6(11):e27075.
23. Wark B, Lundstrom BN, Fairhall A. Sensory adaptation. *Curr Opin Neurobiol*. 2007;17(4):423-429.
24. Shriver S, Knierim KE, O'Shea JP, et al. Pneumatically driven finger movement: a novel passive functional MR imaging technique for presurgical motor and sensory mapping. *AJNR Am J Neuroradiol*. 2013;34(1):E5-E7.
25. Colloca L, Petrovic P, Wager TD, et al. How the number of learning trials affects placebo and nocebo responses. *Pain*. 2010;151(2):430-439.
26. Carlsson K, Petrovic P, Skare S, et al. Tickling expectations: neural processing in anticipation of a sensory stimulus. *J Cogn Neurosci*. 2000;12(4):691-703.
27. Hsieh JC, Hagermark O, Stahle-Backdahl M, et al. Urge to scratch represented in the human cerebral cortex during itch. *J Neurophysiol*. 1994;72(6):3004-3008.
28. Petrovic P, Carlsson K, Petersson KM, et al. Context-dependent deactivation of the amygdala during pain. *J Cogn Neurosci*. 2004;16(7):1289-1301.
29. Drevets WC, Burton H, Videen TO, et al. Blood flow changes in human somatosensory cortex during anticipated stimulation. *Nature*. 1995;373(6511):249-252.
30. Carlsson K, Andersson J, Petrovic P, et al. Predictability modulates the affective and sensory-discriminative neural processing of pain. *Neuroimage*. 2006;32(4):1804-1814.
31. Lang PJ, Greenwald MK, Bradley MM, Hamm AO. Looking at pictures: affective, facial, visceral, and behavioral reactions. *Psychophysiology*. 1993;30(3):261-273.
32. Petrovic P, Dietrich T, Fransson P, et al. Placebo in emotional processing – induced expectations of anxiety relief activate a generalized modulatory network. *Neuron*. 2005;46(6):957-969.
33. Golkar A, Lonsdorf TB, Olsson A, et al. Distinct contributions of the dorsolateral prefrontal and orbitofrontal cortex during emotion regulation. *PLoS One*. 2012;7(11):e48107.
34. Tracey KJ. The inflammatory reflex. *Nature*. 2002;420(6917):853-859.
35. Andersson U, Tracey KJ. Reflex principles of immunological homeostasis. *Annu Rev Immunol*. 2012;30:313-335.
36. Jernelov S, Lekander M, Blom K, et al. Efficacy of a behavioral self-help treatment with or without therapist guidance for comorbid and primary insomnia – a randomized controlled trial. *BMC Psychiatry*. 2012;12:5.
37. Pacheco-Lopez G, Engler H, Niemi MB, Schedlowski M. Expectations and associations that heal: Immunomodulatory placebo effects and its neurobiology. *Brain Behav Immun*. 2006;20(5):430-446.
38. Albring AL, Wendt S, Benson O, et al. Placebo effects on the immune response in humans: the role of learning and expectation. *PLoS One*. 2012;7(11):e49477.
39. Jensen KB, Kaptchuk TJ, Kirsch I, et al. Nonconscious activation of placebo and nocebo pain responses. *Proc Natl Acad Sci USA*. 2012;109(39):15959-15964.
40. Petrovic P, Kalso E, Petersson KM, et al. A prefrontal non-opioid mechanism in placebo analgesia. *Pain*. 2010;150(1):59-65.
41. Atlas LY, Whittington RA, Lindquist MA, et al. Dissociable influences of opiates and expectations on pain. *J Neurosci*. 2012;32(23):8053-8064.
42. Eippert F, Bingel U, Schoell ED, et al. Activation of the opioidergic descending pain control system underlies placebo analgesia. *Neuron*. 2009;63(4):533-543.
43. Wager TD, Scott DJ, Zubieta JK. Placebo effects on human mu-opioid activity during pain. *Proc Natl Acad Sci USA*. 2007;104(26):11056-11061.
44. Zubieta JK, Stohler CS. Neurobiological mechanisms of placebo responses. *Ann N Y Acad Sci*. 2009;1156:198-210.
45. Finsterbusch JF, Eippert F, Büchel C. Single, slice-specific z-shim gradient pulses improve T2*-weighted imaging of the spinal cord. *Neuroimage*. 2012;59(3):2307-2315.
46. Eippert F, Finsterbusch J, Bingel U, Buchel C. Direct evidence for spinal cord involvement in placebo analgesia. *Science*. 2009;326(5951):404.
47. Sprenger C, Eippert F, Finsterbusch J, et al. Attention modulates spinal cord responses to pain. *Curr Biol*. 2012;22(11):1019-1022.
48. Geuter S, Eippert F, Hindi Attar C, Buchel C. Cortical and subcortical responses to high and low effective placebo treatments. *Neuroimage*. 2013;67:227-236.
49. Jensen KB, Kosek E, Wicksell R, et al. Cognitive behavioral therapy increases pain-evoked activation of the prefrontal cortex in patients with fibromyalgia. *Pain*. 2012;153(7):1495-1503.

CHAPTER 10

Brain Predictors of Individual Differences in Placebo Responding

Leonie Koban, Luka Ruzic, Tor D. Wager
University of Colorado, Boulder CO, USA

BRAIN PREDICTORS OF INDIVIDUAL DIFFERENCES IN PLACEBO RESPONDING

One of the most consistent findings in the literature on placebo effects is their inconsistency across individuals. Placebo responses can relieve pain and improve clinical outcomes for some individuals, but others show little or no improvement following placebo treatment. Whereas placebo response magnitude is likely distributed on a continuum across individuals, individuals have often been dichotomized into placebo 'responders' and 'nonresponders'. Assessed in this way, the proportion of 'placebo responders' varies dramatically across different paradigms, probably because of variation in the strength and type of manipulation involved. Whereas 30–50% of individuals are typically found to experience positive placebo effects in clinical placebo-controlled trials,[1] studies that directly manipulate expectations about pain in healthy participants have shown expectancy effects in as much as 90–100% of participants.[2] In this chapter, we refer to the magnitude of placebo responses on a continuous scale, as there is no conclusive evidence for a bimodal distribution of placebo 'responders' and 'nonresponders'. Whether placebo responses are considered in continuous or binary terms, however, the variability in placebo responses across individuals is real and substantial.

In this chapter, we argue that a placebo response at the level of an individual person is driven by both personal and situational factors. Uncovering the factors that underlie these sources of variation, and how they combine together to create a strong placebo response, is an important goal. It has implications for understanding the nature of placebo effects, for promoting efficiency and reducing bias in placebo-controlled clinical trials, and for the potential application of placebo effects in clinical settings.[3,4] Predicting and understanding placebo responses on an individual level could have direct implications for participant selection in clinical trials and for 'personalized medicine'—the prescription of treatments based on the individual characteristics of the patient.

Because of its importance in both basic and clinical research, a large number of studies have tried to uncover the role of individual differences and personality factors in the placebo response, with mixed outcomes. In the following sections, we first describe these previous attempts to characterize individual differences in personality and brain measures predictive of placebo response, as well as their limitations, with a specific focus on methodologic challenges in brain imaging studies. We then discuss the importance of considering the interactions between stable person-level traits and situational context, often referred to as person × situation-interactions,[3,5] and present possible solutions and avenues for future research on the prediction of placebo responses.

PERSONALITY AND BRAIN PREDICTORS OF PLACEBO ANALGESIA

A Mixed History of Individual Differences in Placebo Response

A long history of research has attempted to link different personality characteristics with placebo responses.[6] Interestingly, over the decades there has been a shift in the literature on traits associated with placebo responses. Whereas early studies have often seen the placebo effect as fundamentally pathologic—e.g. arising from 'hysteria'—and related to traits such as social desirability,[3,7] more recent approaches have related placebo responses to more 'positive' traits and interpersonal factors, such as optimism.

Although suggestibility is still found to be associated with individual placebo outcome,[8,9] several other personality variables that are negatively related to neuroticism and positively to resilience are showing promise as predictors of placebo responses. For example, the trait of optimism has repeatedly been shown to correlate with individual differences in placebo responses.[10–12] A recent study found that ego-resilience and agreeableness were positively predictive of the magnitude of placebo analgesia (PA), and neuroticism was negatively predictive of placebo analgesia (PA).[13] In this vein, putatively 'dopamine-related' personality traits such as behavioral activation and novelty seeking are also positively correlated with individual variability in placebo responses.[14] These findings are consistent with others demonstrating that anxiety has a negative impact on placebo effects (see Table 10.1).[15,16]

In addition to stable personality traits, individual differences in emotional and cognitive states—which can vary substantially depending on the context—have important influences on placebo outcomes. Some of the best predictors of placebo analgesia in clinical settings are a trusting relationship with the physician and a positive attitude towards the treatment.[6] Further, a large number of studies have suggested a crucial role for expectations (i.e. subjectively predicted symptom improvement or pain relief)[2,9,17–20] and desire for pain relief.[1] In other settings, when placebo expectations are formed based on the observation of another person experiencing pain relief, PA has been shown to correlate with individual differences in empathic concern for the observed person.[21]

Although these findings suggest a plausible pattern of traits related to placebo responsiveness, it is important to note that the correlations between trait/state variables and individual differences in placebo analgesia have not been replicated across all studies, and the predictive power of personality measures for placebo response remains relatively low. Arthur Shapiro,[6] a prominent scholar who did much of this work, summarized those early studies this way: 'The number of variables thought to be associated with the placebo effect increased with each study that was reviewed. ...dramatic reports of variables associated with the placebo effect could not be replicated.... Evaluation of these studies clearly indicated to us and others that agreement about what variables were consistently and meaningfully associated with placebo reaction could not be achieved.' This might be in part due to the relatively small sample sizes in some of these studies. Another intrinsic limitation of self-report measures is their subjective nature. Self-reported outcomes, by definition, are limited to processes to which patients have introspective access, and they are colored by incentives, decision biases and heuristics, and cultural display rules.[22] In addition, from a conceptual perspective, if placebo responses are driven by situational variables, it is not clear whether personality traits alone should be able to predict a substantial amount of the variance in placebo response magnitude. Previous accounts have highlighted the importance of situational factors and person × situation-interactions for placebo effects, as will be discussed in more detail below.[3,5]

Recent Advances from Brain Imaging Studies

In the past few years, a growing set of studies has focused on physiologic measures, especially neuroimaging-based measures of neurobiologic processes, as potential predictors of individual differences in PA. These studies have advanced our understanding of the brain mechanisms underlying placebo effects. Moreover, they can also shed light on the neurophysiologic systems that contribute to individual variability in placebo responses.

To date, functional MRI (fMRI) studies of placebo have primarily used pain as the modality of interest. Brain responses to pain are relatively well characterized and involve activation of a widespread set of cortical and subcortical areas, including primary somatosensory areas, insula, cingulate cortex, and thalamus.[23–26] Several studies, described below, have used activity in these characteristic 'pain processing' brain regions as an indicator of brain processing of painful events, and tested whether and how they are modulated during placebo analgesia. In addition, placebo-induced activation outside these regions has been interpreted in terms of supporting context and affective processes that contribute to the creation of PA.

Several fMRI studies converge on the observation that placebo-induced relief of pain is paralleled by *decreased* activation of target regions involved in the processing of nociceptive input.[17,22,27–33] In addition, PA and expectation of pain relief are also related to *increased activation* in another set of brain regions—thought to underlie endogenous pain regulation along with other processes—including orbitofrontal cortex, dorsolateral prefrontal cortex, perigenual anterior cingulate cortex, and periaqueductal gray (PAG).[17,22,29,32,34,35] While direct evidence regarding the specific roles of these prefrontal activations remains scarce, several authors have suggested that they might reflect expectation of pain relief or the maintenance of context representations that are used to create the placebo effect.[4,34,36,37]

Can we use functional brain imaging data to predict individual differences in PA? Evidence regarding brain correlates related to individual differences in PA comes from a number of studies (summarized in Tables 10.1 and 10.2, as well as in Fig. 10.1 in the color section)

TABLE 10.1 Overview of Recent Studies Reporting Brain Correlates of Individual Differences in Placebo Analgesia (PA). N = Number of Participants. Abbreviations are Explained in the Footnotes

Study	N	Personality	Brain measure	Brain correlates	Pain
Eippert et al 2009[40]	48	–	fMRI	rACC–PAG coupling	Heat
Elsenbruch et al 2012[88]	36	–	fMRI	DLPFC, SII, vPCC, Thal	Rectal
Geuter et al 2013[39]	40	–	fMRI	DLPFC, SII, Thal, aIns, dACC, sgACC	Heat
Hashmi et al 2012[42]	30	–	FC–fMRI	VMPFC–Ins	Back pain
Koban et al 2012[56]	20	–	ERP	LPFC, SMA	Heat
Kong et al 2006[89]	16	–	fMRI	DLPFC, IFG, ACC, cerebellum, fusiform gyrus, parahippocampal	Heat
Kong et al 2009[90]	24	–	fMRI	RSFG, RIFG	Heat
Kong et al 2013[43]	46	–	FC–fMRI, fMRI	Insula, precentral, DLPFC, fronto–parietal network–rACC connectivity	Heat
Kotsis et al 2012[91]	36	–	fMRI	RpIns, RPCC, RACC, LsFG, LThal LSTG, RMTG, LMTG, LSMA, RPut, RCer, RCaud, LIFG/LaIns	Rectal
Lieberman et al 2004[92]	15	–	PET	RvlPFC, dACC	Rectal (IBS)
Lui et al 2010[37]	31	–	fMRI	RMFG, RIFG	Laser
Lyby et al 2011[93]	33	Stress, fear of pain	ERP	N2, P2	Heat
Peciña et al 2013[13]	50	Ego-resiliancy, agreeableness, neuroticism, altruism, straight-forwardness, angry hostility	PET (Opioid)	dACC, PAG	Saline
Schweinhardt et al[14] 2008	22	BAS, TCI	sMRI (VBM)	Ins/TC, vStr, MFG, pIns	Saline
Scott et al 2007[35]	14, 30	MID task	fMRI, PET (DA, Opioid)	NAC	Saline
Scott et al 2008[49]	20	–	PET (Opioid, DA)	Opioid: sgACC, rACC, oFC, aIns, pIns, NAC, rAmy, PAG. DA: NAC, vPut, RvCaud	Saline
Stein et al 2012[45]	24	–	White matter DTI	rACC, RDLPFC	Heat
Wager et al 2004[32]	24, 23	–	fMRI	DLPFC, OFC	Shock, heat
Wager et al 2011[4]	47	–	fMRI	Antic OFC, DLPFC, Pre–SMA, reduced S2 area	Shock, heat
Watson et al 2009[12]	11	–	fMRI	aMCC, PCC, postcentral gyrus	Laser
Wiech et al 2010[94]	16	–	FC–fMRI	aIns–LMCC, aIns–MCC	Laser
Zubieta et al 2006[95]	19	–	PET	DLPFC, rACC, Ins, NAC	Saline

fMRI, functional magnetic resonance imaging; FC–fMRI, functional connectivity fMRI; PET, positron emission tomography; sMRI (VBM), structural MRI (voxel-based morphometry); DA, dopamine; DTI, diffusion tensor imaging; rACC–PAG, rostral anterior cingulate cortex–periaqueductal gray; DLPFC, dorsolateral prefrontal cortex; SII, second somatosensory area; vPCC, ventral posterior cingulate cortex; thal, thalamus; ins, insular cortex; p-ins, posterior insular cortex; aIns, anterior insular cortex; dACC, dorsal anterior cingulate cortex; sgACC, subgenual anterior cingulate cortex; VMPFC–Ins, ventromedial prefrontal cortex–insula; LPFC, lateral prefrontal cortex; SMA, supplementary motor area; IFG, inferior frontal gyrus; ACC, anterior cingulated cortex; RSFG, right superior frontal gyrus; RIFG, right inferior frontal gyrus; ERP, event-related potentials; PAG, periaqueductal gray; Ins/TC, insula/temporal cortex vStr, ventral striatum; MFG, middle frontal gyrus; NAC, nucleus accumbens; OFC, orbitofrontal cortex; rAmy, right amygdala; vPut, ventral putamen; RvCaud, right ventral caudate; RDLPFC, right dorsolateral prefrontal cortex; aMCC, anterior midcingulate cortex; PCC, posterior cingulate cortex; LMCC, left midcingulate cortex.

TABLE 10.2 Legend for Studies and Contrasts Depicted in Figure 10.1. Abbreviations are Explained in the Footnotes

Number	Study	Contrast	Modality
1	Wager et al 2004,[32] Study 1	(Control – placebo) during pain correlated with placebo analgesia	fMRI
2	Wager et al 2004,[32] Study 1	(Placebo – control) during anticipation correlated with placebo analgesia	fMRI
3	Wager et al 2004,[32] Study 2	(Control – placebo) during early pain correlated with placebo analgesia	fMRI
4	Wager et al 2004,[32] Study 2	(Control – placebo) during late pain correlated with placebo analgesia	fMRI
5	Lieberman et al 2004[92]	(Before – after) placebo treatment correlated with symptom improvement	PET
6	Kong et al 2006[89]	(After – before) placebo – (placebo – control site) correlated with placebo analgesia	fMRI
7	Watson et al 2009[12]	(Control – placebo) correlated with placebo analgesia	fMRI
8	Wager et al 2011[4]	(Placebo – control) during anticipation correlated with placebo analgesia	fMRI
9	Wager et al 2011[4]	(Control – placebo) during anticipation correlated with placebo analgesia	fMRI
10	Wager et al 2011[4]	(Control – placebo) during pain correlated with placebo analgesia	fMRI
11	Eippert et al 2009[40]	(Placebo – control) – (saline – naloxone) correlated with placebo analgesia; brainstem	fMRI
12	Stein et al 2012[45]	FA correlated with placebo analgesia	DTI
13	Hashmi et al 2012[42]	(Decreasing back pain – persisting back pain) functional connectivity with LdlPFC at baseline	FC–fMRI
14	Hashmi et al 2012[42]	(Persisting back pain – decreasing back pain) functional connectivity with LdlPFC at baseline	FC–fMRI
15	Hashmi et al 2012[42]	(Decreasing back pain – persisting back pain) HF power at baseline	fMRI (oscillation)
16	Hashmi et al 2012[42]	(Persisting back pain – decreasing back pain) LF power at baseline	fMRI (oscillation)
17	Schweinhardt et al 2008[14]	GM density correlated with placebo analgesia	VBM
18	Zubieta et al 2006[95]	Opioid increase correlated with experienced/expected analgesia ratio	PET opioid
19	Elsenbruch et al 2012[88]	(Control – placebo) during anticipation correlated with placebo analgesia	fMRI
20	Elsenbruch et al 2012[88]	(Control – placebo) during pain correlated with placebo analgesia	fMRI
21	Elsenbruch et al 2012[88]	(Responders (control – placebo) – nonresponders (control – placebo)) during anticipation	fMRI
22	Elsenbruch et al 2012[88]	(Responders (control – placebo) – nonresponders (control – placebo)) during pain	fMRI
23	Kotsis et al 2012[91]	(Nonresponders > responders) during anticipation	fMRI
24	Kotsis et al 2012[91]	(Nonresponders > responders) during pain	fMRI
25	Wiech et al 2010[94]	(Pain threshold for (high – low) threat stimuli) correlated with functional connectivity with aIns during anticipation	FC–fMRI
26	Wiech et al 2010[94]	(Pain threshold for (high – low) threat stimuli) correlated with functional connectivity with aIns during stimulation	FC–fMRI
27	Geuter et al 2013[39]	(Strong placebo – control) during early pain correlated with PA	fMRI
28	Geuter et al 2013[39]	(Control – weak placebo) during early pain correlated with PA	fMRI
29	Geuter et al 2013[39]	(Strong placebo – control) during late pain correlated with PA	fMRI
30	Geuter et al 2013[39]	(Control – weak placebo) during late pain correlated with PA	fMRI
31	Lui et al 2010[37]	(Placebo – control) during anticipation correlated with PA	fMRI
32	Kong et al 2013[43]	(Good responders (control – placebo) – poor responders (control – placebo)) during pain	fMRI
33	Kong et al 2013[43]	(Placebo – control) correlates with PA	fMRI

TABLE 10.2 Legend for Studies and Contrasts Depicted in Figure 10.1. Abbreviations are Explained in the Footnotes — Cont'd

Number	Study	Contrast	Modality
34	Kong et al 2013[43]	Functional connectivity with sensory–motor network correlated with PA	fMRI
35	Kong et al 2013[43]	Functional connectivity with right frontoparietal network correlated with PA	fMRI
36	Atlas et al 2010[2]	Brain activity mediators (positive) of cue effects on pain. Increases during pain	fMRI
37	Atlas et al 2010[2]	Brain activity mediators (negative) of cue effects on pain. Decreases during pain	fMRI

fMRI, functional magnetic resonance imaging; PET, positron emission tomography; FA, fractional anisotrophy; DTI, diffusion tensor imaging; FC–fMRI, functional connectivity fMRI; LDLPFC, left dorsolateral prefrontal cortex; VBM, voxel-based morphometry; PA, placebo analgesia.

using different brain-imaging techniques and modalities, including functional and structural MRI, PET, and electrophysiology.

In an extensive re-analysis of two previously described experiments,[4] we used brain activity during expectation and pain experience to predict individual differences in PA and placebo-induced decreases in putative 'pain processing' brain regions. We used both standard regression analysis, which involves correlating activity voxel-by-voxel with individual differences in PA, and multivariate LASSO ('least absolute shrinkage and selection operator'[38]) regression, which uses brain activation patterns across multiple voxels in order to identify regions that are predictive of placebo response magnitude.

What predicts subjective reports of placebo analgesia: pain-related activity or activity during pain-anticipation? Counter-intuitively, reductions in pain-processing brain regions during placebo were not correlated with individual differences in reported analgesia. Rather, brain activity during the anticipation of pain best predicted PA. Specifically, we found that PA was predicted by increased anticipatory activity on- vs. off-placebo in dlPFC, OFC, parietal cortex, and cerebellum, as well as decreased somatosensory activity during pain anticipation.[4] Intriguingly, a different network, comprising ventral cerebellum, perigenual cingulate, and anterior cingulate cortex, was predictive of individual differences in reduced activation of the pain-processing brain regions.[4] The only regions found to be predictive of both outcomes—reported PA and reduced activity in pain-processing regions—were pre-SMA and lateral prefrontal cortex. These findings suggest that different outcome measures may reflect different components of the placebo response—and, consequently, different measures may predict reductions in pain and pain-related brain processing. Similarly, another recent study found that subjective pain relief was correlated with increased activation of dlPFC and rostral ACC for a 'strong' (expensive) placebo paired with high pain relief during conditioning phase, as well as with decreased activation of dorsal ACC, anterior insula, and thalamus in a 'weak' (cheap) placebo condition (lower pain relief during the conditioning phase).[39] While placebo treatments clearly have antinociceptive effects[40,41]—even potentially modulating spinal cord signaling[40]—the reduction in nociceptive processing might not be the most important factor driving reductions in self-reported pain in all conditions. In this case, neuroimaging also provides alternative outcomes for identifying placebo responders—e.g. those with reduced activity in pain-responsive regions—which may ultimately help the field to move beyond the exclusive reliance on self-reported outcomes and disentangle the complex influences on self-report.

Prediction of chronic pain and symptom severity in patient populations based on neuroimaging data is important for clinical purposes, but is only beginning to be systematically studied. A recent study investigated how functional connectivity was related to analgesic responses in patients suffering from chronic back pain during a 2-week placebo treatment.[42] Baseline (pre-treatment) functional connectivity in BOLD signal between dorsomedial PFC and bilateral insula cortices was predictive of reduced pain two weeks later, as tested in an independent set of patients. Specifically, patients with persistent pain (i.e. non-responders) showed higher connectivity between dmPFC and insula than did those whose pain resolved (responders), and the authors therefore suggested that a higher negative affect may prevent placebo-induced pain improvements.[42] In addition, high-frequency activity in the left dlPFC and connectivity of this region to other brain areas, including the midcingulate cortex, was higher in resolving versus persistent pain. The authors report that, together, the two connections (dmPFC-insula and LdlPFC-midcingulate cortex) differentiated resolving from persistent pain with 90% accuracy. This unique attempt to predict pain persistence in patients suggests several promising strategies for future research. For example, as in many clinical studies, it was not possible to ascertain whether the patients with resolving pain improved because of an active placebo process or spontaneous healing and recovery. Studies with a no-treatment arm could help assess how much of the overall improvement is due to the placebo itself.

Resting-state functional connectivity of prefrontal areas might also be predictive for how expectations influence pain in healthy participants. A recent study investigated the relationship between resting-state connectivity and cue-mediated changes in pain experience.[43] These authors report that increased baseline resting-state connectivity of the right fronto-parietal network with the rostral ACC was positively related to the magnitude of expectancy effects on pain. Connectivity between the somatosensory network and the cerebellum during rest was also positively associated with reduced pain in the cue-based expectation task. Further, they showed that decreases in pain-processing brain regions and lateral prefrontal cortex during painful stimulation covaried with changes in pain ratings.[43]

Several studies have also correlated measures of brain *structure* with placebo analgesia. Schweinhardt et al,[44] using voxel-based morphometry, found that gray matter density in nucleus accumbens (NAc), insula, and dlPFC correlated with larger PA. PA was also positively correlated with an average across several putatively 'dopamine-related' traits, including novelty seeking and behavioral activation, and these traits were in turn correlated with structural differences in NAc and dlPFC.[44] Given these multiple links between personality, brain structure, and placebo outcomes, it would be interesting to investigate in future studies whether specific traits mediate the link between structural characteristics and PA, and how variables from different levels of observation could be integrated in multivariate predictive models.

Additional support for brain structural predisposition for placebo responsiveness comes from a recent diffusion tensor imaging (DTI) study.[45] These authors measured both local white matter anisotropy and structural connectivity—using a tract-finding algorithm—between *a priori* cortical and subcortical regions of interest. Individual PA was related to increased 'white matter integrity' (higher fractional anisotropy) in the vicinity of rostral ACC and in left dlPFC, as well as to stronger fiber tract connections of these two regions with the periaqueductal gray (PAG).[45] These results further highlight the importance of cortical, 'top-down' modulatory prefrontal regions, and their connections to the descending pain inhibitory system in the brainstem for endogenous pain regulation.

In contrast to MRI, molecular imaging using positron emission tomography (PET) allows the measurement of the activity of specific neurotransmitters and neuropeptides, and their relationship with PA. A number of pharmacologic and brain-imaging studies have provided evidence for a role for opioidergic activity in OFC, dlPFC, PAG, rostral ACC, and other regions in the creation of PA (e.g.[33,40,41,46–49]). Accordingly, several studies have investigated individual differences in the opioidergic system, suggesting that higher μ-opioid activity (reduced [11-C] carfentanil binding), especially in the area of nucleus accumbens (NAc), predicts stronger placebo responses.[49] A recent study (Peciña et al[13]) extends these findings by relating opioid activity to personality measures. Peciña et al reported that different subscales of agreeableness correlated positively, and neuroticism correlated negatively, with individual differences in both PA and placebo-induced mu-opioid activity in NAc, subgenual and dorsal ACC, OFC, dlPFC, insula, PAG, and thalamus. Opioidergic activity in dorsal ACC and PAG also correlated with reduced plasma cortisol levels during pain (Peciña et al[13]). This study cannot establish a causal relationship between placebo and cortisol; the placebo condition always came second, so overall reductions in the placebo condition could be due to the placebo or natural recovery processes. Nonetheless, the finding of individual differences suggests a relationship between the central opioid and neuroendocrine systems.

Another important neurotransmitter associated with PA is dopamine (DA). Scott et al[49] showed that, in addition to opioidergic activity, individual differences in DA activity in NAc were correlated with subjective PA. In another study from the same group, individual differences in DA activity in NAc during pain anticipation, as well as NAc activity in an unrelated reward anticipation task, were correlated with PA.[35] Together, these findings further support the idea that the dopaminergic system and reward prediction mechanisms play an important role in PA (see also Schweinhardt et al[14]).

Whereas imaging studies are powerful in revealing the brain regions associated with pain processing and their modulation by placebo effects, electrophysiology and magnetoencephalography have much higher temporal resolution, and their usefulness in placebo research has not yet been fully exploited. Several studies measuring event-related potentials (ERPs) elicited by painful laser pulses have shown that placebo leads to reduction in the amplitude of laser-evoked potentials, particularly the N2 and/or P2 components.[50–55] These effects indicate a modulation in the neurophysiologic processing of painful stimuli as early as ~200 msec after stimulus onset, which correlated with the subjectively experienced analgesic effect.[15,51,53] The placebo effects on P2 were reduced for participants with high fear of pain, in line with a reduced analgesic effect for highly anxious individuals.[15]

In another recent ERP study, we reasoned that increased prefrontal activations during PA could be functionally related to error monitoring and resolution. During PA, the brain has to adjust to a mismatch (or conflict) between expected and experienced pain, which could be achieved by more general mechanisms involved in conflict and error monitoring.[56] In order to test this hypothesis of a functional relationship between error processing and placebo, we measured ERPs to errors in a go/no-go task during expectation of placebo analgesia. Compared to a matched control condition, placebo analgesia and individual differences in placebo-induced increases in pain tolerance were correlated with an enhanced error-related positivity (Pe).[56] This component peaks around 200 ms after error commission and is related to the affective significance of errors and to post-error changes in regulatory control.[57,58] The Pe effect had sources in the lateral PFC and dorsal mediofrontal cortex/SMA, in line with a critical role of these regions in PA.[4] These results support a functional overlap of prefrontal mechanisms involved in PA and in detecting and adjusting to errors. Our findings are in line with earlier results that have demonstrated additional recruitment of prefrontal areas, especially lateral PFC and OFC, during placebo compared to opioid analgesia.[34] More studies are needed to investigate which other functional processes might be altered during PA, and how individual differences in cognitive control functions and emotion regulation relate to endogenous pain control.

Taken together, different brain imaging methods and experimental approaches in recent years have started to investigate how placebo effects are influenced by individual differences in brain function and structure. Despite the different methodologic approaches and brain-imaging techniques, a relatively coherent pattern of brain correlates of PA is emerging (see Fig. 10.1), and some of these effects have been directly linked to personality traits and mental states. Interestingly, only some of these correlations are found in brain areas that are directly involved in the processing of nociceptive input, such as somatosensory areas (S2 and dorsal posterior insula), dorsal ACC, anterior insula, and medial

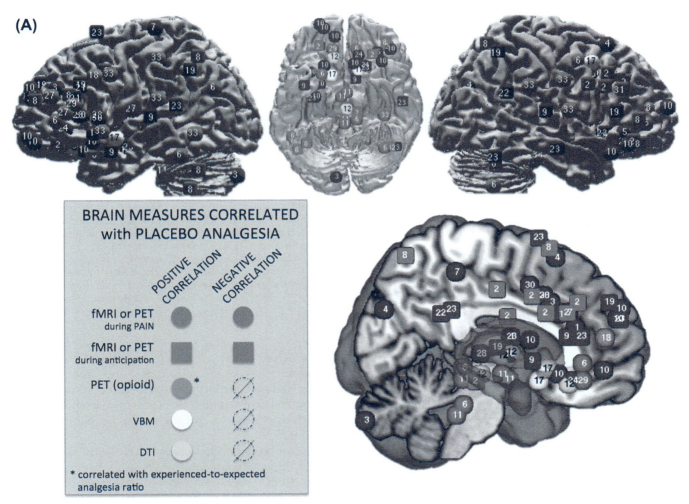

FIGURE 10.1 Brain correlates of individual placebo responses. (A) Coordinates from different studies and contrasts, listed by number in Table 10.2, at which brain measures are reported to correlate with individual differences in PA or to differ between groups of placebo responders and non-responders. Positive and negative correlations from fMRI and PET studies are shown in red and blue, respectively, with spheres and cubes for those points with correlating activity during pain and anticipation, respectively. Magenta spheres denote coordinates at which PET studies found increased opioid activity correlated with the ratio of experienced analgesia to expected analgesia. And locations of gray matter density (VBM) and white matter integrity (DTI) correlates (all positive) with placebo analgesia are marked with yellow and green spheres, respectively.

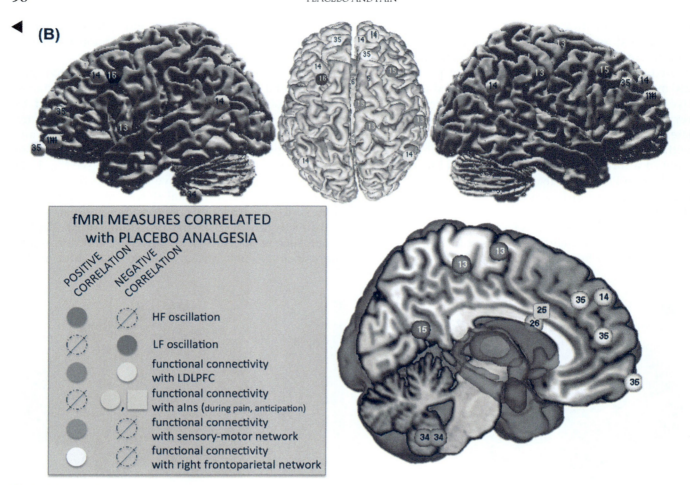

FIGURE 10.1 Cont'd (B) Coordinates at which oscillatory measures and functional connectivity findings from fMRI studies are reported to correlate with placebo analgesia (individual studies and contrasts are listed, by number, in Table 10.2). Included are high frequency band BOLD oscillations (HF) that were positively correlated with placebo analgesia (red); low frequency band BOLD oscillations (LF) that were negatively correlated with placebo analgesia (blue); locations where functional connectivity with left dorsolateral prefrontal cortex (LDLPFC) was positively (magenta) or negatively (cyan) correlated with placebo analgesia; locations where functional connectivity with anterior insula (aIns) during pain (green spheres) or anticipation (green cubes) was negatively correlated with placebo analgesia; locations where functional connectivity with the sensory-motor network was correlated positively with placebo analgesia (orange); and locations where functional connectivity with the right frontoparietal network was correlated positively with placebo analgesia (yellow). *This figure has 2 parts: A and B; both parts are reproduced in color in the color section.*

thalamus. Other correlates of individual differences in PA are found in other brain regions, especially in parietal and prefrontal cortex, as well as brainstem areas (see Fig. 10.1).

One consistent finding across several studies is the involvement of neurophysiologic systems related to valuation and reward-processing, especially NAc/ventral striatum, orbitofrontal areas, and vmPFC.[59–62] These activations have been related to individual differences in reward learning,[35] and to traits including Carver's behavioral activation scale[63] and novelty seeking, which may reflect differences in trait dopaminergic activity across participants.[44] Individuals with increased functional activity and structural integrity in these valuation regions might be especially prone to PA because they efficiently learn to associate medical context and treatment cues with pain relief and reward value. Reward-related activity may also promote opioidergic anti-nociceptive circuits,[64] but might also directly alter subjective evaluation and the affective meaning of pain experience.[61]

Further, as illustrated in Fig. 10.1, one of the most frequently reported correlates of PA and relief expectations is the dlPFC.[4,13,27,29,32,34,36,39,45,56] This region is considered a top-down element of the opioidergic descending pain modulatory system, but also has important roles in cognitive control and conflict resolution.[65,66] As such, the dlPFC might maintain context representations and pain relief expectations during PA,[2,4] which could be crucial to resolve a mismatch between expected pain and actual nociceptive input.[34,56] These ideas are also in line with the findings that opioid activity in this region is positively correlated with resilience and negatively correlated with neuroticism (Peciña et al[13]). Future studies might investigate how prefrontal activity and structural properties relate to other important personality predictors of PA related to expectations and emotion regulation, especially optimism.[67]

Although this research on brain correlates of placebo effects has increased our insight into the mechanisms underlying PA, much more work is needed in order to find reliable predictors of placebo responsiveness and to tailor clinical placebo treatments to individual predisposition and needs. In the following section, we highlight some limitations and methodologic pitfalls that constrain the current research on brain correlates of individual placebo responses.

LIMITATIONS OF STUDIES ON INDIVIDUAL DIFFERENCES IN PA

Thus far, most experiments on brain predictors of placebo responses have not been specifically designed to study individual differences in placebo responses, and most of their findings, although compelling, await replication. One difficulty in comparing results is the variability in experimental procedures and brain-imaging techniques (Fig. 10.1), personality questionnaires, and outcome measures for placebo effects. Given the cost of brain-imaging studies and the need for publishing novel results, replications of previous experiments are often 'conceptual' replications rather than exact ones.[68,69] As a consequence, it is often difficult to determine whether incongruent results are due to differences in experimental design and measures, to lack of statistical power, or to an innate failure to replicate a previous study.[70] This is a very general issue, but one that is amplified for research on brain correlates of individual differences due to the number of tests involved and the diversity of techniques.

A related problem is that not all measures of PA are equivalent, as they might be partly driven by different underlying processes. Differences in outcome measures across studies—e.g. reported pain, tolerance, or pain-related brain activity; or pain in different modalities—might explain some of the inconsistencies. Self-reported PA is likely to be particularly complicated and only partly determined by anti-nociceptive processes. For instance, reductions in pain-related processing were unrelated to individual differences in self-reported PA, and were predicted by a different set of brain regions.[4] In another study, we found no significant correlation between self-reported pain relief and changes in a more implicit measure of PA, pain tolerance temperatures.[56] Future studies might include a comparison of different PA measures, such as pain thresholds, self-report regarding different aspects (e.g. affect and intensity) of pain and relief, and peripheral and neurophysiologic indices of reduced nociceptive processing. An interesting question is whether different outcome measures reflect distinct aspects of PA and could be driven by different processes.

Several other limitations apply to most if not all current studies on brain correlates of individual differences in PA. First, sample sizes in previous studies (typically N < 30, always N < 100, see Table 10.1) have been too small to reliably characterize individual differences. In addition, previous fMRI studies have not been designed to accurately measure how well brain activity can predict placebo response magnitude. This is because stringent multiple comparisons corrections, which are necessary for avoiding false positives, are associated with very low statistical power. Therefore, most true effects will not be detected, and those that are reported suffer from the 'winner's curse'—the selection of significant voxels out of a large set of tests can dramatically inflate the apparent effect sizes.[71–74] Thus, though most brain studies superficially appear to produce strong correlations between PA and brain measures, those effects generally appear much larger than they actually are, and these studies are ultimately inadequate for studying individual differences. If the brain processes that contribute to PA are distributed rather than concentrated in a single anatomic region, the variance explained by any one region is probably small, and can only be detected in much larger samples than currently standard in brain imaging studies.[74]

What can be done in order to overcome some of these problems? First, larger sample sizes are needed to investigate individual differences with sufficient statistical power. Second, echoing recent discussions in general psychology, exact replication studies are needed to test and to challenge current hypotheses and functional models.[68,69] And third, the *use of techniques that combine information from multiple brain regions into a single predictive model* can help overcome both the inflation of apparent effect sizes and some of the power limitations of small sample sizes.

This last point deserves further discussion. Studies investigating individual predictors of PA would benefit from the use of procedures explicitly designed to assess predictive accuracy from multivariate data (e.g. multiple brain voxels and/or multi-modal measures) in an unbiased fashion. The use of *holdout sets* for replication and assessment of predictive accuracy have become commonplace in many fields.[75] Cross-validation is a particular method for selecting holdout sets, in which a model is developed on a *training set* of observations and then accuracy is assessed on an independent *test set*. In cross-validation, the model is developed multiple times on different training sets, so that each observation is part of at least one test set. If correctly applied, cross-validation allows the quantification of how well all the available brain information can predict placebo responses in new patients or participants. Such techniques therefore have a crucial role if we aim at developing tools for predicting outcomes in applied and clinical settings. This procedure represents an important shift away from the use of voxel-wise maps to assess brain correlates of placebo effects, as an *entire brain map* or an *a priori* portion thereof is used to generate an integrated prediction about the outcome (e.g. placebo analgesia), and the predictions are validated on new, independent observations whose 'ground truth' need not be known in advance.

Thus far, there have been two studies using such predictive methods to try to predict placebo responders, one predicting placebo responses in healthy individuals[4] and the other predicting overall improvement on placebo in chronic low back pain.[42] Both report encouraging results, and this method could be more widely used in the future. However, it is important to note that cross-validation and replication are not panaceas, as they only yield truly unbiased (or minimally biased) measures of prediction accuracy the *first time* the test set is evaluated. If different models are developed after 'peeking' at the data in the test sets, the apparent accuracy of the model can be inflated.

In addition to increases in sample size and the use of integrated predictive models, data sharing and pooling across different experiments might also help to identify predictive factors in behavior or brain data.[76] Indeed, several meta-analyses of placebo effects have begun evaluating the consistency of brain findings.[22,27,29]

The Conundrum: Situational Variables Impact Placebo

Even if we follow the recommendations regarding sample size, cross-validation, and other methodologic issues, there is one more fundamental limitation in the study of individual differences in PA. It is the power of the situation, and of interactions between individual predispositions and situational variables.

A necessary prerequisite for the identification of PA predictors based on individual differences is the stability or reliability of placebo responses within individuals across time and situations,[77,78] but very few studies have investigated re-test reliability and situational stability of individual placebo responses. As long as the experimental setting, placebo treatment, and outcome measures are held constant, individual placebo responses can be relatively stable, as indicated by relatively high test-retest reliabilities between $r = 0.6$ and $r = 0.8$,[11,32,78] highlighting the stability of individual predispositions *within* situations.

Yet, individual responses can be highly variable when it comes to the comparison of different situations or experiments. An early study by Liberman[77] compared placebo analgesia in a group of obstetric patients across three different situations: labor pain, post-partum pain, and experimentally induced muscle pain. PA responses were uncorrelated across sessions. Liberman concluded that personality characteristics do not determine individual placebo responses. An important limitation of this study was that the treatment contexts were highly different and the outcome measure of PA very crude, so that the influence of situational factors might have dominated any effect of stable person-level predispositions.

A more recent study by Whalley et al[78] investigated more systematically whether placebo responses are reliable across situations. Whalley and colleagues measured PA across four different trials. They used two differentially labeled, but otherwise identical, placebo creams ('Ibuprofen' and 'Trivaricaine'), which were each tested across two different sessions in the same 71 participants. Whereas they found high correlations across sessions for analgesic responses to the same placebo label (i.e. test-retest reliabilities of $r = 0.60$ and 0.77), the relationship across different labels (brand names) was not significant.[78] These results indicate that even minor changes in placebo treatment have a strong impact on participants' responses, and that individual analgesic outcomes can be very consistent when tested in exactly the same way, but not when tested in a different treatment context. The authors concluded that '[…] the search for a generic placebo responder—one who responds consistently to placebo across different situations—can never be successful, and our failure to find general associations between placebo responses and personality measures is therefore unsurprising.'[78]

This low consistency of individual responses across different treatment settings illustrates that a very important amount of variance in the placebo response is determined by situational factors—though, as we argue below, it does not mean that stable, person-level variables are not important. Previous research has demonstrated that prior experience, type of placebo treatment, and various other factors related to the psychosocial context or the patient–physician relationship impact the placebo response.[1,6,22,79] One example is the influence of prior exposure and learning on placebo effects. For example, a greater number of conditioning trials leads to larger placebo effects, makes them more resistant to extinction,[80] and can even overrule debriefing about the placebo procedure.[81] Deception about the treatment might not even be necessary to elicit a placebo response.[82] On the other hand, it has been shown that suggestion alone can be sufficient to produce placebo effects, albeit weaker than those following conditioning.[46] Together, these findings suggest that there may be multiple routes to placebo analgesia. If we consider these important situational factors, it might come as no surprise that individuals who show strong PA in one situation might experience little relief in another (Colloca, this book, Chapter 15).

SOLUTIONS

To summarize our discussion so far, the review of the previous literature indicates that personality measures and neurophysiologic variables—some of them likely to be stable person-level characteristics, such as measures of brain structure—have some predictive value for individual placebo responses. At the same time, it is becoming increasingly clear that these predictors are not necessarily always replicable or equally important across different experimental procedures or placebo treatments. Situational factors have a powerful role in determining individual placebo responses. How can we reconcile these opposing perspectives and move toward more accurate predictions of individual differences in placebo responses?

We suggest that this is not a new problem, but one of long-standing controversy in social and personality psychology (often referred to as 'person versus situation debate', e.g.[5,70,83,84]) concerning whether more variance in behavior is explained by situational factors or by stable personality traits. As in other debates, the truth lies probably in between—or, more precisely, in interactions, as neither person-level nor situational variables alone are sufficient—and a more fruitful approach lies in the integration of the two opposing views. Mischel (2004)[5] proposed that finding stable personality factors requires taking into account situational variability, and especially person–situation interactions. The temporal stability of personality might even be characterized by how an individual behaves across different situations, which could constitute a relatively stable, person-specific behavioral pattern.[5] For instance, one person could be very outgoing and extraverted when interacting in a small group or with friends, but still be shy in larger group settings and with strangers, whereas another person might enjoy public speaking but be more introverted in smaller groups. Whereas both persons might score intermediate on extraversion when aggregating across situations, their specific personalities would be more finely reflected in the *pattern* of situation-specific individual responses.

From this perspective, placebo effects may be best characterized as driven by complex interactions between individual predispositions such as personality or brain characteristics and situational, contextual factors.[3,5,67,85] In other words, PA requires the right type of person in the right situation, as illustrated in Fig. 10.2. For example, having an optimistic personality is not sufficient for pain relief (Fig. 10.2, P), nor is a positive suggestion alone (Fig. 10.2, S). However, when an optimistic person receives a positive suggestion, he or she may experience a strong analgesic effect (Fig. 10.2, P × S, see also Geers et al[10]). The opposite effect might be observed for a pessimistic personality when encountering a negative suggestion. In this case, even an increase of pain (or a nocebo response) may occur.[67]

This account is in line with the richness of placebo manipulations and the complexity of the mechanisms underlying these effects.[86,87] Whereas one person might respond effectively to a conditioning procedure, another individual may be more susceptible to instructions and conscious expectations of analgesic effects. A third person might be very sensitive to social factors, and therefore respond positively after observing pain relief in another individual[21] or when the treatment is administered by a trusted care provider. Further, individual prior experience with specific treatments might enhance or counteract positive placebo effects by shaping situation-specific expectations of pain relief or increase. For instance, one of the first participants in our early studies[32] dropped out when he learnt that an analgesic cream would be applied to his arm. He told us that he reacts badly to creams, and was sure he was going to get a rash. Unlike for many other participants, the placebo cream was highly aversive, making it unlikely that he would show a placebo response even if he has the right personality, genetics, brain structure, etc. A placebo pill or injection, on the other hand, might have yielded a very different treatment response.

Therefore, an important challenge for future research lies in disentangling the different pathways (e.g. conditioned analgesia versus expectations versus social effects) leading to PA and other kinds of placebo effects, and to refine predictions of placebo responses by taking person × situation-interactions into account (see Fig. 10.2).

HOW CAN BRAIN IMAGING STUDIES FIND BRAIN PREDICTORS OF PA? RECOMMENDATIONS AND CONCLUSIONS

Despite the inconsistency of placebo responses across different situations, brain predictors can have a significant advantage over personality traits such as optimism, agreeableness, or resilience alone. Thus, predictive models using machine learning based on brain activity or functional connectivity might successfully predict PA in new data sets.[4,42] Furthermore, fMRI measures such as dlPFC activity[4] may increase only when the right person is in the right situation, and thus track the product of person × situation-interactions that arise through various means. Functional brain measures might therefore be more

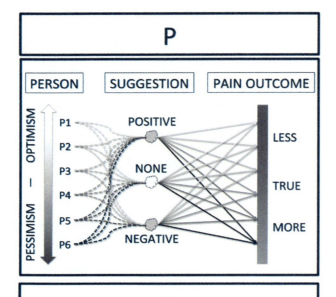
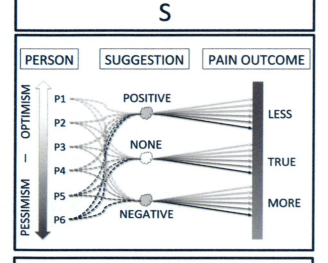
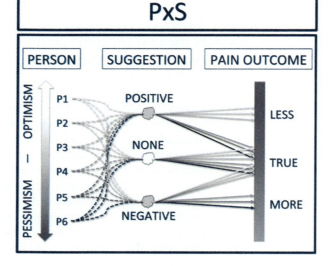

FIGURE 10.2 Illustration of possible roles of person characteristics and situation characteristics with respect to placebo response. Each panel shows diagrammatically, from left to right, a set of individuals with varying trait optimism/pessimism levels (an example of a person-level variable) experiencing a positive, negative, or neutral suggestion about treatment efficacy (a situational variable), and subsequently rating the intensity of a painful stimulus. In the three panels, the set of outcomes are determined by: (P) only the person-level variable; (S) only the situational variable; and (P×S) an interaction between the two variables. In the interaction case, analgesia is shown by the right kind of person in the right situation, and simple effects of each variable need neither exist nor be easily detectable if they do exist. *This figure is reproduced in color in the color section.*

powerful in explaining inter-individual variance than situational or self-report questionnaire measures alone. Brain measures of anatomy, such as gray matter density or fiber tract connectivity, on the other hand reflect relatively stable factors, and might therefore be predictive only when the situation (and participants' perception of it) is held constant. A promising avenue for future research lies in the investigation of how structural brain anatomy and personality factors relate to each other, and how structural factors and functional brain activity interact during PA.

In order to increase our understanding of individual differences in placebo mechanisms and to be able to better predict the magnitude of the analgesic response for individual participants and patients, several approaches might be productively taken. First, in order to detect stable, person-level predictors of PA, individuals' perceptions of the treatment context (the situational factors) must be as identical as possible. Though it may not be possible to completely control situational variables, it is also possible to measure participants' situation-specific feelings, expectations, and attributions and use them to explain variability in PA. Second, it would be productive to jointly manipulate situational and person-level variables, and to study situation × person-interactions[10,15,67] and how they relate to the level of individual brain responses and neuroanatomic characteristics. Finally, larger scale studies will be needed to provide more conclusive evidence regarding the predictability of individual differences in placebo responses. Such responses could include multiple types of outcome variables and multiple types of predictors—e.g. personality questionnaires, multi-modal brain imaging, and genotyping—in order to obtain a comprehensive picture of how people and situations interact to produce placebo effects.

Acknowledgments

This work was supported by R01MH076136 (TDW) and by a postdoctoral fellowship from the Swiss National Science Foundation to LK.

References

1. Price DD, Finniss DG, Benedetti F. A comprehensive review of the placebo effect: recent advances and current thought. *Annu Rev Psychol.* 2008;59:565-590.
2. Atlas LY, Bolger N, Lindquist MA, Wager TD. Brain mediators of predictive cue effects on perceived pain. *J Neurosci.* 2010;30(39):12964-12977.
3. Shapiro AK, Struening EL, Shapiro E. The reliability and validity of a placebo test. *J Psychiatr Res.* 1979;15(4):253-290.
4. Wager TD, Atlas LY, Leotti LA, Rilling JK. Predicting individual differences in placebo analgesia: contributions of brain activity during anticipation and pain experience. *J Neurosci.* 2011;31(2):439-452.
5. Mischel W. Toward an integrative science of the person. *Annu Rev Psychol.* 2004;55:1-22.
6. Shapiro AK, Shapiro E. *The Powerful Placebo: from Ancient Priest to Modern Physician.* Baltimore: Johns Hopkins University Press; 1997.
7. Beecher HKS, Keats A, Mosteller F, Lasagna L. The effectiveness of oral analgesics (morphine, codeine, acetylsalicylic acid) and the problem of placebo 'reactors' and 'non-reactors'. *J Pharmacol Exp Ther.* 1953;109(4):393-400.
8. De Pascalis V, Chiaradia C, Carotenuto E. The contribution of suggestibility and expectation to placebo analgesia phenomenon in an experimental setting. *Pain.* 2002;96(3):393-402.
9. Morton DL, El-Deredy W, Watson A, Jones AK. Placebo analgesia as a case of a cognitive style driven by prior expectation. *Brain Res.* 2010;1359:137-141.
10. Geers AL, Kosbab K, Helfer SG, et al. Further evidence for individual differences in placebo responding: an interactionist perspective. *J Psychosom Res.* 2007;62(5):563-570.
11. Morton DL, Watson A, El-Deredy W, Jones AK. Reproducibility of placebo analgesia: Effect of dispositional optimism. *Pain.* 2009;146(1-2):194-198.
12. Watson A, El-Deredy W, Iannetti GD, et al. Placebo conditioning and placebo analgesia modulate a common brain network during pain anticipation and perception. *Pain.* 2009;145(1-2):24-30.
13. Peciña, M, Azhar, H, Love, TM, et al. Personality Trait Predictors of Placebo Analgesia and Neurobiological Correlates. *Neuropsychopharmacology.* 2013;38(4):639-646. doi:10.1038/npp.2012.227.
14. Schweinhardt P, Kuchinad A, Pukall CF, Bushnell MC. Increased gray matter density in young women with chronic vulvar pain. *Pain.* 2008;140(3):411-419.
15. Lyby PS, Aslaksen PM, Flaten MA. Is fear of pain related to placebo analgesia? *J Psychosom Res.* 2010;68(4):369-377.
16. Ober K, Benson S, Vogelsang M, et al. Plasma noradrenaline and state anxiety levels predict placebo response in learned immunosuppression. *Clin Pharmacol Ther.* 2012;2:220-226.
17. Atlas LY, Whittington RA, Lindquist MA, et al. Dissociable influences of opiates and expectations on pain. *J Neurosci.* 2012;32(23):8053-8064. doi:10.1523/JNEUROSCI.0383-12.2012.
18. Kirsch I. Response expectancy as a determinant of experience and behavior. *Am Psychologist.* 1985;40:1189-1202.
19. Price DD, Milling L, Kirsch I, et al. An analysis of factors that contribute to the magnitude of placebo analgesia in an experimental paradigm. *Pain.* 1999;83:147-156.
20. Vase L, Robinson M, Verne G, Price D. Increased placebo analgesia over time in irritable bowel syndrome (IBS) patients is associated with desire and expectation but not endogenous opioid mechanisms. *Pain.* 2005;115(3):338-347.
21. Colloca L, Benedetti F. Placebo analgesia induced by social observational learning. *Pain.* 2009;144(1-2):28-34.
22. Wager TD, Fields H. Placebo analgesia. In: Textbook of Pain. In press.
23. Apkarian AV, Bushnell MC, Treede RD, Zubieta JK. Human brain mechanisms of pain perception and regulation in health and disease. *Eur J Pain.* 2005;9(4):463-484.
24. Bushnell MC, Duncan GH, Hofbauer RK, et al. Pain perception: is there a role for primary somatosensory cortex? *Proc Natl Acad Sci USA.* 1999;96(14):7705-7709.
25. Iannetti GD, Mouraux A. From the neuromatrix to the pain matrix (and back). *Expl Brain Res.* 2010;205(1):1-12.
26. Rainville P, Duncan GH, Price DD, et al. Pain affect encoded in human anterior cingulate but not somatosensory cortex. *Science.* 1997;277(5328):968-971.
27. Martina Amanzio, Fabrizio Benedetti, Porro Carlo A, et al. Activation likelihood estimation meta-analysis of brain correlates of placebo analgesia in human experimental pain. *Hum Brain Map.* 2011. doi:10.1002/hbm.21471.
28. Eippert F, Finsterbusch J, Bingel U, Buchel C. Direct evidence for spinal cord involvement in placebo analgesia. *Science.* 2009;326(5951):404.
29. Meissner K, Bingel U, Colloca L, et al. The placebo effect: advances from different methodological approaches. *J Neurosci.* 2011;31(45):16117-16124.
30. Petrovic P, Kalso E, Petersson KM, Ingvar M. Placebo and opioid analgesia – imaging a shared neuronal network. *Science.* 2002;95(5560):1737-1740.
31. Price DD, Craggs J, Verne G, et al. Placebo analgesia is accompanied by large reductions in pain-related brain activity in irritable bowel syndrome patients. *Pain.* 2007;127:63-72.
32. Wager TD, Rilling JK, Smith EE, et al. Placebo-induced changes in FMRI in the anticipation and experience of pain. *Science.* 2004;303(5661):1162-1167.
33. Zubieta J, Bueller J, Jackson L, et al. Placebo effects mediated by endogenous opioid activity on mu-opioid receptors. *J Neurosci.* 2005;25(34):7754-7762.
34. Petrovic P, Kalso E, Petersson KM, et al. A prefrontal non-opioid mechanism in placebo analgesia. *Pain.* 2010;150:59-65.
35. Scott D, Stohler C, Egnatuk C, et al. Individual differences in reward responding explain placebo-induced expectations and effects. *Neuron.* 2007;55(2):325-336.
36. Krummenacher P, Candia V, Folkers G, et al. Prefrontal cortex modulates placebo analgesia. *Pain.* 2010;148:368-374.
37. Lui F, Colloca L, Duzzi D, et al. Neural bases of conditioned placebo analgesia. *Pain.* 2010;151(3):816-824.
38. Tibshirani R. Regression shrinkage and selection via the lasso. *J Roy Statist Soc Series B (Methodological).* 1996;58(1):267-288.
39. Geuter S, Eippert F, Hindi A, et al. Cortical and subcortical responses to high and low effective placebo treatments. *Neuroimage.* 2013;67:227-236.
40. Eippert F, Bingel U, Schoell E, et al. Activation of the opioidergic descending pain control system underlies placebo analgesia. *Neuron.* 2009;63:533-543.
41. Wager T, Scott DJ, Zubieta JK. Placebo effects on human μ-opioid activity during pain. *Proc Natl Acad Sci USA.* 2007;104(26):11056-11061.
42. Hashmi JA, Baria AT, Baliki MN, et al. Brain networks predicting placebo analgesia in a clinical trial for chronic back pain. *Pain.* 2012;153(12):2393-2402.
43. Kong J, Jensen K, Loiotile R, et al. Functional connectivity of the frontoparietal network predicts cognitive modulation of pain. *Pain.* 2013. doi:10.1016/j.pain.2012.12.004.
44. Schweinhardt P, Seminowicz DA, Jaeger E, et al. The anatomy of the mesolimbic reward system: a link between personality and the placebo analgesic response. *J Neurosci.* 2009;29(15):4882-4887.
45. Stein N, Sprenger C, Scholz J, et al. White matter integrity of the descending pain modulatory system is associated with interindividual differences in placebo analgesia. *Pain.* 2012;153(11):2210-2217.
46. Amanzio M, Benedetti F. Neuropharmacological dissection of placebo analgesia: expectation-activated opioid systems versus conditioning-activated specific subsystems. *J Neurosci.* 1999;19(1):484-494.

47. Bingel U, Lorenz J, Schoell E, et al. Mechanisms of placebo analgesia: rACC recruitment of a subcortical antinociceptive network. *Pain*. 2006;120(1-2):8-15.
48. Levine JD, Gordon NC, Fields HL. The mechanism of placebo analgesia. *Lancet*. 1978;2(8091):654-657.
49. Scott DJ, Stohler C, Egnatuk C, et al. Placebo and nocebo effects are defined by opposite opioid and dopaminergic responses. *Arch Gen Psychiatr*. 2008;65(2):220-231.
50. Aslaksen PM, Bystad M, Vambheim SM, Flaten MA. Gender differences in placebo analgesia: event-related potentials and emotional modulation. *Psychosom Med*. 2011;73(2):193-199. doi:PSY.0b013e318 2080d73.
51. Colloca L, Tinazzi M, Recchia S, et al. Learning potentiates neurophysiological and behavioral placebo analgesic responses. *Pain*. 2008;139(2):306-314.
52. Flaten MA, Aslaksen PM, Lyby PS, Bjorkedal E. The relation of emotions to placebo responses. *Phil Trans R Soc Ser B*. 2011;366(1572):1818-1827.
53. Morton DL, Brown CA, Watson A, et al. Cognitive changes as a result of a single exposure to placebo. *Neuropsychologia*. 2010;48(7):1958-1964.
54. Wager TD, Matre D, Casey KL. Placebo effects in laser-evoked pain potentials. *Brain Behav Immun*. 2006;20(3):219-230.
55. Watson A, El-Deredy W, Vogt BA, Jones AK. Placebo analgesia is not due to compliance or habituation: EEG and behavioural evidence. *Neurorep*. 2007;18(8):771-775.
56. Koban L, Brass M, Lynn MT, et al. Placebo analgesia affects brain correlates of error processing. *PLoS One*. 2012;7(11):e49784.
57. Ridderinkhof KR, Ramautar JR, Wijnen JG. To P(E) or not to P(E): a P3-like ERP component reflecting the processing of response errors. *Psychophysiology*. 2009;46(3):531-538.
58. Ullsperger M, Harsay HA, Wessel JR, et al. Conscious perception of errors and its relation to the anterior insula. *Brain Struct Funct*. 2010;214(5–6):629-643.
59. O'Doherty J, Kringelbach ML, Rolls ET, et al. Abstract reward and punishment representations in the human orbitofrontal cortex. *Nat Neurosci*. 2001;4(1):95-102.
60. O'Doherty JP, Dayan P, Friston K, et al. Temporal difference models and reward-related learning in the human brain. *Neuron*. 2003;38(2):329-337.
61. Roy M, Shohamy D, Wager TD. Ventromedial prefrontal-subcortical systems and the generation of affective meaning. *Trends Cogn Sci*. 2012;16:147-156.
62. Schultz W. Neural coding of basic reward terms of animal learning theory, game theory, microeconomics and behavioural ecology. *Curr Opin Neurobiol*. 2004;14(2):139-147.
63. Carver CS, White TL. Behavioral inhibition, behavioral activation, and affective responses to impending reward and punishment: the BIS/BAS Scales. *J Pers Soc Psychol*. 1994;67(2):319-333.
64. Leknes S, Tracey I. A common neurobiology for pain and pleasure. *Nat Rev Neurosci*. 2008;9(4):314-320. doi:10.1038/nrn2333.
65. Miller EK, Cohen JD. An integrative theory of prefrontal cortex function. *Annu Rev Neurosci*. 2001;24:167-202.
66. Nee DE, Wager TD, Jonides J. Interference resolution: insights from a meta-analysis of neuroimaging tasks. *Cogn Affect Behav Neurosci*. 2007;7(1):1-17.
67. Geers AL, Helfer SG, Kosbab K, et al. Reconsidering the role of personality in placebo effects: dispositional optimism, situational expectations, and the placebo response. *J Psychosom Res*. 2005;58(2):121-127.
68. Makel MC, Plucker JA, Hegarty B. Replications in psychology research: how often do they really occur? *Perspect Psychol Sci*. 2012;7(6):537-542.
69. Pashler H, Harris CR. Is the replicability crisis overblown? Three arguments examined. *Perspect Psychol Sci*. 2012;7(6):531-536.
70. Mischel W. *Personality and Assessment*. New York: Wiley; 1968.
71. Kriegeskorte N, Lindquist MA, Nichols TE, et al. Everything you never wanted to know about circular analysis, but were afraid to ask. *J Cereb Blood Flow Metab*. 2010;30(9):1551-1557. doi:10.1038/jcbfm.2010.86.
72. Lieberman MD, Berkman ET, Wager TD. Correlations in social neuroscience aren't voodoo: commentary on Vul et al. (2009). *Perspect Psychol Sci*. 2009;4(3):299-307.
73. Vul E, Harris C, Winkielman P, Pashler H. Puzzlingly high correlations in fMRI studies of emotion, personality, and social cognition 1. *Perspect Psychol Sci*. 2009;4(3):274-290.
74. Yarkoni T. Big correlations in little studies: inflated fMRI correlations reflect low statistical power—commentary on Vul et al. (2009). *Perspect Psychol Sci*. 2009;4(3):294-298.
75. Hastie T, Tibshirani R, Friedman J, Franklin J. The elements of statistical learning: data mining, inference and prediction. *Mathemat Intelligen*. 2005;27(2):83-85.
76. Yarkoni T, Poldrack RA, Van Essen DC, Wager TD. Cognitive neuroscience 2.0: building a cumulative science of human brain function. *Trends Cogn Sci*. 2010;14(11):489-496. doi:10.1016/j.tics.2010.08.004.
77. Liberman R. An experimental study of the placebo response under three different situations of pain. *J Psychiatr Res*. 1964;33:233-246.
78. Whalley B, Hyland M, Kirsch I. Consistency of the placebo effect. *J Psychosom Res*. 2008;64(5):537-541.
79. Benedetti F. *Placebo Effects: Understanding the Mechanisms in Health and Disease*. Oxford: Oxford University Press; 2009.
80. Colloca L, Petrovic P, Wager TD, et al. How the number of learning trials affects placebo and nocebo responses. *Pain*. 2010;151(2):430-439.
81. Schaefer S, Wager TD. *Separable correlates of experience and belief in placebo analgesia*. New Orleans: Paper presented at the Neuroscience; October 13-17, 2012.
82. Kaptchuk T, Friedlander E, Kelley J, Sanchez M. Placebos without deception: a randomized controlled trial in irritable bowel syndrome. *PLoS One*. 2010;5(12):e15591.
83. Epstein S, O'Brien EJ. The person–situation debate in historical and current perspective. *Psychol Bull*. 1985;98(3):513-537.
84. Kenrick DT, Funder DC. Profiting from controversy: lessons from the person-situation debate. *Am Psychologist*. 1988;43(1):23-34.
85. Gelfand DM, Gelfand S, Rardin MW. Some personality factors associated with placebo responsivity. *Psychol Rep*. 1965;17(2):555-562.
86. Benedetti F. Mechanisms of placebo and placebo-related effects across diseases and treatments. *Annu Rev Pharmacol Toxicol*. 2007. doi:10.1146/annurev.pharmtox.48.113006.094711.
87. Benedetti F, Mayberg HS, Wager TD, et al. Neurobiological mechanisms of the placebo effect. *J Neurosci*. 2005;25(45):10390-10402.
88. Elsenbruch S, Kotsis V, Benson S, et al. Neural mechanisms mediating the effects of expectation in visceral placebo analgesia: an fMRI study in healthy placebo responders and nonresponders. *Pain*. 2012;153(2):382-390.
89. Kong J, Gollub RL, Rosman IS, et al. Brain activity associated with expectancy-enhanced placebo analgesia as measured by functional magnetic resonance imaging. *J Neurosci*. 2006;26(2):381-388.
90. Kong J, Kaptchuk TJ, Polich G, et al. Expectancy and treatment interactions: a dissociation between acupuncture analgesia and expectancy evoked placebo analgesia. *NeuroImage*. 2009;45(3):940-949.
91. Kotsis V, Benson S, Bingel U, et al. Perceived treatment group affects behavioral and neural responses to visceral pain in a deceptive placebo study. *Neurogastroenterol. Motil*. 2012;24(10):935–e462.
92. Lieberman MD, Jarcho JM, Berman S, et al. The neural correlates of placebo effects: a disruption account. *NeuroImage*. 2004;22(1):447-455.
93. Lyby PS, Aslaksen PM, Flaten MA. Variability in placebo analgesia and the role of fear of pain – an ERP study. *Pain*. 2011;152(10):2405-2412.
94. Wiech K, Lin C, Brodersen KH, Bingel U, et al. Anterior insula integrates information about salience into perceptual decisions about pain. *J Neurosci*. 2010;30(48):16324-16331.
95. Zubieta J, Yau W, Scott DJ, Stohler CS. Belief or need? Accounting for individual variations in the neurochemistry of the placebo effect. *Brain Behav Immun*. 2006;20(1):15-26.

CHAPTER 11

Placebo Responses, Antagonistic Responses, and Homeostasis

Magne Arve Flaten

Department of Psychology, Norwegian University of Science and Technology, Trondheim, Norway

PLACEBO RESPONSES AND HOMEOSTASIS

The placebo response consists of physiological and psychological reactions to information that a treatment has been administered. Such information can be provided verbally, or by presenting conditioned stimuli that have been previously associated with active treatment. These ways of generating placebo responses have been described by the expectancy and the classical conditioning theories of placebo effects (see Ch 16). However, expectancies and learning do not act in a vacuum, and the state of the organism to which information is administered should be considered. The homeostatic level of the organism, it is argued here, should be taken into account to understand the occurrence, amplitude, and duration of placebo responses. Homeostasis refers to the maintenance of physiological and psychological variables within specified limits. When homeostasis is disturbed, feedback mechanisms are activated to bring these variables back within normal limits. Placebos may be among the stimuli that activate such feedback mechanisms. That is, placebo responses may serve to bring systems back to homeostasis, or to maintain homeostasis in the face of stimuli that predict deviations from this set point or normal level of a variable. If so, placebos could be useful in decreasing the impact of such deviations from homeostasis, in the areas of pain reduction and treatment of drug addiction.

Deviations from homeostasis occur in a variety of situations, two examples being symptoms associated with illness, and withdrawal from drug addiction. The most common reason for seeking medical treatment is pain, a state that the organism is motivated to reduce. Pain is usually a signal of some underlying problem, but the pain itself can take on the status of a disorder in its own right. Organisms in pain are undergoing a deviation from their homeostatic condition, and pain reduction is an example of re-establishment of homeostasis. Pain reduction can be achieved pharmacologically, psychologically, or in other ways, and one method of reducing pain is the administration of placebos.

Placebo responses have similarities to the responses seen in individuals in withdrawal from drugs of addiction.[1] Reduction in withdrawal symptoms in drug users represents another example of re-establishing homeostasis. However, if the individual is already at homeostasis, i.e. not in drug withdrawal, then the placebo response should be weak or absent. In fact, *compensatory responses* can be seen when healthy individuals are exposed to placebos, leading to drug tolerance, i.e. a reduction in the response to the drug.[2] Thus, the effects of expectancy and classical conditioning in placebo responding, it is argued, depend on whether or not the individual is at homeostasis, and on the degree of homeostatic disturbance.

THEORETICAL BACKGROUND

Pavlov[3] described an experiment in which morphine was repeatedly administered to dogs. As with all drugs, morphine has several effects, such as sleep, salivation, and vomiting. This was known before Pavlov conducted his experiment, and the novel finding in Pavlov's experiment was that the dog salivated and vomited when seeing the experimenter, before the injection of morphine. Thus, the drug seemed to act as an unconditioned stimulus (US), and the experimenter as the conditioned stimulus (CS) that reliably signalled morphine's effect. Siegel[4] has given a broad historical exposition of the work of Pavlov and others on conditioning with pharmacological stimuli.

Drugs are stimuli[5] that can enter into associations with other stimuli, as described by the principles of classical conditioning. Administration of drugs can be considered conditioning trials, because of signals that are present prior to the administration of the drug, or prior to the occurrence of the effect of the drug. However, there are some important differences between classical conditioning with external sensory stimuli and conditioning with pharmacological stimuli.

Firstly, the unconditioned stimulus is the drug's effects in the central nervous system.[6,7] The drug is not the unconditioned stimulus. Likewise, the peripheral effects of the drug are not the unconditioned stimulus. What should be considered the unconditioned stimulus is the effect of the drug in the central nervous system. Secondly, drugs are distributed throughout the organism and elicit many different effects, some of which may mistakenly be taken for unconditioned responses. Thirdly, the onset latency, duration, and intensity of drug effects are different from those of unconditioned stimuli more commonly used, e.g. electric shock or aversive noise.

Shepard Siegel showed that conditioned stimuli that signalled injections of morphine (or other drugs), lead to accelerated development of tolerance, compared to groups of animals that received un-signaled morphine injections. The observation of rapid and pronounced loss of drug effect was termed associative tolerance, because it could not be explained by physiological processes, and indicated that effects of drugs could be reduced by learning processes. Associative tolerance was observed across laboratories and with different drugs. The theoretical basis of the observed associative tolerance, however, was debated, and Siegel's theory that associative tolerance was due to compensatory conditioned responses received some criticism. Eikelboom and Stewart,[6] in an influential paper (see also Ramsay and Woods[7]), argued that the unconditioned effect of a drug is its action on the central nervous system (Fig. 11.1). Thus, administering a drug and merely observing the subsequent response, does not ensure that the observed response is the proper unconditioned response. All drugs have peripheral effects, e.g. reduction in body temperature after administration of morphine. When the lower temperature is registered in the central nervous system, compensatory mechanisms are activated to restore body temperature to normal values. The input to the central nervous system is the signal to increase temperature, and this is the proper unconditioned stimulus. Thus, what seemingly was a compensatory conditioned increase in body temperature is actually a morphine-agonistic conditioned response, according to this account.

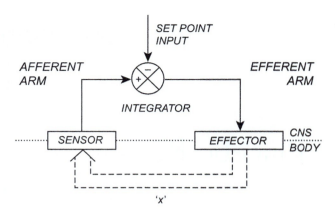

FIGURE 11.1 Diagram of a feedback system that regulates the variable 'X'. The sensor registers the level of 'X' and generates a stimulus that is sent to the integrator, where the stimulus from 'X' is compared to the set-point. When 'X' and the set-point are about equal, no output is sent from the integrator. When 'X' and the set-point are not equal, the integrator sends a signal to the effectors, which act to return 'X' to the normal level and reduce the imbalance between 'X' and the set-point. Reprinted from Eikelboom and Stewart[6] with kind permission from the publisher.

CLASSICAL CONDITIONING AND PAIN

Conditioned hypoalgesia (a placebo effect) and conditioned hyperalgesia (possibly a nocebo effect) have both been observed in several studies on conditioned stimuli that have been paired with the administration of a painkiller. It has been debated why sometimes a hyperalgesic and at other times an analgesic conditioned response has been observed.[6,7] The argument proposed here is that homeostatic regulation can explain at least some instances of these conditioned responses. Pain represents a deviation from homeostasis, and the conditioned effect on pain of signals that a painkiller has been administered may differ between subjects that are in pain or pain-free subjects.

Studies in which drugs have been administered to healthy pain-free animals are mainly studies on tolerance development, and the studies were not performed to investigate placebo effects. However, because they have different aims, studies on tolerance and on placebo effects differ in whether the subjects were in pain or not and whether the drug was administered to reduce pain. This is relevant to whether a conditioned hypo- or hyperalgesia is observed, e.g. whether an analgesic was administered or not, and whether conditioned stimuli were paired with changes in painful stimulation.

CONDITIONING WITH ADMINISTRATION OF PAINKILLERS TO PAIN-FREE SUBJECTS

Siegel and colleagues have performed a series of studies in which they have repeatedly paired an environmental conditioned stimulus with administration of morphine to healthy rats,[8,9] or paired a small dose of morphine, hypothesized to act as a CS, with a larger dose.[10,11] These studies have documented the development of a hyperalgesic CR. In all experiments the morphine was administered to pain-free animals. Testing for pain thresholds were performed after administration of morphine, e.g. 30 minutes after, as in,[12] or in the presence of the CS that signaled morphine, after several conditioning trials. Thus, in these studies morphine most likely induced a deviation from homeostasis, and did not reduce a homeostatic disturbance such as, e.g. pain.

Sherman et al[13] did a particularly interesting study for the argument presented in this chapter. They administered morphine repeatedly; administration was signaled by being administered in a distinctive room (the conditioned stimulus). In this experiment the authors observed the development of tolerance to the analgesic effects of morphine in rats. To investigate extinction of conditioned morphine tolerance, the distinct room where the morphine was administered was repeatedly *not* paired with injections of morphine. During extinction, different groups of rats were exposed to either 52.5°C, to which rats flick their tails to avoid, or about 23°C hot-plate temperatures. Interestingly, extinction of tolerance was observed only in rats that received the 52.5°C temperatures (Experiment 1); it was more pronounced in rats exposed to the warm plate compared to those exposed to the cool plate (Experiment 2). Thus, an extinction process that counteracted associative tolerance and reduced pain developed only in the rats exposed to painful stimulation after morphine administration. If the placebo response is a conditioned response that acts to re-establish homeostasis, and in a feed-forward way to avoid homeostatic perturbations, then these findings are in accordance with a homeostatic hypothesis of placebo responding.

Administration of a cholecystokinin antagonist disrupted morphine tolerance,[11] suggesting that associative tolerance to the hyperalgesic effect of morphine was mediated via cholecystokinin. Therefore, a morphine-compensatory response is expected, according to the present hypothesis that placebos correct deviations from homeostasis, and compensatory responses were indeed observed.

CONDITIONING WITH THE ADMINISTRATION OF PAINKILLERS AS THE UNCONDITIONED STIMULUS TO INDIVIDUALS IN PAIN

This procedure is similar to the one above, with the important difference that morphine or another painkiller is repeatedly administered *to individuals in pain*. Thereafter, an inert treatment is provided that in all ways is similar to the painkiller. Amanzio and Benedetti[14] applied painful stimulation to otherwise healthy subjects across two sessions of classical conditioning, and subsequently infused morphine or the nonsteroid anti-inflammatory drug (NSAID) ketorolac. Both drugs increased pain tolerance, and both drugs supported a conditioned response of analgesia. Interestingly, the placebo analgesic CR was antagonized by naloxone when morphine was used as the painkilling drug, suggesting that the placebo CR involved activation of an opioidergical system. However, when an NSAID was used as the painkiller, the conditioned placebo response was not reduced by naloxone, suggesting that the conditioned placebo response was not mediated by release of endogenous opioids. The same group confirmed the finding of a nonopioid mediated conditioned hypoalgesic response after conditioning with ketorolac, and furthermore showed that the nonopioid placebo response was completely blocked by a cannabinoid antagonist.[15] Thus, different systems are activated by different drug unconditioned stimuli, which seem to also be involved in the conditioned placebo response, and a single system cannot explain all occurrences of conditioned hypoalgesic responses.

Taken together, these findings show that conditioning with painkilling drugs such as morphine can support a conditioned hypoalgesic response when the participants are in pain, i.e. when the participants' homeostasis is disturbed. No human study has, to my knowledge, investigated whether conditioning with morphine as the unconditioned stimulus in pain-free subjects would support an antagonistic response, as it should if homeostatic processes are involved in placebo responses.

CONDITIONING WITH REDUCTION IN, OR ABSENCE OF, PAIN AS THE US

However, pain need not be reduced by a painkiller. In the first study that investigated the role of classical conditioning in placebo effects, Voudouris et al[16] used a procedure in which the subjects first went through a pretest with presentations of painful stimulation that was identical for all groups in the experiment. Thereafter, a placebo treatment was administered, and pain levels were surreptitiously reduced. Thus, administration of the placebo, the CS, was paired with reduced pain, the US. When subsequently tested, pain levels were lower in this group, compared to a control condition where the placebo had not been applied and pain had not been lowered. Thus, when the placebo treatment is paired with a reduction in pain, a conditioned response of lower pain is observed. Several later studies on conditioning as a way of inducing placebo analgesia have used similar procedures, with similar results.

The placebo does not need to be a pill, capsule, cream or other treatment often used in the treatment of pain. Colloca and collaborators[17,18] administered repeated pairings of a green light with absence of pain, and a red light with painful stimulation; a yellow light served as a control condition against which changes in pain levels could be compared. When the painful stimulation was presented in the presence of the green light, pain levels were lower compared to pain reported during the yellow light, indicative of a placebo analgesic response to an initially neutral stimulus paired with lower pain levels.

Taken together, these studies show that conditioned stimuli signaling a reduction in or absence of pain, reduce the pain reported to a painful stimulus. As the conditioned response is a reduction in pain, or some process that results in reduced pain, this may be seen as a way of reducing the perturbation of homeostasis that the painful stimulus exerts.

CONDITIONING WITH AN INCREASE IN PAIN AS THE US

In addition to the two groups that received pairings of a placebo cream with decreases in pain stimulation, Voudouris et al[6] included two groups that received pairings of a placebo cream with an *increase* in pain as the US during conditioning trials. Both of these groups displayed an increase in pain to a standard stimulus presented after conditioning.

As noted above, the placebo does not need to be a treatment often used in the treatment of pain. Colloca et al[18] administered repeated pairings of a green light with absence of pain, red light with more painful stimulation, and yellow light with no change in painful stimulation, respectively. Painful stimulations of the same intensity were subsequently presented in the presence of each of the three conditions. During the red light, pain levels were higher compared to pain during the yellow light, indicative of a nocebo hyperalgesic response.

The findings from the studies reviewed above suggest that analgesic responses are observed when the CS is paired with a reduction in pain. If a CS is paired with a painkiller, in the absence of pain, then a hyperalgesic response may be observed, possibly due to homeostatic processes compensating for the effect of the painkilling drug.

No studies have systematically investigated the role of compensatory and agonistic conditioned responses in the area of pain in humans. However, several studies have been performed, using social drugs as stimuli, to investigate the roles of these processes in humans.

Caffeine

The author has performed several studies with caffeine effects as the unconditioned stimulus. Caffeine is a drug that increases physiological and psychological arousal by inhibiting adenosine receptors in the cerebral cortex and the reticular formation (e.g. the locus coeruleus), areas of importance to arousal regulation.[19,20] Adenosine acts presynaptically to inhibit the action of excitatory transmitters related to arousal, such as glutamate, norepinephrine, and dopamine. By antagonizing the inhibitory effect of adenosine, caffeine has indirect excitatory effects, which can be observed peripherally as sympathomimetic effects, including increases in blood pressure, skin conductance, startle responses, plasma epinephrine, and free fatty acids, and increased feelings of alertness and energy.[21-24] Because of its stimulating effects, caffeine is used in combination with substances that induce relaxation or sleepiness as side effects, e.g. some analgesics, to counteract these side effects. Aqueous oral caffeine is absorbed rapidly and reaches maximum plasma concentration in 30–40 minutes.[25]

Caffeine can act as a reinforcer, in that a novel tasting drink containing caffeine is better liked than the same drink containing placebo.[26] Tolerance develops to the subjective and physiological effects of caffeine.[27,28] In habitual caffeine users, withdrawal effects such as sleepiness, tiredness, weakness, and reduced alertness, all indicative of reduced arousal, can be observed after a period of caffeine abstinence of 12 hours or longer.[23]

Reinforcement, tolerance, and withdrawal effects indicate that long-term caffeine use can lead to dependency. In these respects, caffeine is similar to other drugs of addiction, with the important exception being that caffeine in relatively small doses is not harmful to humans (barring pre-existing conditions such as a seizure disorder). Like other drugs of addiction, caffeine has specific physiological effects, acts on identified neurotransmitter systems, is voluntarily administered, and is negatively reinforcing after the addiction has been established,[29] i.e. intake of caffeine will reduce abstinence symptoms like headache, tiredness and irritability. That is, caffeine addiction may be driven by the ability of caffeine to decrease withdrawal symptoms. The presence of these withdrawal symptoms in the absence of caffeine may be the result of the homeostatic set point for a variety of systems having been reset due to repeated exposure to the drug. A new set point, referred to as an allosteric set point,[30] can affect biochemical systems all the way down to the expression of genes that regulate both presynaptic and postsynaptic processes.

Placebo responses may be seen in both recovery of homeostasis and maintenance of homeostasis. As the drug user repeatedly uses their drug, the normal homeostatic set point for a variety of systems is altered to become an allosteric set point.[30] The biochemical systems now require the drug to maintain that allosteric set point. An example would be the downregulation of presynaptic endorphin production that occurs as autoreceptors are repeatedly activated by heroin. Over time, the presynaptic production of endorphins decreases, leading to a deficiency after the heroin is metabolized. The subjective effect of this is withdrawal, a problem that can be solved temporarily by the injection of more heroin. This becomes a positive feedback system, further downregulating endorphin production and requiring a higher dose of the drug to overcome (i.e. physiological tolerance). During drug rehabilitation, endorphin production slowly up-regulates, approaching the previous homeostatic set point, although this can take weeks to occur. After rehabilitation, the biochemical systems have re-established their prior levels, but the recovered addict still has a conditioning history with which to contend. Exposure to stimuli that have been previously related to the drug (e.g. a syringe or candle) can activate conditioned compensatory mechanisms[31] that can induce craving, a component of withdrawal. This craving is an attempt to maintain homeostasis in the face of an expected drug injection. For the addict, the short-term solution to this problem is reinjection of heroin, resulting in the fairly high relapse rate in recovered addicts.

These conditioned stimuli can be identified in any case of drug addiction in which the administration of the drug is ritualized or involves the repeated use of the same or similar stimuli. For example, a small pink packet of artificial sweetener can gain conditioning properties for a caffeine addict who consumes caffeine in the form of coffee on a regular basis. That pink packet may have no effect on a non-user of coffee, it may induce a slight arousal in an abstained coffee drinker (a drug-like placebo response), and it may induce craving in a former caffeine addict (a compensatory placebo response). It is interesting to note that a given stimulus (for example, the sight of a fine white powder) can be neutral to a non-addict, can reduce withdrawal in an abstaining addict, and can induce craving in a recovered addict. The reduction in withdrawal that occurs early in rehabilitation represents a drug-like response that is an attempt to return to homeostasis, based on the deficiencies in biochemical systems that have been induced by the repeated exposure to the drug. After these biochemical systems have once again achieved their previous levels (homeostasis), the craving is a conditioned compensatory response, an attempt to maintain homeostasis in the face of a predicted deviation based on the conditioning relationship between the CS and the drug.

Withdrawal from a drug leads to a deviation from homeostasis, and stimuli signaling that the drug is about to be administered activate homeostatic mechanisms. By the same token, the presence of illness or disease constitutes a deviation from homeostasis.

One limitation of some experimental models of placebo effects is the use of healthy volunteers as participants.[32–35] Such individuals differ from a clinical population, since the placebo or drug is administered to alter the participants' normal psychological or physiological state. A healthy participant may, for example, be told that the drug will have an effect on 'stomach activity'.[36,37] Thus, the drug is not administered to change symptoms such as pain, airway resistance, or elevated blood pressure, and, thereby, to restore normal function. It should be noted, however, that the evidence base for antagonistic responses induced by verbal information alone is weak. By including the concept of homeostasis on which placebos work, it is suggested that information about drug effects should have different and maybe even opposite effects in individuals in need of treatment, compared to healthy volunteers. Therefore, it is important to understand the differential impact of placebos on people who are or are not at homeostasis.

Compensatory and Drug-Like Conditioned Responses

It is hypothesized that placebo responses occur in individuals who are in pain, have a symptom or disease, suffer from stress or distress, or are in withdrawal from an addictive drug; that is, placebo responses will be seen when the person is not at homeostasis. Placebo responding may therefore be viewed as a way of re-establishing homeostasis. In his law of initial values, Wilder[38] stated that a physiological reaction is dependent on the initial value of the physiological parameter to be measured. Thus, the same stimulus can have an excitatory or inhibitory effect, or no effect at all, depending on the initial state of the organism. This idea is central to the homeostatic theory of placebo effects. If homeostasis is retained, then placebo responses should be absent or minimal.[1,39] This general theory of placebo responses can be operationalized and tested: if placebo responses depend on a deviation from homeostatic levels, then participants who display caffeine withdrawal symptoms and report lower levels of arousal after abstinence should display the largest placebo response to information that caffeine was administered. Caffeine withdrawal involves feelings of decreased vigor and increased fatigue,[40] and these symptoms can be seen even after a brief period of time, equivalent to missing one's morning coffee.[41] Feelings of decreased vigor and increased fatigue are indications of decreased arousal, and we suggest that decreased arousal represents a deviation from homeostatic levels. A deviation from homeostatic levels motivates the correction of the deviation by, in this case, ingesting caffeine. We suggest that a deviation from homeostatic levels can be corrected by presentation of stimuli associated with caffeine, and by the belief that caffeine has been administered and will have certain activating effects.

Caffeine in the Study of Active Placebo

Caffeine is well suited to study active placebo effects because it has nonspecific sympathomimetic effects and is easily detected.[42] Of special importance to the study of active placebo is the drug discrimination procedure, by which participants learn to discriminate a drug from other drugs or placebo,[27] based on the 'discriminative stimulus complex', the subjective feelings or perceptions to which the drug gives rise. Different drugs generate different discriminative stimulus complexes, and participants can be trained to differentiate caffeine from other drugs by reinforcing the identification of one substance as caffeine and others as not-caffeine. Because the active placebo hypothesis is based on the subjective experience of drug effects or side-effects, the drug discrimination method can be used in the study of active placebo effects.

The use of placebos to re-establish homeostasis would be therapeutically similar to cue extinction, or programmed extinction, a procedure in which the conditioned stimuli (cues) associated with a drug are repeatedly presented in the absence of the drug UCS, in hopes of desensitizing the drug user to these conditioned stimuli. Over time, the conditioning power of the drug-related stimuli extinguishes, bringing those stimuli back to the level of neutrality that they possessed before conditioning occurred. The fact that programmed extinction is not very effective in some cases may be due to the fact that the cues produce drug-like placebo responses in abstaining drug users, through conditioning. The cues may then provide comfort by decreasing withdrawal, and this will interfere with extinction.

In support of the homeostatic theory of placebo effects, Rogers et al[26] found that a taste paired with caffeine was preferred to a taste paired with placebo, but only if the participants were caffeine-deprived prior to the experiment. If the participants had already had their morning caffeine dose, then the conditioned flavor preference did not develop. Thus, if the caffeine did not act to reinstate homeostasis, it did not serve as an unconditioned stimulus, and the conditioned flavor preference was not observed.[43] Moreover, people who were not caffeine-users did not develop a liking for the caffeinated drink,[44] suggesting that they were already at an optimal level of arousal, and either caffeine had no effect in these non-users, or it had unpleasant effects due to hyperarousal. Large doses of coffee can lead to feelings of nervousness, anxiety, and tension[28] that are associated with hyperarousal. People with anxiety disorders often avoid caffeine-containing substances, and caffeine could have adverse effects in caffeine avoiders. In support of this hypothesis, Blumenthal et al[45] found that presentation of caffeine-associated stimuli to low-users did not generate conditioned arousal, but tended instead to elicit a compensatory response of hypoarousal that counteracted the caffeine response.

Several studies on the caffeine model of the placebo response have been conducted by the author. Flaten and Blumenthal[1] investigated whether stimuli reliably associated with the effects of caffeine, namely, the smell and taste of coffee, elicited conditioned arousal. This was investigated in a within-participants design (N=21) where participants received either 0 or 2 mg kg^{-1} body weight (bw) caffeine, mixed in either decaffeinated coffee or orange juice (2 caffeine × 2 solution balanced placebo design). The juice without caffeine was the control condition, and the unconditioned effect of caffeine was investigated by adding 2 mg kg^{-1} caffeine to orange juice. The conditioned placebo response was investigated in the decaffeinated coffee condition. Finally, the combination of the conditioned placebo response and the unconditioned drug response could be seen in the condition where coffee with 2 mg kg^{-1} caffeine was administered. All participants drank at least two cups of regular coffee per day, and had done so for at least one year. This ensured that they had experienced several hundred pairings of the taste and smell of coffee with the effect of caffeine. Physiological arousal was measured as increased skin conductance and startle responses, and psychological arousal and stress was assessed. Blood pressure and heart rate were also recorded.

The results showed that caffeine increased several of the arousal indexes, but most important, increased psychological and physiological arousal was also seen after decaffeinated coffee, compared to the control condition. These findings support the hypothesis that presentation of caffeine-associated stimuli increases arousal, a conditioned placebo response, in deprived caffeine-users. The finding that decaffeinated coffee elicited increased physiological and subjective arousal fits well with predictions from the pharmacological conditioning model.[46]

However, the conditioned arousal to caffeine-associated stimuli could be mediated via expectancy.[47] This was tested in a follow-up study[48] which also included information about caffeine content as a third factor; participants were told that they did or did not receive caffeine in their coffee or orange juice. The design was a 2 Caffeine (0, 2 mg kg^{-1}) × 2 Solution (Coffee, orange juice) × 2 Information (told caffeine, told not-caffeine) within-participants factorial design. It was hypothesized that telling people that they were given caffeine should evoke the expectancy that caffeine would have an effect, an expectancy that should not occur when participants were told that they would not receive caffeine in their drink. If the conditioned arousal was mediated via expectancy, then the verbal information should play a large role in modulating the increased arousal seen after decaffeinated coffee. Physiological and psychological arousal was assessed as described by Flaten and Blumenthal.[1] The results indicated that the information played a role, because calmness was decreased and skin conductance levels were increased when participants were informed that they received caffeine in their drink. The verbal information about caffeine content interacted with the type of drink, coffee or orange juice, so that contentedness increased after decaffeinated coffee, especially when participants were informed that they had received caffeine in the coffee. Thus, the conditioned stimulus of the smell and taste of coffee and the verbal information about caffeine content acted in the same direction to produce larger increases in contentedness than either factor alone. However, presentations of decaffeinated coffee also increased alertness and startle reflexes, and this increase was not modified by the type of information provided. Thus, the increased arousal after coffee seemed mainly to have been mediated via classical conditioning and only partly by expectancy.

A third (unpublished) study investigated the relationship of conditioned arousal to the unconditioned arousal generated by caffeine. We know from classical conditioning research that there is a positive correlation between the strength of the unconditioned response and the strength of the conditioned response. A more intense unconditioned stimulus, e.g. an air puff to the eye, supports a stronger conditioned response, e.g. more frequent and larger eyeblink responses.[49,50] The data from Flaten and Blumenthal[1] and Mikalsen et al[48] were re-analyzed with correlation analyses. Since these designs involved a condition where the unconditioned caffeine response was elicited (juice with 2 mg kg^{-1} caffeine) and a condition where the conditioned response was elicited (decaffeinated coffee), responses in these conditions were correlated to provide information about the relationship between the amplitude of the responses in these two conditions. Orange juice without caffeine constituted the control condition. The data illustrate a positive correlation between conditioned and unconditioned arousal. Specifically, there are significant positive correlations between skin conductance responses to caffeine and skin conductance response to decaffeinated coffee. The same is true for alertness.

A fourth study[39] investigated expectancies by asking coffee-drinkers how they expected to feel after one and two cups of coffee. It was found that coffee-drinkers expected coffee to increase arousal in a dose-dependent manner. However, when the participants actually were given one and two cups of decaffeinated coffee, the placebo response was weak. Thus, a strong expectancy for increased arousal was not translated into a strong placebo response when the placebo was actually presented. This underscores the importance of classical conditioning, and indicates that conscious expectancies are not sufficient to elicit a placebo response. In addition, there was a positive correlation between the (conditioned) placebo response and the unconditioned drug response, supporting the (unpublished) finding of Flaten and Blumenthal. This also fits well with the conditioning theory of placebo effects, which postulates that strong drug effects should support strong placebo responses.

One consideration in understanding these prior studies is the fact that all participants were regular coffee drinkers who were required to abstain from caffeine for 12 hours before entering the experimental session. Therefore, these results are based on caffeine consumers who were not at homeostasis at the time of testing.

Previous studies relied on the conditioning history of the participant (regular coffee drinkers). However, conditioning in humans with pharmacological substances can also be generated in the laboratory, where a neutral stimulus is repeatedly paired with a drug in the laboratory. Compared to using naturalistic conditioned stimuli like the taste and smell of coffee, this procedure has the advantage of allowing one to study the development of the conditioned response, because control of the parameters of the conditioned and unconditioned stimulus (the effect of the drug) is better. For example, Flaten et al[2] used a pharmacological conditioning procedure in healthy volunteers, with 700 mg of the muscle relaxant carisoprodol (a prescription drug in Norway that is dispensed under the name Soma in the United States) as the unconditioned stimulus. Carisoprodol inhibits eyeblink reflex magnitude, which was used as the dependent variable. The conditioned stimulus was the administration of the capsules in the laboratory for the CSP group, whereas the control group received placebo capsules (lactose, 700 mg) in the laboratory, and carisoprodol at home. Thus, the laboratory was paired with the carisoprodol administration in the CSP group, but not in the control group. Carisoprodol was given to the control group at home to ensure that both groups had received the same amount of drug to control for habituation or sensitization to carisoprodol.[50] The CSP group received three pairings of the laboratory with the effect of carisoprodol, and the control group received placebo three times in the laboratory, each session spaced exactly one week apart. In each session, eyeblink EMG reflexes to puffs of air to the face were recorded (Fig. 11.2). The results showed that eyeblink EMG was inhibited by 700 mg carisoprodol, but the effect of carisoprodol slightly decreased across sessions, suggesting that tolerance developed. In the fourth session, the CSP group received placebo capsules in the laboratory, i.e. the conditioned stimulus. The control group, on the other hand, received carisoprodol for the first time in the laboratory. In the fourth session, an increase in eyeblink EMG was seen in the CSP group, i.e. the opposite of the response to carisoprodol. This was evidence that a conditioned compensatory response was elicited in this group, and it was hypothesized that this conditioned response antagonized the carisoprodol response, causing tolerance. This hypothesis was supported by the finding of a stronger carisoprodol response in the fourth session in the control group compared to the CSP group. The carisoprodol response was stronger in the control group because no conditioned compensatory response was elicited in this group.

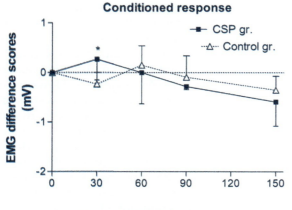
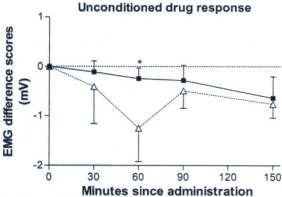

FIGURE 11.2 Top panel: mean EMG difference scores after administration of placebo in session 4 in the CSP group and in session 3 in the control group. The increased blink reflexes in the CSP group compared to the control group at 30 minutes after placebo administration suggest that a carisoprodol-opposite conditioned response was elicited in the CSP group. Bottom panel: mean EMG difference scores after administration of carisoprodol in both groups, in session 3 in the CSP group and session 4 in the control group. Decreased blink reflexes in the control group relative to the CSP group at 60 minutes after drug administration show that carisoprodol had a stronger effect in the control group, again suggesting that a carisoprodol-opposite response was elicited in the CSP group. Reprinted from Flaten et al[2] with permission from the publisher.

These findings fit well with the homeostatic theory discussed above: none of the participants in Flaten et al[2] had medical problems and none reported or used drugs for muscle pain, for which carisoprodol is often prescribed. Thus, the muscle relaxing effect of carisoprodol would represent a deviation from homeostasis in these healthy participants. That homeostatic mechanisms played a role in this experiment was also indicated by the development of tolerance in the CSP group over repeated presentations, which was mediated via a conditioned compensatory response. This shows that conditioned stimuli paired with drugs or events that disrupt homeostasis can elicit conditioned responses that counteract the homeostasis-disrupting effect of the drug or event.[51-53] Thus, seeing the placebo response in terms of an organism that tries to maintain or return to a homeostatic state can explain empirical findings that so far have resisted integration into the expectancy or conditioning theories of placebo effects. Finally, these conditioned compensatory responses could be seen in the placebo response (increased reactivity) that was in the opposite direction of the drug response in the CSP group, an attempt to maintain homeostasis in the conditioned threat to that homeostasis.

Taken together, these data indicate that a conditioned placebo response is not elicited in participants who are at an optimal level of homeostasis. Placebo responses to information that caffeine has been administered should not be seen in non-deprived caffeine users, or in caffeine non-users. Only when homeostasis is disturbed should stimuli paired with substances or events that alleviate this disturbance (e.g. caffeine) generate placebo responses. In fact, compensatory responses to drug-associated stimuli could be seen in participants who are at homeostasis, and such compensatory responses could be responsible for acute or chronic tolerance to drug effects.[54]

Alcohol

Conditioned responses to cues for administration of ethanol have been extensively investigated.[55] These have been motivated by the search for how environmental stimuli may contribute to intoxication, and by a search for how environmental stimuli may affect ethanol tolerance.[56]

The findings of Gundersen et al[55] are of particular interest. They used BOLD fMRI to investigate the effect of alcohol on brain activation in social drinkers, and used a balanced placebo design to control for alcohol-related expectations. Thus, one half of the participants consumed a soft-drink without alcohol before the scanning, and half of these were informed correctly about the drink's content; the other half were informed incorrectly that an alcoholic beverage was consumed. The other half of the participants consumed the soft-drink with alcohol before the scanning, and half of these were informed correctly about the drink's content; the other half were informed incorrectly that a soft-drink was consumed. Interestingly, alcohol decreased neuronal activation, mostly in the dorsal anterior cingulate cortex and in prefrontal areas, while expectancy, i.e. when subjects were told that they got alcohol but received placebo, increased neuronal activation in the same areas. Thus, alcohol intoxication and expectancy had opposite effects on neuronal activation. While the response to placebo is not a *conditioned* compensatory response, it may still be interpreted as a compensatory response to the effects of alcohol.

ACTIVE PLACEBO

A perspective on the placebo response that has received some attention but little empirical study is the hypothesis that drugs can act as active placebos,[32,57,58] meaning that the placebo response can be potentiated or increased by the presence of drug cues. A small drug dose can act as a conditioned stimulus for a larger dose of the same drug.[59]

The changes in arousal induced by verbal information can be potentiated by the administration of drugs, illustrating the active placebo effect. This was investigated by Flaten et al,[32] a study in which participants were told that they had received a stimulant drug, a relaxant drug, or an inactive substance (the control group). Half of the participants in each group received 525 mg carisoprodol (a relaxant), and the other half received 525 mg lactose (placebo). The placebo response to information alone could be investigated in the participants who received information about the drug together with capsules containing placebo. The interaction of the placebo response with the drug response could be seen in the groups who received the same information, but also received active drug. These groups were used to test the hypothesis that an active drug could potentiate the placebo response (the active placebo effect). Blood samples were drawn to determine carisoprodol absorption, and startle reflexes, skin conductance responses, and subjective measures of arousal were also recorded, both before administration of the capsules, and at various points from 15 to 130 minutes after administration of the capsules. Startle reflexes and skin conductance responses were decreased in the group that was told that it had received a relaxant. Subjective tension, indicating arousal, was increased in the group that was told that it had received a stimulant drug. Thus, the information alone generated a placebo response. For the skin conductance and startle reflex data, the placebo response was not potentiated by the administration of carisoprodol. However, the increased tension induced by information that a stimulant had been administered was greatly potentiated by administration of the muscle relaxant carisoprodol. This finding indicates that placebo responses can be augmented by administration of a drug, even if the drug has actions that are opposite to those of the placebo response. Finally, absorption of carisoprodol was faster in the group who were informed that it had received a relaxant: carisoprodol serum levels were significantly higher in this group compared to the other two groups that received carisoprodol. This interesting finding may be due to increased parasympathetic activity in this group, due to being told that it had received a relaxant drug. This hypothesis was investigated in Flaten et al.[60] In relation to the present point of view, the same processes of autonomic activation or deactivation could also modulate the absorption of caffeine.

There are problems with the active placebo hypothesis: because different drugs generate different internal stimulus complexes,[61,62] different drugs seem to interact with instructions in ways that are difficult to predict.[63] Second, individual differences play a role because the active placebo response depends on the anxiety level of the person receiving the drug. For example, informing participants that a drug was a relaxant had opposite effects on those high and low in anxiety.[57] Finally, Mitchell et al[64] found that providing incorrect information about drug effects, e.g. telling participants that a placebo was administered, but administering the stimulant d-amphetamine instead, increased anxiety. This indicates that a mismatch between the expected drug effect (i.e. the internal drug discriminative stimulus) and the experienced drug effect could lead to increased arousal and anxiety.

COMPENSATORY RESPONSES AND THE NOCEBO EFFECT

There may appear to be similarities between the concept of compensatory responses and that of nocebo effects. Compensatory responses are opposite to and counteract the drug response, and increase tolerance to drugs, and this is not beneficial for treatment of symptoms and diseases, of course. Nocebo effects are the increase in symptoms after information that a treatment will increase, e.g. pain, that likewise is not beneficial for treatment. Solomon's[65] opponent-process theory of motivation deals with this issue. According to this theory, two reactions are elicited by any stimulus or drug: the unconditioned response, and a compensatory reaction, or A and B processes in Solomon's terminology. After repeated administration of the stimulus or drug, tolerance or habituation may be observed, and this is hypothesized to result from a gradually stronger compensatory reaction. The compensatory reaction can be viewed as an instance of a homeostatic process that returns the organism to the normal pre-drug state. Solomon's theory provides a framework in which to understand nocebo reactions as homeostatic processes.

Another perspective on the nocebo reaction comes from the work of Benedetti et al,[66] where nocebo effects in the form of hyperalgesia are the result of anxiety. Negative emotions do in many cases increase pain.[67] In this view, nocebo reactions are not seen as compensatory, homeostatic reactions resulting from negative feedback mechanisms. Opposite to this, nocebo reactions are seen as resulting from positive feedback loops, where increased anxiety leads to increased pain that subsequently leads to increased anxiety, i.e. the opposite of homeostatic control and compensatory mechanisms.

In the present context, compensatory conditioned responses could represent the opposite of placebo responses, and are called nocebo responses. The central argument is that placebo responses can be observed in patients suffering from a symptom, or in healthy volunteers where a symptom is induced. Compensatory reactions, on the other hand, are seen in healthy volunteers, in homeostasis without any relevant symptom, but to whom a drug is administered.

SUMMARY AND CONCLUSIONS

Both drug-antagonistic and -agonistic reactions can be observed to signals that a painkiller will be administered. Theories for the prediction of whether agonistic or antagonistic conditioned responses will occur have been put forth, and this chapter has discussed some of these models. A model taking into account the homeostatic level of the organism at the time of drug presentation may explain when drug antagonistic and drug agonistic responses are observed.

References

1. Flaten MA, Blumenthal TD. Caffeine-associated stimuli elicit conditioned responses: an experimental study of the placebo effect. *Psychopharmacology*. 1999;145:105-112.
2. Flaten MA, Simonsen T, Waterloo KK, Olsen H. Pharmacological classical conditioning in humans. *Hum Psychopharmacol Clin Exp*. 1997;12:369-377.
3. Pavlov IP. *Conditioned Reflexes*. New York: Dover Publications; 1927.
4. Siegel S. Explanatory mechanisms for placebo effects: Pavlovian conditioning. In: Guess HA, Kleinman A, Kusek JW, Engel LW, eds. *The Science of the Placebo*. London: BMJ Books; 2002:133-157.
5. Catania AC. Discriminative stimulus functions of drugs: Interpretations. In: Thompson T, Pickens R, eds. *Stimulus Properties of Drugs*. New York: Appleton-Century-Crofts; 1971:149-155.
6. Eikelboom R, Stewart J. Conditioning of drug-induced physiological responses. *Psychol Rev*. 1982;89:507-528.
7. Ramsay DS, Woods SC. Biological consequences of drug administration: implications for acute and chronic tolerance. *Psychol Rev*. 1997;104:170-193.
8. Siegel S. Morphine analgesic tolerance: its situation specificity supports a Pavlovian conditioning model. *Science*. 1976;193:323-325.
9. Siegel S. Classical conditioning, and opiate tolerance and withdrawal. In: Balfour DJK, ed. *Psychotropic Drugs of Abuse. International Encyclopedia of Pharmacology and Therapeutics*. Elmsford, New York: Pergamon Press; 1990:59-85.
10. Sokolowska M, Siegel S, Kim JA. Intra-administration associations: conditional hyperalgesia elicited by morphine onset cues. *J Exp Psychol: Anim Behav Proc*. 2002;28:309-320.
11. Kim JA, Siegel S. The role of cholecystokinin in conditional compensatory responding and morphine tolerance in rats. *Behav Neurosci*. 2001;115:704-709.
12. Siegel S, Sherman JE, Mitchell D. Extinction of morphine tolerance. *Learn Motiv*. 1980;11:289-301.
13. Sherman JE, Proctor C, Strub H. Prior hot plate exposure enhances morphine analgesia in tolerant and drug-naive rats. *Pharmacol Biochem Behav*. 1982;17:229-232.
14. Amanzio M, Benedetti F. Neuropharmacological dissection of placebo analgesia: expectation-activated opioid systems versus conditioning-activated specific subsystems. *J Neurosci*. 1999;19:484-494.
15. Benedetti F, Amanzio M, Rosato R, Blanchard C. Nonopioid placebo analgesia is mediated by CB1 cannabinoid receptors. *Nature Med*. 2011;17:1228-1230.
16. Voudouris NJ, Peck CL, Coleman G. Conditioned placebo responses. *J Pers Soc Psychol*. 1985;48:47-53.
17. Colloca L, Benedetti F. How prior experience shapes placebo analgesia. *Pain*. 2006;124:126-133.
18. Colloca L, Petrovic P, Wager TD, et al. How the number of learning trials affects placebo and nocebo responses. *Pain*. 2010;151:430-439.
19. Daly JW. Mechanism of action of caffeine. In: Garattini S, ed. *Caffeine, Coffee, and Health*. New York: Raven Press; 1993:97-150.
20. Nehlig A, Daval JL, Debry G. Caffeine and the central nervous system: mechanisms of action, biochemical, metabolic and psychostimulant effects. *Brain Res Rev*. 1992;17:139-170.
21. Benowitz NL, Jacob III P, Mayan H, Denaro C. Sympathomimetic effects of paraxanthine and caffeine in humans. *Clin Pharmacol Ther*. 1995;58:684-691.
22. Flaten MA. Caffeine-induced arousal modulates somatomotor and autonomic differential classical conditioning in humans. *Psychopharmacol*. 1998;135:82-92.
23. Griffiths RR, Woodson PP. Caffeine physical dependence: a review of human and laboratory animal studies. *Psychopharmacol*. 1988;94:437-451.
24. Mosqueda-Garcia R, Robertson D, Robertson RD. The cardiovascular effects of caffeine. In: Garattini S, ed. *Caffeine, Coffee, and Health*. New York: Raven Press; 1993:157-176.
25. Blanchard J, Sawers SJA. Comparative pharmacokinetics of caffeine in young and elderly men. *J Pharmacokin Biopharmaceut*. 1983;11:109-126.
26. Rogers PJ, Richardson NJ, Elliman NA. Overnight caffeine abstinence and negative reinforcement of preference for caffeine-containing drinks. *Psychopharmacol*. 1995;120:457-462.
27. Evans SM, Griffiths RR. Caffeine tolerance and choice in humans. *Psychopharmacol*. 1991;108:51-59.
28. Griffiths RR, Mumford GK. Caffeine reinforcement, discrimination, tolerance and physical dependence in laboratory animals and humans. In: Schuster CR, Kuha MJ, eds. *Handbook of Experimental Pharmacology*. Heidelberg: Springer Verlag; 1996:315-341.
29. Dack C, Reed P. Caffeine reinforces flavor preference and behavior in moderate users but not in low caffeine users. *Learn Motiv*. 2009;40(1):35-45.
30. Koob GF, Le Moal M. Drug addiction, dysregulation of reward, and allostasis. *Neuropsychopharmacol*. 2001;24:97-129.
31. Siegel S, Ramos BMC. Applying laboratory research: drug anticipation and the treatment of drug addiction. *Exp Clin Psychopharmacol*. 2002;10(3):162-183.
32. Flaten MA, Simonsen T, Olsen H. Drug-related information generates placebo and nocebo responses that modify the drug response. *Psychosom Med*. 1999;61:250-255.
33. Lyerly SB, Ross S, Krugman AD, Clyde DJ. The effects of instructions upon performance and mood under amphetamine sulphate and chloral hydrate. *J Abn Soc Psychol*. 1964;68:321-327.
34. Jensen MP, Karoly P. Motivation and expectancy factors in symptom perception: a laboratory study of placebo effects. *Psychosom Med*. 1991;53:144-152.
35. Penick SB, Fisher S. Drug-set interaction: psychological and physiological effects of epinephrine under differential expectations. *Psychosom Med*. 1965;27:177-182.
36. Sternbach RA. The effects of instructional set on autonomic responsivity. *Psychophysiol*. 1964;1:67-72.
37. Meissner K. Effects of placebo interventions on gastric motility and general autonomic activity. *J Psychosom Res*. 2009;66(5):391-398.

38. Wilder J. The law of initial values in neurology and psychiatry. *J Nerv Ment Dis*. 1957;125:73-86.
39. Flaten MA, Åsli O, Blumenthal TD. Expectancies and responses to caffeine-associated stimuli. *Psychopharmacol*. 2003;169:198-204.
40. Silverman K, Evans SM, Strain EC, Griffiths RR. Withdrawal syndrome after the double-blind cessation of caffeine consumption. *N Engl J Med*. 1992;327:1109-1114.
41. Lane JD. Effects of brief caffeinated-beverage deprivation on mood, symptoms, and psychomotor performance. *Pharm Biochem Behav*. 1997;58:203-208.
42. Griffiths RR, Evans SM, Heishman SJ, et al. Low-dose caffeine discrimination in humans. *J Pharm Exp Ther*. 1990;252:970-978.
43. Yeomans MR, Spetch H, Rogers PJ. Conditioned flavour preference negatively reinforced by caffeine in human volunteers. *Psychopharmacol*. 1998;137:401-409.
44. Richardson N, Rogers PJ, Elliman NA. Conditioned flavour preferences reinforced by caffeine consumed after lunch. *Physiol Behav*. 1996;60:257-263.
45. Blumenthal TD, Gambill J, Burnett T, et al. Absence of placebo responses in the absence of caffeine-related conditioning history. *Psychophysiology*. 1999;36(suppl):S34.
46. Ader R. Processes underlying the placebo effects. The preeminence of conditioning. *Pain Forum*. 1997;6:56-58.
47. Montgomery GH, Kirsch I. Classical conditioning and the placebo effect. *Pain*. 1997;72:107-113.
48. Mikalsen A, Bertelsen B, Flaten MA. Effects of caffeine, caffeine-associated stimuli, and caffeine-related information on physiological and psychological arousal. *Psychopharmacol*. 2001;157:373-380.
49. Ross LE, Hunter JJ. Habit strength parameters in eyelid conditioning as a function of UCS intensity. *Psychol Rec*. 1959;9:103-107.
50. Smith MC. CS-US interval and US intensity in classical conditioning of the rabbit's nictitating membrane response. *J Comp Phys Psychol*. 1968;66:679-687.
51. Robbins SJ, Ehrman RN. Designing drug conditioning studies in humans. *Psychopharmacol*. 1992;106:143-153.
52. Poulos CX, Cappell H. Homeostatic theory of drug tolerance: a general model of physiological adaptation. *Psychol Rev*. 1991;98:390-408.
53. Siegel S, Allan LG. Learning and homeostasis: drug addiction and the McCollough effect. *Psychol Bull*. 1998;124:230-239.
54. Siegel S, Baptista AS, Kim JA, et al. Pavlovian psychopharmacology: the associative basis of tolerance. *Exp Clin Psychopharmacol*. 2000;8:276-293.
55. Gundersen H, Specht K, Gruner R, et al. Separating the effects of alcohol and expectancy on brain activation: an fMRI working memory study. *Neuroimage*. 2008;42:1587-1596.
56. Birak KB, Higg S, Terry P. Conditioned tolerance to the effects of alcohol on inhibitory control in humans. *Alcohol Alcoholism*. 2011;46:686-693.
57. Dinnerstein AJ, Lowenthal M, Blitz B. The interaction of drugs with placebos in the control of pain and anxiety. *Persp Biol Med*. 1966;10:103-117.
58. Kirsch I, Sapirstein G. Listening to prozak but hearing placebo: a meta-analysis of antidpressant medication. *Prevent Treat*. 1998;1: article 0002a.
59. McDonald RV, Siegel S. Intra-administration associations and withdrawal symptoms: morphine-elicited morphine withdrawal. *Exp Clin Psychopharmacol*. 2004;12:3-11.
60. Flaten MA, Simonsen T, Åsli O. The effect of stress on absorption of acetaminophen. *Psychopharmacol*. 2006;185:471-478.
61. Colpaert FC, Niemegeers CJE, Janssen PAJ. Theoretical and methodological considerations on drug discrimination learning. *Psychopharmacologia*. 1976;46:169-177.
62. Flaten MA, Simonsen T, Olsen H, et al. A study of active placebo. *Psychopharmacol*. 2004;176:426-434.
63. Dinnerstein AJ, Halm J. Modification of placebo effects by means of drugs – effects of aspirin and placebos on self-rated moods. *J Abn Psychol*. 1970;75:308-314.
64. Mitchell SH, Laurent CL, deWit H. Interaction of expectancy and the pharmacological effects of d-amphetamine: subjective effects and self-administration. *Psychopharmacol*. 1996;125(4):371-378.
65. Solomon RL. The opponent process theory of acquired motivation. The cost of pleasure and the benefit of pain. *Am Psychol*. 1980;35: 691-712.
66. Benedetti F, Amanzio M, Casadio C, et al. Blockade of nocebo hyperalgesia by the cholecystokinin antagonist proglumide. *Pain*. 1997;71:135-140.
67. Rhudy JL, Williams AE, McCabe KM, et al. Emotional control of nociceptive reactions (ECON): do affective valence and arousal play a role? *Pain*. 2008;136:250-261.

CHAPTER 12

Placebo Analgesia, Nocebo Hyperalgesia, and Acupuncture

Jian Kong, Randy L. Gollub

Psychiatry Department, Massachusetts General Hospital and Harvard Medical School, MA, USA

Humans have used acupuncture treatment and placebo treatment (or more accurately, medical rituals)[1,2] for thousands of years. Only in modern times has the effectiveness of acupuncture come into question. The methodology and philosophy underlying acupuncture has become an obstacle to acceptance by practitioners of modern medicine because acceptance by those who practice 'Western' medicine requires evidence that the therapeutic effects of acupuncture treatment are superior to that of a placebo form of treatment. As a result, we have begun to wonder if acupuncture is simply a very effective form of placebo treatment or if acupuncture produces specific effects in addition to what are its increasingly evident placebo effects.

Acupuncture is a component of the 'Traditional Chinese Medicine' (TCM) system. The basic theory of acupuncture involves meridians and acupuncture points (acupoints). Meridians are believed to be channels within which Qi, or life force, flows. Acupuncture points are found primarily at specific locations along the meridians. It is worth noting that some practicing acupuncturists also use points which are not on the meridians, such as 'extraordinary' points with specific therapeutic properties, 'Ashi' (tender/reflexive) points unique to each individual patient, and auricular (ear) points. Based on TCM/acupuncture theory, stimulating acupoints with needles, Moxa (slowly burning mugwort herb) or other tools can regulate the flow of Qi and blood to keep the functions of the body in harmony.

Placebo, from the Latin term meaning 'I shall please', is an inert treatment for a disease or other medical condition. In their book *The powerful placebo: From Ancient Priest to Modern Physician*, Shapiro and Shapiro state that the history of medical treatment is essentially the history of the placebo effect,[1] implying a long history of placebo practice. It's worth noting that, in the book, the authors also used acupuncture as an example of such placebo practice, stating: 'controlled studies have failed to confirm its effectiveness'. The use of the term 'placebo' in a medical context, describing inert treatments used to comfort a patient, dates from at least the end of the 18th century.[3] Mainstream interest in placebo effects only began with the adoption of the randomized controlled trial (RCT), in which investigators found significant improvement in the placebo control group of some studies.[3]

Theoretically at least, both acupuncture and placebo may work by activating a 'self-regulation' system in the body, thus at some level, the pathways involving the two treatment modalities may share common mechanisms. For instance, studies have shown that endogenous opioids and cholecystokinin are involved in both acupuncture analgesia[4,5] and placebo analgesia.[6,7] Under the umbrella of self-regulation, shared mechanisms between the two may further blur the distinction between acupuncture and placebo.

Conversely, while there is no evidence to suggest that acupuncture treatment exacerbates or causes pain or illness, with the exception of adverse effects such as pneumothorax, bacterial/viral infections that are due to improper acupuncture practice.[8] We have shown in an experimental paradigm that sham acupuncture treatment can be used in the opposite way to elicit a nocebo effect. The term 'nocebo effects' refers to adverse events produced by negative expectations and represent the negative side of placebo effects.[9] Unlike placebo effects, which have drawn the attention of both the scientific community and the public, nocebo effects have received relatively scant attention from the field of neuroscience[9] (for more details please also see the chapter on nocebo in this book).

In this chapter, we start by elucidating the complex relationship between acupuncture and the placebo effect that has been illuminated by the practicalities of conducting research investigations in this field. In particular, we explore the complexity of defining an inert (sham) acupuncture treatment that respects traditional acupuncture theory. Then, we review the results of research studies that have begun to investigate the dissociation and interaction between positive and negative placebo effects and acupuncture treatment. Finally, we suggest future directions of research in this field.

IS ACUPUNCTURE A FORM OF PLACEBO TREATMENT?

Long ago, the Ancient Chinese acknowledged the effects of expectancy and belief in healing. For instance, in the Yellow Emperor's Inner Classic (Huang Di Nei Jing), the oldest canonical classic of Chinese medicine written in the first century BCE, it states: 'if a patient does not consent to therapy with positive engagement, the physician should not proceed as the therapy will not succeed' (SuWen Chapter 11). It seems reasonable to conclude that from the very inception of this treatment modality, experienced clinicians recognized the essential role of patient expectations and participation in the procedural rituals to obtain a salubrious outcome.

To date, acupuncture has been studied in about 1000 different RCTs.[10–12] In these trials, it is not uncommon for sham acupuncture, a placebo form of acupuncture, to induce positive therapeutic effects similar to those of verum acupuncture. While individual patient outcomes vary widely, both verum and sham groups usually demonstrate superiority and clinical benefits when compared with wait list control groups.[12–18]

In a landmark clinical trial on acupuncture treatment of chronic low back pain, Cherkin and colleagues[19] compared the treatment effects of individualized acupuncture, standardized acupuncture, simulated acupuncture, and usual care. After eight weeks of treatment, mean pain-related dysfunction scores for the individualized, standardized, and simulated acupuncture groups improved by 4.4, 4.5, and 4.4 points respectively, compared with 2.1 points for those receiving usual care. The results from this study raised more questions than they answered. *Why can sham acupuncture produce such a powerful placebo effect? What is the best predictor of acupuncture treatment outcome? What is the best inert sham acupuncture treatment? What is/are the most crucial component(s) of acupuncture?*

In an attempt to find predictors of acupuncture response, Linde and colleagues[20] reanalyzed the results from several RCTs of acupuncture treatment for chronic pain and found that expectation of relief was the only factor that correctly predicted outcome. In support of this hypothesis, Kaptchuk and colleagues[21] found that relative to sham acupuncture treatment delivered to patients with limited treatment provider engagement, sham acupuncture delivered with augmented context (more empathy, attention and rituals from the acupuncturist) increased expectation of a good outcome and indeed produced a significant increase in clinical benefit. Taken together, these studies suggest that many aspects of acupuncture treatment, including needle penetration, empathic touch, and other related ritual procedures as well as patients' belief/expectancy, may all contribute to a large effect in sham acupuncture treatment.

Due to contradictory results from individual clinical trials, investigators attempted to pool data from multiple studies to estimate the effect size of acupuncture treatment to reduce chronic pain. Initial meta-analytic reviews tended to arrive at negative conclusions (e.g.[11]), while more recent reviews tend to draw more positive conclusions (e.g.[8]). This may be due to improvements in the quality of clinical trial methods in acupuncture research. In the most recent meta analysis, Vickers and colleagues[22] pooled the individual subject data from 29 high-quality RCTs (17922 patients in total) to estimate the efficacy of acupuncture treatment on four chronic pain conditions: back and neck pain, osteoarthritis, chronic headache, and shoulder pain. Results indicate that acupuncture treatment was superior to both sham and no-acupuncture control for all four pain conditions, and further that the effect sizes were similar across pain conditions when outlier trials were excluded. These results provide strong evidence that acupuncture is more than a placebo (Fig. 12.1). Nevertheless, the results also indicate the differences between true and sham acupuncture are relatively modest, implying non-specific effects are important contributors to acupuncture therapeutic effects.

An important observation from these reviews is the heterogeneity in the efficacy of acupuncture treatment for different pain disorders, i.e. reporting a positive conclusion for acupuncture treatment for chronic pain conditions such as low back pain, osteoarthritis and neck pain, but a negative conclusion for other chronic pain conditions such as fibromyalgia.[8] Thus, although there is little truly convincing evidence that acupuncture is effective in reducing all types of clinical pain, for some specific chronic pain disorders the evidence for acupuncture effectiveness is very strong. Further investigation of the disorder specificity of acupuncture efficacy may enhance our understanding of mechanisms underlying both acupuncture and chronic pain disorders.

In summary, although the sources of heterogeneity in the efficacy of acupuncture treatment on chronic pain are not clear, there is strong evidence indicating that acupuncture can produce a moderately greater effect than

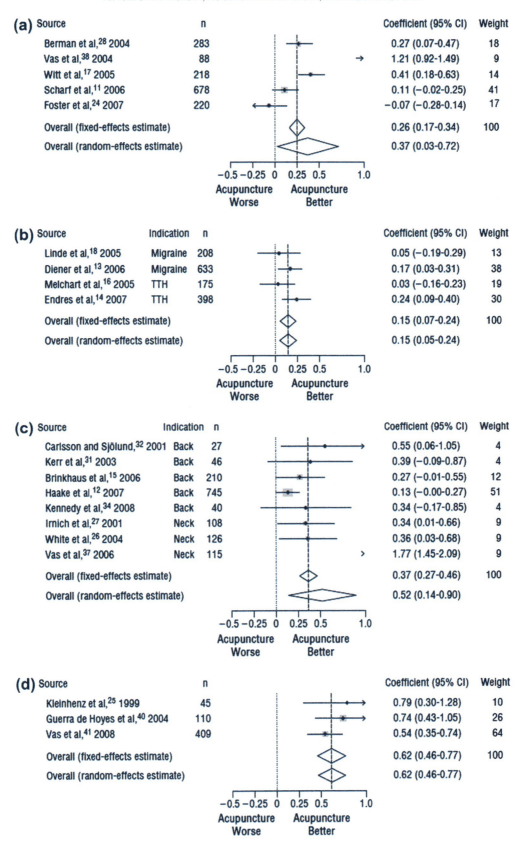

FIGURE 12.1 Forest plots for the comparison of true and sham acupuncture across different chronic pain disorders from Vivkers et al.[22] (A) Osteoarthritis; (B) chronic headache; (C) musculoskeletal pain; (D) shoulder pain.

sham treatment for specific chronic pain conditions. This conclusion coupled with the very low incidence of risks and side effects calls for the incorporation of acupuncture treatment into management of these chronic pain disorders.

CHALLENGES AND ISSUES IN PLACEBO/SHAM ACUPUNCTURE STUDIES

Placebo effects are psychobiologic events attributable to the overall therapeutic context in which an inert treatment is given.[3] One of the greatest challenges in acupuncture research is the difficulty in finding an appropriate inert acupuncture treatment. This is partly due to the lack of a clear definition of acupuncture and limitations in our understanding of the underlying mechanisms of acupuncture treatment.

For example, although accurately stimulating acupoints is historically an important aspect of effective acupuncture, studies have shown that similar treatment effects could be produced by stimulating areas outside of the boundaries of the proper acupoints or meridians.[23] These findings raise other important questions. *If administration of acupuncture to an area that is not an acupoint produces treatment effects, should we regard acupuncture as having a placebo effect? Or should we be re-working our current definition of genuine acupuncture acupoints to include those other areas?* For instance, an area of active research on its own is the investigation of innovative hypotheses regarding the mechanisms of how acupuncture needle manipulation of connective tissue translates into homeostatic biomechanical signaling of neuronal, immunological and endocrine systems.[24-29]

In addition to acupuncture point specificity, we should also be re-thinking our current definition of what constitutes a proper acupuncture tool. Acupoints can be stimulated by 'superficial' stimulation, that is, stimulation without needle insertion. Indeed, there is an entire system of treatment, known as Jin Shin or acupressure, that uses light touch of the treater's fingertips over the acupoints as the 'tool'.[30] For another example, transcutaneous electrical nerve stimulation (TENS), which involves a patch that electrically stimulates a specific area of skin, could be considered a form of acupuncture in which the acupuncture tool (the TENS patch) is stimulating a large acupoint area.[5] Ancient traditional Chinese literature lists nine forms of acupuncture tools, including both sharp needles that penetrate the skin at a small, specific acupoint, and smooth-tipped needles that stimulate larger acupoint areas. The ancient literature calls all nine types of acupuncture tools 'needles', regardless of whether they puncture the skin (Neijing, SuWen, Chapter 1). Today, smooth-tipped needles that produce treatment effects are generally regarded as having produced a placebo effect; however, traditional Chinese acupuncture would most likely regard the use of smooth-tipped needles as one of the multiple, genuine forms of acupuncture.

Currently, investigators use a variety of tools as sham acupuncture devices for research, including tooth picks with guide tubes, fake or minimum TENS, and different kinds of sham needles. The sham acupuncture needle developed by Streitberger and colleagues has become popular in acupuncture research. This validated device[31-34] consists of a sham needle inserted through a small, tape-covered plastic ring. The needle retracts into its casing when pressed on the skin, similar to the action of a retractable stage knife. A second sham needle, developed by a Japanese group,[35] is gaining in popularity. The tip of this double-blind non-penetrating sham needle simply presses against the skin.[35] The appearance and feel of these needles are designed to be indistinguishable to the treater from real acupuncture needles, which do penetrate the skin. Thus, both patients and acupuncturist can be blinded in clinical trials.

The general assumption about using placebo needles is that non-penetrating needle stimulation cannot produce treatment effects. As noted above, according to Traditional Chinese Medicine, this assumption is not necessarily true. The advantage of using placebo needles is the ability to mimic genuine acupuncture procedures in all aspects of clinical practice. To avoid the potential limitations of non-acupoints and sham needles, some investigators have attempted to combine the two methods, using sham acupuncture on non-acupoints, which may be a more appropriate, albeit still imperfect, control.

SUBJECTIVE AND OBJECTIVE MEASUREMENTS IN ACUPUNCTURE AND PLACEBO STUDIES

Outcome measurement is a crucial component of evaluating treatment effects. In placebo studies, it is not uncommon to find a discrepancy between objective and subjective improvements from placebo treatments.[36-41] For instance, in a within-subject, repeated measures, crossover study published in the *New England Journal of Medicine*, the authors[41] randomly assigned 46 patients with asthma to active treatment with an albuterol inhaler, a placebo inhaler, sham acupuncture, or no intervention. At each visit, spirometry was performed repeatedly over a period of 2 hours. Maximum forced expiratory volume in 1 second (FEV(1)) was measured, and patients' self-reported improvement in breathing ratings were recorded. The results showed that albuterol resulted in a 20% increase in FEV(1), as compared with approximately 7% increases with each of the other three interventions. However, patients' reports of improvement after the intervention did not differ significantly for the albuterol inhaler (50% improvement), placebo inhaler (45%), or sham acupuncture (46%).

Similarly, in another study, Fregni and colleagues[42] investigated the acute effect of levodopa, placebo pill or sham transcranial magnetic stimulation on the motor function of patients with Parkinson's disease on three different occasions using a crossover design. They found that only the levodopa treatment resulted in objectively measured motor function improvement, yet patient's subjective report of motor function improvement after the two different placebo treatments was equal to that of the levodopa.

CONTRIBUTION OF NEUROIMAGING TO ACUPUNCTURE AND PLACEBO/NOCEBO RESPONSE

Both because of this need for objective measurements and because of the clear role of the brain in mediating the effects of acupuncture and placebo, investigators are turning to imaging methods, such as functional magnetic resonance imaging (fMRI) and positron emission tomography (PET). Technical improvements in fMRI due to more powerful magnets, increasingly sophisticated imaging hardware, and, in particular, the development of new experimental paradigms and data analysis methods, allow the investigation of neural events as dynamic processes within the whole brain. Both the spatial and temporal aspects of neural activity underlying placebo and acupuncture treatment can be explored. Technical advances in PET imaging not only provide tools for investigating brain metabolism, blood flow changes, and other non-selective markers of neural activity, but also for investigating whole-brain determinants of specific receptor-binding distributions in fully conscious humans. Such progress enables us to indirectly assess neurotransmitter changes associated with placebo analgesia.

Several functional neuroimaging placebo acupuncture studies have been published.[43–45] The results are reminiscent of earlier non-imaging studies in that they have shown significant differences between verum and sham acupuncture in objectively measured imaging markers of brain activation in the setting of no significant differences between the two treatments with regard to subjective measurements of treatment effect.

In a previous review,[46] we posed a theoretical framework for interpreting the results of the neuroimaging literature of placebo analgesia. According to this framework, placebo treatment may exert an analgesic effect on at least three stages of pain processing, by (1) influencing pre-stimulus expectation of pain relief, (2) modifying pain perception, and (3) distorting post-stimulus pain rating. Based on the framework and on our subsequent studies,[46–50] we speculate that some level of bias/distortion in subjective pain ratings may account for this outcome, as in cognitive neuroscience, subjective reports can be significantly biased on account of previous experience and expectation.[51,52] In a recent study,[50] we found a placebo-related analgesia effect evoked by visual cues to be significantly correlated with hippocampal activation during pain rating. (We used placebo-related analgesia here because there is no placebo administered in this study. However, in both placebo analgesia evoked by placebo treatment and placebo-related analgesia evoked by a visual cue, the psychosocial context plays a key role.[53])

Over the past few years, we have conducted several studies to investigate both the placebo and nocebo effects produced by acupuncture. One advantage of studying sham acupuncture is that it is novel and unfamiliar to many individuals in western culture, unlike pills, ointments, or injections. This minimizes the variance in subjective response introduced by prior conditioning. These studies also directly compared placebo analgesia with acupuncture analgesia and investigated how expectancy can modulate the effect of acupuncture treatment on pain.

Sham Acupuncture Evoked Placebo Effect

One challenge of studying placebo analgesia in healthy subjects is that expectancy evoked by verbal cues is, in general, weak compared to that in patients.[54–56] This is particularly true in the United States where acupuncture is a novel treatment method and thus there is no prior positive conditioning from previous healing experiences. Thus, the aim of our first study[57] was to test whether sham acupuncture could elicit detectable placebo analgesia in healthy subjects.

To overcome the limitation of weak placebo effect in healthy, acupuncture-naïve subjects, we applied an expectancy manipulation model[47,48,58,59] to boost the analgesic effect of sham acupuncture. This conditioning method has been widely used in neuroimaging placebo analgesia studies.[60–64] Specifically, we performed a three-session study in which all subjects participated in two behavioral testing sessions and one fMRI scanning session, separated by a minimum of three days.

The first behavioral session was used to determine appropriate stimulus intensities for each subject, to minimize anticipatory anxiety, and to train subjects to rate a noxious thermal pain stimulus using the Sensory Box and Affective Box 0–20 scales.[65–67]

The second behavioral session was designed to manipulate the subjects' expectancy of acupuncture analgesia. At the beginning of the session, subjects viewed a figure depicting how acupuncture meridian lines connect a set

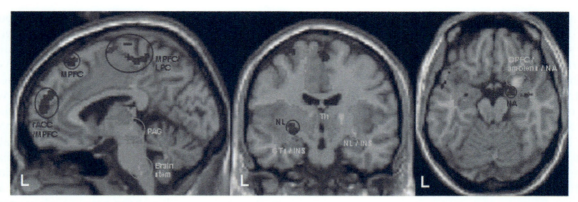

FIGURE 12.2 Representative brain regions involved in expectancy (blue) and acupuncture treatment (green) from ANOVA analysis across four groups.[44] The red color indicates the mask of high pain minus low pain across four groups. L indicates left side of the brain, R indicates right side of the brain. rACC, rostral anterior cingulate cortex; MPFC, medial prefrontal cortex; LPC, paracentral lobule; PAG, periaqueductal gray; NL, lentiform nucleus; INS, insula; OPFC, orbital prefrontal cortex; NA, amygdala. *This figure is reproduced in color in the color section.*

of acupuncture points according to Traditional Chinese Medicine. Subjects were then told that according to acupuncture theory and previous reports, acupuncture would only produce an analgesic effect on the side of the arm where needles were placed ('treatment' side), but not on the other side of the arm where there were no needles (control side). Finally, subjects were told that they would be receiving identical sequences of painful heat stimuli before and after treatment so that if they were 'good' acupuncture responders, they would feel less pain on the treated (sham acupuncture) than on the untreated (control) side of the arm after the acupuncture treatment. In reality, the information they were told is likely not true (Fig. 12.2).

Next, three identical sequences of painful stimuli were applied to areas on both sides of the arm. Sham acupuncture using Streitberger sham needles[31–34] applied at non acupoints was then performed for 5 minutes. After treatment, the same pain sequences were applied to the same areas on the control side of the arm, but surreptitiously lowered temperatures were used when the sequences were applied to the areas on the sham acupuncture treatment side (the 'treated' side of the arm). This gave them an unmistakable 'analgesic' experience. After this manipulation, all subjects reported greater expectation for pain relief following acupuncture treatment.

Session three took place inside the fMRI scanner. Subjects were told that the exact procedures of Session 2 would be repeated while the subject was inside the scanner to investigate the brain networks involved in acupuncture analgesia. In reality, subjects only received the same pre-treatment stimuli sequences and sham acupuncture treatment on the same side of their arm as they did in Session 2. In contrast to Session 2, this time after treatment, subjects received the sequence with lowered temperature stimulation on the first area of the treated side of the arm and instead received identical sequences with unchanged temperatures on the remaining areas of both the treated and untreated sides of their arm. Thus we were able to analyze fMRI scans of the subjects collected while they experienced exactly identical, calibrated noxious heat stimuli, where the only difference was that they expected to feel less pain after treatment on spots that were on the 'treated' side of their arm.

Sham acupuncture with the manipulation reduced subjective pain rating (pre minus post) significantly more on the treated side of the arm compared to the control side. When the contrast that subtracts the fMRI signal difference (pre minus post-treatment) during application of the noxious stimulus on the treated side of the arm from the same difference on the control side (e.g. placebo (post-pre)—control (post-pre)) was calculated, highly localized significant differences were observed in brain activation in the bilateral rostral anterior cingulate cortex (rACC), lateral prefrontal cortex, right anterior insula, and left inferior parietal lobule. Interestingly, the pattern of brain activity changes differ from a similar previous expectancy enhanced, placebo analgesia fMRI study[60] that found decreased activity in pain-sensitive regions such as the thalamus, insula, and ACC when comparing the response to noxious stimuli applied to control and placebo cream-treated areas of the skin. Our results suggest that placebo analgesia may be configured through multiple brain pathways and that different placebo modalities, pain stimuli paradigms and other experimental details may all influence the final results.

Sham Acupuncture Evoked Nocebo Effect

Previous studies suggest that nocebo effects, sometimes termed 'negative placebo effects', contribute appreciably to a variety of medical symptoms,[68,69] adverse events in clinical trials and medical care,[70–73] and public health 'mass psychogenic illness' outbreaks.[74] For instance, Ko and colleagues[75] found patients who received a beta-blocker and patients who received a placebo reported comparable levels of common side effects, including depressive symptoms, fatigue, and sexual dysfunction.

In previous studies, neurophysiologic theories to account for nocebo effects have been tested in human subjects. For instance, Benedetti and colleagues[76] found that while proglumide, a nonspecific cholecystokinin antagonist, could counteract nocebo-induced hyperalgesia, the opioid antagonist naloxone had no effect on nocebo responses. In another study,[77] they found that while the benzodiazepine diazepam could block both nocebo hyperalgesia and hypothalamic–pituitary–adrenal (HPA) hyperactivity, proglumide could block only the former. The results indicate that proglumide may act specifically on the CCK-mediated link between pain and anxiety that may or may not be the same site of action of the benzodiazepine-mediated anti-nociception. These studies underscore the complexity of these neural systems and the need for further work.

In an fMRI study from our group,[78] the neural mechanism of nocebo hyperalgesia evoked by sham acupuncture treatment was investigated using a variation on the modified expectancy model described above. The main difference between this study[78] and the placebo analgesia study discussed above[57] is that, in this study, subjects were told at the beginning of the session that acupuncture can sometimes produce hyperalgesia and we increased the pain stimulus temperature on the treated side of the forearm after treatment to create the expectation of pain worsening, instead of decreasing the stimulus temperature in the manipulation session on the treated side to enhance the expectancy of pain relief. The nocebo hyperalgesia effect of sham acupuncture treatment was then tested while subjects were in the fMRI scanner (Session 3).

We found that after administering sham acupuncture treatment, subjective pain intensity ratings increased significantly more in areas of the arm associated with the nocebo treatment compared with the control areas where no expectancy conditioning manipulation was performed. fMRI analysis of hyperalgesic nocebo responses to identical calibrated noxious stimuli showed signal increases in brain regions, including bilateral dorsal ACC, insula, medial frontal gyrus, orbital prefrontal cortex, superior parietal lobule and hippocampus; and right claustrum/putamen, and lateral prefrontal gyrus. These results are consistent with the interpretation that nocebo hyperalgesia is mediated by activity in a network referred to as the affective–cognitive pain pathway (medial pain system) and further, that the left hippocampus may play an important role in this process. The crucial role of hippocampus in negative expectancy and nocebo has been supported by a subsequent study from another group,[79] in which the analgesic effect of the potent mu opioid analgesic, remifentanil, was both positively and negatively manipulated using a clever, within-subject design. After an initial conditioning session, the subjects were tested in the fMRI scanner. The negative conditioning was sufficiently robust to completely abolish the analgesic effect of the drug, and the hippocampus is involved in the modulation process.

How Expectancy can Modulate Acupuncture Effects

In a study with a 2×2 design (treatment by expectancy),[43,80] we combined the expectancy manipulation model described above with brain imaging (fMRI) to investigate (1) the underlying mechanism of acupuncture and placebo analgesia, and (2) whether expectancy can modulate acupuncture treatment effects.

In this three-session study, we used Session 1 for training and familiarity. In Session 2, subjects were randomized to one of four groups: verum acupuncture with high expectancy (VH), verum acupuncture with low expectancy (VL), placebo acupuncture with high expectancy (PH), and placebo acupuncture with low expectancy (PL). Subjects in the high-expectancy groups received an expectancy manipulation following treatment with verum or sham acupuncture. Subjects in the low-expectancy groups did not receive an expectancy manipulation but did go through otherwise identical procedures. In other words, the high-expectancy groups were conditioned to expect acupuncture treatment to produce analgesia by the surreptitiously lowered intensity post-treatment stimuli, while the low-expectancy groups were more purely testing the effects of the physical aspects of the acupuncture treatment by minimizing the placebo component. In Session 3, subjects entered the MRI scanner to investigate fMRI signal change and subjective pain rating changes after the expectancy manipulation.

Changes in subjective pain ratings in all four groups were analyzed via a 2×2×2 mixed model analysis of variance (ANOVA). We found a significant main effect on which side subjects were told would be affected by treatment, with subjects reporting greater pain reduction on the 'treated' side than on the control side of the arm. There was also a significant interaction between what they were told (treated or control) and expectancy. There were no other significant main effects or interactions.[44]

For the fMRI data, we first performed an ANOVA analysis on the treated side of the four groups (a matching ANOVA on the subjective pain rating changes showed a significant main effect for expectancy level (high > low), but not for treatment mode (verum or sham acupuncture), nor the interaction of expectancy and treatment mode.

The brain regions involved in the main effects of treatment mode (verum acupuncture vs sham acupuncture) include bilateral PAG, thalamus; left insula, pons/medulla oblongata, and inferior frontal cortex/insula; and right lentiform nucleus/insula.[44] Most of these regions are involved in pain perception and modulation,[81,82] indicating that acupuncture can directly modulate pain perception (Fig. 12.2).

The brain regions involved in the main effects of expectancy (high expectancy vs low expectancy) included bilateral MPFC/rostral ACC, precentral gyrus and medial prefrontal cortex/paracentral lobule, left primary somatomotor, posterior insula/operculum, and superior frontal gyrus, and right amygdala. Most of these regions are involved in emotion modulation and cognitive control of pain (Fig. 12.2).

To investigate whether expectancy enhances the treatment effect of verum acupuncture, a between-group comparison of pre- and post-treatment fMRI signal differences between the verum acupuncture with high expectancy and verum acupuncture with high expectancy groups on treated (HE/LE) and control sides of the arm were conducted separately. On the treated side of the arm, verum acupuncture in the VH group produced significantly greater fMRI signal decreases than did verum acupuncture in the VL group in brain regions including bilateral rACC/medial prefrontal cortex, medial prefrontal cortex, left dorsolateral prefrontal cortex and orbital prefrontal cortex. No region was above threshold for the opposite comparison. On the control side, no region was above threshold for the comparison of VH > VL. The only region above threshold for the opposite comparison, VL > VH, was a small area in the left subgenual cortex.[44] Thus, expectancy significantly enhanced the treatment effect of verum acupuncture as indicated by both subjective pain rating and fMRI signal changes.

It is worth noting that this dissociation between subjective and objective measurements between real and sham acupuncture is also observed in chronic pain patients. A recent study conducted by Harris and colleagues[45] compared both short- and long-term effects of traditional Chinese verum acupuncture versus sham acupuncture treatment on *in vivo* mu opioid receptor (MOR) binding availability in patients with fibromyalgia. Results showed that after 4 weeks of treatment, verum acupuncture evoked (1) short-term increases in MOR binding potential in the cingulate (dorsal and subgenual), insula, caudate, thalamus, and amygdale, and (2) long-term increases in MOR binding potential in the cingulate (dorsal and perigenual), caudate, and amygdala. These short- and long-term effects were absent in the sham group where small reductions were observed. Long-term increases in MOR binding potential following verum acupuncture were also associated with greater reductions in clinical pain. These findings suggest that different MOR processes may mediate clinically relevant analgesia effects for acupuncture and sham acupuncture.

As we have described previously, this significant modulatory effect of expectancy is not specific to acupuncture. The recent study[79] investigating expectancy modulation of the analgesia effect of remifentanil found a significant 'dose' effect of expectancy, such that positive expectancy can significantly enhance the analgesic effect of remifentanil, and negative expectancy can abolish the analgesic effect of the drug. Interestingly, although the brain fMRI signal changes are not exactly the same as reported in our acupuncture studies (not surprising given the multiple differences in experimental paradigms), their study also found that positive expectancy is associated with increased activity in rACC, and negative expectancy is associated with increased activity in the hippocampus, which is consistent with our previous placebo[82] and nocebo[78] studies, respectively (for more details please also see Chapter 13).

In summary, these results indicate that different brain mechanisms are involved in verum and sham acupuncture analgesia; and further that these pathways are uniquely different from those mediating nocebo hyperalgesia. As a peripheral-central modulation, acupuncture needle manipulation may inhibit incoming noxious afferent activity, while as a top-down modulation, expectancy (placebo) may work through cognitive control and emotional neural circuitry. In addition, we also found that expectancy modulates the treatment effect of acupuncture, as measured by both fMRI signal and subjective pain rating changes.

Brain Network Related to Acupuncture Stimulation

In recent decades, investigators have also explored what happens during acupuncture needle stimulation. Acupuncture needle stimulation has been found to evoke changes in an extensive brain network, with reports of both fMRI signal increases and decreases.[83–87] This has been a unique aspect of acupuncture research. Treatment procedures of other therapeutic methods, such as administration of pain pills or application of anesthetic cream, have generally not been studied. The rational for investigating brain activity changes during acupuncture needle stimulation is the belief that elucidating the brain networks involved in acupuncture needling will eventually increase our understanding of acupuncture's therapeutic effects.

Pariente and colleagues compared brain responses to three modes of stimulation, verum acupuncture, sham acupuncture with Streitberger needle stimulation, and overt sham skin pricking, to directly assess the neural activity associated with the treatment in patients with osteoarthritis using PET imaging.[88] They found that verum acupuncture evoked more activation in the insula ipsilateral to the site of needling compared to the site of placebo intervention. Both genuine acupuncture and sham acupuncture evoked greater activation than skin prick (no expectation of a therapeutic effect) in the right dorsolateral prefrontal cortex, ACC, and midbrain. These results indicate that verum acupuncture has a specific physiologic effect that differs from the effects evoked by patients' expectation.

In the 2×2 expectancy study discussed above,[43,80] we used continuous electroacupuncture stimulation (EAS) (2 Hz) for about 20 minutes in total. One challenge of analyzing fMRI data on continuous EA is that traditional block or event-related design/data analysis is not appropriate for long-duration stimulation. To overcome this issue, we posed the concept of EAS state and utilized functional connectivity data analysis methods typically applied to resting state fMRI data.

Using the coordinates of the PAG (identified by the main effect of acupuncture) as a seed, we[89] investigated differences in functional connectivity between subjects receiving the verum and sham acupuncture with different expectancy levels. The PAG was chosen for analysis because it is well known that the PAG plays a role in pain modulation,[90,91] and because the PAG has been found to play an important role in acupuncture analgesia,[92] and also because a previous analysis conducted using this same data set found that fMRI signal changes in response to calibrated heat pain were inhibited at the PAG after verum acupuncture treatment, but were not influenced by the subject's level of expectancy of treatment analgesia.[44] We found that, compared with sham EAS, EAS state showed significantly enhanced functional connectivity between the PAG and left posterior cingulate cortex & precuneus, and reduced functional connectivity between the PAG and right anterior insula.

In another study[93] that utilizes a data-driven functional connectivity data analysis method called independent component analysis (ICA), we found that EA can enhance the functional connectivity between the executive control network and PAG, which, consistent with our previous findings, indicates that EA can produce analgesia through the descending pain control system.[44,89] We also found that high expectancy is associated with strong functional connectivity in areas associated with emotion and memory retrieval (such as the amygdala and parahippocampus), suggesting that expectancy may produce analgesia through emotion modulation. Finally, we found that there is a significant interaction between EA and expectancy in the anterior insula, an area of the 'salience network' that integrates information about the significance of stimulation into perceptual decision-making.[94]

These results suggest that the functional connectivity changes evoked by expectancy are different from functional connectivity changes during EAS state. This provides new information on our understanding of the mechanisms underlying verum and sham acupuncture analgesia.

Overall, studies have indicated that verum acupuncture modulates pain perception differently than sham acupuncture and that acupuncture stimulation itself is associated with complicated functional changes in brain networks. Clearly, much more work is needed on this topic.

SUMMARY AND FUTURE DIRECTIONS

Clinical trials suggest that verum acupuncture can be moderately more effective than placebo acupuncture treatment for some chronic pain conditions. Nevertheless, non-specific effects provide a significant contribution to the therapeutic benefits of acupuncture. With the aid of brain-imaging tools, our understanding of the underlying mechanisms of both verum and sham acupuncture has been significantly enhanced. It seems that different mechanisms are responsible for the analgesic effect of verum acupuncture and placebo acupuncture treatment. We must, however, realize that both verum and sham acupuncture are very complicated phenomena, and we have a long way to go before we can harness and combine the two to obtain the best clinical outcome. To reach this goal, several key questions need to be answered.

1. Source of individual variability in response to acupuncture treatment? For a long time, we have known that individuals' responses to both placebo verum and sham acupuncture vary significantly, but the underlying mechanisms for this variability remain to be answered. It may be that a multitude of factors, including personal traits, differences in brain structure, brain response during anticipation or the pain or other symptoms, genetic or epigenetic factors, are responsible.
2. Relationship between different modalities of placebo treatment (for instance, placebo pills versus placebo injections versus sham acupuncture) and schools of acupuncture treatment (for instance, manual acupuncture versus electroacupuncture)?
3. Is our response to placebo and acupuncture treatment a state or trait characteristic, and does it vary across different conditions or symptoms?
4. Relationship between the specific and non-specific effects of particular treatments? Does an individual who tends to respond to verum acupuncture treatment also tend to respond to sham acupuncture treatment?
5. Experimental studies in healthy subjects indicate that expectancy manipulation significantly boosts placebo effects. Are we allowed to boost the placebo or non-specific effect of treatment to

enhance the cumulative effects of acupuncture treatments on patients in clinics? What ethical issues need to be taken into account when we boost placebo effect in clinics, if and when deception is involved? It remains challenging to maintain a balance between maximizing the benefit to patients (through some level of deception) and truthful disclosure in acupuncture practice.[95]

References

1. Shapiro AK, Shapiro E. *The Powerful Placebo: From Ancient Priest to Modern Physician*. Baltimore: The Johns Hopkins University Press; 1997.
2. Kaptchuk TJ. Placebo studies and ritual theory: a comparative analysis of Navajo, acupuncture and biomedical healing. *Philos Trans R Soc Lond B Biol Sci*. 2011;366(1572):1849-1858.
3. Finniss DG, Kaptchuk TJ, Miller F, Benedetti F. Biological, clinical, and ethical advances of placebo effects. *Lancet*. 2010;375(9715):686-695.
4. Mayer DJ, Prince DD, Rafii A. Antagonism of acupuncture analgesia in man by the narcotic antagonist naloxone. *Brain Res*. 1977;121:368-372.
5. Han JS. Acupuncture: neuropeptide release produced by electrical stimulation of different frequencies. *Trends Neurosci*. 2003;26:17-22.
6. Levine JD, Gordon NC, Fields HL. The mechanism of placebo analgesia. *Lancet*. 1978;2(8091):654-657.
7. Benedetti F, Arduino C, Amanzio M. Somatotopic activation of opioid systems by target-directed expectations of analgesia. *J Neurosci*. 1999;19(9):3639-3648.
8. Ernst E, Lee MS, Choi TY. Acupuncture: does it alleviate pain and are there serious risks? A review of reviews. *Pain*. 2011;152(4):755-764.
9. Colloca L, Finniss D. Nocebo effects, patient-clinician communication, and therapeutic outcomes. *JAMA*. 2012;307(6):567-568.
10. Ernst E. Acupuncture – a critical analysis. *J Intern Med*. 2006;259(2):125-137.
11. Linde K, Vickers A, Hondras M, et al. Systematic reviews of complementary therapies – an annotated bibliography. Part 1: acupuncture. *BMC Complement Altern Med*. 2001;1:3.
12. Kaptchuk TJ. *The Web that has no Weaver: Understanding Chinese Medicine*. Chicago: Contemporary Books (McGraw-Hill); 2000.
13. Kaptchuk TJ. Acupuncture: theory, efficacy, and practice. *Ann Intern Med*. 2002;136(5):374-383.
14. Leibing E, Leonhardt U, Koster G, et al. Acupuncture treatment of chronic low-back pain – a randomized, blinded, placebo-controlled trial with 9-month follow-up. *Pain*. 2002;96:189-196.
15. Brinkhaus B, Witt CM, Jena S, et al. Acupuncture in patients with chronic low back pain: a randomized controlled trial. *Arch Intern Med*. 2006;166(4):450-457.
16. Haake M, Muller HH, Schade-Brittinger C, et al. German acupuncture trials (GERAC) for chronic low back pain: randomized, multicenter, blinded, parallel-group trial with 3 groups. *Arch Intern Med*. 2007;167(17):1892-1898.
17. Melchart D, Streng A, Hoppe A, et al. Acupuncture in patients with tension-type headache: randomised controlled trial. *BMJ*. 2005;331(7513):376-382.
18. Linde K, Streng A, Jurgens S, et al. Acupuncture for patients with migraine: a randomized controlled trial. *JAMA*. 2005;293(17):2118-2125.
19. Cherkin DC, Sherman KJ, Avins AL, et al. A randomized trial comparing acupuncture, simulated acupuncture, and usual care for chronic low back pain. *Arch Intern Med*. 2009;169(9):858-866.
20. Linde K, Witt CM, Streng A, et al. The impact of patient expectations on outcomes in four randomized controlled trials of acupuncture in patients with chronic pain. *Pain*. 2007;128(3):264-271.
21. Kaptchuk TJ, Kelley JM, Conboy LA, et al. Components of placebo effect: randomised controlled trial in patients with irritable bowel syndrome. *BMJ*. 2008;336(7651):999-1003.
22. Vickers AJ, Cronin AM, Maschino AC, et al. Acupuncture for chronic pain: individual patient data meta-analysis. *Arch Intern Med*. 2012;172:1-10.
23. Han JS. Acupuncture analgesia: areas of consensus and controversy. *Pain*. 2011;152(suppl 3):S41-S48.
24. Langevin HM, Bouffard NA, Badger GJ, et al. Subcutaneous tissue fibroblast cytoskeletal remodeling induced by acupuncture: evidence for a mechanotransduction-based mechanism. *J Cell Physiol*. 2006;207(3):767-774.
25. Langevin HM, Churchill DL, Fox JR, et al. Biomechanical response to acupuncture needling in humans. *J Appl Physiol*. 2001;91:2471-2478.
26. Langevin HM, Churchill DL, Wu J, et al. Evidence of connective tissue involvement in acupuncture. *FASEB J*. 2002;16(8):872-874.
27. Langevin HM, Konofagou EE, Badger GJ, et al. Tissue displacements during acupuncture using ultrasound elastography techniques. *Ultrasound Med Biol*. 2004;30(9):1173-1183.
28. Langevin HM, Sherman KJ. Pathophysiological model for chronic low back pain integrating connective tissue and nervous system mechanisms. *Med Hypotheses*. 2007;68(1):74-80.
29. Langevin HM, Storch KN, Cipolla MJ, et al. Fibroblast spreading induced by connective tissue stretch involves intracellular redistribution of alpha- and beta-actin. *Histochem Cell Biol*. 2006;125(5):487-495.
30. Lee EJ, Frazier SK. The efficacy of acupressure for symptom management: a systematic review. *J Pain Symptom Manage*. 2011;42(4):589-603.
31. Kleinhenz J, Streitberger K, Windeler J, et al. Randomised clinical trial comparing the effects of acupuncture and a newly designed placebo needle in rotator cuff tendinitis. *Pain*. 1999;83(2):235-241.
32. Streitberger K, Kleinhenz J. Introducing a placebo needle into acupuncture research. *Lancet*. 1998;352:364-365.
33. White P, Lewith G, Hopwood V, Prescott P. The placebo needle, is it a valid and convincing placebo for use in acupuncture trials? A randomised, single-blind, cross-over pilot trial. *Pain*. 2003;106:401-409.
34. Kong J, Fufa DT, Gerber AJ, et al. Psychophysical outcomes from a randomized pilot study of manual, electro, and sham acupuncture treatment on experimentally induced thermal pain. *J Pain*. 2005;6(1):55-64.
35. Takakura N, Yajima H. A double-blind placebo needle for acupuncture research. *BMC Complement Altern Med*. 2007;7:31.
36. van Leeuwen JH, Castro R, Busse M, Bemelmans BL. The placebo effect in the pharmacologic treatment of patients with lower urinary tract symptoms. *Eur Urol*. 2006;50(3):440-452; discussion 453.
37. Nickel JC. Placebo therapy of benign prostatic hyperplasia: a 25-month study. Canadian PROSPECT Study Group. *Br J Urol*. 1998;81(3):383-387.
38. de Jong PJ, van Baast R, Arntz A, Merckelbach H. The placebo effect in pain reduction: the influence of conditioning experiences and response expectancies. *Int J Behav Med*. 1996;3(1):14-29.
39. Feather BW, Chapman CR, Fisher SB. The effect of a placebo on the perception of painful radiant heat stimuli. *Psychosom Med*. 1972;34(4):290-294.
40. Kelley JM, Boulos PR, Rubin PAD, Kaptchuk TJ. Mirror, mirror on the wall: placebo effects that exist only in the eye of the beholder. *J Eval Clin Pract*. 2009;15:292-298.
41. Wechsler ME, Kelley JM, Boyd IO, et al. Active albuterol or placebo, sham acupuncture, or no intervention in asthma. *N Engl J Med*. 2011;365(2):119-126.

42. Fregni F, Boggio PS, Bermpohl F, et al. Immediate placebo effect in Parkinson's disease – is the subjective relief accompanied by objective improvement? *Eur Neurol*. 2006;56(4):222-229.
43. Kong J, Kaptachuk TJ, Polich G, et al. Expectancy and treatment interactions: a dissociation between acupuncture analgesia and expectancy evoked placebo analgesia. *Neuroimage*. 2009;45: 940-949:PMID: 19159691.
44. Kong J, Kaptchuk TJ, Polich G, et al. An fMRI study on the interaction and dissociation between expectation of pain relief and acupuncture treatment. *Neuroimage*. 2009;47(3):1066-1076:PMID: 19501656.
45. Harris RE, Zubieta JK, Scott DJ, et al. Traditional Chinese acupuncture and placebo (sham) acupuncture are differentiated by their effects on mu-opioid receptors (MORs). *Neuroimage*. 2009;47(3):1077-1085.
46. Kong J, Kaptchuk TJ, Polich G, et al. Placebo analgesia: findings from brain imaging studies and emerging hypotheses. *Rev Neurosci*. 2007;18(3–4):173-190.
47. Montgomery GH, Kirsch I. Classical conditioning and the placebo effect. *Pain*. 1997;72:107-113.
48. Price DD, Milling LS, Kirsch I, et al. An analysis of factors that contribute to the magnitude of placebo analgesia in an experimental paradigm. *Pain*. 1999;83:147-156.
49. Amanzio M, Benedetti F, Porro CA, et al. Activation likelihood estimation meta-analysis of brain correlates of placebo analgesia in human experimental pain. *Hum Brain Mapp*. 2011; doi:10.1002/hbm.21471.
50. Kong J, Jensen K, Loiotile R, et al. Functional connectivity of frontoparietal network predicts cognitive modulation of pain. *Pain*. 2013. http://dx.doi.org/10.1016/j.pain.2012.1012.1004.
51. Mesulam MM. From sensation to cognition. *Brain*. 1998;121: 1013-1052.
52. Miller EK, Cohen JD. An integrative theory of prefrontal cortex function. *Annu Rev Neurosci*. 2001;24:167-202.
53. Benedetti F. Mechanisms of placebo and placebo-related effects across diseases and treatments. *Annu Rev Pharmacol Toxicol*. 2008;48:33-60.
54. Roberts AH, Kewman DG, Hevell M. The power of nonspecific effects in healing implication for psychosocial and bioligical treatments. *Clin Pysychol Rev*. 1993;12:375-391.
55. Charron J, Rainville P, Marchand S. Direct comparison of placebo effects on clinical and experimental pain. *Clin J Pain*. 2006; 22(2):204-211.
56. Colloca L, Tinazzi M, Recchia S, et al. Learning potentiates neurophysiological and behavioral placebo analgesic responses. *Pain*. 2008;139(2):306-314.
57. Kong J, Gollub RL, Rosman IS, et al. Brain activity associated with expectancy-enhanced placebo analgesia as measured by functional magnetic resonance imaging. *J Neurosci*. 2006;26(2):381-388.
58. Voudouris NJ, Peck CL, Coleman G. The role of conditioning and verbal expectancy in the placebo response. *Pain*. 1990;43(1): 121-128.
59. De Pascalis V, Chiaradia C, Carotenuto E. The contribution of suggestibility and expectation to placebo analgesia phenomenon in an experimental setting. *Pain*. 2002;96(3):393-402.
60. Wager TD, Rilling JK, Smith EE, et al. Placebo-induced changes in FMRI in the anticipation and experience of pain. *Science*. 2004;303:1162-1167.
61. Bingel U, Lorenz J, Schoell E, et al. Mechanisms of placebo analgesia: rACC recruitment of a subcortical antinociceptive network. *Pain*. 2006;120(1-2):8-15.
62. Watson A, El-Deredy W, Iannetti GD, et al. Placebo conditioning and placebo analgesia modulate a common brain network during pain anticipation and perception. *Pain*. 2009;145(1-2):24-30.
63. Wager TD, Scott DJ, Zubieta JK. Placebo effects on human mu-opioid activity during pain. *Proc Natl Acad Sci USA*. 2007;104(26): 11056-11061.
64. Eippert F, Bingel U, Schoell ED, et al. Activation of the opioidergic descending pain control system underlies placebo analgesia. *Neuron*. 2009;63(4):533-543.
65. Gracely RH, Dubner R, McGrath PA. Narcotic analgesia: fentanyl reduces the intensity but not the unpleasantness of painful tooth pulp sensations. *Science*. 1979;203:1261-1263.
66. Gracely RH, McGrath PA, Dubner R. Ratio scales of sensory and affective verbal pain descriptors. *Pain*. 1978;5:5-18.
67. Gracely RH, McGrath PA, Dubner R. Validity and sensitivity of ratio scales of sensory and affective verbal pain descriptors: manipulation of affect by diazepam. *Pain*. 1978;5:19-29.
68. Barsky AJ, Borus JF. Functional somatic syndromes. *Ann Intern Med*. 1999;130(11):910-921.
69. Barsky AJ, Saintfort R, Rogers MP, Borus JF. Nonspecific medication side effects and the nocebo phenomenon. *JAMA*. 2002;287(5): 622-627.
70. Kaptchuk TJ, Stason WB, Davis RB, et al. Sham device v inert pill: randomised controlled trial of two placebo treatments. *BMJ*. 2006;332(7538):391-397.
71. Myers MG, Cairns JA, Singer J. The consent form as a possible cause of side effects. *Clin Pharmacol Ther*. 1987;42(3):250-253.
72. Roscoe JA, Hickok JT, Morrow GR. Patient expectations as predictor of chemotherapy-induced nausea. *Ann Behav Med*. 2000;22(2): 121-126.
73. Reuter U, Sanchez del Rio M, Carpay JA, et al. Placebo adverse events in headache trials: headache as an adverse event of placebo. *Cephalalgia*. 2003;23(7):496-503.
74. Jones TF, Craig AS, Hoy D, et al. Mass psychogenic illness attributed to toxic exposure at a high school. *N Engl J Med*. 2000;342(2): 96-100.
75. Ko DT, Hebert PR, Krumholz HM. Review: beta-blockers increase fatigue and sexual dysfunction but not depression after myocardial infarction. *ACP J Club*. 2003;138(1):30; author reply 30.
76. Benedetti F, Amanzio M, Casadio C, et al. Blockade of nocebo hyperalgesia by the cholecystokinin antagonist proglumide. *Pain*. 1997;71(2):135-140.
77. Benedetti F, Amanzio M, Vighetti S, Asteggiano G. The biochemical and neuroendocrine bases of the hyperalgesic nocebo effect. *J Neurosci*. 2006;26(46):12014-12022.
78. Kong J, Gollub RL, Polich G, et al. A functional magnetic resonance imaging study on the neural mechanisms of hyperalgesic nocebo effect. *J Neurosci*. 2008;28(49):13354-13362:PMCID: PMC2649754.
79. Bingel U, Wanigasekera V, Wiech K, et al. The effect of treatment expectation on drug efficacy: imaging the analgesic benefit of the opioid remifentanil. *Sci Transl Med*. 2011;3(70):70ra14.
80. Kong J, Kaptchuk TJ, Webb JM, et al. Functional neuroanatomical investigation of vision-related acupuncture point specificity – a multisession fMRI study. *Hum Brain Mapp*. 2009;30(1):38-46.
81. Kong J, Loggia ML, Zyloney C, et al. Exploring the brain in pain: activations, deactivations and their relation. *Pain*. 2010;148: 257-267:PMID: 20005043.
82. Kong J, White NS, Kwong KK, et al. Using fMRI to dissociate sensory encoding from cognitive evaluation of heat pain intensity. *Hum Brain Mapp*. 2006;27(8):715-721.
83. Kong J, Ma L, Gollub RL, et al. A pilot study of functional magnetic resonance imaging of the brain during manual and electroacupuncture stimulation of acupuncture point (LI-4 Hegu) in normal subjects reveals differential brain activation between methods. *J Altern Complement Med*. 2002;8(4):411-419.
84. Hui KK, Liu J, Makris N, et al. Acupuncture modulates the limbic system and subcortical gray structures of the human brain: evidence from fMRI studies in normal subjects. *Hum Brain Mapp*. 2000;9:13-25.
85. Hui KK, Liu J, Marina O, et al. The integrated response of the human cerebro-cerebellar and limbic systems to acupuncture stimulation at ST 36 as evidenced by fMRI. *Neuroimage*. 2005;27(3):479-496.

86. Napadow V, Kettner N, Ryan A, et al. Somatosensory cortical plasticity in carpal tunnel syndrome – a cross-sectional fMRI evaluation. *Neuroimage.* 2006;31(2):520-530.
87. Kong J, Gollub RL, Webb JM, et al. Test-retest study of fMRI signal change evoked by electroacupuncture stimulation. *Neuroimage.* 2007;34(3):1171-1181:PMCID: PMC1994822.
88. Pariente J, White P, Frackowiak RS, Lewith G. Expectancy and belief modulate the neuronal substrates of pain treated by acupuncture. *Neuroimage.* 2005;25(4):1161-1167.
89. Zyloney CE, Jensen K, Polich G, et al. Imaging the functional connectivity of the Periaqueductal Gray during genuine and sham electroacupuncture treatment. *Mol Pain.* 2010;6:80.
90. Fields H. State-dependent opioid control of pain. *Nat Rev Neurosci.* 2004;5(7):565-575.
91. Kong J, Tu PC, Zyloney C, Su TP. Intrinsic functional connectivity of the periaqueductal gray, a resting fMRI study. *Behav Brain Res.* 2010;211(2):215-219.
92. Zhao ZQ. Neural mechanism underlying acupuncture analgesia. *Prog Neurobiol.* 2008;85(4):355-375.
93. Wey HY, Gollub R, Kong J. *The modulation effect of expectancy on electro-acupuncture stimulation state*:Paper presented at: 18th Annual Meeting of the Organization for Human Brain Mapping; 2012:Beijing.
94. Wiech K, Lin CS, Brodersen KH, et al. Anterior insula integrates information about salience into perceptual decisions about pain. *J Neurosci.* 2010;30(48):16324-16331.
95. Brody H, Colloca L, Miller FG. The placebo phenomenon: implications for the ethics of shared decision-making. *J Gen Intern Med.* 2012;27(6):739-742.

CHAPTER 13

The Relevance of Placebo and Nocebo Mechanisms for Analgesic Treatments

Ulrike Bingel

NeuroImage Nord, Department of Neurology, University Medical Center Hamburg-Eppendorf, Hamburg, Germany

PLACEBO AND NOCEBO IN PAIN TREATMENTS: BEHAVIORAL EVIDENCE

The Role of Expectation in Analgesic Treatments

It has been a longstanding clinical notion dating back to reports from ancient Greek literature (Platon, Charmides) that the individual's beliefs and expectations can significantly influence the therapeutic benefit and adverse effects of a pharmacologic (or other specific) treatment for pain. The recent scientific interest in placebo and nocebo phenomena, and their psycho-neurobiologic underpinnings, has rekindled awareness of the fact that, inevitably, active pharmacologic treatments also include interacting physiologic and psychosocial components. The crucial influence of expectation on therapeutic outcome is best illustrated in the so-called open/hidden drug paradigm (see Fig. 13.1).[1] In this paradigm, identical concentrations of the same drug are administered under two conditions: an *open condition*, in which the patient is aware of the time-point at which the medication is administered by a health-care provider—and of the intended treatment outcome (e.g. analgesia)—and a *hidden condition*, in which the patient is unaware of the medication being administered by a computer-controlled infusion. The comparison of both conditions allows for the dissociation of the genuine pharmacodynamic effect of the treatment (hidden treatment) and the additional benefit of the psychosocial context in which the treatment is provided. The difference in outcome following the administration of the expected and unexpected therapy can be seen as the 'placebo' (psychologic) component, even though no placebo is given. Studies based on an open/hidden paradigm have revealed that psychosocial factors, such as the awareness of a drug being given, can considerably enhance its analgesic effect.[2] Conversely, the hidden administration attenuates the analgesic effect of nonsteroidal anti-inflammatory drugs to non-significance, and even the effects of opioids are substantially reduced by hidden application.[3] As a result of a hidden application, the drug dosage had to be doubled to achieve the same result as during open application.

Intriguingly, this phenomenon is not limited to analgesics, as similar effects have also been reported for pharmacotherapy in other clinical conditions, such as Parkinson's disease and anxiety disorder, or drug addiction.[1,4,5] Findings from these studies using the open/hidden drug paradigm are supported by studies that have explicitly modulated the expectancy regarding a given drug by verbal instruction.[6-8]

Results from these studies using open/hidden drug paradigms are complemented by observations from clinical trials showing that the efficacy of an analgesic treatment under test differs considerably depending on the trial type and randomization scheme used in the study, as these clearly influence the patients' expectation of actually receiving an analgesic drug and subsequent pain relief. For instance, the analgesic effect of paracetamol is almost twice as high in so-called comparator trials, i.e. randomized controlled trials (RCTs) that test paracetamol against another analgesic, such as naproxen,[9,10] compared to trials, where the same dosage of paracetamol is compared against placebo. The critical difference between these two types of trial is that, in the comparator trial, the patient has a 100% probability (and subsequent expectation) of receiving an analgesic, but only 50% when paracetamol is compared to placebo. It is worth noting that substantial effects of expectation on the analgesic effect can also be observed in non-pharmacologic therapies for pain, such as acupuncture.[11]

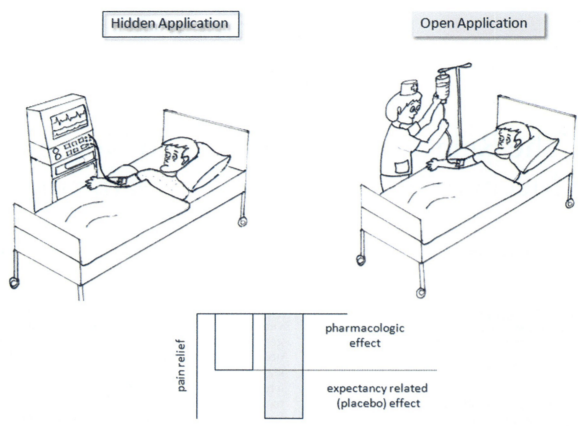

FIGURE 13.1 Open–hidden drug paradigms. The pivotal role of expectation in mediating placebo responses is best illustrated by the so-called 'open–hidden' drug paradigm. In this paradigm, identical concentrations of active drugs are administered by a doctor in a visible way (open condition) or in a hidden condition, in which the patient is unaware of the timing of the medication administration (e.g. using a computer-controlled infusion). This permits dissociation of the pure pharmacodynamic effect of the treatment (hidden treatment) from the additional benefit of the psychologic context that comes from knowing that treatment is being administered. The difference between the outcomes following the administration of the expected and unexpected therapy can be seen as the 'placebo' or psychologic component, even though no placebo treatment has been used.

While the above examples highlight the impact of positive expectations on analgesic outcome, the detrimental influence of negative expectations on drug response has, for instance, been demonstrated in a behavioral experimental study by Dworkin and colleagues who reported a reversal of analgesia by nitrous oxide in dental pain when participants expected the drug to increase awareness of bodily sensations.[12]

Two additional studies underline the power of negative expectations regarding the 'side effects' of treatments or diagnostic procedures. In an early study by Daniels et al, patients undergoing lumbar puncture were randomly assigned to two groups. One group was not informed about the potential incidence of post-puncture headache, the other group was informed about this potential side effect. Post-puncture headache was assessed at 4 and 24 hours after the lumbar puncture. Intriguingly, about 50% (7/15) of the informed group, but less than 10% of the non-informed group developed a headache after the procedure.[13] A similar observation has been reported by Varelmann and colleagues who studied how the wording of information about potential side effects of a local anaesthetic affects the treatment tolerance. One hundred and forty healthy women at term gestation requesting neuraxial analgesia were randomized into a positive-information group ('We are going to give you a local anesthetic that will numb the area and you will be comfortable during the procedure') or into a group that was provided with the standard form of information often used in the clinical setting ('You are going to feel a big bee sting; this is the worst part of the procedure'). Pain was assessed immediately after the injection of the local anaesthetic using verbal analogue scale scores of 0 to 10. The reported pain scores were significantly lower in the positive-information group.

The crucial contribution of negative expectations on the development of unwanted effects during pain treatments is further supported by data from clinical trials. Several recent meta-analyses have investigated the frequency, severity and quality of unwanted side effects that occurred in the placebo arm of randomized placebo-controlled trials of pharmacologic treatments for different chronic pain conditions. Papadopoulos and Mitsikostas for instance, who looked at neuropathic pain trials, report an alarming rate of side effects under placebo (nocebo effects) that affected about 50% of the patients[14]; 6% of the patients who had been assigned to the placebo arm even dropped out of the study due to intolerable 'drug'-related side effects. A meta regression analysis on study-related and patient-related variables revealed a higher frequency of nocebo responses in male patients. The relevance of verbal instruction and other information given to the patient is illustrated by the fact that these side effects in the placebo arm reflect typical side-effect patterns expected in the drug group. Placebo groups of tricyclic antidepressant trials, for instance, show higher rates of adverse events compared with placebo groups of trials testing selective serotonin reuptake inhibitors.[15] These analyses highlight the fact that the patients' individual beliefs and negative expectations regarding a drug's effect (nocebo effects) substantially contribute to the development of side effects that adversely affect adherence and efficacy, and at worst, lead to discontinuation of treatment. Alarmingly, experimental approaches support the notion that such nocebo responses are easier to induce than placebo responses. Colloca and co-workers found that negative conditioning and verbal suggestions induce equally significant nocebo hyperalgesic responses, whereas in the case of placebo analgesia, conditioning elicits larger reductions in pain than do verbal suggestions alone.[16] These findings suggest that nocebo responses may be elicited faster than placebo responses to protect the body from dangerous and negative outcomes.

In sum, behavioral studies have convincingly shown that the effect of drug treatment is determined not only by the pharmacologic profile of the drug. Rather, the individual's expectation substantially modulates the overall efficacy and tolerability of drugs and other specific pain treatments as well as the quality and severity of side effects associated with the treatment.

The Role of Conditioning and Prior Experience in Analgesic Treatments

As outlined in the chapter by Colloca in this book, conditioning is another key mechanism underlying placebo and nocebo responses. Several studies have investigated placebo effects by reinforcing expectations through a placebo manipulation in which the inert treatment is paired with lowered pain stimuli so that patients come to experience and expect pain relief.[16–18] This procedure typically evokes much stronger and more stable placebo analgesic effects compared to verbal suggestion alone and is therefore often used in experimental paradigms of placebo analgesia. It is important to emphasize that any pharmacologic (or other specific) treatment for pain can induce conditioned responses that may then modulate future responses to the same treatment.

In pharmacologic conditioning (see Fig. 13.2), after repeated associations between a neutral stimulus (e.g. a syringe/infusion/capsule containing an inert substance) and an active drug (unconditioned stimulus), the neutral stimulus may become able, by itself, to elicit a response characteristic of the unconditioned stimulus. Intriguingly, pharmacologic conditioning has been shown to produce placebo responses largely independent of verbal suggestion or expectation, both in animals and humans.

For instance, healthy volunteers who have repeatedly received an opioid infusion during an ischemic arm pain model will also develop analgesia when receiving a saline infusion that is explicitly given as a control condition.[19] Remarkably, the neurobiology underlying the conditioned response seems to vary depending on the specific substance used for conditioning. When morphine was used for pharmacologic conditioning, the conditioned analgesic response can be blocked by the administration of the opioid-antagonist naloxone, indicating the involvement of the endogenous opioid system in the conditioned response. In contrast, when analgesia is conditioned using the nonopioid ketorolac, the conditioned analgesic response is not blocked by naloxone. According to recent evidence, conditioned analgesia after exposure to the nonopioid ketorolac involves the endogenous cannabinoid systems, as the CB1 cannabinoid receptor antagonist rimonabant can block these responses.[20]

These phenomena are not restricted to conditioned placebo analgesia, as similar pharmacologic conditioning effects have also been observed in the motor, hormone and immune systems.[21] Particularly fascinating are conditioned responses in the immune system which does not underlie voluntary control. Repeated pairing of the immunosuppressive drug cyclosporin A (US, unconditioned stimulus), which inhibits both IL-2 and IFN-γ, with strawberry milk (CS, conditioned stimulus) induces conditioned responses in which the strawberry milk alone (without cyclosporin A) is capable of inhibiting both IL-2 and IFN-γ.[22]

Taken together, these behavioral observations highlight the crucial relevance of psychosocial effects such as expectancy induced by verbal suggestion, prior experience, or a combination of both, to drug efficacy.

UNDERSTANDING THE NEURAL MECHANISMS UNDERLYING THE EFFECTS OF EXPECTATION AND LEARNING ON DRUG EFFICACY

Given that functional neuroimaging has successfully been used to characterize pain processing, the modulation of pain by cognitive factors, and analgesia, the same techniques should be capable of unravelling the neural

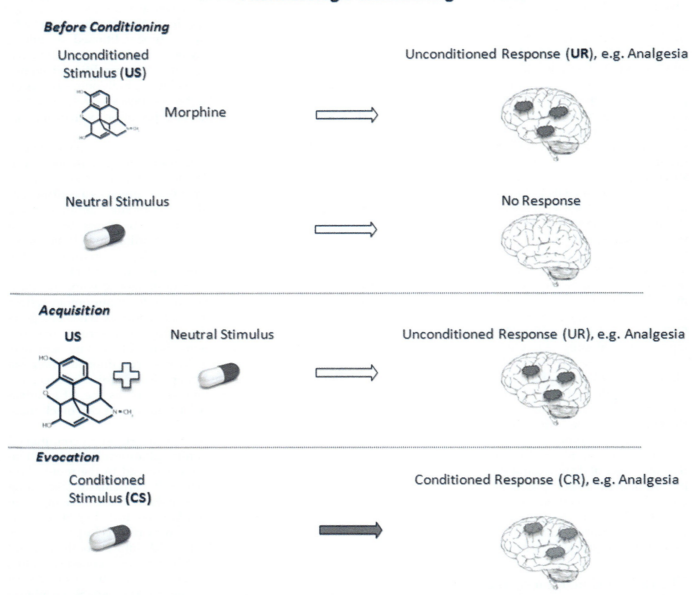

FIGURE 13.2 Pharmacologic conditioning. In the context of behavioral conditioning (B), the unconditioned stimulus (e.g. a pharmacologic agent) is inducing a response in the CNS (unconditioned response/UR); a neutral stimulus (e.g. environmental stimuli, an inert substance) is inducing no such response. During the acquisition phase, the neutral stimulus is paired with the unconditioned stimulus (US). After one or several pairings of the neutral stimulus with the US, the neutral stimulus becomes the conditioned stimulus (CS). During evocation, the CS is now able to mimic the effects formally induced by the US. *This figure is reproduced in color in the color section.*

and neurochemical mechanisms by which expectations modulate pharmacologically induced analgesia.

In a first study addressing this issue, we recently investigated the effect of expectancy modulation of opioid analgesia using fMRI.[23] In this study, the potent opioid analgesic remifentanil was administered to healthy volunteers under three conditions: without expectation of analgesia (hidden application), with expectancy of a positive analgesic effect, and with negative expectancy of analgesia, i.e. expectation of hyperalgesia. Importantly, fMRI was used to study the efficacy of the opioid (i.e. to exclude a report bias) and to elucidate the underlying neural mechanisms. Our results showed that positive treatment expectancies doubled the analgesic benefit of remifentanil (Fig. 13.3). Negative treatment expectation interfered with the analgesic potential of remifentanil to the extent that its analgesic effect was completely abolished.

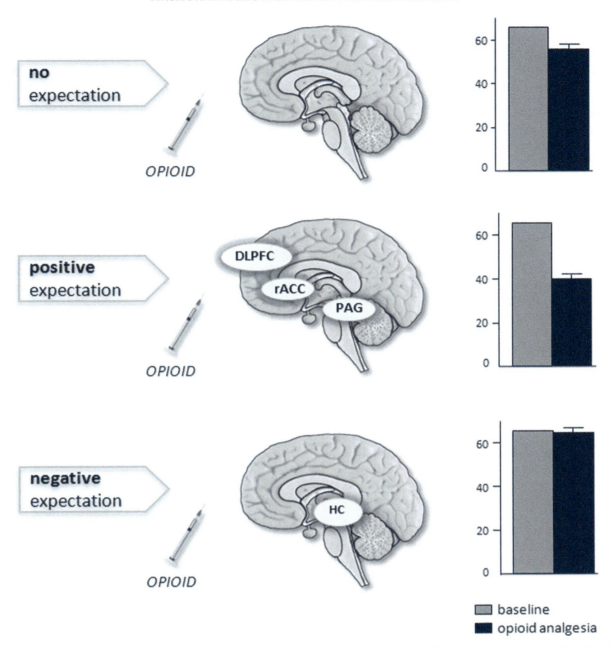

FIGURE 13.3 Influence of expectations on opioid-induced analgesia. The opioid remifentanil led to a significant reduction of pain when participants were not aware of the time-point of drug application, reflecting the pharmacologic effect with no expectations (top row). The analgesic effect was, however, doubled when participants were informed about application onset and expected a reduction of pain (middle row). Conversely, the drug effect was completely abolished in the condition where participants expected the drug to exacerbate pain (bottom row). (See Bingel et al[23] for details.) DLPFC, dorsolateral prefrontal cortex; rACC, rostral anterior cingulate cortex; PAG, periaqueductal gray; HC, hippocampus. *This figure is reproduced in color in the color section.*

The first part of our study, showing that psychosocial factors, such as awareness of a drug being given, can considerably enhance the overall clinical response to a drug,[1] confirmed previous behavioral observations that used hidden versus open application of analgesics. Importantly, the fMRI data obtained in this study allowed for demonstrating that the expectancy-related changes in pain perception were paralleled by significant alterations in the neural response to noxious thermal stimulation in core brain regions of the pain- and opioid-sensitive brain networks, such as the thalamus, mid-cingulate cortex and primary somatosensory cortex (SI). Activity in these brain areas has consistently been shown to scale with the intensity of nociceptive input and resultant pain perception,[24,25] and may therefore serve as an objective index of analgesic efficacy. This study therefore provided strong evidence that context-related differences in reported analgesia, as observed here and in previous studies, are not the result of report bias.

With respect to underlying mechanisms, we found that the individual benefit from positive treatment expectancy during remifentanil analgesia was associated with activity in the descending pain modulatory system, including cingulo-frontal and subcortical brain areas that are known to contribute to both opioid and placebo analgesia.

In contrast, negative expectancy that abolished the analgesic effect of the opioid was associated with reduced activity in the subgenual anterior cingulate cortex (sgACC). This response pattern suggests that both positive and negative expectancy use a key component of the descending pain modulatory control system, but in opposite ways. Furthermore, we found that negative expectancy was selectively associated with increased activity in the hippocampus and medial prefrontal cortex. These brain areas have previously been implicated in the exacerbation of pain by mood and anxiety in patients as well as in healthy controls.[26,27] Interestingly, activity in medial frontal areas and hippocampus has also been observed in a recent study on the nocebo hyperalgesic effects during sham acupuncture.[28] While the contribution of the descending pain modulatory system to various types of pain modulation is well characterized, the role of the hippocampus to negative expectations on both pain and treatment outcome requires further attention. Negative treatment expectancy in our study produced a significant increase in anxiety. This is in line with the existing evidence that anxiety represents a powerful modulator in nocebo hyperalgesia,[29] most likely via activation of the endogenous cholecystokinin (CCK) system.[30] The CCK peptide is a known pronociceptive, anxiogenic neurotransmitter that has been found in some of the key structures of the descending pain modulatory system, including the PAG.[31]

These first experimental data on the expectancy modulation of opioid analgesia substantiate the significant contribution of cognitive factors to the overall benefit from pharmacologic treatments. Even though the applied neuroimaging technique used in this study does not allow for detecting interactions at the receptor level, these data demonstrate that pharmacologic and psychologic factors, such as an individual's expectation, ultimately converge at the neuronal level and can substantially improve or abolish the net-analgesic effect of a potent analgesic. Similar interactions between pharmacodynamic and psychologic effects on regulatory brain mechanisms have been reported for the administration of methylphenidate in cocaine-addicted patients.[5] Together with evidence from experimental placebo and nocebo studies, these observations indicate that the effects of expectancy and conditioning converge with pharmacologic effects at the very same biologic systems, involving distinct CNS and peripheral physiologic mechanisms.

Implications for Clinical Practice

The improved understanding of the psychology and neurobiology underlying placebo and nocebo effects and their substantial contribution to the efficacy and tolerability of pharmacologic treatments has far-reaching implications for future research and clinical practice. However, to date, the potential of placebo mechanisms is far from being systematically exploited in daily clinical care.

From a clinical perspective, maximum drug efficacy is desirable, irrespective of whether the improvements are based on specific pharmacologic effects, on placebo mechanisms, or on a combination of both. Accordingly, while placebo responses should be controlled and reduced to improve assay sensitivity in clinical trials,[32] in the clinical setting the mechanisms underlying placebo responses (i.e. treatment outcome expectation, conditioning/learning processes, physician–patient relationship) should be systematically exploited to maximize treatment benefits.

MODULATING EXPECTATIONS TO OPTIMIZE ANALGESIC OUTCOME

Treatment expectations are shaped by various factors, including prior experiences with physicians and treatments. Particularly in patients with chronic diseases, treatments often fail repeatedly. Frustration inevitably mounts and may result in negative expectancies for future treatments. Furthermore, the negative mood states that occur in patients with chronic disease[33] may themselves generate negative treatment expectations and increased anxiety. In these situations, drugs with biologically plausible intrinsic actions compete with the negative treatment expectancies of the patient. Because both processes activate similar target brain regions, negative expectations can modulate, or in the worst case, completely abolish, the drug effects and clinical outcome.[23] The underestimation of the influence that psychologic states have on the pharmacodynamics of a drug might therefore, inadvertently, contribute to the frequent failure of clinical translation of drugs that show target engagement in preclinical studies, especially when drugs are developed for the treatment of chronic illness.

Patients' outcome expectations are malleable and can be systematically altered by instruction. This has been demonstrated after myocardial infarction where the modification of expectations resulted in improved function, earlier return to work, and improved quality of life.[34] However, more research is needed to investigate how expectations can be assessed and modified in the context of complex medical settings. Promising tools are brief psychologic interventions that can be used by doctors and nurses under daily routine conditions to optimize patients' expectations. Because inadequate and overly optimistic expectations might be as detrimental as negative expectations, the modification of outcome expectations should be adjusted to the individual patient's expectation. Patients with inadequate expectations (overly negative or over-optimistic cognitions) could undergo re-attribution training to more positive and realistic expectations. In addition, adequate expectations should be consolidated by strengthening the cognitive and emotional impact of positive treatment results. Influencing beliefs about outcome by the careful use of language and, importantly, provision of appropriate information regarding the expected drug effect, should be considered as an important feature of any pharmacologic treatment. Indeed, this is already done by some physicians. However, the observation that in the United States, 50% of patients leave after an office visit without an adequate understanding of what the physician has told them,[35] highlights a need to improve this element of the patient–physician interaction if we are to improve treatment outcomes.

It is worth noting that many variables can determine whether expectation interventions are useful. Accordingly, psychologic, physiologic, and medical predictors for successful expectation interventions must be defined and evaluated for adequate patients' selection (see also 'interindividual differences,' below). Further medical setting variables and information provided through leaflets, consent forms, etc. should be optimized to support the development of positive outcome expectations.

EXPLOITING LEARNING MECHANISMS TO OPTIMIZE ANALGESIC OUTCOME

Classical (Pavlovian) conditioning of drug responses is another promising tool to improve treatment outcome (for a review see Doering and Rief[36]). However, drug intake is rarely analyzed under the perspective of associative learning processes. Experimental studies have shown that learned placebo responses can be used to induce analgesia.[19,20] Reframing continuous drug intake as a learning process opens a new avenue for treatment optimization with the aim of reducing drug dosage, unwanted side effects and treatment costs while maximizing treatment efficacy. New application strategies have been suggested that comprise a learning phase, in which the active drug is administered, and the subsequent maintenance phase, in which a placebo is intermittently applied.[32,36–38] Using this 'partial reinforcement' scheme, drug efficacy can be maintained with a reduced drug dosage. The feasibility of such an approach has not yet been shown for pain, but the 'proof of principle' has been demonstrated in psoriasis treated with corticosteroids,[39] in attention deficit hyperactivity disorder (ADHD),[40] the suppression of cough[41] and other clinical conditions. Although these studies showed positive short-term effects, the potential for reducing negative consequences of long-term drug applications requires further investigation.

The expectation-related and conditioning-based approaches should be combined with approaches to optimize physician–patient interactions, as characteristics of the physician–patient relationship have been shown to predict outcome in various medical conditions, including pain.[42] For irritable bowel syndrome, for instance, switching from a brief, technical interaction style to an empathic, emotionally warm interactive style has been shown to increase the efficacy of a (placebo) intervention from 42% to 82%.[43] (For further details regarding the practical implications of these lines of research on the treatment of chronic pain patients see also Ch 26.)

Implications for Clinical Trial Designs

Our current understanding of the role of psychosocial components in relation to drug efficacy and tolerability also require reconsidering the design of clinical trials. The question is not only whether a drug shows superior efficacy over an inert placebo pill, but also what are the optimal contextual conditions that maximize the analgesic effect and minimize side effects of pain medications by integrating the effects of expectation, learning and active pharmacologic treatment in a real-life scenario? New trial designs could be developed that explore the disease and drug-specific treatment context that optimally enhances the overall treatment outcome after the general efficacy of a drug has been demonstrated in comparison to placebo using conventional RCTs. Such optimized, enriched clinical trial arms may provide a better estimate for the potential of a drug under optimal clinical context conditions.[32,44]

FUTURE AIMS AND CHALLENGES

Additive vs. Interactive Effects?

One of the crucial yet unanswered questions is whether placebo responses and pharmacologically induced analgesia combine in an additive or interactive manner. Depending on the specific pharmacologic agent, pharmacologically induced analgesia and the distinct endogenous cascades triggered by placebo mechanisms may combine in an additive manner for one substance such as opioids,[45] but in an interactive fashion in another substance (Fig. 13.4).

Future studies involving different methodologies and designs should aim at unraveling the effects of expectations and learning on drug action at a receptor/molecular level, and determine whether the effects of expectation and drug effect combine in an additive or interactive manner. Furthermore, if expectations have the potential to substantially modify drug effects, the interface between drug effects and cognitive effects is a promising target for future pharmacologic developments. However, in order to allow for the development of specific drugs, a more detailed understanding of the downstream neurobiologic cascades is required. Molecular imaging approaches, such as positron emission tomography (PET), with distinct pain- and analgesia-related ligands (e.g. of the opioid, cannabinoid or gabaergic system) in combination with MRI (PET-MR), or MR-spectroscopy, may help to unravel these effects.

Predictors for Placebo and Nocebo Effects and Their Impact on Analgesic Treatments

Both the susceptibility to analgesic drugs and to placebo/nocebo mechanisms varies considerably between individuals. It therefore seems reasonable to also assume substantial inter-individual differences for their combined (or interactive) effect. To date, specific factors that contribute to this variability have not been identified. However, insights from pharmacology and cognitive neuroscience predict that genetics, disease-specific factors and the personality of the individual are important influential variables that shape the individual response. This is, for example, illustrated in an open/hidden study of local anesthesia in patients suffering from Alzheimer's disease.[46] In this patient group the loss of placebo-related neurobiologic functions reduced overall treatment efficacy of the local anesthetic, with dose increases necessary to achieve adequate analgesia. Specifically, those Alzheimer patients who displayed reduced connectivity of the prefrontal lobes, which are crucially involved in the initiation of placebo analgesia, experienced less additional pain relief from an open compared with a hidden application of lidocaine. This study demonstrates that the individual contribution of placebo mechanisms to therapeutic outcome is critically determined by the neurobiologic make-up of an individual, and underscores the necessity to adjust drug treatment approaches depending on the individual predisposition for placebo responses.

Other pathologic or normal variations in brain function or structure are likely to modulate the influence of placebo/nocebo mechanisms on pharmacologically induced analgesia in similar ways. Evidence from diffusion tensor imaging, for instance, shows that—also in healthy human volunteers—individual brain anatomy predicts the individual capacity to develop placebo analgesia.[47] The search for predictor variables in placebo responses and their influence on pharmacologic analgesia is still at a very early stage. Many of the available studies included only small sample sizes, which might explain the often inconclusive results.[48] Thus, future large-scale experimental and clinical studies have to address this important issue of biologic and psychologic predictors of placebo responsiveness that will be a major research challenge in the next decade.

CONCLUSION

Experimental and clinical evidence demonstrates that placebo and nocebo mechanisms substantially affect the efficacy and tolerability of analgesic treatments. The systematic exploitation of expectancy and learning mechanisms promises to fundamentally improve analgesic treatments in daily clinical routine. Understanding the neurobiology of placebo and nocebo modulations of analgesic drug efficacy represents a new and promising avenue of research. Instead of studying the effect of one of them in isolation by controlling for the other, it is time to unravel how both mechanisms combine at the neural and physiologic level. A more detailed neurobiologic understanding of their potential interaction promises to ultimately optimize treatment outcome

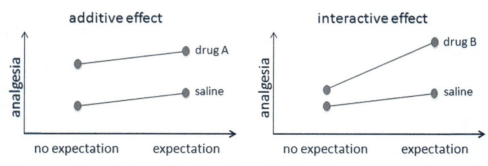

FIGURE 13.4 Additive versus interactive drug effects. Depending on the drug, pharmacologically induced mechanisms and cognition (e.g. expectation) may combine in an additive manner for one substance (left), but in an interactive fashion in another substance (right).

and push forward the development of personalized treatment strategies.

References

1. Amanzio M, Pollo A, Maggi G, Benedetti F. Response variability to analgesics: a role for non-specific activation of endogenous opioids. *Pain*. 2001;90(3):205-215.
2. Levine JD, Gordon NC. Influence of the method of drug administration on analgesic response. *Nature*. 1984/1985;312(5996):755-756.
3. Colloca L, Lopiano L, Lanotte M, Benedetti F. Overt versus covert treatment for pain, anxiety, and Parkinson's disease. *Lancet Neurol*. 2004;3(11):679-684.
4. Colloca L, Lopiano L, Lanotte M, Benedetti F. Overt versus covert treatment for pain, anxiety, and Parkinson's disease. *Lancet Neurol*. 2004;3(11):679-684.
5. Volkow ND, Wang GJ, Ma Y, et al. Expectation enhances the regional brain metabolic and the reinforcing effects of stimulants in cocaine abusers. *J Neurosci*. 2003;23(36):11461-11468.
6. Metrik J, Rohsenow DJ, Monti PM, et al. Effectiveness of a marijuana expectancy manipulation: piloting the balanced-placebo design for marijuana. *Exp Clin Psychopharmacol*. 2009;17(4):217-225.
7. Lyerly SB, Ross S, Krugman AD, Clyde DJ. Drugs and placebos: the effects of instructions upon performance and mood under amphetamine sulphate and chloral hydrate. *J Abnorm Psychol*. 1964;68:321-327.
8. Kirk JM, Doty P, De Wit H. Effects of expectancies on subjective responses to oral delta9-tetrahydrocannabinol. *Pharmacol Biochem Behav*. 1998;59(2):287-293.
9. Skovlund E, Fyllingen G, Landre H, Nesheim BI. Comparison of postpartum pain treatments using a sequential trial design. I. Paracetamol versus placebo. *Eur J Clin Pharmacol*. 1991;40(4):343-347.
10. Skovlund E, Fyllingen G, Landre H, Nesheim BI. Comparison of postpartum pain treatments using a sequential trial design: II. Naproxen versus paracetamol. *Eur J Clin Pharmcol*. 1991;40(6):539-542.
11. Linde K, Witt CM, Streng A, et al. The impact of patient expectations on outcomes in four randomized controlled trials of acupuncture in patients with chronic pain. *Pain*. 2007;128(3):264-271.
12. Dworkin SF, Chen AC, LeResche L, Clark DW. Cognitive reversal of expected nitrous oxide analgesia for acute pain. *Anesth Analg*. 1983;62(12):1073-1077.
13. Daniels AM, Sallie R. Headache, lumbar puncture, and expectation. *Lancet*. 1981;1(8227):1003.
14. Papadopoulos D, Mitsikostas DD. A meta-analytic approach to estimating nocebo effects in neuropathic pain trials. *J Neurol*. 2012;259(3):436-447.
15. Mora MS, Nestoriuc Y, Rief W. Lessons learned from placebo groups in antidepressant trials. *Philos Trans R Soc Lond B Biol Sci*. 2011;366(1572):1879-1888.
16. Colloca L, Sigaudo M, Benedetti F. The role of learning in nocebo and placebo effects. *Pain*. 2008;136(1-2):211-218.
17. Voudouris NJ, Peck CL, Coleman G. Conditioned response models of placebo phenomena: further support. *Pain*. 1989;38(1):109-116.
18. Colloca L, Benedetti F. How prior experience shapes placebo analgesia. *Pain*. 2006;124(1-2):126-133.
19. Amanzio M, Benedetti F. Neuropharmacological dissection of placebo analgesia: expectation-activated opioid systems versus conditioning-activated specific subsystems. *J Neurosci*. 1999;19(1):484-494.
20. Benedetti F, Amanzio M, Rosato R, Blanchard C. Nonopioid placebo analgesia is mediated by CB1 cannabinoid receptors. *Nat Med*. 2011;17(10):1228-1230.
21. Benedetti F, Pollo A, Lopiano L, et al. Conscious expectation and unconscious conditioning in analgesic, motor, and hormonal placebo/nocebo responses. *J Neurosci*. 2003;23(10):4315-4323.
22. Goebel MU, Trebst AE, Steiner J, et al. Behavioral conditioning of immunosuppression is possible in humans. *FASEB J*. 2002;16(14):1869-1873.
23. Bingel U, Wanigasekera V, Wiech K, et al. The effect of treatment expectation on drug efficacy: imaging the analgesic benefit of the opioid remifentanil. *Sci Transl Med*. 2011;3(70):70ra14.
24. Tracey I, Mantyh PW. The cerebral signature for pain perception and its modulation. *Neuron*. 2007;55(3):377-391.
25. Apkarian AV, Bushnell MC, Treede RD, Zubieta JK. Human brain mechanisms of pain perception and regulation in health and disease. *Eur J Pain*. 2005;9(4):463-484.
26. Ploghaus A, Narain C, Beckmann CF, et al. Exacerbation of pain by anxiety is associated with activity in a hippocampal network. *J Neurosci*. 2001;21(24):9896-9903.
27. Schweinhardt P, Kalk N, Wartolowska K, Chessell I, Wordsworth P, Tracey I. Investigation into the neural correlates of emotional augmentation of clinical pain. *Neuroimage*. 2008;40(2):759-766.
28. Kong J, Gollub RL, Polich G, et al. A functional magnetic resonance imaging study on the neural mechanisms of hyperalgesic nocebo effect. *J Neurosci*. 2008;28(49):13354-13362.
29. Colloca L, Benedetti F. Nocebo hyperalgesia: how anxiety is turned into pain. *Curr Opin Anaesthesiol*. 2007;20(5):435-439.
30. Benedetti F, Amanzio M, Vighetti S, Asteggiano G. The biochemical and neuroendocrine bases of the hyperalgesic nocebo effect. *J Neurosci*. 2006;26(46):12014-12022.
31. Hebb AL, Poulin JF, Roach SP, et al. Cholecystokinin and endogenous opioid peptides: interactive influence on pain, cognition, and emotion. *Prog Neuropsychopharmacol Biol Psychiatry*. 2005;29(8):1225-1238.
32. Rief W, Bingel U, Schedlowski M, Enck P. Mechanisms involved in placebo and nocebo responses and implications for drug trials. *Clin Pharmacol Ther*. 2011;90(5):722-726.
33. Edwards RR, Bingham 3rd CO, Bathon J, Haythornthwaite JA. Catastrophizing and pain in arthritis, fibromyalgia, and other rheumatic diseases. *Arthritis Rheum*. 2006;55(2):325-332.
34. Petrie KJ, Cameron LD, Ellis CJ, et al. Changing illness perceptions after myocardial infarction: An early intervention randomized controlled trial. *Psychoso Med*. 2002;64:580-586.
35. Bodenheimer T. The future of primary care: transforming practice. *N Engl J Med*. 2008;359(20):2086-2089.
36. Doering BK, Rief W. Utilizing placebo mechanisms for dose reduction in pharmacotherapy. *Trends Pharmacol Sci*. 2012;33(3):165-172.
37. Ader R, Cohen N. Behaviorally conditioned immunosuppression. *Psychosom Med*. 1975;37(4):333-340.
38. Colloca L, Miller FG. How placebo responses are formed: a learning perspective. *Philos Trans R Soc Lond B Biol Sci*. 2011;366(1572):1859-1869.
39. Ader R, Mercurio MG, Walton J, et al. Conditioned pharmacotherapeutic effects: a preliminary study. *Psychosom Med*. 2010;72:192-197.
40. Sandler AD, Glesne CE, Bodfish JW. Conditioned placebo dose reduction: a new treatment in attention-deficit hyperactivity disorder? *J Dev Behav Pediatr*. 2010;31(5):369-375.
41. Leech J, Mazzone SB, Farrell MJ. The effect of placebo conditioning on capsaicin-evoked urge-to-cough. *Chest*. 2012;142(4):951-957.
42. Koudriavtseva T, Onesti E, Pestalozza IF, et al. The importance of physician–patient relationship for improvement of adherence to long-term therapy: data of survey in a cohort of multiple sclerosis patients with mild and moderate disability. *Neurol Sci*. 2012;33(3):575-584.
43. Kaptchuk TJ, Kelley JM, Conboy LA, et al. Components of placebo effect: randomised controlled trial inpatients with irritable bowel syndrome. *Br Med J*. 2008;336(7651):999-1003.

44. Enck P, Bingel U, Schedlowski M, Rief W. The placebo response in medicine: minimize, maximize or personalize? *Nature Rev Drug Disc*. 2013 (in press).
45. Atlas LY, Whittington RA, Lindquist MA, et al. Dissociable influences of opiates and expectations on pain. *J Neurosci*. 2012;32(23): 8053-8064.
46. Benedetti F, Arduino C, Costa S, et al. Loss of expectation-related mechanisms in Alzheimer's disease makes analgesic therapies less effective. *Pain*. 2006;121(1-2):133-144.
47. Stein N, Sprenger C, Scholz J, et al. White matter integrity of the descending pain modulatory system is associated with interindividual differences in placebo analgesia. *Pain*. 2012;153(11): 2210-2217.
48. Kaptchuk TJ, Kelley JM, Deykin A, et al. Do 'placebo responders' exist? *Contemp Clin Trials*. 2008;29(4):587-595.

CHAPTER 14

How Placebo Responses are Formed: From Bench to Bedside

Luana Colloca

National Center for Complementary and Alternative Medicine (NCCAM), National Institutes of Health, Bethesda, MD, USA, National Institute of Mental Health, National Institutes of Health, Bethesda, MD, USA, Department of Bioethics, Clinical Center, National Institutes of Health, Bethesda, MD, USA

INTRODUCTION

Current research has pointed to very different neurobiologic pathways serving to mediate the formation of placebo responses depending on the medical condition of subjects and the outcomes investigated. There are relatively few comprehensive theories about how beliefs and psychosocial messages are decoded to form a placebo response.[1,2] The most extensively accepted theories are expectation and conditioning, involving both conscious and unconscious information processing. The question remains: how can this confluence of theories and diverse views with regard to the placebo response on a neurobiologic level, enable pain analgesia in patients?[3] Moreover, can we exploit the placebo phenomenon in a safe and controlled manner? Delving into these questions naturally requires a thorough understanding of the formation of behavioral and biologic placebo changes as a coherent phenomenon while taking into account important implications for laboratory research and medical care. In addition, there are still many unknowns when it comes to dealing with placebo analgesia, even if recent human and nonhuman research has impressively increased our knowledge of neurobiologic mechanisms underlying placebo effects in different medical conditions and physiologic processes.

This chapter orchestrates common themes in the placebo literature with the aim of articulating a unified account of the phenomenon through a learning perspective within the context of pain analgesia. The core of this chapter focuses on behavioral and neurobiologic evidence of placebo (and nocebo) responses formed by processing verbal instructions, conditioning and social cues (observation and interactions). Verbal, conditioning, and social cues are decoded and processed by the brain, creating dynamic expectations that, in turn, shape pain perception. We present a general account of the concept of expectations as central to the formation of placebo responses, and develop speculations relating to the evolution of placebo responses. We suggest interpreting, critiquing and formally modeling the existing experimental and clinical research on placebo (and nocebo) effects in terms of expectations formed through different kinds of learning. This approach is promising for future laboratory investigation and translational patient-oriented placebo research within the domain of pain analgesia.

INSTRUCTIONAL LEARNING

Numerous studies have explored the role of verbal suggestions in eliciting placebo responses, as well as examining the impact of the various associated cognitive and emotional elements involved. Although the contributions of verbal suggestions and conditioning are often intertwined, it is important to acknowledge the distinction between these two forms of process with regard to top-down modulation of pain and placebo responses. In this section, the focus is on research examples that have demonstrated a placebo response by suggestions of a benefit from treatment via persuasive verbal communication. The converse is also inherently possible, that is, a verbal suggestion of harm invoking a nocebo response.

Henry Knowles Beecher, the father of randomized controlled trials (RCTs), was able to demonstrate the profound ability of placebos to relieve pain. Beecher, facing a depleted morphine supply, surreptitiously gave saline as a substitute to morphine to wounded soldiers. Remarkably, soldiers who were administered saline responded in a similar fashion to those soldiers who were given morphine. This extraordinary consequence is one of the first concrete examples of using top-down modulation of pain such as verbal suggestions invoking a powerful placebo response in pain analgesia.

Similar results have been observed in modern clinical settings of postoperative pain[4] as well as visceral and somatic pain.[5,6] In the former, thoracotomized patients were treated with buprenorphine on request for three consecutive days, together with a basal intravenous infusion of saline solution while three different sets of verbal instructions were given to the patients. The first group was told nothing about the analgesic effect (natural history). The second group was told that the basal infusion was either a powerful painkiller or a placebo according to a double-blind paradigm. The third group was informed that the basal infusion was a potent painkiller (full-deceptive administration). The placebo response was measured by recording the doses of buprenorphine requested over the 3-day treatment. Buprenorphine requests decreased in the double-blind group by 20.8% compared with the natural history group, and by 33.8% in the deceptive administration group.[4] There are a couple of immediate implications of this research study. First, once more there is distinct evidence of verbal suggestions contributing to an analgesic effect. Second, there is a clinically significant reduction of buprenorphine in the third group as compared to the second group, thus revealing a dramatic variation in analgesia as a consequence of differences in instructions communicated to patients. With regard to visceral pain, patients with irritable bowel syndrome (IBS) exposed to painful rectal balloon distension under no treatment, rectal placebo, or rectal lidocaine conditions showed a higher analgesia when the active rectal lidocaine was administered along with the information that they were given an agent which is known to significantly reduce pain in some patients as compared to the patients who were informed that they may receive either an active pain-reducing medication or an inert placebo agent.[5,6]

The conclusions drawn from these studies are not in isolation. Indeed, early experimental studies of Wolf indicate that placebo effects could mimic, mask, enhance or prevent beneficial responses to active drugs.[7] Earlier studies by Luparello and co-workers[8,9] indicated significant increases in airway resistance in nearly half the asthmatic patients under investigation when they inhaled a nebulized saline solution along with the information that it was an allergen with irritant properties.[5,6] The same patients were able to reverse airway obstruction by inhaling the identical substance presented as a medicine with beneficial effects on asthma.[8] Similarly, the effects of the bronchoconstrictor carbachol, when administered along with the information that this drug is a bronchoconstrictor, were higher than when subjects were told that the drug was a bronchodilator.[9] Since the results of these experiments were published, additional systematic experimental evidence has supported the intuition that verbally induced expectations can markedly influence the response to drugs and other interventions. Coupled with this research is the observation that different sets of verbal instructions result in the modulation (enhancement or reversal, or both) of a variety of clinical outcomes and specific perceptions. In another study, verbal suggestions produced different outcomes in healthy subjects randomized to receive decaffeinated coffee under two different verbal descriptions: group 1, in which participants were told that they would receive either regular or decaffeinated coffee according to a double-blind design; and group 2, in which decaffeinated coffee was presented as regular coffee. Placebo responses were higher in group 2 compared to group 1, suggesting a difference in expectation.[10]

The results derived from research on placebo effects naturally extend to the nocebo. To this end, research on the nocebo has indicated that communication and verbally induced expectations can also produce negative responses, termed *nocebo effects*.[11,12] Negative information given verbally once can induce nocebo responses as strong as those induced by direct experience of negative outcomes.[13,14] After informing healthy subjects about the hyperalgesic effect of a treatment, the subjects perceived pain with both low and high non-painful tactile stimuli associated with a green light along with negative verbal suggestion, with or without pre-conditioning. Nocebos produced allodynic effects, whereby non-painful tactile stimuli become painful. In addition, low-intensity painful stimuli were perceived as high-intensity stimuli after negative verbal suggestion, with or without pre-conditioning, indicating that nocebos can also induce hyperalgesic effects, whereby low-intensity painful stimuli are perceived as high-intensity stimuli (Fig. 14.1).[13] These results have important implications for daily practice. For a more comprehensive overview, see some related chapters in this book.[15–17] Since negative verbal suggestions are powerful in eliciting a nocebo response, it follows that nocebos (aversive responses) exist in the organism to enhance perceptual processing which, in turn, helps initiate potentially defensive behavioral reactions.

In line with these findings, Rodriguez-Raecke et al performed experiments wherein contextual information was given once at the beginning of the investigation, indicating that repeated painful stimulations over several days will increase pain sensation from day to day. The researchers discovered that this procedure elicited

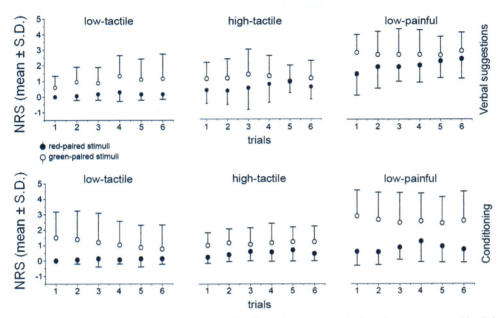

FIGURE 14.1 Informing healthy subjects about the hyperalgesic effect of a treatment led to the perception of both low- and high-tactile stimuli as painful. Also, low-painful stimuli were perceived as higher intensity painful stimuli. Similar effects were observed after a conditioning procedure, thus indicating that, in the presence of a stressor, expectations derived from verbal suggestions produced effects comparable to those induced by direct prior experience. *Data from Colloca et al.[13]*

an immediate hyperalgesic effect in healthy subjects and interfered with the long-term natural course of pain habituation over 8- and 90-day periods at the level of the brain opercular region.[14] At the neurochemistry level, verbal suggestions of hyperalgesia trigger the activation of cholecystokinin pathways[18] with a possible involvement of dopaminergic and opioidergic systems as well.[19]

Collectively, these research results provide valuable evidence of the effectiveness of verbal suggestions in promoting placebo and nocebo responses in the context of pain. In the following section, the impact of conditioning on placebo and nocebo responses is explored.

ASSOCIATIVE LEARNING

Pavlov's famous experiments demonstrated that dogs would salivate (conditioned response, CR) in response to a bell (conditioned stimulus, CS) that had previously been paired with the administration of food (unconditioned stimulus, US).[20] The formation of these conditioned responses in the dogs suggests that these dogs learned that the ringing bell implied food, hence the salivary reaction upon hearing the bell. An obvious question is whether these results could be extended to patients in a clinical setting, especially as a means of augmenting pain analgesia? Extending the classic conditioning concept in the clinical sphere translates into the intuition that visual, tactile, and gustatory stimuli associated with the ingestion of a medication can become CSs through their repeated association with the delivery of a US in the form of active medication. Moreover, placebos given along with the presentation of CSs and subsequently without the USs could potentially elicit CRs that are similar to the response to medication. Thus, classic conditioning has been the prevalent paradigm to explain the genesis of (unconscious) placebo responses by learning mechanisms. The initial support for a classic conditioning interpretation of the placebo effect arises from some studies in animals.[21,22] Herrnstein demonstrated after 14 pairings of scopolamine with sweetened milk, that the presentation of the pure sweetened milk alone (CS) caused scopolamine-like alteration of behavior (depression of rates of a lever-pressing task). Ader and Cohen found that pairing a novel saccharine-flavored solution with the immunosuppressant cyclophosphamide induced immunosuppression in rats presented with saccharine solution alone. Rats that received two doses of cyclophosphamide during the conditioning phase showed greater conditioned immunosuppression than those given one dose;[21] hence, the stronger the US, the more robust the CR is. Evidence of conditioned placebo responses in the immune system has been extensively documented in human models with promising clinical implications.[23,24] More recently, Guo et al. investigated the effect of prior pharmacologic opioid and nonopioid exposure in mice by using a model of a hot-plate test.[25] Conditioned cues were paired with either opioid agonist morphine hydrochloride or nonopioid aspirin. Placebo analgesic responses evoked by morphine-based pharmacologic conditioning were completely antagonized by naloxone. By contrast, after aspirin conditioning the placebo responses were not blocked by naloxone.[25] Therefore, the responses evoked were either naloxone-reversible or naloxone-insensitive, depending on previous drug exposure, confirming parallel early results in humans.[26] Some more recent human studies adopting a pharmacologic conditioning with drugs such as immunosuppressor cyclosporin A,[27,28] dopamine-agonist apomorphine,[29] benzodiazepine receptor agonist midazolan and antagonist flumazenil,[30] found that the conditioned placebo responses mimic drug effects, supporting the notion that such placebo responses depend on the kind of drug exposure that is originally performed.[31] Based on these and similar research studies, it is not surprising then that the analysis of the placebo effect in terms of learning has for the most part been discussed with respect to classic conditioning and unconscious processes.[22,23,32–35]

In systems such as the immune and endocrine systems, which are not accessible consciously, and do not differ, other than in degree of complexity, from those in nonhuman animals, classic conditioning plays a crucial role in eliciting a placebo response. For example, Benedetti et al observed that the pharmacologic exposure to a serotonin agonist of $5\text{-HT}_{1B/1D}$ receptors, stimulating growth hormone (GH) and inhibiting cortisol secretion in humans, produced similar responses when the drug was replaced by a placebo treatment. Furthermore, these responses were not influenced by verbal instructions,[36] thus providing evidence of a demarcation between verbal suggestions and conditioning.

It is worth mentioning that the traditional concept of conditioning based on the acquired ability of one stimulus (CS) to evoke the original response by means of pairing with the US may only partially explain conditioned response in humans. Rescorla described conditioning as the learning of relations among events so as to allow the organism to represent its environment.[37] According to this definition, pairing and contiguity remain a central concept, but conditioning depends on both the information that the CS provides about the US and the learning of relations among events.[37–40]

In laboratory settings, numerous studies have found that conditioning is more effective when verbal suggestions of benefit are also provided.[13,33,41-47] Voudouris and colleagues tested the effects of verbal suggestion and conditioning manipulations.[47] Healthy participants underwent an iontophoretic pain stimulation after the establishment of a baseline pain tolerance level. All participants attended four sessions on four consecutive days. The experimental design consisted of verbal suggestion manipulation (session 1), pain-test (session 2), conditioning manipulation (session 3), and pain-test 2 (session 4). In the first session, half the participants were informed that a topical cream was a powerful painkiller (expectation of analgesia), and would provide pain relief, and the other half were informed that the cream was neutral (no expectation). In the second session, half of the participants received a neutral cream (placebo) and the other half were given none. In the third session, half the participants in each of the verbal suggestion and no-verbal-suggestion groups received conditioning in which the pain stimulus was reduced after the cream was applied, while the other half received the same pain stimulus. Thus, group 1 received a combined verbal suggestion and conditioning manipulation; group 2 received verbal suggestion alone; group 3 conditioning alone; and group 4 was the control. The findings revealed an enhancement of placebo response in groups 1 and 3. Thus, conditioning was more effective than verbal suggestion in eliciting placebo analgesia.[47] We studied the contribution of both verbal suggestions and prior experience via conditioning, at the level of both N1 and biphasic N2–P2 components of scalp laser-evoked potentials (LEPs), which presented the advantage to explore early central nociceptive processing noninvasively. N1 is generated in the second somatosensory (SII) area, while N2–P2 is a biphasic negative–positive complex obtained at the vertex which originates in the bilateral operculo-insular areas and in the cingulate gyrus. Scalp-LEP components are modulated by top-down mechanisms and placebo manipulations.[48–50] We found that verbal suggestions of benefit and conditioning induced a decrease in amplitude of the N2–P2 complex, but not of the earlier N1 component. Verbal suggestions induced LEP changes occurring without subjective perception of pain decrease, while N2–P2 amplitude reductions induced by the conditioning procedure were associated with a strong reduction of subjective pain. Overall, the experience of an analgesic effect, via conditioning produced a substantial reduction of LEP amplitudes as compared to the natural history group and verbal suggestions (Fig. 14.2).[51]

By manipulating the number of CS–US pairings, we have found in our own studies that the persistence of placebo and nocebo responses was firmly connected to the length of exposures to effective (and ineffective) treatments. The increase in the number of associations from 1 to 4 sessions of conditioning resulted in a higher magnitude of placebo and nocebo responses and their resistance to extinction of the ensuing responses (Fig. 14.3).[52]

The experience of mastery also strongly impacts expectations and on subsequent behavior. In this regard, Price and colleagues applied a pain stimulus and placebo cream to healthy volunteers under two experimental conditions: A (strong placebo) and B (weak placebo), and a control agent, C.[45] During the conditioning procedure, the intensity of the stimulus was decreased 67% in A and 17% in B. There was no reduction in C (control). Expectancies were also assessed after the trials of intensity manipulation. In comparison to the pre-treatment

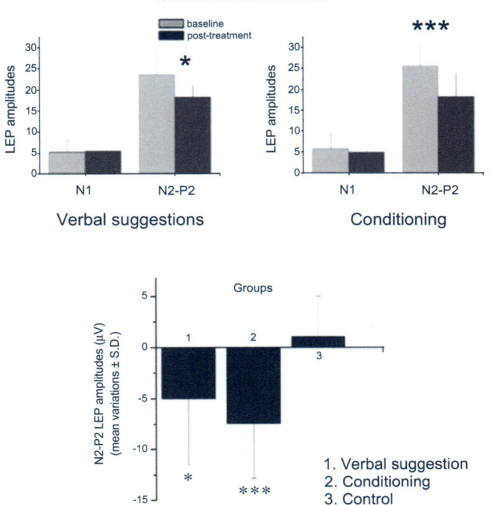

FIGURE 14.2 Verbal suggestions of analgesia and conditioning modulated N2–P2 amplitude reductions. Subjects in the conditioning procedure experienced an analgesic effect (Group 2), and produced a larger pain reduction as compared to subjects in the verbally induced group (Group 1) and subjects in the no-treatment group (Group 3). This result suggests that the perception of treatment effectiveness via conditioning is a crucial factor for shaping the central nociceptive processing of placebo analgesia. *Data from Colloca et al.*[51]

condition, subjects experienced a large and small reduction, as well as no reduction in the three areas on the ventral part of the forearm, which had corresponding labels of A, B, and C.[45] The findings were congruent with what the subjects previously perceived, thus the intensity of 'US' likely graded different expectation levels so that graded placebo analgesic effects were obtained.

In general, learning via prior experience has been increasingly emphasized as a modulating factor of the placebo effect[53] owing to the awareness that a previous direct experience of benefit via pharmacologic or biologically significant cue exposure can powerfully change behavior and clinical outcomes. Prior positive experience induces increased analgesic responses of a subsequent placebo, whereas ineffective previous experiences negatively influence the formation of placebo effects.[41] We performed in one group a simulation of an effective treatment (exposure to a surreptitious reduction of intensity of painful stimulations, treatment A) to create expectations of benefit. After this learning procedure, the administration of a placebo induced robust analgesic responses. A second group of subjects of the same study received an ineffective treatment (no reduction of intensity of painful stimulation, treatment B) which resulted in no placebo analgesic effects. After a time lag of 4–7 days, both the groups were shifted to either the effective or ineffective procedure according to a within cross-over design. Interestingly, the placebo analgesic responses following an effective procedure in the first group were significantly high showing no extinction over time. In the second group, the experience of ineffectiveness was capable of altering negatively the effects of the subsequent effective procedure, as shown in Figure 14.4. These findings suggest that placebo analgesia is finely tuned by prior experience (either positive or negative), and that the effect of initial treatment

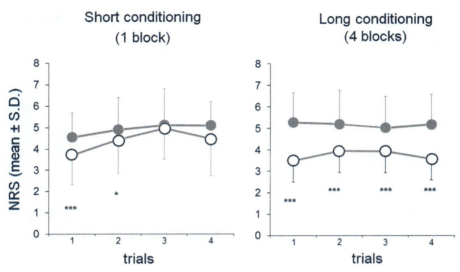

FIGURE 14.3 The use of CS, reinforced on 100% of trials, produced two distinct patterns of learned placebo analgesic effects. The persistence of placebo analgesic responses was strongly related to the number of trials in the acquisition phase (1 versus 4 blocks). The long conditioning paradigm resulted in the formation of sustained placebo analgesic responses lasting for the entire experimental session. Conversely, the short conditioning paradigm induced responses of small magnitude and duration. *Data from Colloca et al.[52]*

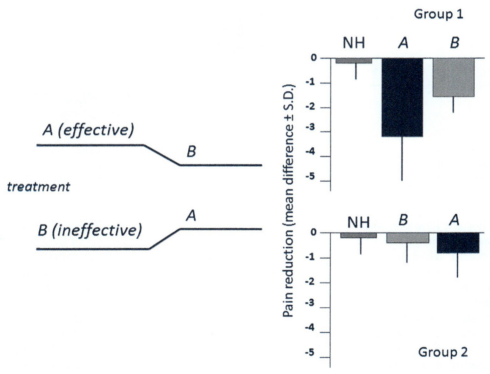

FIGURE 14.4 Placebo analgesic effects in sequential trials. Prior positive experience induces increased analgesic responses of a subsequent placebo, whereas ineffective previous experiences negatively influence the formation of placebo effects. The scheme shows the magnitude of placebo analgesic effects in healthy subjects as a result of being exposed first to an effective (A) or ineffective (B) procedure according to a within cross-over design. Group 1 was exposed to an effective treatment first, while Group 2 received a placebo manipulation after an ineffective treatment. After a time lag of 4–7 days, both groups were shifted to either effective or ineffective procedures. Interestingly, placebo analgesic effects increased in Group 1 as compared to Group 2. In addition, a treatment initially perceived as ineffective, when given after a treatment perceived as effective, resulted in a high analgesic effect, suggesting that the sequence of treatments influences the relative magnitude of analgesic responses. *Data from Colloca and Benedetti.[41]*

influences the magnitude of placebo responses after several days.[54]

Thus, research on the formation of placebo responses via conditioning and learning experiences has important implications for clinical practice. Ader and colleagues have recently provided proof-of-concept evidence that a partial schedule of pharmacologic reinforcement consisting of reduced cumulative amounts of corticosteroids associated with placebos, was effective in suppressing symptoms of psoriasis comparable to the reduction in symptoms induced by a full-dose treatment.[55] Thus, learned placebo responses following the exposure to drugs may be successfully exploited in routine clinical practice by integrating placebos in schedules of reinforcement, so that conditioned stimuli acquire properties and characteristics of USs. These effects may become part of the pharmacotherapeutic protocol preserving therapeutic benefits while side effects are reduced.[3,56]

SOCIAL LEARNING

The previous sections have examined in detail, forms of learning that have shown to be capable of evoking placebo and nocebo responses and, in turn, reducing or increasing, respectively, pain analgesia. However, beyond direct first-hand experience, people can also learn by observing others. Likewise, placebo responses can be formed by means of observational learning in a social context without any deliberate reinforcement.

By investigating the role of social observation in placebo analgesia, Colloca and Benedetti demonstrated that healthy subjects can learn to form placebo analgesic effects by observing the experience of a demonstrator, suggesting that the information drawn from social learning may establish a self-projection into the future outcome.[57] Observationally induced effects exhibited no extinction, indicating implicit acquisition and retention of behavioral output. The magnitude of observationally induced placebo responses was similar to those induced by directly experiencing the benefit through a conditioning procedure in which subjects were exposed to first-hand experience of analgesia. Interestingly, the more pronounced observationally induced placebo responses were found in those subjects who presented higher empathy scores. This result suggests a link between the ability to modify behaviors following mere observation, the formation of placebo analgesia responses, and empathy (Fig. 14.5).

These observations emphasize that social interactions are potential cues to induce expectations of benefit and activate specific mechanisms.

In terms of nocebo effects, recently Mazzoni et al reported observationally induced responses in a model of nocebo mass psychogenic illness.[58] Healthy subjects were asked to inhale a sample of normal air which was presented as a product containing a suspected environmental toxin known to cause headache, nausea, itchy skin and drowsiness. Half of the healthy subjects also observed an actor who inhaled the product. Those who watched another person displaying signs of illness reported a significant increase in the four described symptoms, suggesting that social learning might also play a role in clinical contexts.[58] It is plausible that these results may be extended to pain and hyperalgesia.

Attempts to analyze observationally induced effects within the associative learning framework have been made in human and nonhuman models. In particular,

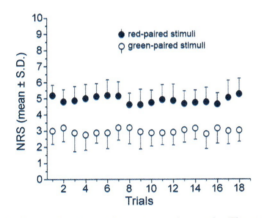 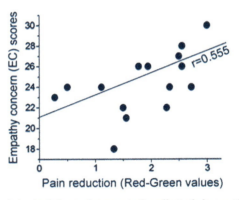

FIGURE 14.5 Placebo analgesia, social learning and empathy. The graph in the left panel presents the effect of observation on pain reports. Healthy subjects were tested for placebo analgesic responses after observing a demonstrator. The demonstrator rated green-associated stimuli as analgesic and red-associated stimuli as painful. After the observation, subjects perceived the green-associated stimuli as less painful, although the pain intensities for both colors were the same. Pain perception was assessed by means of a Numerical Rating Scale (NRS) ranging from 0 = no pain to 10 = maximum imaginable pain. The analgesic effects were correlated with empathy concern scores. The higher the scores in empathy, the higher the analgesic effects (B). *Data from Colloca and Benedetti.*[57]

in animals, studies on observational aversive learning in rats fail to find blocking, latent inhibition, and overshadowing—three well-documented features of classical conditioning.[59] Conversely, studies in humans reported classic conditioning features for social aversive learning, including overshadowing and blocking,[60] implying that observation might serve as a US. Beyond these attempts to categorize observational learning, the ability to alter behaviors without any practice and direct reinforcement is part of the large repertoire of prosocial behaviors (e.g. ability to share another's feelings, imitation, mimicry) which allows humans and nonhumans to draw information from their social environment and make inferences about a future outcome.[61]

The power of social learning in placebo health-promoting processes is demonstrated by studies focusing on contextual cues in clinical settings and the patient–provider relationship. Aspects of conditioning, instructional learning, and observational learning are likely to combine in the clinical encounter to promote socially induced placebo responses. The corresponding contributions of the various psychosocial components can be disentangled by comparing the effects of medication administered in the manner of routine clinical practice with a hidden administration of the same medication (for a review see[62]) and the effects of business-like placebo responses versus an augmented patient–doctor relationship.[63] In other words, the goal is to compare and contrast a system with common psychosocial components and inter-personal relationships intact versus a system with one or more psychosocial components removed or reduced. As an example of the impact of social interaction in pain analgesia, studies compared two groups of patients hospitalized after surgery. The first group received an injection of analgesic drugs administered by a physician and were told that this injection contained a powerful painkiller, which should produce pain relief in a few minutes. The second group, who received analgesic medication from a preprogrammed infusion machine without being told when they would be given the medication, required a much higher dose of medication to reduce pain by 50% than those in the first group.[64,65]

Patients with irritable bowel syndrome enrolled in a run-in phase of a randomized trial comparing verum and sham acupuncture reported greater symptom relief when they received an augmented sham acupuncture intervention arm consisting of a longer and more empathetic initial conversation with the practitioner as compared to a more business-like sham intervention, whereby communication between practitioner and patient was reduced to a minimum. Both sham acupuncture groups reported superior outcomes relating to symptoms than those in a no-treatment waiting-list group. When augmented by supportive communication, the ritual of treatment can produce an enhanced and sustained placebo response in a difficult-to-treat patient population.

These findings make some inroads into revealing the cognitive and emotional appraisal of a situation based on human social interactions, in terms of the generation and maintenance of placebo effects.

EXPECTATIONS

Expectations are central to the formation of placebo responses and can derive from information learnt from instructions, via personal experiences and from social and observational cues. Such expectations are connected with individual beliefs and emotions, dynamically updated over time and graded by the perceived likelihood of an outcome such as placebo analgesia (Fig. 14.6).

This perspective is in line with the semiotic theory developed by Charles Peirce.[1,2] Iconic, indexical, or symbolic signs convey information that is processed by a person. These signs can be vehicles of placebo responses. When a placebo response occurs, the interpreted sign is the therapeutic agent and the response involves learning processes, through which the patient processes and responds to the information coded in the signs. Cognition shapes the interpretation of the signs, producing positive and negative outcomes. As Kirsch suggested, expectations can be elicited by explicit (e.g. suggestions of positive or negative outcomes) and implicit processes (e.g. individual previous experience).[66] This observation suggests that it is important to avoid any strict dichotomy between conditioning and expectation mechanisms, as the former involves information processing by which a subject anticipates (i.e. expects) a future event, which may or may not be conscious.[67] Conversely, expectations formed on the basis of explicit instructions are often

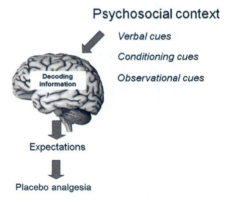

FIGURE 14.6 The diagram shows how verbal, conditioning, and observational cues—all components of the psychosocial context surrounding the patient and any treatments—are finely decoded by the brain to form expectations which are pivotal to placebo analgesia. *Adapted from Colloca and Miller.[1]*

associated with unconscious prior experience and thus involve different degrees of awareness.

When a perception, such as pain relief is consciously accessible, verbal instructions become a crucial modulator of placebo effects. However, it is worth noting that both nocebo hyperlagesic responses and placebo analgesic responses can be also triggered by stimuli presented outside of conscious awareness. By contrast, hormonal conditioned placebo responses are shaped by unconscious conditioning but are not affected by verbal instructions and such an event cannot be experienced and perceived by human cognition (e.g. growth factor secretion).[36]

Defining the boundaries of consciousness along with conscious and unconscious expectancies in forming placebo responses is a challenging field of research, which requires more investigation. Some authors have defined expectations as consciously accessible mental entities,[68,69] but others have suggested that unconscious expectations also exist.[70] If conditioning may be understood as a process generating expectations and conditioned responses in humans and animals without being mediated by consciousness, it follows that expectations are not necessarily conscious. However, it is reasonable to assume that, by and large, the closer the phylogenetic distance to humans, the larger the role of conscious cognition and the smaller the role of unconscious processes.

EVOLUTIONARY PRINCIPLES BEHIND PLACEBO ANALGESIA

In evaluating the hypothesis that the placebo effect predominantly relies on learning, it is worth asking the reasons for human and nonhuman ability to release endogenous substances with health-promoting effects in the context of nature and how these processes may have evolved.[71] In an attempt to reconstruct the evolutionary meaning of the placebo responses, we briefly rely on a combination of the innate capacity of humans to enhance their adaptation to the environment by learning, social behaviors, naturalistic and ecologic contexts, and potential genetic contributions via selection.

The ability to modulate endogenous systems and heal by learning presumably would enhance survival, and hence, placebo analgesia may be favored directly by natural selection. In particular, genetic make-up and social contexts may interact in controlling behaviors and promoting adaptation. As reflected in the research findings on pain and other domains, the propensity of humans and nonhumans to be conditioned as well as the potential for placebo interventions to modify disease outcomes by means of classic conditioning, may also be part of their biologic heritage. Pavlovian conditioning which humans share with other species and relatively simple invertebrates is one of the most powerful mechanisms underlying placebo analgesia.

Interestingly, the ability to imitate and learn from the behaviors of others appears advantageous because this mechanism is a rapid form of learning, saving cost and time, in terms of effort and risk, leading the animal or individual to acquire appropriate information without risks associated with trial-and-error learning.[61,72-74] Therefore, social learning may be an evolutionary adaption that prevents errors, selects strategies useful for survival, and facilitates the formation of placebo analgesic responses.

Viewed as being activated by conditioning and observational processing, the placebo effect would represent a form of learned responses that antedated the emergence of language. However, human and social environment linguistic abilities evolved to recode and re-represent individual and social experiences. Patients who are given inert treatments along with the verbal suggestion that they are powerful remedies have shown an improvement in a variety of symptoms. In the doctor–patient relationship, the way in which doctors inform patients about diagnoses and/or prognoses may result not only in good compliance with treatment and patients' satisfaction but also in placebo responses promoting healing processes. Moreover, social interactions seem to be important determinants of the formation of placebo responses, suggesting that these responses may represent a byproduct of altruism and social solidarity, nurturance and inter-animal practices of grooming.[75] Monkeys used grooming to build up alliances and are selective in allocation of grooming, likely recognizing in this practice an honest indicator of time invested and a form of reciprocal altruism. Also relevant in humans is the prolonged process of dependency in infancy and childhood that gives more salience to parental nurturing, thus laying a strong foundation for projecting the relief of suffering that children receive from their parents' intervention in interactions with healers. This observation may explain why some internal mechanisms of symptoms' relief so often take the intervention of a healer[71] and are not activated spontaneously when an animal or an individual is at rest and is doing what is needed to avoid further injury to the organism. Patients are inevitably not only confronted with their own expectations and experiences but are strongly influenced by beliefs of their families, peers, clinicians, and cultural elements. Finally, cultural processes may facilitate the spread of adaptive knowledge over generations by being able to recognize vital life skills to cultivate successful social relations, and prosocial behaviors. Thus, it is possible to argue that the human placebo effect is based on an innate ability to learn with obvious survival value, which is made use of in healing in light of the human situation of prolonged dependency, social interactions, and features of cultural evolution. An important aspect related to the genesis of placebo effects is the idea that placebo responders can be predicted by specific genetic polymorphisms and brain network activation. For example, serotonin-related gene polymorphisms have been found to influence the individual placebo response in social anxiety, both at the behavioral and neural levels (as indicated by amygdala activity during a stressful public speaking task).[76] Genetic polymorphisms modulating monoaminergic tone (catabolic enzymes catechol-O-methyltransferase and monoamine oxidase A) have been related to the degree of placebo responsiveness in a major depressive disorder.[77] The COMT val158met functional single nucleotide variant has also been associated with placebo analgesic responses in patients suffering from IBS, and has been suggested as a potential predictor of the benefit from a supportive doctor–patient relationship.[78] Moreover, brain networks have been shown as powerful tools for predicting placebo responsiveness in both healthy individuals[79,80] and patients in pain.[81]

Overall, research clarifying the relation between specific genetic polymorphisms and brain placebo effects paves the way for predicting the ability to activate endogenous modulation of pain, gathering knowledge regarding the role of evolution in determining this phenomenon and promising a better approach for pain management.

CONCLUSION

This chapter has explored a wealth of research serving to elucidate the mechanisms responsible for activating placebo and nocebo responses. In particular, learning (and associated mechanisms) has been demonstrated to be a key mediator of expectations and placebo responses. A large body of research has been formally systemized here, integrating behavioral and neurobiologic literature in terms of information processing, reframing the placebo effect as a learning phenomenon. The evidence points to learning processes ultimately guiding the changes in behaviors and expectations that, in turn, lead to the formation of placebo responses. Viewing the placebo effect via a learning perspective may foster scientific investigation, promoting a deeper and better knowledge of the phenomenon in health care. It is patently clear that the ramifications of such knowledge are of paramount importance to the study of pain management, given the potential capacity of the placebo and nocebo responses in affecting pain outcomes in diverse physiologic and pathologic conditions.

CONFLICTS OF INTEREST

There are no conflicts of interest.

Acknowledgments

This research is supported by the Intramural Research Program of the National Center for Complementary and Alternative Medicine (NCCAM), the National Institute of Mental Health (NIMH), and the Clinical Center, National Institutes of Health.

The opinions expressed are those of the author and do not necessarily reflect the position or policy of the National Institutes of Health, the Public Health Service, or the Department of Health and Human Services.

References

1. Colloca L, Miller FG. How placebo responses are formed: a learning perspective. *Philos Trans R Soc Lond B Biol Sci*. 2011;366(1572): 1859-1869.
2. Miller FG, Colloca L. Semiotics and the placebo effect. *Perspect Biol Med*. 2010;53(4):509-516.
3. Colloca L, Miller FG. Harnessing the placebo effect: the need for translational research. *Philos Trans R Soc Lond B Biol Sci*. 2011;366(1572):1922-1930.
4. Pollo A, Amanzio M, Arslanian A, et al. Response expectancies in placebo analgesia and their clinical relevance. *Pain*. 2001;93(1):77-84.
5. Vase L, Robinson ME, Verne GN, Price DD. Increased placebo analgesia over time in irritable bowel syndrome (IBS) patients is associated with desire and expectation but not endogenous opioid mechanisms. *Pain*. 2005;115(3):338-347.
6. Verne GN, Robinson ME, Vase L, Price DD. Reversal of visceral and cutaneous hyperalgesia by local rectal anesthesia in irritable bowel syndrome (IBS) patients. *Pain*. 2003;105(1-2):223-230.
7. Wolf S. Effects of suggestion and conditioning on the action of chemical agents in human subjects; the pharmacology of placebos. *J Clin Invest*. 1950;29(1):100-109.
8. Luparello T, Lyons HA, Bleecker ER, McFadden Jr ER. Influences of suggestion on airway reactivity in asthmatic subjects. *Psychosom Med*. 1968;30(6):819-825.
9. Luparello TJ, Leist N, Lourie CH, Sweet P. The interaction of psychologic stimuli and pharmacologic agents on airway reactivity in asthmatic subjects. *Psychosom Med*. 1970;32(5):509-513.
10. Kirsch I, Weixel LJ. Double-blind versus deceptive administration of a placebo. *Behav Neurosci*. 1988;102(2):319-323.
11. Benedetti F, Lanotte M, Lopiano L, Colloca L. When words are painful: unraveling the mechanisms of the nocebo effect. *Neuroscience*. 2007;147(2):260-271.
12. Colloca L, Finniss D. Nocebo effects, patient-clinician communication, and therapeutic outcomes. *JAMA*. 2012;307(6):567-568.
13. Colloca L, Sigaudo M, Benedetti F. The role of learning in nocebo and placebo effects. *Pain*. 2008;136(1-2):211-218.
14. Rodriguez-Raecke R, Doganci B, Breimhorst M, et al. Insular cortex activity is associated with effects of negative expectation on nociceptive long-term habituation. *J Neurosci*. 2010;30(34):11363-11368.
15. Moussata D, Goetz M, Gloeckner A, et al. Confocal laser endomicroscopy is a new imaging modality for recognition of intramucosal bacteria in inflammatory bowel disease in vivo. *Gut*. 2011;60(1): 26-33.
16. Kambal AA, De'Ath HD, Albon H, et al. Endovenous laser ablation for persistent and recurrent venous ulcers after varicose vein surgery. *Phlebology*. 2008;23(4):193-195.
17. Bentley DE, Watson A, Treede RD, et al. Differential effects on the laser evoked potential of selectively attending to pain localisation versus pain unpleasantness. *Clin Neurophysiol*. 2004;115(8): 1846-1856.
18. Benedetti F, Amanzio M, Vighetti S, Asteggiano G. The biochemical and neuroendocrine bases of the hyperalgesic nocebo effect. *J Neurosci*. 2006;26(46):12014-12022.

19. Scott DJ, Stohler CS, Egnatuk CM, et al. Placebo and nocebo effects are defined by opposite opioid and dopaminergic responses. *Arch Gen Psychiatr*. 2008;65(2):220-231.
20. Pavlov IP. Conditioned Reflexes: An Investigation of the Physiological Activity of the Cerebral Cortex. In: Anrep GV, ed. London: Oxford University Press; 1927.
21. Ader R, Cohen N. Behaviorally conditioned immunosuppression. *Psychosom Med*. 1975;37(4):333-340.
22. Herrnstein RJ. Placebo effect in the rat. *Science*. 1962;138:677-678.
23. Ader R. The role of conditioning in pharmacotherapy. In: Harrington A, ed. *The Placebo Effect: An Interdisciplinary Exploration*. Cambrige, MA: Harvard UP; 1997:138-165.
24. Schedlowski M, Pacheco-Lopez G. The learned immune response: Pavlov and beyond. *Brain Behav Immun*. 2010;24(2):176-185.
25. Guo JY, Wang JY, Luo F. Dissection of placebo analgesia in mice: the conditions for activation of opioid and non-opioid systems. *J Psychopharmacol*. 2010;24(10):1561-1567.
26. Amanzio M, Benedetti F. Neuropharmacological dissection of placebo analgesia: expectation-activated opioid systems versus conditioning-activated specific subsystems. *J Neurosci*. 1999;19(1):484-494.
27. Goebel MU, Hubell D, Kou W, et al. Behavioral conditioning with interferon beta-1a in humans. *Physiol Behav*. 2005;84(5):807-814.
28. Goebel MU, Trebst AE, Steiner J, et al. Behavioral conditioning of immunosuppression is possible in humans. *FASEB J*. 2002;16(14):1869-1873.
29. Benedetti F, Colloca L, Torre E, et al. Placebo-responsive Parkinson patients show decreased activity in single neurons of subthalamic nucleus. *Nat Neurosci*. 2004;7(6):587-588.
30. Petrovic P, Dietrich T, Fransson P, et al. Placebo in emotional processing – induced expectations of anxiety relief activate a generalized modulatory network. *Neuron*. 2005;46(6):957-969.
31. Watson AD, Pitt Ford TR, McDonald F. Blood flow changes in the dental pulp during limited exercise measured by laser Doppler flowmetry. *Int Endod J*. 1992;25(2):82-87.
32. Siegel S. Explanatory mechanisms for placebo effects—Pavlovian conditioning. In: Guess HA, Kleinman A, Kusek JW, Engel LW, eds. *The Science of the Placebo: Toward an Interdisciplanary Research Agenda*. London: BMJ Books; 2002.
33. Voudouris NJ, Peck CL, Coleman G. Conditioned placebo responses. *J Pers Soc Psychol*. 1985;48(1):47-53.
34. Wickramasekera I. A conditioned response model of the placebo effect predictions from the model. *Biofeedback Self Regul*. 1980;5(1):5-18.
35. Wickramasekera I. A conditioned response model of the placebo effect: predictions from the model. In: White L, Tursky B, Schwartz GE, eds. *Placebo: Theory, Research, and Mechanisms*. New York: Guilford Press; 1985:255-287.
36. Benedetti F, Pollo A, Lopiano L, et al. Conscious expectation and unconscious conditioning in analgesic, motor, and hormonal placebo/nocebo responses. *J Neurosci*. 2003;23(10):4315-4323.
37. Rescorla RA. Pavlovian conditioning: it is not what you think it is. *Am Psychologist*. 1988;43:151-160.
38. Reiss S. Pavlovian conditionig and human fear: an expectancy model. *Behav Ther*. 1980;11(3):380-396.
39. Rescorla RA, Wagner AR. A theory of Pavlovian conditioning: variation in the effectiveness of reinforcement and nonreinforcement. In: Black AH, Prokasy WF, eds. *Classical Conditioning II: Current Research Theory*. New York: Appleton-CenturyCrofts; 1972.
40. Shanks DR. Learning: from association to cognition. *Annu Rev Psychol*. 2010;61:273-301.
41. Colloca L, Benedetti F. How prior experience shapes placebo analgesia. *Pain*. 2006;124(1-2):126-133.
42. Pollo A, Carlino E, Benedetti F. The top-down influence of ergogenic placebos on muscle work and fatigue. *Eur J Neurosci*. 2008;2:379-88.
43. Klinger R, Soost S, Flor H, Worm M. Classical conditioning and expectancy in placebo hypoalgesia: a randomized controlled study in patients with atopic dermatitis and persons with healthy skin. *Pain*. 2007;128(1-2):31-39.
44. Montgomery GH, Kirsch I. Classical conditioning and the placebo effect. *Pain*. 1997;72(1-2):107-113.
45. Price DD, Milling LS, Kirsch I, et al. An analysis of factors that contribute to the magnitude of placebo analgesia in an experimental paradigm. *Pain*. 1999;83(2):147-156.
46. Voudouris NJ, Peck CL, Coleman G. Conditioned response models of placebo phenomena: further support. *Pain*. 1989;38(1):109-116.
47. Voudouris NJ, Peck CL, Coleman G. The role of conditioning and verbal expectancy in the placebo response. *Pain*. 1990;43(1):121-128.
48. Watson A, El-Deredy W, Vogt BA, Jones AK. Placebo analgesia is not due to compliance or habituation: EEG and behavioural evidence. *Neuroreport*. 2007;18(8):771-775.
49. Wager TD, Matre D, Casey KL. Placebo effects in laser-evoked pain potentials. *Brain Behav Immun*. 2006;20(3):219-230.
50. Jones A, Brown C, El-Deredy W. How does EEG contribute to our understanding of placebo response. Chapter 5 of this book.
51. Colloca L, Tinazzi M, Recchia S, et al. Learning potentiates neurophysiological and behavioral placebo analgesic responses. *Pain*. 2008;139(2):306-314.
52. Colloca L, Petrovic P, Wager TD, et al. How the number of learning trials affects placebo and nocebo responses. *Pain*. 2010;151(2):430-439.
53. Porro CA. Open your mind to placebo conditioning. *Pain*. 2009;145(1-2):2-3.
54. Colloca L. Learned placebo analgesia in sequential trials: what are the Pros and Cons? *Pain*. 2011;152(6):1215-1216.
55. Ader R, Mercurio MG, Walton J, et al. Conditioned pharmacotherapeutic effects: a preliminary study. *Psychosom Med*. 2010;72(2):192-197.
56. Doering BK, Rief W. Utilizing placebo mechanisms for dose reduction in pharmacotherapy. *Trends Pharmacol Sci*. 2012;33(3):165-172.
57. Colloca L, Benedetti F. Placebo analgesia induced by social observational learning. *Pain*. 2009;144(1-2):28-34.
58. Mazzoni G, Foan L, Hyland ME, Kirsch I. The effects of observation and gender on psychogenic symptoms. *Health Psychol*. 2010;29(2):181-185.
59. White DJ, Galef JBG. Social influence on avoidance of dangerous stimuli by rats. *Anim Learn Behav*. 1998;26:433-438.
60. Lanzetta JT, Orr SP. Influence of facial expressions on the classical conditioning of fear. *J Pers Soc Psychol*. 1980;39(6):1081-1087.
61. Iacoboni M. Imitation, empathy, and mirror neurons. *Annu Rev Psychol*. 2009;60:653-670.
62. Colloca L, Lopiano L, Lanotte M, Benedetti F. Overt versus covert treatment for pain, anxiety, and Parkinson's disease. *Lancet Neurol*. 2004;3(11):679-684.
63. Kaptchuk TJ, Kelley JM, Conboy LA, et al. Components of placebo effect: randomised controlled trial in patients with irritable bowel syndrome. *BMJ*. 2008;336(7651):999-1003.
64. Amanzio M, Pollo A, Maggi G, Benedetti F. Response variability to analgesics: a role for non-specific activation of endogenous opioids. *Pain*. 2001;90(3):205-215.
65. Benedetti F, Amanzio M, Maggi G. Potentiation of placebo analgesia by proglumide. *Lancet*. 1995;346(8984):1231.
66. Kirsch I. Response expectancy as a determinant of experience and behavior. *Am Psychologist*. 1985;40(11):1189-1202.
67. Jensen KB, Kaptchuk TJ, Kirsch I, et al. Nonconscious activation of placebo and nocebo pain responses. *Proc Natl Acad Sci USA*. 2012;109(39):15959-15964.
68. Kirsch I. Conditioning, expectancy, and the placebo effect: comment on Stewart-Williams and Podd. *Psychol Bull*. 2004;130(2):341-343; discussion 344-345.

69. Stewart-Williams S, Podd J. The placebo effect: dissolving the expectancy versus conditioning debate. *Psychol Bull.* 2004;130(2): 324-340.
70. Hahn RA. The nocebo phenomenon: concept, evidence, and implications for public health. *Prev Med.* 1997;26(5 Pt 1):607-611.
71. Miller FG, Colloca L, Kaptchuk TJ. The placebo effect: illness and interpersonal healing. *Perspect Biol Med.* 2009;52(4):518-539.
72. Heyes CM. Social learning in animals: categories and mechanisms. *Biol Rev Camb Philos Soc.* 1994;69(2):207-231.
73. Byrne RW, Bates LA. Sociality, evolution and cognition. *Curr Biol.* 2007;17(16):R714-R723.
74. Paukner A, Suomi SJ, Visalberghi E, Ferrari PF. Capuchin monkeys display affiliation toward humans who imitate them. *Science.* 2009;325(5942):880-883.
75. Fruteau C, Voelkl B, van Damme E, Noe R. Supply and demand determine the market value of food providers in wild vervet monkeys. *Proc Natl Acad Sci USA.* 2009;106(29):12007-12012.
76. Furmark T, Appel L, Henningsson S, et al. A link between serotonin-related gene polymorphisms, amygdala activity, and placebo-induced relief from social anxiety. *J Neurosci.* 2008;28(49): 13066-13074.
77. Leuchter AF, McCracken JT, Hunter AM, et al. Monoamine oxidase a and catechol-o-methyltransferase functional polymorphisms and the placebo response in major depressive disorder. *J Clin Psychopharmacol.* 2009;29(4):372-377.
78. Hall KT, Lembo AJ, Kirsch I, et al. Catechol-O-methyltransferase val158met polymorphism predicts placebo effect in irritable bowel syndrome. *PLoS One.* 2012;7(10):e48135.
79. Wager TD, Atlas LY, Leotti LA, Rilling JK. Predicting individual differences in placebo analgesia: contributions of brain activity during anticipation and pain experience. *J Neurosci.* 2011;31(2): 439-452.
80. Lui F, Colloca L, Duzzi D, Anchisi D, et al. Neural bases of conditioned placebo analgesia. *Pain.* 2010;151(3):816-824.
81. Hashmi JA, Baria AT, Baliki MN, et al. Brain networks predicting placebo analgesia in a clinical trial for chronic back pain. *Pain.* 2012;153(12):2393-2402.

CHAPTER 15

Methodologic Aspects of Placebo Research

Magne Arve Flaten[1], Karin Meissner[2], Luana Colloca[3,4,5]

[1]Department of Psychology, Norwegian University of Science and Technology, Trondheim, Norway,
[2]Institute of Medical Psychology, Ludwig-Maximilians-University Munich, Munich, Germany,
[3]National Center for Complementary and Alternative Medicine (NCCAM), National Institutes of Health, Bethesda, MD, USA, [4]National Institute of Mental Health (NIMH), National Institutes of Health, Bethesda, MD, USA,
[5]Department of Bioethics, Clinical Center, National Institutes of Health, Bethesda, MD, USA

METHODOLOGY OF STUDIES INVESTIGATING PLACEBO ANALGESIA AND NOCEBO HYPERALGESIA

Placebo analgesic and nocebo hyperalgesic responses are due to expectations, in some form, that a treatment will reduce or enhance pain. The design of a study on placebo and nocebo effects needs to ensure that any change in pain is due to the expectations, and to control for other factors that may reduce or increase pain. Thus, the observation that pain is reduced after administration of a placebo is not proof of a placebo analgesic response, as the relief in pain could be due to e.g. natural variation in pain, regression to the mean, response bias, reporting errors, other medication, and a number of other factors.

Over time, pain and many other pathologic conditions show spontaneous variations and fluctuations of symptoms that are known as natural history.[1] Indeed, relapses and remissions can occur in the absence of any treatment. If a research subject or a patient takes a medication (e.g. aspirin or placebo) just before his/her discomfort starts to vanish, he/she may believe that the medication was effective but that decrease would have occurred anyway. Importantly, this effect should not be considered as a placebo effect, but rather a spontaneous remission. Spontaneous remissions can easily lead to misinterpretation of the cause–effect relationship. It is worth noting that in order to determine a placebo effect, scientists and trialists should look for a difference between the natural history and the placebo groups.

Another variable that can be erroneously confused with placebo effects is the regression to the mean, a statistical effect which accounts for changes in subsequent measurements. A variable, A, will tend to move closer to the center of its distribution from initial to later measurements. This is a property of all measurements subject to random error. Subsequent measurements then tend to be lower, because of regression to the mean, even if no biologically or psychologically mediated placebo effects are present.[2] If a research subject or a patient receives an initial clinical assessment when his/her pain is near its highest intensity, then the pain level is likely to be lower when they return for a second pain assessment. In this case also, the improvement or worsening in pain experience should not be attributed to placebo (or nocebo) effects. The unique, reliable way to establish to what proportion an observed improvement is due to the placebo effect is to compare a group receiving a placebo to a group receiving no treatment.

A further source of confusion is represented by particular kinds of error made by the patient and/or physician, i.e. false-positive (and -negative) errors. This concept has been explained in the framework of signal detection theory that was developed to model errors in the detection of ambiguous signals, such as a symptom.[3] The ambiguity of symptom intensity may lead to biases following verbal suggestions and other placebo manipulations that can occur even subconsciously. A patient might report that he/she feels better after a given medication by erroneously detecting a symptomatic relief (e.g. false-positive errors). False-positive (or -negative) errors are common in medical decision making, both by physicians who diagnose patients' symptoms and by patients who report the severity of their own symptoms. The false-positive errors and scaling biases can be significantly reduced by using objective physiologic measurements (e.g. brain activity) along with pain reports. In clinical practice and clinical trials, the Hawthorne effect, namely, the effect of being under study, may be a source of biases as well. It is likely that a research subject or a patient, once included in a study or trial, perceives pain reductions merely because of study participation.

INDUCED PAIN AND CLINICAL PAIN

Placebo effects are most often observed as changes in a symptom or disease, although some studies have shown that expectations may change base levels of psychologic and physiologic processes, e.g. arousal.[4,5] However, in most studies a symptom such as pain is induced, and in some studies clinical pain has been monitored. The advantage with experimentally induced pain is that the experimenter has control over the painful stimulus, and may apply the same painful stimulus to all participants, e.g. all subjects may receive the same temperature and duration of stimulation. Although the stimulus is identical for all participants, reported pain levels may still vary. Alternatively, the experimenter can calibrate the stimulus to the same perceived pain levels for all participants. This last method reduces error variance and is preferred by many investigators. It also has the advantage that pain levels may be calibrated to, e.g. a 5 on a 10 cm visual analogue scale, that allows the observation of both increases and decreases in pain levels.

Pain can be induced experimentally in a variety of ways, but an overview of these methods is outside the scope of this chapter; reviews of experimental induction of pain can be found in Gracely.[6] It has been suggested that placebo analgesia is observed in tonic pain, and not in phasic pain. However, several studies using event-related potentials, where painful stimulation has an abrupt onset and a duration of less than 0.1 seconds, have shown reliable placebo analgesic responding, both at the subjective and cortical level.[7–9] Placebo analgesia has also been observed during application of the sub-maximum tourniquet test, where duration of the experiment can be up to 1 hour, depending on the pain tolerance of the participant. Thus, placebo analgesia can be observed during application of painful stimuli of short duration with stimuli of longer duration of up to 1 hour. Clinical pain is also subject to modulation by expectations of pain relief, and controlled experiments have shown placebo analgesic responses with a duration of more than 4 hours.[10]

The effect of a placebo on chronic pain and postoperative pain has been investigated in a few studies aimed specifically at investigating placebo analgesia. Benedetti's group has showed that postoperative pain can be reduced by administration of a placebo.[11] Kupers et al[12] observed large placebo analgesic responses in a female patient with chronic back pain, and Charron et al[13] found placebo analgesic responses in patients with chronic low-back pain. Petersen et al[14] found no effects of a placebo manipulation on spontaneous pain in patients with neuropathic pain, but evoked pain was reduced in the same patients after a placebo manipulation.

QUANTIFICATION OF PAIN

Pain is a subjective experience, and there are several ways of quantifying the experience, and of recording the physiologic correlates of pain. The visual analog scale (VAS) is a 100 mm long line anchored by two statements: the subject is to cross the line at the left end if no pain is experienced, and the right end of the line should be crossed if pain is extreme, intolerable, or is the worst pain imaginable. The point where the subject has crossed the line is recorded in millimeters and is indicative of pain levels. The VAS may also be computerized, allowing for continuous recordings of pain, or to be used in situations when the ordinary VAS may be difficult to use, as under dental procedures or during scanning of the brain. The numerical rating scale (NRS) is conceptually similar to the VAS, but it involves the vocal expression of a number that corresponds to the pain level. The NRS has, as the VAS, a range from 0 to 10 (or 100), where 0 represents no pain and 10 (or 100) maximum pain.

The two major dimensions of pain are its intensity and its unpleasantness,[15] often referred to as the sensory and affective dimensions of pain. Both are recorded in the same way on the VAS or NRS. Pain is a sensation, with defined receptors and sensory pathways leading via the thalamus to cortical areas, as other somatosensory senses. However, pain is also an unpleasant feeling, and painful stimulation also activates areas in the brain that are involved in emotional reactions. Pain intensity and pain unpleasantness are often highly correlated in experimentally induced pain, but the correlation may decrease under some conditions, as in dental pain,[16] where pain unpleasantness can be high, whereas the intensity of pain may be lower. Subjective pain can also be expressed across several other dimensions, e.g. with the McGill Pain Questionnaire, but in studies of placebo analgesia interest is often in pain intensity or pain unpleasantness.

Pain threshold is the lowest stimulation that is judged as painful, and is assessed by a series of ascending and descending stimulation intensities. Pain tolerance, on the other hand, is a behavioral measure of pain where the time from onset of painful stimulation till the participant terminates the stimulus is recorded. Pain tolerance is often used with the cold pressor test or the submaximum tourniquet technique, and robust placebo analgesic responses have been observed in pain tolerance.[17,18]

Clinical pain can be quantified via measurements of the size of the painful area. Placebo analgesia has been associated with a reduction in the hyperalgesic area.[14]

RESPONSE BIAS

Pain report may be influenced by situational characteristics. Aslaksen et al[19] (see also[20]) showed that male subjects reported lower pain to female experimenters compared to male subjects who reported pain to male experimenters. Pain reported to females was about 50% lower than pain reported to male experimenters. The physiologic response to painful stimulation, on the other hand, was similar in both males and females. Thus, the presence of a female experimenter may have induced a bias in males towards reporting lower pain. Response bias is a serious threat towards the reliability of pain measurement, and thus to placebo analgesia.

Allan and Siegel[3] have proposed that placebo responding may be understood in the light of signal detection theory. According to this account, the pain report is influenced by two factors: the signal-to-noise ratio, and the consequences of reporting that pain was reduced or not. Thus, how well can the patient detect a change in pain, compared to the pain levels before taking the medication? In the case of fluctuating pain levels this can be difficult. Furthermore, when medication for pain has been administered, there is expectancy on the part of the patient and the physician that pain will be reduced. However, reporting that pain has not been reduced may have negative consequences for the patient: it may appear that the patient questions the physician's authority, or the patient may be classified as a complainer. Thus, administration of the medication introduces a bias towards reporting lower pain. The reporting of lower pain, even when pain levels are not lower, is termed a 'false-positive error,' as the subject reports that a signal (lower pain) has been detected, while the signal is not there.

Not many studies have followed-up on this hypothesis, but the few studies that have[21] (see also[22,23]) have provided support for the idea that the information that a painkiller has been administered changes the response criterion for what is defined by the patient as a painful stimulus, but does not change the experience of pain.

To bypass response bias, the recording of physiologic correlates of pain has been used. Painful stimuli with short rise times and short duration have been shown to reliably elicit event-related potentials (ERPs) in the electroencephalogram.[24] The painful stimuli generate a negative component, termed N2, followed by a positive component, the P2, that both correlate with reported pain. Several studies[7–9,25] have shown that the P2 component is reduced under placebo analgesia. This observation is important for two reasons. First, the reduced P2 component indicates that the brain's response to the painful stimulation is reduced. Second, placebo analgesia involves biologic processes that reduce the pain signal prior to reaching the brain. Consequently, the observation of reduced P2 amplitude shows that placebo analgesia is not due to response bias alone.

Studies employing functional magnetic resonance imaging (fMRI), positron emission tomography (PET), and brain mapping (e.g. laser-evoked potentials) have also found evidence of reduced activation of brain areas and scalp-responses involved in the processing of pain, which is further support that placebo analgesia is due to brain processes that reduce pain nociception and the experience of pain in the brain, and is not merely due to response biases.[26–28] Although there are limitations related to the brain, imaging and mapping techniques have extensively contributed to recent advances in the field of pain and placebo (see[29,30] and the first part of this book).

DESIGN

To ensure internal validity, i.e. that the reduction in pain observed in studies of placebo analgesia is due to expectations and not to other factors, a minimum of two groups or conditions needs to be included in the experimental design. For example, in the case of a pain study, the placebo group receives the experimental pain, and an inactive treatment along with the information that the given treatment is an effective painkiller (Fig. 15.1). Thus, expectations of pain reduction are induced in this group. A natural history control group receives the same type of pain, but neither a placebo treatment nor a verbal suggestion of pain relief is provided. The only difference between the two groups is hypothesized to be the expectation of reduced pain in the placebo group. Thus,

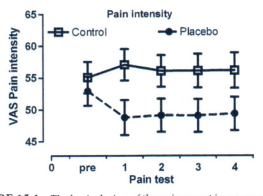

FIGURE 15.1 The basic design of the pain report in a group receiving repeated painful stimulation (Control group) and a group receiving the same painful stimulation, but who received placebo capsules with information that they contained a powerful painkiller (Placebo group). The capsules were administered after the pre-test to this group. *Reprinted from Aslaksen and Flaten[31] with kind permission from the publisher.*

	REDUCED PAIN	NOT REDUCED PAIN
MEDICATION	Correct decision	False decision
NO MEDICATION	False positive	Correct decision

FIGURE 15.2 After administration of active drug or placebo, said to reduce pain, the patient is asked whether the treatment reduced pain or not. The decision the patient makes depends on the discriminability of changes in the pain, and on the consequences of reporting lower pain or not.[3] *Adapted from Siegel and Allan.*

a placebo analgesic response is observed if pain levels are lower in the placebo group as compared to pain levels observed in the natural history group.

The minimal design for the study of placebo analgesia, therefore, involves two groups: one group with no expectations of treatment effects, where pain is recorded but not treated or in other ways modulated; and a second group that receives positive or negative verbal instructions that the treatment will reduce or enhance pain, but where an inactive placebo is administered. Many of the early studies on the response to placebos did not take this into account, as the concept of 'placebo effect' was not well discussed and understood at that time.

A three-armed design with a natural history group, a placebo group, and a group receiving the study drug has been employed in some clinical drug trials.[32,33] This design allows the comparison of the placebo group with a natural history group. However, the design is not suited to study placebo responses, as the subjects in the placebo group are informed that they will receive either a placebo or a drug. Thus, expectations in the placebo group will be weak in this study design, with a resultant lower placebo response. This observation was documented in a meta-analysis by Hróbjartsson and Gøetzsche[32,33] that showed small placebo analgesic responses in three-armed clinical trials where patients, who received a placebo, were told that they may or may not receive a painkiller. A meta-analysis by Vase et al,[5] on the other hand, showed significantly larger placebo analgesic responses when the research participants were told that they received a powerful painkiller, although they received a placebo.

A robust way to isolate the placebo effect as a context effect is achieved with the open–hidden paradigm,[35,36] also termed the overt–covert paradigm.[10] The term refers to the modality of administration of a treatment that can be doctor-initiated versus machine-initiated therapy. The former is the classical situation of routine medical practice whereby an active treatment is administered to the patient, who is aware that a medical therapy is being administered. The latter condition – the hidden administration – consists of administration of the active treatment while the patient is completely unaware that it is being given; the treatment is administered by means of a computer-controlled infusion pump that is preprogrammed to deliver the treatment at the required time. This paradigm has proven to be reliable in studying the placebo effect in the field of experimental and clinical pain, anxiety and Parkinson's disease (for a review see Colloca et al[10]). By comparing open–hidden administrations of medications it is possible to highlight the active role of cognition in therapeutic outcomes Fig. 15.2).

The role of cognition can be explored by using a balanced placebo design as well. Introduced by Ross et al,[8] the balanced placebo design refers to an orthogonal manipulation in which instructions (told drug versus told placebo) and drug administered (received drug versus received placebo) are carefully manipulated. This design is particularly interesting for investigating placebo effects and verbally induced expectations, although it requires elements of deception and some subjects will receive placebos. By using this design, Flaten et al[5] showed that carisoprodol, a centrally acting muscle relaxant, acted either as a relaxant or a stimulant, depending on the combination of verbal suggestion associated with the administration of treatment. Patients given a muscle relaxant and told that the administered drug was a stimulant reported greater muscle tension than did those who were told that it was a relaxant. Similarly, when an aerosolized, active bronchoconstrictor (carbachol) was administered to asthmatic subjects, it produced more airway resistance and dyspnea in patients who were told that it was a bronchoconstrictor than in those who were told it was a bronchodilator.[38,39] Keltner et al[40] demonstrated different patterns of brain activity in relation to pain under different verbal suggestions.

WITHIN-SUBJECTS VERSUS BETWEEN-SUBJECTS DESIGNS

In a between-subjects design, subjects are randomized to either the natural history group or the placebo group, and the placebo analgesic effect is the mean difference in pain levels between the groups after administration of the placebo. A less used design is the within-subjects design, where all subjects are run in both the natural history and the placebo conditions. The advantage with the within-subjects design is that of statistical power, since error variance is reduced, and fewer subjects may be needed. However, the most important conceptual advantage with this form of design is that a placebo response can be computed, for each subject, as the difference between that subject's pain levels in the natural history and placebo conditions. The individual placebo response cannot be computed in a between-subjects experiment. Reduced pain observed after administration of a placebo is not evidence of a placebo response, as other factors, like natural variation in pain levels or habituation, may reduce pain. It has been argued that the term 'placebo response' should be used for the response at the individual level, whereas the term 'placebo effect' should be used for between-groups comparison, and this nomenclature is used in the present chapter. When the interest is in predicting individual differences in placebo response, and to identify placebo responders and non-responders, the within-subjects design should be considered.

On the other hand, the within-subjects design introduces some problems. The order of presentation of the natural history and placebo conditions has been found to affect placebo analgesia. When the placebo condition has been presented before the natural history condition, placebo effects have been small or absent.[41,42] When the natural history condition has been presented before the placebo condition, significant placebo effects have been observed. The effect of order of presentations is due to higher levels of stress or negative emotions on the first day of the experiment that increase pain on that day. After habituation to the experimental procedures, that decreases emotional arousal, pain levels are lower. It is recommended that subjects are familiarized with the experimental procedures prior to the start of the experiment, to avoid habituation across the experimental conditions. Another problem with the within-subjects design is subject drop-out, as the experiment is run across two or more days. Running both conditions on the same day is problematic, since the placebo condition will have to be run before the natural history condition for half the subjects, and there may be carry-over effects that may reduce pain in the natural history condition, that could reduce any difference in pain between the conditions.

THE PRE-TEST

The pre-test refers to the application of painful stimulation prior to introduction of the placebo treatment, to ensure that the response levels are similar in all the groups. Although the pre-test is not a necessity in experimental design,[43] its use is seldom questioned and it is very commonly used in studies on placebo effects.

RESEARCHERS' AND SUBJECTS' PERCEPTION OF THE TREATMENT ALLOCATION

Gracely et al[44] showed that when the person administering a placebo knew that the capsule could be a placebo, the effect on pain was smaller compared to a placebo administered by a person who believed she was administering a painkiller. Thus, the knowledge that the experimenter has about the treatment to be administered can affect the patient or research participant. Thus, it is common in many experiments on placebo analgesia that some participants in the placebo group receive an active drug, and these participants are subsequently removed from the data analyses. The experimenter will not know which subjects have received the active drug and which have received the placebo. As indicated by Gracely et al,[44] if the experimenter has knowledge that only a placebo is administered, this may reduce placebo effects.

Furthermore, placebo effects may be induced by cues in the experimental procedures. Levine and Gordon[37] showed that placebo analgesia could be induced by a hidden infusion of saline, performed in another room by the experimenter, in the absence of explicit information. These placebo analgesic responses were of the same magnitude as placebo responses induced by the experimenter sitting at the patient's bedside. When the hidden saline infusions were made by a preprogrammed infusion pump, placebo analgesic responses were not observed. They concluded that 'even the most subtle cues can elicit a placebo response' (Levine and Gordon[37 p. 755]). Galer et al[45] have shown that the patients' response to treatment was associated with physician expectations of treatment effects, even if these expectations had not been made known to the subjects. It was concluded that physicians could subtly communicate their expectations of treatment outcome, which, in turn, could influence patient response. It has also been reported that higher physiologic arousal prior to administration of electric shock, compared to arousal measured prior to administration of an aversive tactile stimulus,[46] was observed even if no programmed information was available to the subjects prior to the aversive stimulus. A similar finding was reported in Flaten et al,[18] where nurses were informed that one group of subjects would receive a small dose of a less-effective painkiller and another group would receive a larger dose of an effective painkiller. In a pre-test prior to administration of the drug and painful stimulation, and before any verbal information was provided to the participants, stress levels were higher in the participants that were about to receive the less-effective medication. Thus, information available to the experimenter can affect the subject's behavior before that information has explicitly been transmitted to the subject, maybe through changes in the tone of voice or facial expression. This is not a placebo effect because the decreased pain levels were not due to the subjects' expectations. Thus, the experimenter should not be informed about which group the research participant belongs to when performing the pre-test as this may modulate pain levels even prior to administration of the placebo. This situation may not be possible in within-subjects designs, however, where the experimenter will know which condition is to be run before the second session.

SINGLE-BLIND VERSUS DOUBLE-BLIND DESIGNS

The double-blind design ensures that the research participant and the experimenter are blind to whether the participant received active treatment or placebo. The design is the gold standard in studies of effects of drugs, or other treatments where this design can be applied. The concept of double-blind in studies of placebo effects can, however, mean other things. In some studies the experimenter is blinded regarding the hypotheses.[47,48] This is recommendable, but uncommon. This form of blinding ensures that the experimenter does not induce a bias in the pain report that may mistakenly be interpreted as a placebo analgesic response. Knowledge of the experimental hypothesis can affect the behavior of the experimenter so that the pain report may differ between the natural history and placebo groups.

INDUCTION OF PLACEBO ANALGESIA BY CLASSIC CONDITIONING: METHODOLOGICAL ISSUES

Placebo effects may be induced by verbal information alone, but have been found to be stronger when the participant has experienced that the treatment is effective.[25,49,50] This procedure was first introduced by Voudouris et al.[51] The experience that a drug is effective is induced via a classic conditioning procedure that has a minimum of three phases: in a pre-test painful stimulation is presented, which acts as a baseline against which the participant can compare subsequently induced pain. In the second phase, the conditioning phase, inactive treatment is administered, often a cream applied to the bodily site where the pain is administered, that is followed by painful stimulation where the intensity of stimulation is surreptitiously reduced compared to the pre-test. The placebo treatment is the conditioned stimulus, and the reduced pain after application of the placebo is the unconditioned stimulus. The pairing of the conditioned and unconditioned stimuli gives the impression that the reduced pain is due to the placebo treatment. A control group does not receive pairings of the placebo cream with reduced pain. In the post-test, the cream is again applied in the placebo group to induce an expectation of reduced pain, and both groups receive the same levels of painful stimulation, often the same level as in the pre-test. The participants are often informed verbally that the placebo cream is a powerful painkiller, but in some experiments this information has not been provided and a placebo analgesic response has still been observed after application of the cream.[51]

Two central issues in this type of design are: how much should the painful stimulation be reduced from the pre-test to the conditioning phase, and what intensity the pain stimulations in the post-test should have. A large reduction from the pre-test to the conditioning phase will generate expectations that the placebo is a powerful painkiller. However, if the pain in the post-test is similar to that in the pre-test there may be a mismatch between the expectation of low levels of pain, and the actual pain experienced. The mismatch could weaken expectations about the analgesic effect of the placebo, and reduce placebo analgesia. Thus, some studies have used a level of painful stimulation in the post-test that is intermediate between the pre-test and the conditioning phase.[52] However, robust placebo analgesia has been observed in this form of design when pain levels are identical in the pre- and post-tests, and there is little support for the hypothesis that a mismatch between expected and experienced pain reduces placebo analgesia.[53] In support of this, data from our laboratory, where the same painful stimulations were used in the first and third phases, showed a positive correlation between the magnitude of reduced pain from the first to the second phase, and the placebo analgesic response. The advantage of having identical pain stimulations in the pre- and post-test is that it allows for the computation of the Group by Test interaction. This interaction cannot be assessed if pain stimuli are of less intensity in the post-test compared to the pre-test, and only the main effect of Group can be computed.

An alternative way of inducing placebo and nocebo responses via classic conditioning has been developed by Colloca and Benedetti.[50] The basic idea is to pair neutral conditioned stimuli, e.g. lights of different wavelengths, with painful unconditioned stimuli of different intensities (see Fig. 15.3). Participants were informed that a green light displayed on a computer screen will indicate activation of the electrode pasted on their middle finger which, in turn, would induce analgesia by virtue of a sub-threshold stimulation. Conversely, a red light will indicate that the electrode is not activated, thus they would experience a red-light-associated painful stimulus (which serves as control). The language was simplified for the participant, and they were told 'When the green light is on, there will be a stimulus sent to your middle finger so that you will feel either no pain or less pain. On the other hand, when you see the red light, then the stimulus to the finger is turned off so that you will feel pain.' This approach provides evidence of the possibility to study analgesic placebo responses even when no placebo medication is provided by manipulating expectations and prior experience.

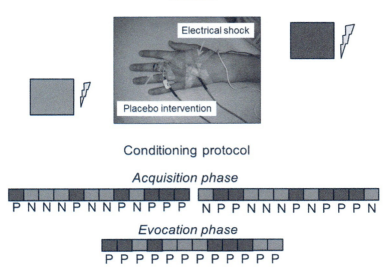

FIGURE 15.3 Experimental paradigm to investigate the role of conditioning and prior experience. Participants were informed that a green light displayed on a computer screen indicates activation of the electrode pasted on their middle finger which, in turn, would induce analgesia by virtue of a sub-threshold stimulation. Conversely, a red light indicates that the electrode is not activated, so that they would experience pain (which serves as control). The intensity of stimulation was manipulated to give the experience of analgesia in association with the presentation of a green light during the acquisition phase of conditioning (P = pain; N = no pain). No intensity manipulation was performed during the evocation phase when red and green lights were set at the control level of pain. Placebo responses were calculated as the differences between pain reports under red and green stimuli in the evocation phase. Adapted from Colloca and Benedetti (2006).[49] *This figure is reproduced in color in the color section.*

MEASUREMENT OF EXPECTATIONS

The concept of expectation was most thoroughly discussed by Kirsch.[54] Expectations are central to the placebo effect, but are difficult to measure in ways that do not affect the subsequent pain report. In most studies on placebo analgesia, expectations are not measured. Some studies have asked the subjects to indicate, before the placebo takes effect, their expected pain levels, i.e. expected pain after treatment is rated, on the same scale that actual pain is measured.[55,56] If the subject is asked to indicate the pain level she expects after administration of the placebo, that may affect the pain actually reported. To avoid this problem, some studies have asked subjects to estimate the expected reduction in pain, on a scale from 0 to 100%, prior to administration of the placebo.[13,57] Other researchers have asked subjects retrospectively, after the experiment, to rate the expectations held after placebo administration.[58] By using this method, correlations with actual pain are lower compared to when expectations are assessed before the placebo takes effect. Others have asked subjects about how much painkillers have reduced pain in the past, as past experiences may determine present expectations towards painkillers. However, correlations between reported previous effects of painkillers and the placebo analgesia observed in the experiment are typically low.[47]

Expectations have at least two different dimensions, which can affect placebo analgesia:[54] the first dimension is how certain the subject is that a painkiller has been administered, and the second is how much the subject believes that the painkiller will reduce pain. Most studies that have monitored expectations have recorded the second type. However, Bjørkedal and Flaten[58] investigated both dimensions by asking subjects after the post-test, how certain they were that they had received a painkiller, and how much they had expected the painkiller to reduce their pain. There were correlations between certainty of having received a painkiller and placebo analgesia, but not between degree of expected pain reduction and actually reduced pain.

CONCLUSION

The validity of research on placebo analgesia depends on the researcher taking into account the factors reviewed and discussed above. Although a number of issues important for the study of the field have been identified and solved, at least partly, some methodologic issues are in need of more work. The same can also be said of the concept of expectation. There is no commonly accepted way of measuring expectations. On the other hand, important issues regarding response bias and the design of studies have been clarified. However, the field is developing rapidly, and a chapter on the methodologic issues written in 10 years will likely have many new issues not discussed here.

Acknowledgment

The opinions expressed by LC are those of the author and do not necessarily reflect the position or policy of the National Institutes of Health, the Public Health Service, or the Department of Health and Human

Services. LC thanks the intramural program of NCCAM and NIMH for their support.

References

1. Fields HL, Levine JD. Pain – mechanisms and management. *Western J Med.* 1984;141(3):347-357.
2. Benedetti F. *Placebo Effects: Understanding the Mechanisms in Health and Disease.* New York: Oxford University Press; 2009.
3. Allan L, Siegel S. A signal detection theory analysis of the placebo effect. *Eval Health Prof.* 2002;25:410-420.
4. Flaten MA. Information about drug effects modify arousal. An investigation of the placebo response. *Nord J Psychiatry.* 1998;52:147-151.
5. Flaten MA, Simonsen T, Olsen H. Drug-related information generates placebo and nocebo responses that modify the drug response. *Psychosom Med.* 1999;61:250-255.
6. Gracely RH. Studies of pain in human subjects. In: McMahon SB, Koltzenburg M, eds. *Wall and Melzack's Textbook of Pain.* 5th ed. Philadelphia: Elsevier; 2006:267-290.
7. Wager TD, Matre DB, Casey KL. Placebo effects in laser-evoked pain potentials. *Brain Behav Immun.* 2006;20(3):219-230.
8. Watson A, El-Deredy W, Vogt BA, Jones AKP. Placebo analgesia is not due to compliance or habituation: EEG and behavioural evidence. *Neurorep.* 2007;18:771-775.
9. Aslaksen PM, Bystad M, Vambheim SM, Flaten MA. Gender differences in placebo analgesia: event-related potentials and emotional modulation. *Psychosom Med.* 2011;73:193-199.
10. Colloca L, Lopiano L, Lanotte M, Benedetti F. Overt versus covert treatment for pain, anxiety, and Parkinson's disease. *Lancet Neurol.* 2004;3(11):679-684.
11. Pollo A, Amanzio M, Arslanian A, et al. Response expectancies in placebo analgesia and their clinical relevance. *Pain.* 2001;93(1):77-84.
12. Kupers R, Maeyaert J, Boly M, et al. Naloxone-insensitive epidural placebo analgesia in a chronic pain patient. *Anesthesiol.* 2007;106(6):1239-1242.
13. Charron J, Rainville P, Marchand S. Direct comparison of placebo effects on clinical and experimental pain. *Clin J Pain.* 2006;22:204-211.
14. Petersen GL, Finnerup NB, Nørskov KN, et al. Placebo manipulations reduce hyperalgesia in neuropathic pain. *Pain.* 2102;153:1292–1300.
15. Price DD. *Psychological Mechanisms of Pain and Analgesia.* Washington DC: IASP Press; 1999.
16. Hunsbeth P, Flaten MA. The effect of extended and positive information on pain induced by dental procedures. *Psychosom Med.* 2013 (in press.).
17. Amanzio M, Benedetti F. Neuropharmacological dissection of placebo analgesia: expectation-activated opioid systems versus conditioning-activated specific subsystems. *J Neurosci.* 1999;19(1):484-494.
18. Flaten MA, Aslaksen PM, Simonsen T, et al. Cognitive and emotional factors in placebo analgesia. *J Psychosom Res.* 2006;61:81-89.
19. Aslaksen PM, Myrbakk IN, Høifødt RS, Flaten MA. The effect of experimenter gender on autonomic and subjective responses to pain stimuli. *Pain.* 2007;129:260-268.
20. Kallai L, Barke A, Voss U. The effects of experimenter characteristics on pain in women and men. *Pain.* 2004;112:142-147.
21. Clark WC. Sensory-decision theory analysis of the placebo effect on the criterion for pain and thermal sensitivity. *J Abnorm Psychol.* 1969;74:363-371.
22. Clark WC, Goodman JS. Effects of suggestion on d´ and Cx for pain detection and pain tolerance. *J Abnorm Psychol.* 1974;83:364-372.
23. Clark WC, Mehl L. Thermal pain: a sensory decision theory analysis of age and sex on d´, various response criteria, and 50% pain threshold. *J Abnorm Psychol.* 1971;78:202-212.
24. Granovsky Y, Granot M, Nir RR, Yarnitsky DJ. Objective correlate of pain perception by contact-heat pain potentials. *J Pain.* 2008;9(1):53-63.
25. Colloca L, Tinazzi M, Recchia S, et al. Learning potentiates neurophysiological and behavioral placebo analgesic responses. *Pain.* 2008;139(2):306-314.
26. Wager TD, Rilling JK, Smith EE, et al. Placebo-induced changes in FMRI in the anticipation and experience of pain. *Science.* 2004;303(5661):1162-1167.
27. Petrovic P, Kalso E, Petersson KM, Ingvar M. Placebo and opioid analgesia – imaging a shared neuronal network. *Science.* 2002;295(5560):1737-1740.
28. Zubieta JK, Bueller JA, Jackson LR, et al. Placebo effects mediated by endogenous opioid activity on mu-opioid receptors. *J Neurosci.* 2005;25(34):7754-7762.
29. Colloca L, Benedetti F, Porro CA. Experimental designs and brain mapping approaches for studying the placebo analgesic effect. *Eur J Appl Physiol.* 2008;102(4):371-380.
30. Meissner K, Colloca L, Bingel U, et al. The placebo effect: advances from different methodological approaches. *J Neurosci.* 2011;31(45):16117-16124.
31. Aslaksen PM, Flaten MA. The roles of physiological and subjective stress in the effectiveness of a placebo on experimentally induced pain. *Psychosom Med.* 2008;70:811-818.
32. Hróbjartsson A, Gøetzsche PC. Is the placebo powerless? An analysis of clinical trials comparing placebo with no treatment. *N Engl J Med.* 2001;344(21):1594-1602.
33. Hróbjartsson A, Gøetzsche PC. Placebo interventions for all clinical conditions. *Cochrane Database Syst Rev.* 2010(1):CD003974.
34. Vase L, Petersen GL, Riley 3rd JL, Price DD. Factors contributing to large analgesic effects in placebo mechanism studies conducted between 2002 and 2007. *Pain.* 2009;145(1-2):36-44.
35. Levine JD, Gordon NC. Influence of the method of drug administration on analgesic response. *Nature.* 1984;312(5996):755-756.
36. Benedetti F, Pollo A, Lopiano L, et al. Conscious expectation and unconscious conditioning in analgesic, motor, and hormonal placebo/nocebo responses. *J Neurosci.* 2003;23(10):4315-4323.
37. Ross S, Krugman AD, Lyerly SB, Clyde DJ. Drugs and placebos: a model design. *Psychol Rep.* 1962;10:383-392.
38. Luparello T, Lyons HA, Bleecker ER, McFadden ER. Influences of suggestion on airway reactivity in asthmatic subjects. *Psychosom Med.* 1968;30(6):819-825.
39. Luparello TJ, Leist N, Lourie CH, Sweet P. The interaction of psychologic stimuli and pharmacologic agents on airway reactivity in asthmatic subjects. *Psychosom Med.* 1970;32(5):509-513.
40. Keltner JR, Furst A, Fan C, et al. Isolating the modulatory effect of expectation on pain transmission: a functional magnetic resonance imaging study. *J Neurosci.* 2006;26(16):4437-4443.
41. Lyby PS, Aslaksen PM, Flaten MA. Is fear of pain related to placebo analgesia? *J Psychosom Res.* 2010;68:369-377.
42. Lyby PS, Aslaksen PM, Flaten MA. Variability in placebo analgesia and the role of fear of pain-an ERP study. *Pain.* 2011;152(10):2405-2412.
43. Campbell DT, Stanley JC. *Experimental and Quasi-Experimental Designs for Research.* Boston: Houghton Mifflin Company; 1966.
44. Gracely RH, Dubner R, Wolskee PJ, Deeter WR. Placebo and naloxone can alter post-surgical pain by separate mechanisms. *Nature.* 1983;306(5940):264-265.
45. Galer BS, Schwartz L, Turner JA. Do patient and physician expectations predict response to pain-relieving procedures? *Clin J Pain.* 1997;13:348-351.
46. Grillon C, Baas JP, Lissek S, et al. Anxious responses to predictable and unpredictable aversive events. *Behav Neurosci.* 2004;118:916-924.
47. Johansen O, Brox J, Flaten MA. Placebo and nocebo responses, cortisol, and circulatingbeta-endorphin. *Psychosom Med.* 2003;65(5):786-790.

48. Walach H, Schmidt S, Dirhold T, Nosch S. The effects of a caffeine placebo and suggestion on blood pressure, heart rate, well-being and cognitive performance. *Int J Psychophysiol*. 2002;43(3):247-260.
49. Colloca L, Benedetti F. How prior experience shapes placebo analgesia. *Pain*. 2006;124:126-133.
50. Colloca L, Sigaudo M, Benedetti F. The role of learning in nocebo and placebo effects. *Pain*. 2008;136(1-2):211-218.
51. Voudouris NJ, Peck C, Coleman G. Conditioned placebo responses. *J Personality Soc Psychol*. 1985;48(I):47-53.
52. Eippert F, Bingel U, Schoell ED, et al. Activation of the opioidergic descending pain control system underlies placebo analgesia. *Neuron*. 2009;63(4):533-543.
53. Arntz A, Hopmans M. Underpredicted pain disrupts more than correctly predicted pain, but does not hurt more. *Behav Res Ther*. 1998;36(12):1121-1129.
54. Kirsch I. *How Expectations Change Experience*. Washington DC: APA Press; 1999.
55. Price DD, Milling LS, Kirsch I, et al. An analysis of factors that contribute to the magnitude of placebo analgesia in an experimental paradigm. *Pain*. 1999;83(2):147-156.
56. Vase L, Robinson ME, Verne GN, Price DD. Increased placebo analgesia over time in irritable bowel syndrome (IBS) patients is associated with desire and expectation but not endogenous opioid mechanisms. *Pain*. 2005;115(3):338-347.
57. Scott DJ, Stohler CS, Egnatuk CM, et al. Individual differences in reward responding explain placebo-induced expectations and effects. *Neuron*. 2007;55(2):325-336.
58. Bjørkedal E, Flaten MA. Interaction between expectancies and drug effects: an experimental investigation of placebo analgesia with caffeine as an active placebo. *Psychopharmacol*. 2011;215:537-548.

CHAPTER 16

Balanced Placebo Design, Active Placebos, and Other Design Features for Identifying, Minimizing and Characterizing the Placebo Response

Paul Enck, Katja Weimer, Sibylle Klosterhalfen
Department of Psychosomatic Medicine, University Hospital Tübingen, Tübingen, Germany

INTRODUCTION

Because of the difficulties of reliably identifying placebo responders and predicting placebo response rates in clinical and experimental trials, different methodologic approaches have been tried when testing (novel) drugs against placebo. At the same time, but to a varying degree, these and other designs have also been used to unravel some of the mechanisms behind the placebo response. This chapter discusses clinical and experimental designs that may allow characterization and quantification of the placebo response in respective settings; it will not focus exclusively on designs for better discrimination between the drug response and the placebo response in clinical trials.[1] While many of the cited references refer to pain and placebo analgesia—because this is the most widely studied placebo phenomenon[2]—other significant contributions derive from psychiatry, neurology, and internal medicine.[3,4]

We will introduce two distinctly different approaches to allow a better discrimination and quantification of drug and placebo effects: by manipulating the information provided to patients and participants, and by manipulating the timing of drug action; we will present examples for both from the current literature. A third and more hypothetical approach—the free choice paradigm—is currently lacking empirical evidence that it will work under clinical conditions (see Table 16.1).

The following sections will briefly discuss (1) the apparent dichotomy between minimizing and maximizing the placebo response, (2) the assumptions of the 'additive model,' (3) the goals of the so-called 'balanced placebo design' (BPD) and two potential alternatives to the BPD, (4) the 'balanced cross-over design' (BCD), (5) a 'delayed response test' (DRT), (6) the advantages and disadvantages of the use of 'active placebos' in different medical subspecialties, (7) the need for effective blinding, (8) the pitfalls of waiting-list controls and a potential alternative, (9) a placebo-controlled trial design without ethical concerns, and (10) ethical aspects that are inherent in all placebo research.

MINIMIZE VERSUS MAXIMIZE

While for randomized controlled trials (RCTs) minimizing the placebo response and optimizing the drug–placebo difference remains the gold standard for new compounds to be developed and selected for drug treatment,[1] the situation is different in experimental settings when it comes to investigating the placebo response and its interaction with drug effects: here, maximizing the placebo response may be a valid goal, because only with substantial variation in the placebo response between subjects may the psychosocial and physiologic predictors (mediators, moderators) of its size and nature become detectable.[5]

The 'hidden treatment'[6] is a valid example of how to generate high placebo responses that may even be used in patients without ethical limitations: patients/volunteers receive a drug but are not informed when the drug is provided. In comparison to an open administration of the drug, which maximizes expectancy of symptom improvement, the hidden application allows the effects of expectancy to be separated from the 'true' effects of the drug, as has been shown in a number of examples.[7]

TABLE 16.1 Design Alternatives to Separate the Drug Response from the Placebo Response

Manipulation	Name of design
Of patient/volunteer information	Balanced placebo design
	Balanced cross-over design
Of drug application time	Hidden treatment
	Delayed release medication
Choice between drug and placebo	Free choice paradigm

Table 16.2 lists a variety of traditional and novel design features—usually employed in RCT—that may also be used in experimental settings to minimize the placebo response and allow a better discrimination between drug and placebo responses.

Most of these measures apply to RCTs during drug development, especially in Phase III trials aimed at optimizing drug–placebo differences (also called 'assay sensitivity').

Approaches that identified and excluded placebo responders at an early stage of an RCT (e.g. by placebo run-in periods, repeated treatment phases with re-randomization) could not prevent placebo responders at a later phase of the RCT. Randomized run-in and withdrawal periods appear to be more promising but so far have only rarely been tested for their efficacy to improve assay sensitivity.

Cross-over designs do not seem to optimize assay sensitivity either. If different treatment periods apply to

TABLE 16.2 Strategies used to Minimize the Placebo Response and/or to Optimize Drug–Placebo Differences

Design features	Effects
Cross-over trials	Potentiate the risk of learning (conditioning) and carry-over effects and increase the PR[8]
Use of placebo run-in phases	Exclusion of responders does not prevent PR during the medication phase[9]
Increase the number of trial arms	Increases acceptance due to higher chances of receiving medication but increases the PR[10]
Single-case (N=1) studies	Requires very high number of subjects, but the PR is difficult to assess[11]
Using active placebos	Difficult for most drugs but would substantially increase drug-placebo differentiation[12]
Using a comparator drug	Increases the acceptance to enter a drug study but also substantially increases the PR[13]
Repetitive treatment periods	Re-randomization does not prevent PR and do not predict PR during all periods[14]
Use of drug let-in phases	Helps identify drug non-responders and augments drug–placebo differences but is biased[15]
Randomized run-in/withdrawal	May identify some but not all PR, allows better discrimination between drug and placebo[16]
Adaptive dosing	Placebo responders tend to demand fewer dose adjustments, enrichment of groups increases the PR[17]
Preference design	Reduces disappointment and drop-outs specifically in the placebo arm, optimal for CER[18]
Step-Wedge design (waiting list)	Induces disappointment and drop-out, PR may show improvement during waiting[19]
Zelen design	Allows 'no-treatment control' without randomization, improved separation of PR from natural course[20]
Cluster randomization	Randomization of health service delivery units that either provide drug or placebo[21]
OTHER MEASURES	
Patient reported outcomes (PRO)	PROs instead of physician reports may decrease the PR to some extend[22]
Using biomarkers instead of PROs	Biomarkers of drug efficacy may reduce the PR, biomarkers of the PR would be even better[23]
Standardize symptom severity	Placebo responders are frequently patients with less severe symptoms at study entry[24]
Control for center effects	Standardizing recruitment and separating it from study conductance may reduce the PR[25]
Personality profiling of PR	Is possible, but currently no valid psychometric test exists that reliably predicts the PR[26]
Control for patient expectations	Seldom done, may eliminate patients with inappropriate expectations and high PR[27]
Increase medication adherence	Better adherence is associated with higher responses in the placebo arm[28]
Ensure effective blinding	Incomplete blinding is frequent and the reason for drop-out in the placebo arm[29]
Responder analysis	Allows post-hoc identification of the responder to endpoint definitions[30]

the same patients, carry-over effects occur, and within-subject variability of the placebo response hinders an adequate interpretation of trial results. Cross-over trials have been questioned because treatments in the first phase may generate conditioning effects during the second phase. They also may lead to un-blinding of the study due to perceived differences in side effects. While the risk of un-blinding could be controlled for by using 'active placebos' (see below) that mimic side effects of the drug under test, active placebos are difficult to develop and are therefore used only occasionally in a few clinical conditions, e.g. in the treatment of depression.

Recently, approaches to identifying drug responders rather than placebo responders during a run-in phase, or pre-selection of patients with previous exposure to a similar drug, were reported. This may improve assay sensitivity by increasing the drug–placebo difference, although this carries the risk of biasing and selected drug indication. Any previous treatment will affect a new therapy via conditioning and positive or negative expectation; therefore, documentation of the patient's history of disease, the prior management of the disease, and the patient's previous participation in RCTs need to become part of the baseline documentation in any RCT (see Enck et al[1] for further discussion and references).

THE 'ADDITIVE MODEL' ASSUMPTIONS

RCTs usually determine the efficacy of the drug by subtracting the placebo response (in the placebo arm) from the response in the drug arm. The implicit assumption of this 'additive model'[31] is that, in both the drug and the placebo arms, drug-nonspecific effects (that include the placebo effect) are equal (Fig. 16.1). This model reflects a general assumption in almost all placebo-controlled drug trials that have been performed since its dawn in the 1940s[32]. Interestingly the underlying hypothesis that the placebo response is equal in size, irrespective of whether an active drug or a placebo was given, has never been thoroughly tested.

The model has been questioned on theoretical grounds[33] but also on the basis of empirical evidence.[34] One argument against the assumption is that 'active' placebos (see below) may produce different placebo effects as compared to inactive placebos.[35]

THE BALANCED PLACEBO DESIGN

The so-called 'balanced placebo design' (BPD) was traditionally used in the testing for expectancy effects of frequently consumed everyday drugs such as caffeine, nicotine and alcohol,[36] more recently also with drugs such as marijuana.[37]

While one half of the study sample receives placebo, the other half receives the drug. Half of each group receives correct information while the other half receives false information on the nature of their study condition (drug or placebo) immediately prior to drug testing. This allows differentiation between the 'true' drug effect (those receiving the drug, but told that they received placebo) and the 'true' placebo effect (those receiving placebo, but told that they received the drug) (Fig. 16.2).

The central concept of the design is—as in the 'hidden treatment' paradigm—to separate the 'true' effects of

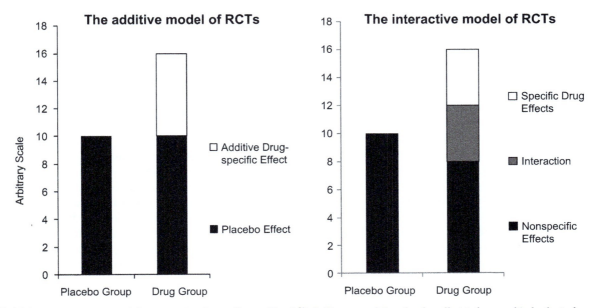

FIGURE 16.1 The 'additive model' assumption (according to Kirsch[31]) (left) assumed the placebo effect to be equal in both study arms, while it may be that in the drug arm the placebo may be smaller or larger, depending on interactive effects of placebo and drug. *(adopted from Enck et al[1])*

		Information	
		Medication	Placebo
Application	Medication	1: true positive	2: false negative
	Placebo	3: false positive	4: true negative

FIGURE 16.2 The 'balanced placebo design' (BPD): all participants are told that they are participating in a double-blind parallel-group design study. After drug intake, and immediately before testing, half of the participants in each group are given false and correct information on what they received. *(adopted from Enck et al[45])*

a drug from the effects of expectancy that occur when participants and patients are given a pill and told that it may or may not contain the active compound. It has been shown that the likelihood of receiving the active treatment determines the size of both the drug response and the placebo response in clinical trials[38] but also in experimental settings: the greater the likelihood of active treatment, the higher the response to both the drug and the placebo, solely attributable to the increased expectancy.[39] Maximal response difference between drug and placebo is achieved with a 50% chance when the chances of receiving either drug or placebo are equalized. This is thought to be associated with maximal reward activity in the brain, e.g. with maximal dopamine release in subthalamic neurons.[40]

The 'non-additive model' according to Kirsch[31] (see above, Fig. 16.2) can be tested in the following way. If the difference between Group 1 (drug plus placebo effect) and Group 2 (true drug effect) is unequal to Group 4 (true placebo effect), the non-additivity assumption is correct.

A variant of the BPD is the 'half BPD' in which all participants are given placebos, but half of them are told that they are receiving the drug; this is a common design in current placebo research as it does not require approval for performing a drug study where the legal stakes are usually higher. However, effective double-blinding of such a study is difficult unless—as in a recent test in our laboratory[41]—the participants and the experimenter(s) conducting the study are made to believe that they are participating in a full BPD. In this study, male and female smoking and non-smoking volunteers were investigated for the effects of an assumed nicotine-containing chewing gum for the effects on neurocognitive functions (go/no-go task). A significant interaction of all three factors (information, smoking status, gender) was found, indicating that the information that they had received nicotine shortened reaction times in smoking women and non-smoking men but increased reaction times in non-smoking women and smoking men.[41]

One of the pitfalls of the BPD is the fact that all participants are informed (either correctly or falsely), prior to testing, whether and what they have received. In sceptical participants (especially in medical students), this may raise doubts about the truth of the information provided and may require additional measures, such as a reliable explanation as to why the information is being given at all. This is usually done by informing them that once the drug is active, the information as to whether and what they received may no longer be relevant; however, the participants' acceptance of such information is difficult to prove prior to the test, and its testing afterwards may be subject to other biases.

THE BALANCED CROSS-OVER DESIGN

In an attempt to overcome the serious limitations of the BPD, we designed another strategy that may account for some of the BDP limitations. In this case, participants are divided into four groups, and all are told that they are participating in a conventional randomized double-blinded and placebo-controlled cross-over trial, in which they will receive both the drug and the placebo on two different occasions in a randomized and double-blinded order. However, only groups 2/3 will be exposed to drug and placebo in a balanced way; that is, half the participants will receive the drug first and the placebo on the second occasion, while the other half will receive placebo first and then the drug. Participants in Group 1 will receive the drug twice, and those in group 4 will receive placebo twice instead (Fig. 16.3). In this case, groups 2 and 3 represent the conventional drug trial, assuming the 'additive model' for drug and placebo effects. In group 1, the minimal value of both measures represents the 'true' drug effect (plus other nonspecific effects), and the difference between both is the expectancy component of the drug response. In group 4, the maximum value should represent the 'true' placebo effect (plus other nonspecific effects), and the difference between both values should be the expectancy component of the placebo response. Comparing these expectancy effects between groups 1 and 4 allows us to test

		First medication application	
		drug	placebo
Second medication application	drug	1: drug - drug	2: placebo - drug
	placebo	3: drug - placebo	4: placebo placebo

FIGURE 16.3 The 'balanced cross-over design' (BCD): all participants are told that they are participating in a double-blind cross-over design study and will receive both drug and placebo; this is true for groups 2 and 3, while in groups 1 and 4 they receive twice the drug and the placebo, respectively. *(adopted from Enck et al[45])*

whether the expectancy component (the placebo effect) is equal under drug and placebo conditions—which is the assumption of the 'additive model'. All other non-specific factors are assumed to be equally effective in all the groups.

The balanced cross-over design (BCD) has one important methodologic limitation. As with other cross-over designs, interference of learning effects needs to be kept in mind,[42,43] and any adaptation or habituation between measurement 1 and measurement 2 should be minimized, e.g. by increasing the time interval between the two. Its ethical limitation (deception) is similar to that of the BPD, with the exception that participants may receive a drug twice but expect to receive it only once; any risk involved in such a repetition of drug application would exclude the BCD from use, and it can be used in patients only when the deception is authorized.[44]

To our knowledge, this design has not so far been used in any experimental research, whether related to a drug or to the placebo response. An ongoing study in our laboratory testing the effects of a nicotine patch on neurocognitive performance in healthy volunteers[45] will show its applicability and limitations.

THE 'DELAYED RESPONSE' TEST

Another unique experimental approach manipulating drug timing has—to our knowledge—never been explored in any placebo testing; this approach assumes that a drug can elicit its action at a predefined time point hours after ingestion, via either a coating technology[46] or a radio-transmitted mechanical capsule technology[47] for controlled release of the compound. Such technology has been used in many clinical conditions, such as diabetes and Alzheimer's dementia. In this case, a three-arm design (Fig. 16.4) would allow a very elegant proof of the additive versus non-additive model of the placebo response.

All participants receive the same information: that they will receive either a drug or placebo in a double-blinded fashion, and no information is given about the

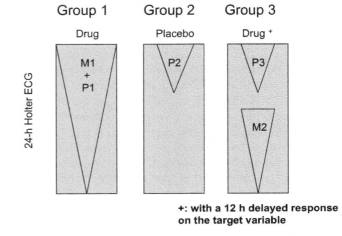

+: with a 12 h delayed response on the target variable

Assumption: M1 = M2 and P2 = P3   Hypothesis: M1+ P1 ≠ M2 + P3

FIGURE 16.4 The 'delayed response' design; M1 and M2 stand for medication response, P1 and P2 for placebo response; the 'additive model' by Kirsch[31] assumes that P1 = P2. Under the further assumptions that M1 = M2 and P2 = P3, the hypothesis of the 'additive model' is falsified if (M1 + P1 ≠ M2 + P3). *(adopted from Enck et al[45])*

timing of drug action; instead, a rationale ('cover story') is provided for prolonged drug action monitoring, e.g. for 24 hours.

Group 1 will receive the drug with immediate action, group 2 the respective placebo. Group 3 receives the delayed response medication, e.g. with drug release after 12 hours. To confirm the 'non-additive model', the placebo responses (P1, P2) should not be identical, the compound effects of drug and placebo (M1 + P1; M2 + P3) should also not be the same, but the drug effects (M1, M2) and placebo effects (P2, P3) should be equal (see Fig. 16.4 for the formula).

A variant of such a design that intended to elucidate the drug response in a clinical trial in Parkinson's disease was recently described[48]: patients in the placebo arm are planned to switch from placebo to drug at some time point during the trial unbeknown to the patient and physician, but in this case pre-treatment with placebo may affect the later drug treatment by conditioning procedures.[42] A better way of separating drug and placebo effects may be randomized run-in and withdrawal periods (see Table 16.2, above).

While the DRT has never been used in a laboratory setup for assessment of the placebo response, preliminary approval of the underlying hypothesis may be drawn from clinical studies that have used the delayed release technology in the treatment of patients, e.g. in 5-ASA treatment of chronic inflammatory bowel diseases[49]; in these studies, the placebo response should be overall lower as compared to immediate release medication. Unfortunately, the number of placebo-controlled studies using such technology is so far rather low to allow a reliable meta-analytic comparison.

ACTIVE PLACEBOS

Active placebos mimic the side effects of a drug under investigation without inducing its main effect in clinical trials. Active placebos in experimental research induce side effects which make the volunteer believe that he/she has received active treatment (e.g. a pain medication); this may be achieved by any perceivable effect following a placebo application, e.g. by skin, olfactory, gustatory and other signals that are easy to induce and do not interfere with the function under test. Interestingly, active placebos have rarely been used, either in clinical trials or in experimental placebo research. Among the few experimental studies that tested active placebos in comparison with inactive ones, Rief et al[35] recently showed that adding pepper to an otherwise inert nasal placebo spray increased the response rate (placebo analgesia) from under a 50:50 chance to 100%.

In clinical trials, active placebos are difficult to develop and are therefore used only occasionally in a few clinical conditions, e.g. in the treatment of depression.[50] A 2004 Cochrane meta-analysis[51] reported only nine studies on 751 patients with depression, all conducted/published between 1961 and 1984. In all of these cases, the 'active placebo' was atropine compared with amitryptilin or imipramine, and all but one study used a parallel-group design. While the overall effect size was in favor of active treatment, it was small compared with placebo-controlled trials using inactive placebos, indicating that unblinding effects may inflate the efficacy of antidepressants in trials using inert placebos.

All this argues strongly in favor of the use of active placebos. Without active placebos it is often difficult to decide whether drugs have beneficial effects because of their genuine action, or because they induce side effects that trigger positive outcome expectations. Albeit active placebos likely increase the placebo response, they may help to reduce false-positive results in drug testing.

Appropriate 'active placebo' procedures beyond drug treatment raise a number of other issues. We will briefly discuss the situation with surgery, acupuncture and other device-driven therapies, as well as physical therapy.

Surgery

While an 'active placebo' appears difficult to develop for drug treatment, it appears to be less problematic for medicinal interventions such as surgery. 'Sham surgery'[52] leaves visible marks (scars) and induces side effects (pain, wound healing), forcing patients to believe that they underwent true surgery. The few well-conducted sham surgery trials for knee orthoarthritis[53] and for Parkinson's disease[54] showed equal efficacy of surgery and sham surgery; however, they have raised ethical concerns[55] as the placebo procedure itself carries substantial risks (risks that are different from placebo administration in drug trials) such as wound infections and anesthesia risks.[56]

Acupuncture

While in the past, clinical efficacy of acupuncture was regarded as predominantly driven by placebo effects, recent RCTs and their results have changed this view[57] and attributed analgesic effects of acupuncture to neurophysiologic mechanisms.

A commonly used sham procedure in acupuncture trials is 'minimal acupuncture' by inserting a needle either at or near a conventional needling point and providing low-grade stimulation, e.g. with a lower intensity electrical stimulation or by only inserting the needle. It seems to be a specifically valid ethical argument to perform acupuncture trials in acutely ill patients because therapy is not withheld; only the degree of intensity of the therapy is different. This procedure resembles medicinal studies where a low (and likely ineffective) dose is compared with higher dosages of the drug under investigation.

As is evident from published trials,[58] minimal acupuncture is often as effective as true acupuncture. In the review by Lundberg et al,[59] the six studies that used this control condition yielded similar, though somewhat lower, response rates compared with acupuncture—but overall substantial improvement in clinical conditions. The largest acupuncture trial so far, the German Acupuncture Trial for Chronic Low Back Pain (GERAC)[60] that included more than 1000 patients, revealed that both acupuncture and sham acupuncture were equally effective and superior to conventional treatment alone.

Lundberg et al,[59] and the same group in other papers,[61,62] argue that minimal acupuncture is 'not a valid placebo control' due to the physiologic effects that minimal acupuncture procedures are able to elicit. However, as long as the correct conclusion is drawn from the study, i.e. that efficacy is confirmed only if the difference between both study arms is significant, the use of

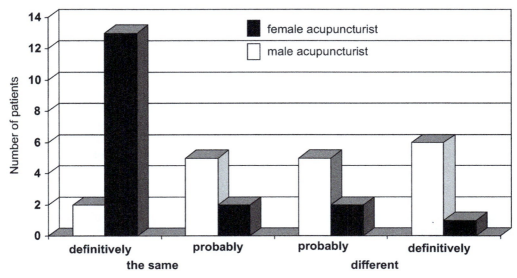

FIGURE 16.5 Estimation of acupuncture-naive patients, whether they had received acupuncture and sham acupuncture on two occasions in a randomized sequence. Thirty-six patients (18 females) were stimulated by a female and a male practitioner in a balanced fashion. Significantly more patients believed that they had received true acupuncture on both occasions when the acupuncturist was female (black bars). *(adopted from White et al[68])*

minimal acupuncture requires only higher efficacy of the true procedure. It is a conservative statistical argument (against overestimating the acupuncture efficacy) if the control condition simulates as many features of the true procedure as possible.

To overcome the need for an appropriate sham acupuncture procedure in respective trials, Streitberger et al[63] and subsequently others developed specific sham acupuncture needles by either blunting[64] or by blunting and shortening true acupuncture needles,[40] or by using needles of similar appearance that do not penetrate the skin but retract telescopically into the needle handle, invisible to the patient.[65] These devices have been validated for efficacy and specifically for blinding[66,67] and appear to work properly. In the study by White et al,[68] the 'Streitberger needle' was able to hide the assignment to the true and sham acupuncture group in acupuncture-naive participants (Fig. 16.5).

A similar situation exists with other 'technical' interventions such as transcutaneous electrical nerve stimulation (TENS), laser stimulation, and transcranial electrical or magnetic stimulation where non-effective or less effective stimuli and/or stimulation of adjacent body parts can be used as sham treatment targets. It has also been shown that 'sham devices' produce a greater placebo response than does a placebo pill for the same condition.[69]

Physical Therapy

Development of appropriate 'active' control or sham procedures becomes complicated again with physical therapies (massage, Kneipp applications, gymnastics and similar). In all of these cases, false interventions may be unethical as they may induce harm, and the risk of unblinding of the patient may be high, resulting in non-acceptance of randomization or early withdrawal in the control group.[70]

EFFECTIVE BLINDING

In a systematic review by Boutron et al,[71] blinding methods were classified according to whether they primarily focused on blinding of patients or health-care providers, or the evaluators of treatment outcomes. They identified 819 articles, of which more than 50% described the method of blinding. Methods to avoid unblinding of patients and/or health-care providers involved the use of active placebos, centralized assessment of side effects, patients informed only in part about the potential side effects of each treatment, centralized adapted dosage, or provision of sham results of complementary investigations; blinding of evaluation included the use of video, audiotape, or photography, or adjudication of clinical events.

In a systematic review of 126 trials with different treatment options for low-back pain, Machado et al[72] investigated appropriate and inappropriate blinding procedures, among them 10 acupuncture trials using different sham control strategies.

With respect to acupuncture, only four of the studies assessed by Machado et al[72] assessed whether sham acupuncture can be distinguished from true acupuncture, and only two used acupuncture-naive participants, making it likely that in the others that the assumed 'inertness' and blindness of the sham procedure may have been unmasked. The review by Madsen et al[73] noted that in their 13 meta-analyzed trials none had blinded the acupuncturist.

However, these problems are not specific to acupuncture trials, as with the other non-medicinal treatment options for low-back pain—back school, behavioral treatment, electrotherapy, exercise, heat-wrap therapy, insoles, magnet therapy, massage, neuroreflex therapy, spinal manipulative therapy, traction—similar incomplete blinding problems were noted, while drug trials usually comply with this condition.

While many studies state that they are double-blinded, they rarely report how effective the blinding actually was. In 1986, Ney et al[74] stated that the effectiveness of blinding was assessed in less than 5% of studies conducted between 1972 and 1983. Twenty years later, Hrobjartsson et al[75] identified 1599 blinded randomized studies and found that only 31 (2%) reported tests for the success of blinding. Even then, only 14 of 31 studies (45%) reported that blinding was successful. Boehmer and Yong[76] consequently asked for inclusion of the evaluation of the effectiveness of blinding in RCTs, but this request should also be extended to experimental studies.

NO-TREATMENT AND WAITING-LIST CONTROLS

All RCTs need to control for 'spontaneous variation of symptoms' that occurs with all medical conditions, and especially with chronic diseases, as they are part of the 'nonspecific effects' seen in both arms of drug trials (Fig. 16.6). To separate 'spontaneous variation' from 'placebo responses', a 'no-treatment' control group appears necessary in order to determine how much of the nonspecific effects can be attributed to spontaneous variation and recovery. Because this is rarely done for ethical reasons (see below), the exact size of the contribution of spontaneous variation to the placebo response is known only for minor and benign clinical conditions and may account here for approximately 50% of the placebo effect.[77]

In experimental settings, 'no treatment controls' may serve to control for habituation and sensitization effects that may occur with repetitive stimulation, e.g. in pain and placebo analgesia experiments.

Waiting-list (WL) and 'treatment as usual' (TAU) are common control strategies in all non-medication trials in which an inert 'placebo' treatment is difficult to provide, such as in psychotherapy, physical rehabilitation, surgery, and 'mechanical' interventions (TENS, magnetic stimulation, laser, acupuncture). While some of these therapies have developed their own strategy (e.g. sham surgery, see above), others have to rely on WL and TAU. Their limitations are that patients' expectation of receiving effective therapy is in conflict with being randomized to routine treatment (which most of them will already have experienced) and to delays in the onset of therapy (which may increase the placebo response, but also drop-out rates). This may significantly affect recruitment and compliance in trials, and may lead to biased patient populations in respective studies.[4]

According to a review by Lundeberg et al,[61] of nine trials in migraine (n = 3), low-back pain (n = 3), and osteoarthritis (n = 3), one-third used WL and/or TAU only to control for nonspecific effects. In the meta-analyses of Moffet,[58,78] 22 of 36 studies used TAU and other non-acupuncture therapies, of which most (n = 18) reported significant clinical efficacy without controlling for the placebo effect.

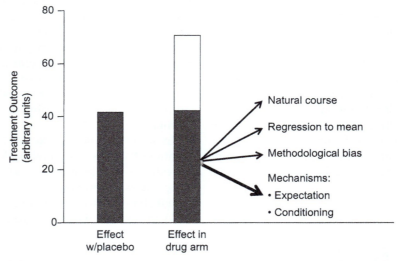

FIGURE 16.6 The 'placebo effect' in both arms of RCTs is thought to be a composite of spontaneous disease variation, regression to the mean, and specific contextual factors that represent the placebo response. *(adopted from Enck et al[1])*

In contrast, 'no treatment' control groups have been mandated by critiques of the current placebo discussion[79,80] to account for spontaneous variation of symptoms in many clinical trials that may falsely be attributed to the placebo response. When they meta-analyzed studies,[77] they found that about half of the placebo response can be attributed to spontaneous remission; this was also true for included pain trials. They also noted that the number of studies using no-treatment controls is low, that they often involve benign clinical conditions (smoking cessation, insomnia), and that they most often include non-medicinal interventions such as psychotherapy and acupuncture.

Evidently, WL controls, as well as TAU, lack credibility to serve as proper control groups in many clinical areas, and certainly do so where pain patients ask for therapy. According to recent meta-analyses[81,82] many drug studies in acute and chronic pain are conducted with comparator drugs rather than with placebos for ethical reasons. However, comparative effectiveness research (CER) studies lack the possibility of assessing the placebo response at all, but such studies have been shown to enhance the drug response (compared with placebo-controlled trials of the same compounds) by 100%—solely by the patient's expectation of receiving a drug.[83]

Recently, novel designs have been developed to account for these disadvantages of WL and TAU in RCTs.[1] Among them, the 'preference design'[19] asks for patients' preference (usually in CER) before patients with no preference are randomized to treatment arms. The step-wedge design[84] randomizes patients to different treatment groups that are stacked (immediate start, beginning after 8 weeks, after 16 weeks etc.) so that waiting becomes less of a disappointment, and waiting time allows assessment of spontaneous variation of symptoms. An even more acceptable strategy for patients is the (classical or modified) Zelen design[20] (Fig. 16.7). This separates recruitment for an observational study for spontaneous symptom variation (the 'no-treatment' control) from randomization for an interventional study, either placebo-controlled or as a CER study.

All of these design features are usually not applicable to experimental studies in the laboratory. In experimental research, 'no treatment' controls have become the standard to evaluate habituation/adaptation to repetitive stimuli.

THE FREE-CHOICE PARADIGM

The free-choice paradigm (FCP) breaks most radically with current traditions in clinical and experimental placebo research by introducing the option of choosing between drug and placebo for the patient/volunteer.[85]

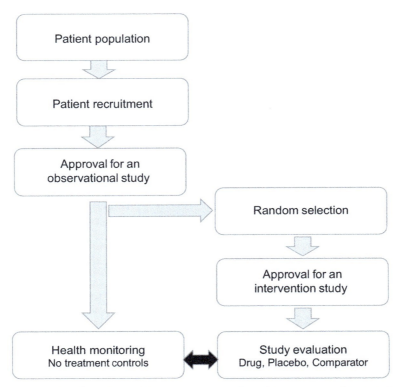

FIGURE 16.7 Schematics of the so-called Zelen design[20] that separates recruitment for an observational study from recruitment for an intervention study. *(adopted from Enck et al[1])*

The design allows volunteers/patients to choose between two pills different in color. They receive the correct information that one contains the drug while the other contains the placebo, but that conditions are double-blinded. In this case there is no obvious deception, and hence ethical limitations are minimal; furthermore, the dependent variable for measuring drug efficacy is the choice of behavior rather than reported symptoms or symptom improvement.

The design does not manipulate the information provided to participants and patients, neither does it manipulate the timing of drug release, both kinds of manipulation being common when novel designs are proposed in experimental studies on the placebo effect in healthy volunteers.[45] It thus avoids ethical concerns (deception) if patients are included. It also increases the number of events that can be used for evaluation of drug efficacy, e.g. superiority of drug over placebo by computing.

One has, however, to make sure that patients indeed select and do not take both pills simultaneously, thus undermining the intention of the design. It further has to be ascertained that technical solutions have been installed to warrant appropriate compliance, to prevent over-dosage, and to monitor drug intake.

Other restrictions may be short-acting effects of the drug, the need for steady drug levels, the effect on symptoms rather than on biochemical disease indicators—and hence symptomatic endpoints rather than disease biomarkers. In this case, the primary outcome measure of drug testing is the 'selection behavior' of patients (Fig. 16.8).

The FCP maybe regarded as a modification of the 'adaptive response design,'[86] the 'early-escape design'[87] and other adaptive strategies.[88] It may offer an alternative approach to common drug test procedures, although its statistics have still to be established.

Other requirements of such an approach may be due to the fact that the patient is allowed to switch to the other condition at any time, hence, the pharmacodynamics of the compound under investigation have to be appropriate, e.g. the speed of action, and the feasibility of on-demand medication. It would, on the other hand, allow assessment of drug efficacy via the choice of behavior rather than with symptomatic endpoints.

As the FCP most radically breaks with some of the current principles of RTC trials, its consequences on the purpose of randomization need to be considered. Randomization (but not yet double-blinding) of patients to either the treatment or the placebo group was introduced with the British Streptomycin Trial in Tuberculosis as late as 1948, more than 100 years after introducing placebo treatment into drug testing.[32] Its purpose is to prevent selection bias, i.e. the biased decision of the treating doctor to allocate one patient to active treatment and the other to placebo treatment. It thereby also balances other potential confounders of treatment efficacy in both treatment arms. Randomization is not necessary to balance the number of patients in all treatment arms, as this is usually decided upon by statistical power analysis on the one hand and on design issues on the other hand (dose range testing, comparator drug testing, enrichment trials etc.).[1] The FCP may eventually end up with quite unequal numbers of those choosing drug or placebo, but this would be its advantage because it would indicate statistical superiority of drug over placebo (see Fig. 16.8), and avoids unnecessary placebo treatment in many patients.[89]

With the FCP, no randomization is needed as all patients have the choice between drug and placebo at predefined time points. Because reasons to alter from one day to the next may vary within and across patients, they need to be assessed continuously, e.g. by symptom diaries, and may be taken as covariates in the efficacy analysis. As with conventional designs, unblinding may be an issue in the FCP: adverse events rather than the course of symptoms may indicate which is the drug and which the placebo, and thus may determine the choice of behavior. This also requires additional data to be monitored. It may also be combined with other design features common in

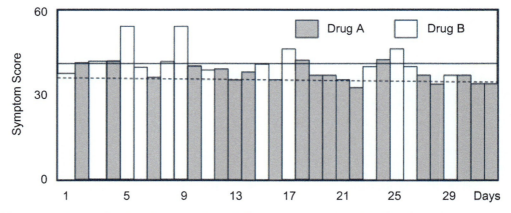

FIGURE 16.8 The 'free-choice paradigm': patients can choose daily between drugs A and B. The efficacy measures are either the average symptom score with A (solid line) and B (dotted line) or number of days with A and B. *(adopted from Enck et al[85])*

conventional trial designs, e.g. active placebos that mimic the side effects of the drug under investigation (see above).

While, in conventional RCTs, unequal dropout rates between the drug and placebo arm mirror superiority of drug over placebo—but also carry the potential for unblinding of the study—the FCP overcomes this disadvantage by substantially increasing the number of 'events' that can be taken as statistical significance (Fig. 16.9).

A procedure similar to the FCP has been used occasionally in optimizing dosage of drugs[86] in clinical trials. Its methodology and statistics in assessing drug superiority over placebo have not been validated, and it is not able to overcome the 'additive model' assumptions[31] by separating expectancy and true drug effects; it may, however, provide a better estimate of both by using regression functions over time per patient rather than average response rates across treatment groups.

ETHICS OF PLACEBO RESEARCH

Conventional RCTs imply that patients are properly informed about their relative chances of receiving placebo, that they do understand its meaning, and that they do agree to it, although it has been shown that this may not always be the case.[90] Even so, RCTs have been questioned for ethical reasons: they withhold effective therapy for a large proportion of the patients, and this is in conflict with the ethical rules of the Declaration of Helsinki.[91] Increasing the number of patients in the active treatment arm (e.g. in enrichment trials, see above) or even testing novel drugs against a comparator drug offers only a 'sham' solution but creates an ethical paradox[92]: by increasing the number of actively treated patients for ethical reasons, more patients need to be enrolled in the study to demonstrate superiority (or non-inferiority in comparator trials), and this again contradicts the Declaration of Helsinki's rule that the least number of patients possible should undergo drug testing in RCTs while all others should receive the best treatment available. This apparent contradiction between ethics on the one hand and methodologic requirements on the other hand is even more evident with the need for 'no-treatment' controls as discussed above: leaving patients without any treatment for methodologic reasons is unacceptable unless TAU is highly effective or other solutions (such as the Zelen design, see above) are applicable.

Planning a drug trial with a fast acting proton pump inhibitor for functional dyspepsia and/or gastrointestinal reflux in 100 patients, to be taken daily for 30 consecutive days (see Fig. 1) Assessment of efficacy by a dichotomous endpoint (improved, not improved) on a daily symptoms score.

Conventional DBRPC	Free choice paradigm (FCP)
50 patients on drug and 50 patients on placebo	100 patients with the daily choice between drug and placebo

The drug is superior to placebo if

- depending on the response in the placebo arm - at least 6 more patients in the drug arm compared to placebo need to be responders. Assuming 42 responders with placebo (beyond this, chi-square test needs Yates correction), 48 need to respond with drug ($\chi^2=4.0$, p=.046). With lower response rates on placebo, up to 10 more patients need to respond to drug than to placebo,	- assuming every choice to be an independent event, based on daily symptom assessment, and random selection would result in a 1500:1500 distribution - 1576 of out 3000 choices need to be in favor of drug compared to 1424 in favor of placebo ($\chi^2=3.853$, p=.05). In theory, each patient therefore needs to select the drug on 16 of the 30 days to make drug superior to placebo,

otherwise the drug is equal to placebo (null hypothesis).

Restrictions and caveats:

High rates of spontaneous remission may occur as drug or placebo response but are indistinguishable without "no treatment" controls. High placebo response rates require more patients to be tested.	Individual patients with a 1:1 choice ratio may have to be excluded because of random choice behavior. Placebo responders are those with more choices of the placebo over drug.

FIGURE 16.9 Fictive draft and calculation of a clinical trial using a conventional double-blinded randomized placebo-controlled (DBRPC) trial and the free-choice paradigm (FCP). *(adopted from Enck et al[85])*

The situation is somewhat different in experimental placebo research: as is obvious, manipulating either timing of drug application, or the information provided with the application of a drug, are deceptive procedures and therefore excluded from use in patients. Ethical provisions are less strict in healthy participants who will not experience harm in experimental settings beyond temporary discomfort (such as experimental pain of short duration). However, it should be noted that ethical standards are equal for studies in healthy people and patients, and patient studies are, for one reason only, to be treated differently: patients usually seek health care and not participation in experimental or clinical trials, while healthy volunteers seek monetary or other reward when responding to calls for participation in an experiment.

A further complicating issue for both experimental and clinical studies is the inclusion of children and adolescents: while we know that the placebo response is higher in children, as compared to adolescents, and in adolescents as compared to adults,[93] current legal and ethical rules for the inclusion of children into randomized, double-blinded and placebo-controlled trials are far from being clear and are, to some degree, even contradictory.

Guidelines of the National Institutes of Health (NIH)[94] and the Food and Drug Administration (FDA)[95] in the USA, and the European Medicines Agency (EMA),[96] all explicitly call for the need to include children into the testing of novel compounds as long as no specific reasons argue against it,[97,98] as drugs should not be used in children without such tests, and their use is based merely on dose adjustments from doses for adults (e.g. based on body weight). The current practice of not including children is reflected in the missing data on placebo response rates and placebo effects in children.[93]

In contrast to the policy of drug approval authorities, the Declaration of Helsinki (DoH) of the World Medical Association (WMA) in its current version (as of 2008)[99] does not explicitly mention children but includes them among all 'incompetent' persons that are not to be included in placebo-controlled trials unless important reasons argue in favor of such inclusion, e.g. no adequate medication available that is approved for use in adults.[100]

Common to both sets of rules is that children can be included into trials only after careful consideration of the risks and benefits.

The ethical limitations of performing placebo experiments in both healthy volunteers and patients may be avoided only by a procedure called 'authorized deception': they are informed prior to carrying out the study that—at some point during the procedure—they will be misinformed about one or another feature of the study design. If they still agree to participate, the requirements of fully informed consent are met. This procedure has been shown not to corrupt data collection in a placebo analgesia experiment.[101] It resembles a clinical analogue procedure called 'patient authorized concealment'[102] when patients authorize their doctors to e.g. not disclose a fatal diagnosis during a clinical work-up.[103]

Finally, it has been shown that full information of a placebo application may still generate substantial clinical improvement when performed properly, and patients do not misunderstand the information that receiving a placebo pill implies no improvement.[104] Such a procedure is, however, unlikely to work in an experiment with healthy volunteers. Any change in e.g. placebo analgesia in a laboratory setting will need to be controlled for habituation/adaptation responses using a 'no-treatment' control group (see Table 16.2, above).

SUMMARY

With growing knowledge of the mechanisms behind the placebo response in drug trials and clinical routine,[68] different methods have been developed over the last 50 years of placebo-interested research to identify, characterize and modulate the placebo response in individuals, but also to minimize it in RCTs.[1] Among the designs that manipulate information, the BPD and the BCD are not applicable to patients without authorized deception. Manipulating timing of the drug, such as in the hidden treatment and the DRT, may be more acceptable but is still limited to experimental settings. In RCTs, 'active placebos' and sham controls for non-drug therapy are difficult to develop, and effective blinding has to be secured. Waiting-list controls and treatment-as-usual are inappropriate control strategies unless combined with novel approaches such as the Zelen design.

Acknowledgment

Supported by a grant from Deutsche Forschungsgemeinschaft (En 50/30-1).

References

1. Enck P, Bingel U, Schedlowski M, Rief M. Minimize, maximize or personalize: what to do with the placebo response in medicine? *Nat Rev Drug Disc*. 2013;12:191-204.
2. Staud R, Price DD. Role of placebo factors in clinical trials with special focus on enrichment designs. *Pain*. 2008;139:479-480.
3. Iovieno N, Papakostas GI. Correlation between different levels of placebo response rate and clinical trial outcome in major depressive disorder: a meta-analysis. *J Clin Psychiatry*. 2012;73: 1300-1306.
4. Enck P, Horing B, Weimer K, Klosterhalfen S. Placebo responses and placebo effects in functional bowel disorders. *Eur J Gastroenterol Hepatol*. 2012;24:1-8.

5. Enck P, Benedetti F, Schedlowski M. New insights into the placebo and nocebo responses. *Neuron*. 2008;59:195-206.
6. Colloca L, Lopiano L, Lanotte M, Benedetti F. Overt versus covert treatment for pain, anxiety, and Parkinson's disease. *Lancet Neurol*. 2004;3:679-684.
7. Benedetti F, Carlino E, Pollo A. Hidden administration of drugs. *Clin Pharmacol Ther*. 2011;90:651-661.
8. Hrobjartsson A, Boutron I. Blinding in randomized clinical trials: imposed impartiality. *Clin Pharmacol Ther*. 2011;90:732-736.
9. Lee S, Walker JR, Jakul L, Sexton K. Does elimination of placebo responders in a placebo run-in increase the treatment effect in randomized clinical trials? A meta-analytic evaluation. *Depress Anxiety*. 2004;19:10-19.
10. US Department of Health and Human Services, Food and Drug Administration. Guidance for Industry. E 10 Choice of control groups and related issues in clinical trials. International Conference on Harmonisation of Technical Requirements for Registration of Pharmaceuticals for Human Use (ICH). <http://www.fda.gov/downloads/Drugs/GuidanceComplianceRegulatoryInformation/Guidances/UCM073139.pdf>; 2001 Accessed 24.06.12.
11. Madsen LG, Bytzer P. Review article: single subject trials as a research instrument in gastrointestinal pharmacology. *Aliment Pharmacol Ther*. 2002;16:189-196.
12. Moncrieff J, Wessely S, Hardy R. Active placebos versus antidepressants for depression. *Cochrane Database Syst Rev*. 2004:CD003012.
13. Woods SW, Gueorguieva RV, Baker CB, Makuch RW. Control group bias in randomized atypical antipsychotic medication trials for schizophrenia. *Arch Gen Psychiatry*. 2005;62:961-970.
14. Rao S, Lembo AJ, Shiff SJ, et al. A 12-week, randomized, controlled trial with a 4-week randomized withdrawal period to evaluate the efficacy and safety of linaclotide in irritable bowel syndrome with constipation *Am J Gastroenterol*. 2012;107:1714-1724 .
15. Iovieno N, Papakostas GI. Does the presence of an open-label antidepressant treatment period influence study outcome in clinical trials examining augmentation/combination strategies in treatment partial responders/nonresponders with major depressive disorder? *J Clin Psychiatry*. 2012;73:676-683.
16. Mallinckrodt C, Chuang-Stein C, McSorley P, et al. A case study comparing a randomized withdrawal trial and a double-blind long-term trial for assessing the long-term efficacy of an antidepressant. *Pharm Stat*. 2007;6:9-22.
17. Staskin DR, Michel MC, Sun F, et al. The effect of elective sham dose escalation on the placebo response during an antimuscarinic trial for overactive bladder symptoms. *J Urol*. 2012;187:1721-1726.
18. Ivanova A, Tamura RN. A two-way enriched clinical trial design: combining advantages of placebo lead-in and randomized withdrawal. *Stat Methods Med Res*. 2011 (in press).
19. King M, Nazareth I, Lampe F, et al. Conceptual framework and systematic review of the effects of participants' and professionals' preferences in randomised controlled trials. *Health Technol Assess*. 2005;9:1-186; iii-iv.
20. Zelen M. A new design for randomized clinical trials. *N Engl J Med*. 1979;300:1242-1245.
21. Weijer C, Grimshaw JM, Eccles MP, et al. Ottawa Ethics of Cluster Randomized Trials Consensus Group. The Ottawa statement on the ethical design and conduct of cluster randomized trials. *PLoS Med*. 2012;9:e1001346.
22. Rief W, Nestoriuc Y, Weiss S, et al. Meta-analysis of the placebo response in antidepressant trials. *J Affect Disord*. 2009;118:1-8.
23. Hrobjartsson A, Gotzsche PC. Placebo interventions for all clinical conditions. *Cochrane Database Syst Rev*. 2010:CD003974.
24. Kirsch I, Deacon BJ, Huedo-Medina TB, et al. Initial severity and antidepressant benefits: a meta-analysis of data submitted to the Food and Drug Administration. *PLoS Med*. 2008;5:e45.
25. Nijs J, Inghelbrecht E, Daenen L, et al. Recruitment bias in chronic pain research: whiplash as a model. *Clin Rheumatol*. 2011;30:1481-1489.
26. Kaptchuk TJ, Kelley JM, Deykin A, et al. Do 'placebo responders' exist? *Contemp Clin Trials*. 2008;29:587-595.
27. Stone DA, Kerr CE, Jacobson E, et al. Patient expectations in placebo-controlled randomized clinical trials. *J Eval Clin Pract*. 2005;11:77-84.
28. Pressman A, Avins AL, Neuhaus J, et al. Adherence to placebo and mortality in the Beta Blocker Evaluation of Survival Trial (BEST). *Contemp Clin Trials*. 2012;33:492-498.
29. Boehmer J, Yong P. How well does blinding work in randomized controlled trials?: a counterpoint. *Clin Pharmacol Ther*. 2009;85:463-465.
30. Cepeda MS, Berlin JA, Gao CY, et al. Placebo response changes depending on the neuropathic pain syndrome: results of a systematic review and meta-analysis. *Pain Med*. 2012;13:575-595.
31. Kirsch I. Are drug and placebo effects in depression additive? *Biol Psychiatry*. 2000;47:733-735.
32. Hill AB. Suspended judgement: memories of the British streptomycin trial in tuberculosis. The first randomized clinical trial. *Contemp Clin Trials*. 1990;11:77-79.
33. Muthén B, Brown HC. Estimating drug effects in the presence of placebo response: causal inference using growth mixture modeling. *Stat Med*. 2009;28:3363-3385.
34. Petrovic P, Kalso E, Petersson KM, et al. A prefrontal non-opioid mechanism in placebo analgesia. *Pain*. 2010;150:59-65.
35. Rief W, Glombiewski JA. The hidden effects of blinded, placebo-controlled randomized trials: An experimental investigation. *Pain*. 2012;153:2473-2477.
36. Kelemen WL, Kaighobadi F. Expectancy and pharmacology influence the subjective effects of nicotine in a balanced-placebo design. *Exp Clin Psychopharmacol*. 2007;15:93-101.
37. Metrik J, Rohsenow DJ, Monti PM, et al. Effectiveness of a marijuana expectancy manipulation: piloting the balanced-placebo design for marijuana. *Exp Clin Psychopharmacol*. 2009;17:217-225.
38. Papakostas GI, Fava M. Does the probability of receiving placebo influence clinical trial outcome? A meta-regression of double-blind, randomized clinical trials in MDD. *Eur Neuropsychopharmacol*. 2009;19:34-40.
39. Lidstone SC, Schulzer M, Dinelle K, et al. Effects of expectation on placebo-induced dopamine release in Parkinson disease. *Arch Gen Psychiat*. 2010;67:857-865.
40. Fiorillo CD, Tobler PN, Schultz W. Discrete coding of reward probability and uncertainty by dopamine neurons. *Science*. 2003;299:1898-1902.
41. Weimer K, Horing B, Stürmer J, et al. Nicotine expectancy differentially affects reaction time in healthy non-smokers and smokers depending on gender. *Exp Clin Psychopharmacol*. 2013;21:181-187.
42. Suchman AL, Ader R. Classic conditioning and placebo effects in crossover studies. *Clin Pharmacol Ther*. 1992;52:372-377.
43. Colloca L, Benedetti F. How prior experience shapes placebo analgesia. *Pain*. 2006;124:126-133.
44. Miller FG, Wendler D, Swartzman LC. Deception in research on the placebo effect. *PLoS Med*. 2005;2:e262.
45. Enck P, Klosterhalfen S, Zipfel S. Novel study designs to investigate the placebo response. *BMC Med Res Methodol*. 2011;11:90.
46. Behzadi SS, Toegel S, Viernstein H. Innovations in coating technology. *Recent Pat Drug Deliv Formul*. 2008;2:209-230.
47. Twomey K, Marchesi JR. Swallowable capsule technology: current perspectives and future directions. *Endoscopy*. 2009;41:357-362.
48. D'Agostino RB. The delayed-start study design. *N Engl J Med*. 2009;361:1304-1306.
49. Fernandez-Becker NQ, Moss AC. Improving delivery of aminosalicylates in ulcerative colitis: effect on patient outcomes. *Drugs*. 2008;68:1089-1103.

50. Edward SJ, Stevens AJ, Braunholtz DA, et al. The ethics of placebo-controlled trials: a comparison of inert and active placebo controls. *World J Surg.* 2005;29:610-614.
51. Moncrieff J, Wessely S, Hardy R. Active placebos versus antidepressants for depression. *Cochrane Database Syst Rev.* 2004:CD003012.
52. Campbell MK, Entwistle VA, Cuthbertson BH, et al. KORAL study group. Developing a placebo-controlled trial in surgery: issues of design, acceptability and feasibility. *Trials.* 2011;12:50.
53. Moseley JB, O'Malley K, Petersen NJ, et al. A controlled trial of arthroscopic surgery for osteoarthritis of the knee. *N Engl J Med.* 2002;347:81-88.
54. Freed CR, Greene PE, Breeze RE, et al. Transplantation of embryonic dopamine neurons for severe Parkinson's disease. *N Engl J Med.* 2001;344:710-719.
55. Frank SA, Wilson R, Holloway RG, et al. Ethics of sham surgery: perspective of patients. *Mov Disord.* 2008;23:63-68.
56. Landau W. What is the risk of sham surgery in Parkinson disease clinical trials? A review of published reports. *Neurology.* 2006;66:1788-1789.
57. Enck P, Klosterhalfen S, Zipfel S. Acupuncture, psyche and the placebo response. *Auton Neurosci.* 2010;157:68-73.
58. Moffet HH. Sham acupuncture may be as efficacious as true acupuncture: a systematic review of clinical trials. *J Altern Complement Med.* 2009;15:213-216.
59. Lundeberg T, Lund I, Sing A, Näslund J. Is placebo acupuncture what it is intended to be? *Evid Based Complement Alternat Med.* 2011;2011:932407.
60. Haake M, Müller HH, Schade-Brittinger C, et al. German Acupuncture Trials (GERAC) for chronic low back pain: randomized, multicenter, blinded, parallel-group trial with 3 groups. *Arch Intern Med.* 2007;167:1892-1898.
61. Lundeberg T, Lund I, Näslund J, Thomas M. The Emperors sham - wrong assumption that sham needling is sham. *Acupunct Med.* 2008;26:239-242.
62. Lund I, Näslund J, Lundeberg T. Minimal acupuncture is not a valid placebo control in randomised controlled trials of acupuncture: a physiologist's perspective. *Chin Med.* 2009;4:1.
63. Streitberger K, Kleinhenz J. Introducing a placebo needle into acupuncture research. *Lancet.* 1998;352:364-365.
64. Tough EA, White AR, Richards SH, et al. Developing and validating a sham acupuncture needle. *Acupunct Med.* 2009;27:118-122.
65. Takakura N, Yajima H. A double-blind placebo needle for acupuncture research. *BMC Complement Altern Med.* 2007;7:31.
66. McManus CA, Schnyer RN, Kong J, et al. Sham acupuncture devices–practical advice for researchers. *Acupunct Med.* 2007;25:36-40.
67. Takakura N, Yajima H. A placebo acupuncture needle with potential for double blinding – a validation study. *Acupunct Med.* 2008;26:224-230.
68. White P, Lewith G, Hopwood V, Prescott P. The placebo needle, is it a valid and convincing placebo for use in acupuncture trials? A randomised, single-blind, cross-over pilot trial. *Pain.* 2003;106:401-409.
69. Kaptchuk TJ, Stason WB, Davis RB, et al. Sham device v inert pill: randomised controlled trial of two placebo treatments. *BMJ.* 2006;332:391-397.
70. Lindström D, Sundberg-Petersson I, Adami J, Tönnesen H. Disappointment and drop-out rate after being allocated to control group in a smoking cessation trial. *Contemp Clin Trials.* 2010;31:22-26.
71. Boutron I, Estellat C, Guittet L, et al. Methods of blinding in reports of randomized controlled trials assessing pharmacologic treatments: a systematic review. *PLoS Med.* 2006;3:e425.
72. Machado LA, Kamper SJ, Herbert RD, et al. Imperfect placebos are common in low back pain trials: a systematic review of the literature. *Eur Spine J.* 2008;17:889-904.
73. Madsen MV, Gøtzsche PC, Hróbjartsson A. Acupuncture treatment for pain: systematic review of randomised clinical trials with acupuncture, placebo acupuncture, and no acupuncture groups. *BMJ.* 2009;338:a3115.
74. Ney PG, Collins C, Spensor C. Double blind: double talk or are there ways to do better research? *Med Hypotheses.* 1986;21:119-126.
75. Hróbjartsson A, Forfang E, Haahr MT, et al. Blinded trials taken to the test: an analysis of randomized clinical trials that report tests for the success of blinding. *Int J Epidemiol.* 2007;36:654-663.
76. Boehmer J, Yong P. How well does blinding work in randomized controlled trials?: a counterpoint. *Clin Pharmacol Ther.* 2009;85:463-465.
77. Krogsbøll LT, Hróbjartsson A, Gøtzsche PC. Spontaneous improvement in randomised clinical trials: meta-analysis of three-armed trials comparing no treatment, placebo and active intervention. *BMC Med Res Methodol.* 2009;9:1.
78. Moffet HH. Traditional acupuncture theories yield null outcomes: a systematic review of clinical trials. *J Clin Epidemiol.* 2008;61:741-747.
79. Hróbjartsson A, Gøtzsche PC. Is the placebo powerless? An analysis of clinical trials comparing placebo with no treatment. *N Engl J Med.* 2001;344:1594-1602.
80. Hróbjartsson A, Gøtzsche PC. Is the placebo powerless? Update of a systematic review with 52 new randomized trials comparing placebo with no treatment. *J Intern Med.* 2004;256:91-100.
81. Saarto T, Wiffen PJ. Antidepressants for neuropathic pain. *Cochrane Database Syst Rev.* 2007:CD005454.
82. Quilici S, Chancellor J, Löthgren M, et al. Meta-analysis of duloxetine vs. pregabalin and gabapentin in the treatment of diabetic peripheral neuropathic pain. *BMC Neurol.* 2009;9:6.
83. Rutherford BR, Sneed JR, Roose SP. Does study design influence outcome? The effects of placebo control and treatment duration in antidepressant trials. *Psychother Psychosom.* 2009;78:172-181.
84. De Allegri M, Pokhrel S, Becher H, et al. Step-wedge cluster-randomised community-based trials: an application to the study of the impact of community health insurance. *Health Res Policy Syst.* 2008;6:10.
85. Enck P, Grundy D, Klosterhalfen S. A novel placebo-controlled clinical study design without ethical concerns – the free choice paradigm. *Med Hypotheses.* 2012;79:880-882.
86. Rosenberger WF, Lachin JM. The use of response-adaptive designs in clinical trials. *Contemp Clin Trials.* 1993;14:471-484.
87. Vray M, Girault D, Hoog-Labouret N, et al. Methodology for small clinical trials. *Therapie.* 2004;59:273-286.
88. Zhang L, Rosenberger WF. Response-adaptive randomization for clinical trials with continuous outcomes. *Biometrics.* 2006;62:562-569.
89. Enck P, Klosterhalfen S, Weimer K, et al. The placebo response in clinical trials: more questions than answers. *Philos Trans R Soc Lond B Biol Sci.* 2011;366:1889-1895.
90. McCann SK, Campbell MK, Entwistle VA. Reasons for participating in randomised controlled trials: conditional altruism and considerations for self. *Trials.* 2010;11:31.
91. Ehni HJ, Wiesing U. International ethical regulations on placebo-use in clinical trials: a comparative analysis. *Bioethics.* 2008;22:64-74.
92. Rief W, Bingel U, Schedlowski M, Enck P. Mechanisms involved in placebo and nocebo responses and implications for drug trials. *Clin Pharmacol Ther.* 2011;90:722-726.
93. Weimer K, Gulewitsch MD, Schlarb AA, et al. Placebo effects in children: a review. *Pediat Res.* 2013; doi: 10.1038/pr.2013.66.
94. Workshop on Ethical and Regulatory Issues in Global Pediatric Trials 2009. *Eunice Kennedy Shriver National Institute of Child Health and Human Development, National Institutes of Health Office of Pediatric Therapeutics.* Rockville, MD: US Food and Drug Administration; September 20-22, 2009.
95. Committee on Drugs, American Academy of Pediatrics. Guidelines for the ethical conduct of studies to evaluate drugs in pediatric populations. *Pediatrics.* 1995;95:286-294.

96. Regulation (EC) No 1901/2006 of the European Parliament and of the Council of 12 December 2006 on medicinal products for paediatric use and amending Regulation (EEC) No 1768/92, Directive 2001/20/EC, Directive 2001/83/EC and Regulation (EC) No 726/2004. Official Journal of the European Union 27.12.2006, L378/1.
97. Miller FG, Wendler D, Wilfond B. When do the federal regulations allow placebo-controlled trials in children? *J Pediatr*. 2003;142:102-107.
98. Derivan AT, Leventhal BL, March J, et al. The ethical use of placebo in clinical trials involving children. *J Child Adolesc Psychopharmacol*. 2004;14:169-174.
99. Yan EG, Munir KM. Regulatory and ethical principles in research involving children and individuals with developmental disabilities. *Ethics Behav*. 2004;14:31-49.
100. Roth-Cline M, Gerson J, Bright P, et al. Ethical considerations in conducting pediatric research. *Handb Exp Pharmacol*. 2011;205:219-244.
101. Martin AL, Katz J. Inclusion of authorized deception in the informed consent process does not affect the magnitude of the placebo effect for experimentally induced pain. *Pain*. 2010;149:208-215.
102. Dowrick CF, Hughes JG, Hiscock JJ, et al. Considering the case for an antidepressant drug trial involving temporary deception: a qualitative enquiry of potential participants. *BMC Health Serv Res*. 2007;7:64.4.
103. Shahidi J. Not telling the truth: circumstances leading to concealment of diagnosis and prognosis from cancer patients. *Eur J Cancer Care (Engl)*. 2010;19:589-593.
104. Kaptchuk TJ, Friedlander E, Kelley JM, et al. Placebos without deception: a randomized controlled trial in irritable bowel syndrome. *PLoS One*. 2010;5:e15591.

CHAPTER 17

Psychological Processes that can Bias Responses to Placebo Treatment for Pain

Ben Colagiuri[1], Peter F. Lovibond[2]

[1]School of Psychology, University of Sydney, Sydney, Australia, [2]School of Psychology, University of New South Wales, Australia

There is a wealth of evidence demonstrating improvement following placebo treatment. Expectancy is one of the primary mechanisms by which placebos are proposed to elicit their effects, whether established via suggestion alone, conditioning, or a combination. However, there are other psychological processes that can bias responses to placebo treatment and that may lead to over- or underestimation of the placebo effect. This chapter reviews three such psychological processes that are commonly cited as potential explanations for changes following placebo treatment for pain, particularly when self-reported outcomes are used (as is often the case for pain), namely, demand characteristics, the Hawthorne effect, and response shift. Understanding how these processes can affect treatment responses is important for evaluating evidence for the placebo effect and estimating the extent to which the placebo effect can be harnessed in order to improve beneficial outcomes and reduce adverse outcomes.

THEORETICAL MODEL

Before reviewing evidence for these three processes, it is worth outlining a theoretical model for how they could contribute to treatment responses following placebo administration. This model is presented in Figure 17.1. It is important to emphasize that this model is in no way intended to be a complete model of the placebo effect. For example, it ignores factors, such as desire, that are reviewed elsewhere in this book (see Ch 20). Instead, the model simply aims to highlight that several processes can influence responses to a placebo treatment, but that not all of the processes constitute a genuine placebo effect.

The model begins with the treatment context, which can lead to expectancies directly, e.g. via suggestion, or indirectly via classic conditioning (see Mitchel et al[1] and Lovibond and Shanks[2] for a detailed discussion of the relationship between conditioning and expectancy). These expectancies have been proposed to produce placebo effects in a number of ways, including unmediated and mediated effects. To account for unmediated expectancy effects, Kirsch[3] explains that under any non-dualist account of psychology, all psychological states must have corresponding physiological states, referred to as the mind–body identity assumption. Adopting this approach means that any change in psychological state, such as changes in expectancy, are accompanied by a physiologic change. In this way, Kirsch[3] argues that expecting pain relief can produce a genuine and unmediated reduction in pain because of the physiological changes that correspond to the reduced expectancy.

Mediated expectancy effects could occur through processes such as behavior change, perceptual change, and emotion change (see Stewart-Williams[4] for a review). Behavioral change occurs when the patient or participant modifies his/her behavior in response to receiving treatment. For example, patients experiencing chronic pain may feel more confident engaging in daily activities after receiving a treatment, and this may lead to improvements in actual pain. Perceptual change refers to changes in the way a patient or participant attends to his/her bodily symptoms and may also be affected by the receipt of treatment. For instance, a patient in chronic pain may attend less to their pain as a result of receiving treatment compared with a patient who is not receiving treatment and this may reduce his/her overall experience of pain. Emotional change refers to the

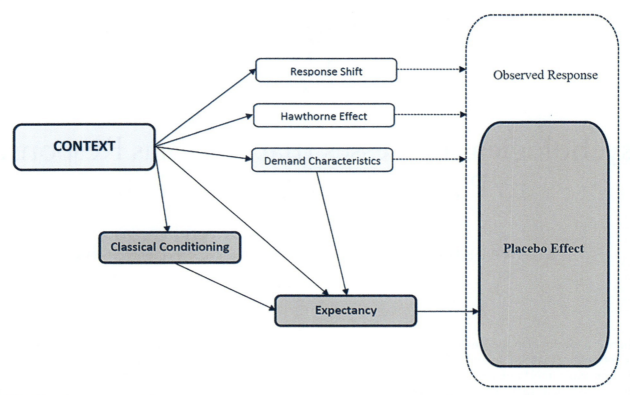

FIGURE 17.1 Theoretical model for how expectancy and other psychological processes can produce responses following placebo treatment. While the context in which placebo (or other) treatment is delivered can produce genuine placebo effects via expectancies, it may also produce responses via other processes such as demand characteristics, the Hawthorne effect, and response shift. All of these effects contribute to the observed response, but only genuine changes that are not artefacts of the context constitute the placebo effect. Dark gray boxes correspond to components of a genuine placebo effect. Light gray boxes correspond to components that could produce a placebo effect, but may also produce bias in the observed response. White boxes correspond to components that only produce bias.

situation when delivery of treatment elicits an increase or decrease in emotions. For example, receiving treatment may reduce a patient's anxiety about his/her pain, which may in turn reduce the intensity of pain.

In addition to direct and indirect effects of expectancy, however, the treatment context could also produce response shift, demand characteristics, and the Hawthorne effect. All of these can lead to changes in the way a symptom is reported in the absence of any real change in the underlying condition via reporting bias. In addition to this, demand characteristics and the Hawthorne effect may produce real changes in the underlying condition if they generate behavior modification, but these effects are generally not considered to constitute genuine placebo effects because they are artefacts to do with simply participating in a study or being under observation. They would not be maintained if the patient or participant was still receiving treatment, but was no longer under observation.

Thus, a key factor in this model is the distinction between the observed response and the placebo effect. The observed response encompasses all changes that occur following placebo treatment, regardless of whether or not they reflect real changes in the underlying condition or are artefacts of the treatment context. The placebo effect is more specific and refers only to genuine changes in the underlying condition that are not artefacts of the treatment context. Seen in this way, it becomes clear that there are many potential factors that can influence responses to placebo treatment and that understanding these factors is important for accurately estimating the magnitude and reliability of the placebo effect.

It is worth noting here that some authors have argued against the use of the term placebo effect in favor of terms such as the 'meaning response',[5] context effects,[6] and the care effect.[7] These approaches aim to highlight the fact that this phenomenon is not constrained to situations in which a dummy treatment, e.g. sugar pill or saline injection, is administered. However, we do not consider the term placebo effect as being limited to such situations and prefer its use, given that it is the term predominantly used throughout medicine and other related areas. In addition, while it is common to separate positive and negative effects into placebo and nocebo effects, respectively, in the current chapter we use the single term 'placebo effect' to refer to all such effects. This is because we consider that the psychological processes discussed below apply equally to positive and negative responses to placebo treatment. Attention now turns to processes that could bias responses to placebo treatment for pain.

DEMAND CHARACTERISTICS

Demand characteristics refer to the possibility that participants and patients may modify their responses in order to conform to their expectancies about the aims of the study or treatment they are taking part in.[8] This can lead to bias when evaluating the effects of a treatment (placebo or otherwise) if the demand characteristics differ between the two groups being investigated. For example, in an open-label trial of a new analgesic, participants receiving treatment would be much more likely to believe that the researchers are expecting them to improve compared with a no-treatment control group. If these demand characteristics encourage participants receiving treatment to report more improvement than those in the control group, then the trial may provide a biased estimate of the true efficacy of the treatment. The same applies to placebo treatment. If a participant receiving placebo treatment is inclined to report more improvement simply because that is what they believe their physician hopes to achieve, then an improvement in the patient's reported pain may occur in the absence of any genuine improvement, and this bias would lead to an overestimation of the placebo effect.

An early demonstration of demand characteristics comes from a study by Orne and Scheibe.[9] In their study, healthy volunteers were led to believe that they were taking part in an experiment on a form of meaning deprivation. Two groups of participants were asked to sit alone in a chamber with a one-way researcher-to-participant mirror for an undisclosed amount of time. In the experimental group, various props were included in order to give the participants the impression that the researchers expected the deprivation would lead to distress and cognitive impairment. As part of this participants in this group were shown a tray of emergency medications and had a panic button which they could press to end the experiment. Participants in the control group were not exposed to these props, but were simply told to knock on the one-way mirror if they wanted to terminate the experiment. Interestingly, participants in the experimental group reported more deprivation symptoms and had poorer cognitive performance on some tests than those in the control group. Given that the actual deprivation was identical across groups, this suggests that increased deprivation symptoms in the experimental group were attributable to the demand characteristics created by the props.

While the study of Orne and Scheibe[9] is considered a classic demonstration of demand characteristics, it is worth evaluating whether or not such an effect should only be considered as a form of bias. The authors themselves proposed that expectancy could account for the increased deprivation symptoms in the experimental group. That is, exposure to the additional cues in the experimental group could have caused that group to expect the deprivation to be more aversive and these expectancies may have caused a genuine increase in deprivation symptoms, either directly or indirectly via emotion change. This would be distinct from the participants in the experimental group simply reporting more deprivation symptoms because they believed that is what the researchers were hoping to find. Thus, the extent to which contextual cues and the demand characteristics they create should be considered a source of bias may depend on the type of effects they create. In the context of the placebo treatment, demand characteristics that produce genuine changes as a result of the expectancies they activate could be considered as an important part of the placebo effect. Demand characteristics that lead participants to adjust their responses, even in the absence of any real change, however, remain a source of bias that contributes to the observed response to a treatment and that could lead to over- or underestimation of the placebo effect.

Given this distinction, it is important to determine the extent to which demand characteristics can bias outcomes in studies on pain, as these studies generally rely on subjective ratings of painful stimuli. Relatively few studies appear to have addressed this question.[10–12] However, those that do exist suggest that pain ratings are amenable to demand characteristics. For example, Roche and colleagues[12] randomized participants undergoing cold pressor-induced pain to receive training in acceptance-based or control-based strategies for dealing with pain and then re-randomized them to either high or low demand characteristics. In this case, high demand characteristics involved close contact with the experimenter and statements such as: 'I need you to do your best for me'. They found a main effect of demand characteristics with participants in the high demand characteristics groups keeping their hand submerged in the cold water for longer than those in the low demand characteristics groups. This effect also appeared to interact with type of treatment, in that the largest difference between high and low demand was in the acceptance-based training groups.

Demand characteristics are a very real possibility when administering placebo treatment. This is because almost all placebo interventions involve clear (albeit often deceptive) suggestion about the aim of the intervention. That is, generally when placebo treatment is administered, there are very clear cues indicating that the clinician or researcher is hoping or expecting the patient or participant to improve and this might encourage reporting of improvements, even in the absence of any real improvement.

At least two studies suggest that demand characteristics can influence pain ratings following placebo administration.[10,11] In these studies, signal detection theory was used to investigate the extent to which placebo administration affected pain ratings and pain sensitivity, separately. Signal detection theory is a method of differentiating a person's ability to discriminate the presence and absence of a stimulus (or different stimulus intensities) from the criterion the person uses to make responses to those stimuli.[13] It can be used to provide an estimate of pain sensitivity, that is, the actual experience of pain including the ability to discriminate between painful stimuli of varying intensities, versus response bias, which reflects the criterion an individual uses for making the behavioral response, i.e. the pain rating. In both studies applying this technique to placebo treatment for pain,[10,11] the placebo treatment reduced pain ratings, but failed to reduce actual pain sensitivity. On this basis the authors argued that the reduction in pain ratings following placebo treatment could be attributable to the demand characteristics created by suggesting that the participants would experience pain relief because, if the placebo treatment actually reduced pain, then both pain ratings and pain sensitivity should have decreased.

These results suggest that demand characteristics could account for a significant proportion of responses to placebo treatment, whether in research or in the clinic. However, it is worth noting that despite being raised as a common potential limitation to research on the placebo effect (and many other areas of psychology), there is fairly limited empirical research on demand characteristics and the circumstances most likely to produce them, especially in non-laboratory settings.[14] Further, in the limited research that does exist, there have been some notable failures of replication. For example, Barabasz and colleagues[15] failed to replicate the study of Orne and Scheibe[9] described above. While such failures of replication could be due to numerous factors, the general lack of empirical research on demand characteristics makes it difficult to estimate the magnitude and reliability of these effects in studies on the placebo effect. Thus, while they are very difficult to rule out, it is possible that demand characteristics are overstated as a potential explanation for observed placebo effects.

THE HAWTHORNE EFFECT

The Hawthorne effect refers to when knowledge of being under observation affects a participant's or patient's behavior/responses. The effect was coined as a result of research conducted from the mid-1920s to the early 1930s at the Hawthorne plant of the Western Electric Factory.[16-19] Textbooks and other sources generally describe this research as a series of experiments aimed at determining the optimum conditions (e.g. lighting, number of breaks, pay structure) for maximizing the output of factory workers with the curious finding that no matter what the conditions were changed to, productivity relative to pre-study baseline increased, even when original baseline conditions were restored following the experimental manipulations.[20] The most common interpretation of this effect is that changes in the workers' productivity were due to the fact that they were under observation, not the actual conditions in which they were working.[20] Mechanisms generally proposed to account for the Hawthorne effect are extra attention, the novelty of the situation, or conformity to the perceived expectations of the researchers.[21]

Given that participants and patients are generally aware that they are under observation when they receive treatment (placebo or otherwise), the Hawthorne effect may have a significant influence on their responses to the treatment. As such, any responses to placebo treatment could occur simply as a result of the participant or patient being under observation, rather than the placebo treatment per se. Some have even suggested that in certain treatment contexts, the Hawthorne effect may be stronger than any genuine placebo effect itself.[22]

However, the Hawthorne effect does not appear to be as robust as is commonly proposed. In fact, the common textbook description and interpretation of the original research at the Hawthorne factory that led to the definition of the effect misrepresents the actual findings.[20,21,23,24] Kompier,[20] for example, conducted a detailed review of the original studies[16-19] and concluded that consistent increases in productivity, regardless of the conditions, are not supported by the data. Further, he argues that even if the research had shown continued increases in productivity, the use of a pre-post design without a control group would limit the validity of the research. A similar analysis is provided by Jones.[23] He analysed the raw output data reported in the original studies[16-19] to determine whether there were any consistent changes in output across the experimental manipulations and concluded that there was slender, if any, evidence for a Hawthorne effect. As such, the existence of a clear Hawthorne effect in the original research appears largely overstated.

Of course, a misrepresentation of the original research does not mean that the Hawthorne effect itself does not exist. It does, however, mean that other evidence is needed in order to establish if and to what extent such effects do occur. To this end, Adair[21] reviewed studies involving explicit controls for the Hawthorne effect, most of which were educational interventions. The control groups attempted to account for 'incidental' components of the intervention, such as extra attention, awareness of being in a study, and the novelty of the situation, which are commonly proposed mechanisms of the Hawthorne effect. Notably, he found only weak evidence that these factors could influence outcomes. Interestingly, however, Adair[21] did argue that if the Hawthorne effect exists, then it is most likely to be a product of the participants' or patients' expectancies about how they should respond, rather than attention or other factors associated with simply being observed. That is, the Hawthorne effect may simply be a case of demand characteristics.

Two studies relevant to placebo effects for pain provide some support for this interpretation. In one, patients undergoing knee arthroscopy were randomized to receive standard pre-surgery information, i.e. with no indication that they were part of a study, or to additional information that they were part of a study on the acceptability of the anesthetic procedure being employed.[25] Participants who received the additional information reported better psychological wellbeing, less knee pain, and fewer adverse effects following the surgery than did those who received the standard information. Because all patients received the same anesthetic, and there was minimal, if any, additional attention given to the experimental group, such factors are unlikely to explain the observed results. Instead, it seems quite possible that participants in the additional information group were responding consistently with the demand characteristics of the scenario.

In the second study, Wolfe and Michaud[26] tracked rheumatoid arthritis patients' disability scores at three points around a sponsored 3-month open-label, phase 4 trial of a US Food and Drug Administration approved treatment: (1) prior to the trial, (2) at the end of the 3 month trial period, and (3) in a survey 8 months following the trial's conclusion, but while the patients were still receiving the study treatment. The critical difference between the latter two assessments was that, at the end of the 3 month trial period, participants knew that they were part of a trial aiming/hoping to demonstrate the treatment's efficacy, whereas the survey 8 months following the trial was not presented as related to the trial itself. Consistent with the notion of demand characteristics influencing outcomes, participants rated their disability as significantly lower at the end of the 3 month trial period compared with the 8 months following this in an apparently unrelated survey, despite the fact that there were no differences in the treatment they were receiving. Of course, a significant limitation to this study is the lack of an appropriate control group. As with the original Hawthorne studies,[16–19] the study of Wolfe and Michaud[26] did not include a group of participants who were unaware that they were under observation with a particular goal. Nonetheless, it points towards demand characteristics as being the most likely factor mediating any effect of knowledge about being under observation.

Overall, then, the evidence for the existence of the Hawthorne effect as a unique phenomenon distinct from other processes, such as demand characteristics, is weak. For the most part, the strength and the prominence of the Hawthorne effect as a potential explanation of responses in situations where participants or patients know they are under observation appears to rest largely upon misrepresentation of the original Hawthorne studies.[20] This suggests that the Hawthorne effect may contribute to the observed response to a treatment much less than is commonly suggested. Nonetheless, the possibility that being under observation creates demand characteristics that influence participants' or patients' behavior/responses means that this knowledge could bias responses following placebo treatment. However, without further research indicating the specific effects of attention or other distinct factors, it may be best to avoid the use of the term 'Hawthorne effect' and, instead, to focus on demand characteristics.

RESPONSE SHIFT

Response shift refers to an internal change in a person's criteria for evaluating the same experience such that even if his/her condition remained the same to an external observer, it would be rated differently by the individual. More formally, Shwartz and Sprangers[27] have defined response shift as 'a change in the meaning of one's self-evaluation of a target construct as a result of: (a) a change in the respondent's internal standards of measurement (i.e. scale recalibration); (b) a change in the respondent's values (i.e. the importance of component domains constituting the target construct) or (c) a redefinition of the target construct (i.e. reconceptualization)' (p. 1532). In relation to placebo treatment for pain, this means that differences in patients' or participants' ratings before and after placebo treatment, may not accurately reflect the actual changes, and this may bias estimates of treatment efficacy in either direction, if the treatment affects the way they evaluate or conceptualize their pain.

Howard[28] provided an early empirical demonstration of response shift in his summary of a series of studies evaluating a communication skills training program aimed at reducing dogmatism in the US Air Force. In the first study,[29] a peculiar effect was observed in which there was an apparent increase in dogmatism from before the training to after the training. However, discussions with the trainees revealed that the most likely reason for the higher ratings of dogmatism post-intervention compared with pre-intervention was that the trainees had changed their perceptions of what constitutes dogmatism as a result of the training. In order to investigate this possibility further, Howard et al[29] developed the Then-Test, which aimed to assess pre-intervention levels of dogmatism retrospectively at the same time post-intervention levels were being assessed to ensure that both were being evaluated by individuals according to the same criteria. They then compared traditional Pre-Post ratings with Then-Post ratings in participants who underwent the communication training program. Interestingly, they found that reduced dogmatism was evident in the Then-Post ratings, but not in the Pre-Post ratings. Importantly, in a post-intervention memory test, participants demonstrated almost perfect memory for their pre-intervention ratings, despite the fact that these ratings differed significantly from their Then-ratings. This suggests that the difference between Pre-Post and Then-Post ratings is unlikely to be attributable to poor recall of the pre-intervention state. Instead, Howard[28] proposed that the training had increased participants' understanding of dogmatism and this increased understanding affected how they evaluated their own dogmatism. Specifically, it caused them to realize that their pre-intervention behavior was more dogmatic than they had previously realized.

More recently, the majority of research into response shift has been to do with assessing quality of life.[30-33] For example, Ring and colleagues[33] investigated response shift in quality of life ratings in edentulous patients receiving high-quality conventional dentures. They assessed quality of life before treatment and then 3 months following treatment using both a standard Pre-Post test and a Then-Post test. While there was no significant difference when comparing Pre-Post scores, the Then-Post test revealed significant improvements in quality of life following treatment. This pattern was as a result of Then-ratings of quality of life being significantly lower than Pre-ratings. The authors attributed this response shift to a re-conceptualization and reprioritization of quality of life as a result of receiving the dentures.

In addition to evidence of response shift in ratings of quality of life, there is also direct evidence of response shift in pain ratings. For example, Razmjou and colleagues[34] examined self-reported pain before and after total knee arthroplasty in patients with degenerative arthritis. Patients completed a Pre-test before their surgery and then both a Post-test and Then-test 6 months following treatment. There was a significant response shift in that patients rated their pain before surgery as higher in the Then-test compared with the Pre-test. As a result, the Then-test indicated a larger improvement as a result of the surgery compared with the Pre-test.

In terms of the placebo effect, these studies suggest that response shift may lead to an underestimation of the efficacy of a placebo treatment, if the treatment leads patients to use higher criteria for rating their health. While this is a real possibility, response shifts in the opposite direction are also possible. Paterson[35] interviewed patients receiving acupuncture for a variety of conditions; they were asked to complete questionnaires on health status three times over a 6-month period. Several of the 23 participants demonstrated a response shift in varying directions. Consistent with the studies described above, one participant reported that having experienced a successful treatment increased her standards for assessing her health. However, another participant reported having rated her daily function in relationship to her inability to do paid work at baseline, but then rating it in relation to her ability to do household tasks 6 months later. Thus, for this participant, there was a substantial increase in reported daily function, but this was entirely attributable to employing different benchmarks across testing times.

It is worth noting here that, despite being a potential source of bias when testing the efficacy of a treatment, response shift could be an important coping mechanism that allows the individual to adapt to different circumstances. This particularly applies to cases when the individual's health status deteriorates. For example, an individual might believe that he or she could never experience a satisfactory quality of life if confined to a wheelchair. However, if that individual experiences a spinal injury resulting in paraplegia, then reconceptualizing and recalibrating his or her criteria in such a way that satisfactory quality of life can still be achieved even with the injury, then this type of response shift will facilitate adjustment to his/her new circumstances.

Nonetheless, one of the key concerns for research on the placebo effect is that response shift could affect participants systematically depending on their treatment allocation. Participants allocated to receive placebo treatment may experience a response shift that leads them to use higher criteria for assessing their pain, say, compared with a no-treatment control group whose criteria remain the same. If this were the case, then the estimate of the placebo effect derived from Pre-Post ratings would underestimate the true placebo effect. Conversely, participants allocated to a no-treatment control group may be disappointed that they are not receiving treatment and this may lead them to reduce their standards for assessing their pain. In this case, the Pre-Post ratings would overestimate the true placebo effect.

Based on the available evidence, it is difficult to estimate the magnitude of response shift in studies on the placebo effect, because, to our knowledge, these studies generally do not incorporate Then-tests. However, as with demand characteristics, response shift effects may be less reliable than is commonly suggested. In a recent meta-analysis of 19 studies, Schwartz and colleagues[36] found only a small effect size when averaging response shift across five health outcomes. Most relevant to the current chapter, the effect size for pain was almost zero, suggesting that response shift effects may be minimal. Further, given that response shift occurs as a result of changes in internal standards for evaluating a symptom, it is much more likely to bias outcomes in studies with more than a day or two in between pre- and post-treatment assessments. It would be hard to imagine, for example, that participants in single-session experimental studies on placebo analgesia with intermixed placebo and control trials[37-41] are undergoing constant response shifts back and forth in line with the trial sequence.

RETURNING TO THE THEORETICAL MODEL

Reviewing the evidence for demand characteristics, the Hawthorne effect, and response shift in the context of the placebo effects, demonstrate that there are a number of factors that could contribute to the observed response to placebo treatment, as shown in Figure 17.1. The demand characteristics of the study or treatment context may create expectancies for certain outcomes that produce a genuine placebo effect. However, they may also lead to biased reporting—whether conscious or not—such that the observed treatment response does not accurately reflect the magnitude of the placebo effect. While the original evidence on which the Hawthorne effect is based seems less consistent than it is generally reported, participation in a study or treatment could also influence the observed response in the absence of a genuine placebo effect, if it creates biased reporting or induces behavioral change that would not be maintained if the participant or patient was not under observation. Response shift may bias the observed treatment response if participants' or patients' criteria for evaluating their symptoms change for pre- to post-intervention such that the observed response over- or underestimates the actual placebo effect.

IMPORTANCE OF OBJECTIVE OUTCOMES

Clearly, delineating the contribution of these psychological processes and of expectancy to the observed treatment response is difficult. However, from the above analysis, it is clear that most of the bias that can occur when evaluating a placebo intervention is associated with self-report. Thus, one obvious way to significantly reduce the possibility of bias is to incorporate objective outcomes. While many symptoms of health and illness are difficult to assess objectively, advances in technology for assessing psychophysiologic functioning are providing increasing opportunities to do so, such as fMRI and EEG. In the last decade, these approaches have been employed increasingly in research on the placebo effect with encouraging results. For example, self-reported placebo analgesia is associated with a reduced P2 component in EEG studies of placebo treatment for laser-evoked potentials, with the magnitude of P2 evoked response itself found to correlate with the intensity of painful stimulation.[42] Similarly, in an fMRI study, Bingel and colleagues[43] found that instruction-induced modulation of the analgesic remifentanil was associated with changes in activity of pain and opioid-sensitive brain regions consistent with the direction of pain modulation. While correlational only, such studies are important for providing objective evidence of placebo-induced analgesia that would be difficult to explain in terms of demand characteristics, the Hawthorne effect, or response shift. As such, wherever possible, future research on the placebo effect should incorporate objective outcomes. In the absence of these, it is difficult to rule out any effect of demand characteristics, the Hawthorne effect, or response shift on reported changes following placebo treatment.

CONCLUSIONS AND FUTURE DIRECTIONS

Demand characteristics, the Hawthorne effect, and response shift are commonly cited mechanisms that could bias estimates of the placebo effect. While reviewing the evidence for these psychological processes suggests that they are very difficult to rule out as potential sources of bias in placebo trials assessing self-reported outcomes, such as pain, it is also clear that there is a general lack of empirical evidence on the reliability and characteristics of these effects. There are a number of ways in which future research could address this problem. Signal detection theory could be incorporated into experimental studies on placebo analgesia in which participants rate a series of painful stimuli with and without placebo. In terms of the Hawthorne effect, direct comparisons of the effects of observation alone versus observation with additional attention and support would help to delineate its mechanisms and the extent to which it should be considered distinct from demand characteristics. In both research and clinical practice involving pre- and post-treatment pain assessments, Then-tests could be used in order to attempt to investigate any response shift. A better understanding of these social processes would facilitate estimating the magnitude of the placebo effect and how it may be used to enhance treatment outcomes. In addition, where possible, researchers should incorporate objective outcomes when investigating the placebo effect, as these may reduce the potential influence of demand characteristics, the Hawthorne effect, and response shift.

References

1. Mitchell CJ, De Houwer J, Lovibond PF. The propositional nature of human associative learning. *Behav Brain Sci.* 2009;32(02):183-198.
2. Lovibond PF, Shanks DR. The role of awareness in Pavlovian conditioning: empirical evidence and theoretical implications. *J Exp Psychol Anim Behav Process.* 2002;28:3-26.
3. Kirsch I. Response expectancy as a determinant of experience and behaviour. *Am Psychol.* 1985;40(11):1189-1202.
4. Stewart-Williams S. The placebo puzzle: Putting together the pieces. *Health Psychol.* 2004;23(2):198-206.
5. Moerman DEP, Jonas WBMD. Deconstructing the Placebo Effect and Finding the Meaning Response. *Ann Intern Med.* 2002;136(6):471-476.
6. Di Blasi Z, Harkness E, Ernst E, et al. Influence of context effects on health outcomes: a systematic review. *Lancet.* 2001;357(9258):757-762.
7. Louhiala P, Puustinen R. Rethinking the placebo effect. *Med Humanit.* 2008;34(2):107-109.
8. Orne MT. On the social psychology of the psychological experiment: with particular reference to demand characteristics and their implications. *Am Psychol.* 1962;17(11):776-783.
9. Orne MT, Scheibe KE. The contribution of nondeprivation factors in the production of sensory deprivation effects: the psychology of the 'panic button'. *J Abnorm Social Psychol.* 1964;68(1):3-12.
10. Clark WC. Sensory-decision theory analysis of the placebo effect on the criterion for pain and thermal sensitivity (d'). *J Abnorm Psychol.* 1969;74(3):363-371.
11. Fernandez E, Turk DC. Demand characteristics unerlying differential ratings of sensory versus affective components of pain. *J Behav Med.* 1994;17(4):375-390.
12. Roche B, Forsyth JP, Maher E. The impact of demand characteristics on brief acceptancy- and control-based interventions for pain tolerance. *Cogn Behav Practice.* 2007;14:381-393.
13. Lloyd MA, Appel JB. Signal detection theory and the psychophysics of pain: an introduction and review. *Psychosom Med.* 1976;38(2):79-94.
14. McCambridge J, de Bruin M, Witton J. The effects of demand characteristics on research participant behaviours in non-laboratory settings: a systematic review. *PLoS One.* 2012;7(6):e39116.
15. Barabasz M, Barabasz A, O'neill M. Effects of experimental context, demand characteristics, and situational cues: new data. *Percept Mot Skills.* 1991;73:83-92.
16. Mayo E. *The Human Problems of an Industrial Civilization.* New York: MacMillan; 1933.
17. Whitehead TN. *The Industrial Worker: a Statistical Study of Human Relations in a Group of Manual Workers.* Cambridge: Harvard University Press; 1938.
18. Roethlisberger FJ, Dickson WJ. *Management and the Worker.* Cambridge: Harvard University Press; 1939.
19. Roethlisberger FJ. *Management and Morale.* Cambridge: Harvard University Press; 1941.
20. Kompier MAJ. The 'Hawthorne effect' is a myth, but what keeps the story going? *Scand J Work Environ Health.* 2006;32(5):402-412.
21. Adair JG. The Hawthorne effect: a reconsideration of the methodological artifact. *J Appl Psychol.* 1984;78:413-432.
22. Berthelot J-M, Le Goff B, Maugars Y. The Hawthorne effect: stonger than the placebo effect? *Joint Bone Spine.* 2011;78:335-336.
23. Jones SRG. Was there a Hawthorne effect? *Amercian J Sociol.* 1992;98(3):451-468.
24. Merrett F. Reflections on the Hawthorne effect. *Educat Psychol.* 2006;26(1):143-146.
25. De Amici D, Klersy C, Ramajoli F, et al. Impact of the Hawthorne Effect in a longitudinal clinical study: The case of anesthesia. *Control Clin Trials.* 2000;21(2):103-114.
26. Wolfe F, Michaud K. The Hawthorne effect, sponsored trials, and the overestimation of treatment effectiveness. *J Rheumatol.* 2010;37(11):2216-2220.
27. Schwartz CE, Sprangers MAG. Methodological approaches for assessing response shift in longitudinal health-related quality-of-life research. *Soc Sci Med.* 1999;48(11):1531-1548.
28. Howard GS. Response-shift bias: a problem in evaluating interventions with pre/post self-reports. *Eval Rev.* 1980;4(1):93-106.
29. Howard GS, Ralph KM, Gulanick NA, et al. Internal invalidity in pretest-posttes self-report evaluations and a reevaluation of retrospective pretests. *Appl Psychol Measure.* 1979;3:1-23.
30. Ahmed SE, Mayo N, Wood-Dauphinee S, et al. Response shift influenced estimates of change in health-related quality of life post-stroke. *J Clin Epidemiol.* 2004;57(6):561-570.
31. Hagedoorn M, Sneeuw KCA, Aaronson NK. Changes in physical functioning and quality of life in patients with cancer: response shift and relative evaluation of one's condition. *J Clin Epidemiol.* 2002;55(2):176-183.
32. Jansen SJT, Stiggelbout AM, Nooij MA, et al. Response shift in quality of life measurement in early-stage breast cancer patients undergoing radiotherapy. *Qual Life Res.* 2000;9(6):603-615.
33. Ring L, Hofer S, Heuston F, et al. Response shift masks the treatment impact on patient reported outcomes (PROs): the example of individual quality of life in edentulous patients. *Health Qual Life Outc.* 2005;3:55.
34. Razmjou H, Yee A, Ford M, Finkelstein JA. Response shift in outcome assessment in patients undergoing total knee arthroplasty. *J Bone Joint Surg Am.* 2006;88(12):2590-2595.
35. Paterson C. Seeking the patient's perspective: a qualitative assessment of EuroQol, COOP-WONCA charts and MYMOP. *Qual Life Res.* 2012;13(5):871-881.
36. Schwartz CE, Bode R, Repucci N, et al. The clinical significance of adaptation to changing health: a meta-analysis of response shift. *Qual Life Res.* 2006;15(9):1533-1550.
37. Colloca L, Benedetti F. How prior experience shapes placebo analgesia. *Pain.* 2006;124(1-2):126-133.
38. Voudouris NJ, Peck CL, Coleman G. The role of conditioning and verbal expectancy in the placebo response. *Pain.* 1990;43(1):121-128.
39. Voudouris NJ, Peck CL, Coleman G. Conditioned response models of placebo phenomena: Further support. *Pain.* 1989;38(1):109-116.
40. Voudouris NJ, Peck CL, Coleman G. Conditioned placebo responses. *J Pers Soc Psychol.* 1985;48(1):47-53.
41. Montgomery GH, Kirsch I. Classical conditioning and the placebo effect. *Pain.* 1997;72(1-2):107-113.
42. Wager TD, Matre D, Casey KL. Placebo effects in laser-evoked pain potentials. *Brain Behav Immun.* 2006;20(3):219-230.
43. Bingel U, Wanigasekera V, Wiech K, et al. The effect of treatment expectation on drug efficacy: imaging the analgesic benefit of the opioid remifentanil. *Sci Transl Med.* 2011;3(70):70ra14.

CHAPTER 18

Against 'Placebo.' The Case for Changing our Language, and for the Meaning Response

Daniel E. Moerman

William E. Stirton Professor Emeritus of Anthropology, University of Michigan-Dearborn, MI, USA

A SUMMARY OF THE ARGUMENT

First, I will provide a quick summary of the argument, which will be followed by a brief review of the relevant data. For many years I have argued that researchers should abandon the phrase 'placebo effect' (or 'placebo response') and replace it with something else; I have proposed 'meaning response' because I believe that it points directly to the mechanisms involved.[1-4] Others, for similar reasons, propose 'context effect'[5] or 'expectation effect'[6] or other similar terms. I do this for two primary reasons. First, a placebo (a perfectly legitimate concept) is an inert substance (a sugar or starch pill, a saline injection) used in place of, or next to, an ordinary drug. By 'inert', we mean a substance that 'doesn't do anything,' that has no effect on human physiology. Given that, the term 'placebo effect/response' is an oxymoron; nothing can't do anything.

Second, however, we all know that after administering inert medications, things often do happen. People often get better, or their pain reports moderate, and they often even show 'side effects.' Given the definition above, there is one thing we can be absolutely certain of, which is that whatever happened was *not* due to the placebo, which was inert (unless we had an incompetent pharmacist). I argue instead that ordinarily such effects are due to the 'meanings' of the placebo, to the understandings of physicians, patients, families, communities, about such drugs/surgeries/treatments, etc. The colors of placebos make a difference, as does their number and administration; as does the cultural context of the care given. Most important, a range of recent studies have shown the same sorts of 'meaning responses' attached to real drugs as, for example, in the various open/hidden experiments of the past few years. In cases where fentanyl works better when administered by a clinician, than when given secretly (via intravenous line), we clearly cannot attribute anything to a 'placebo effect' because there were no placebos given at all to anyone.[7] But we can easily attribute such differences to human interaction, to language, to caring, whatever seems plausible given the details of the situation. It is important to recognize that meaning responses can occur across the breadth of medicine: internal, surgical, manipulative, linguistic. Howard Brody has several times written about the 'meaning model' of medicine, which is quite similar to my approach.[8,9]

Let me be more explicit why I prefer the notion of the meaning response to, first, expectation, and second, conditioning. Expectation is probably a universal human psychologic phenomenon. One anticipates that someone you trust will do what she or he says, or that someone you know well will ordinarily act the same way under certain circumstances that you have observed before. In this, conditioning and expectation are, perhaps, two sides of a coin. The facts at issue—truthful, or customary, behavior—lead you to anticipate something in particular, and behave accordingly yourself, by, say, producing endorphins or some other neurochemical. In these cases, the prior experience with these people acts as the conditioned stimulus, and the subsequent behaviors, the expected ones, can be considered conditioned responses. All of these processes may occur in any human relationships. However, as an anthropologist, I am confident that quite different (culturally constructed) experiences might yield the same, or different, responses in others. And, human behaviors are rarely as simple and unadorned as might be the food given to that rat which then shakes her tail. The shaman might enact a complex expression of a (presumably) ancient myth to evoke the healing he seeks in his patient; and everyone in the room—the shaman, the patient, the families and friends, everyone in town—knows the story in advance, having heard it dozens of times on their mother's knees. Hence it is practically impossible for human beings to avoid a substantial cognitive, that is, meaningful, dimension in any healing rite. The rites will perforce be different; the induction of expectations may be the same, or at least similar. A shaman sucking an evil spirit out of a sick patient makes perfect sense to a Navajo or Eskimo;[10,11] while to a native of Kansas or Alsace, it might seem absurd, and would likely create highly negative responses. While expectations may be involved in all of these places, what evokes them may vary dramatically, and needs to be understood in local terms. I don't deny the psychologic processes, but I am more interested in the 'what' than the 'how.'

A significant implication of this approach is that 'meaning' (or if you prefer, context, or expectation) is clearly not nothing; and it's the nothingness of placebos that causes the huge ethical dilemma of their use. We know that placebos continue to 'work' when they are identified as inert substances; but most believe that placebos must be utilized with some sort of deception. 'Care,' 'language,' 'talk,' 'touch,' need not be used deceptively (indeed it's hard to imagine that they could be).

A BRIEF REVIEW OF THE DATA

Color and Number

The color of pills can make a difference in their effectiveness: in a study done in the Netherlands, red, yellow, and orange pills were perceived to work better as stimulants while blue and green pills were perceived to work better as tranquilizers. The same study showed that of 49 drugs available for treating the nervous system, sedatives were more often green, blue or purple, while stimulants were more often red or orange.[12] In an Italian study, women responded better to blue sleeping tablets than did men, who preferred orange ones.[13] I have recently provided an extensive analysis of this situation elsewhere.[4]

In a study of a large number of clinical trials of anti-ulcer drugs, I have shown that, in 51 trials, patients took four tablets a day (drug or placebo) while in another 28 trials, patients took two pills a day. After 4 weeks, 44% of the four-per-day placebo patients (805 of 1821) showed healed ulcers (on endoscopy) while only 36% of the two-per-day patients (545 of 1504) were healed ($\mu^2 = 21.7$; $p = 0.000$).[14]

Note that a blue nothing should not work better than a red nothing. But we all know that red is different from blue, and has a different range of meanings. And note that $2 \times 0 = 4 \times 0$. That is, four nothings should do no more than two nothings. But everybody knows that 4 means more than 2.

Form

When sumatriptan first became available for the treatment of migraine, it was available only as a subcutaneous injection. Later, it became available as an oral tablet. Ton De Craen and his colleagues have shown that, in a series of trials, 32% of patients taking placebo injections were better after an hour (279 of 862) while 26% of patients taking placebo tablets were better after an hour (222 of 865) a '6.7% difference, 95% CI 2.4–11.0%).'[15] A placebo injection is zero, and a placebo tablet is zero; yet apparently one zero is larger than the other, because this difference, although clinically modest, is highly statistically significant. And everybody knows that injections are bigger, and more powerful, than pills.[16]

Except when they are not. De Craen reported separately the results of European and US trials, but did not compare the differences in drug and placebo outcomes in the two places. The overall difference in the US was 33.6% for sham injection and 22.3% for sham pill, a difference of 11.3%, larger than the overall difference. This difference disappears in the European studies where 27.1% of placebo tablet patients were cured and 25.1% of placebo injection patients were cured. So, injections work better than pills, but only in the United States, not in Europe. That is, there are cultural factors at work here. Researchers regularly note that American medicine is more aggressive than European medicine.[17] And it appears that aggressive medicine works better in the US than in Europe. Even when it's inert.

Culture

In a large review of over 100 controlled trials of H2-receptor-antagonists for peptic ulcer, there were dramatic differences in the number of (endoscopically verified) healed ulcers in placebo-treated patients; the number averaged about 32% but varied from zero to 100%. The highest control group healing rates were in Germany, over 60%. The lowest were in Brazil, at about 6%. There were significant differences between control group healing rates in Germany and their northern neighbors, the Dutch and Danes, where rates were about 20%. At the same time, the German control group improvement rates for anti-hypertensive treatment were among the lowest in the world, with a small increase in blood pressure noted. So, these factors are not generic ones addressing German genes, but cultural ones, affecting their (and everyone else's) ideas and understandings of particular diseases and what they mean.[18]

Hype

There is a lot of advertising of medicines and drugs in western cultures; although direct-to-consumer advertising (DTCA) is rare for prescription medicines (only the US and New Zealand see a lot of it), over the counter (OTC) medicines are widely advertised, and, of course, the bulk of advertising is to doctors, which seems to be common everywhere. Drug company 'detail men' are, today, often fetchingly attractive young women, expensively dressed, carrying drugs packaged more like cosmetics, and bringing the staff a nice lunch.[19] Medical conventions are often more or less fully paid for by drug and device manufacturers who put up lavish displays of their wares, with champagne at 3:00. There is good evidence to show that such detailing and advertising increases the intention to prescribe, even for off-label uses (such detailing is illegal in the US).[20] There is some evidence (called 'weak evidence') that 'DTCA may increase compliance and improve clinical outcomes.'[21] Although little studied, there is some evidence to show increased effectiveness of OTC medicines which have widely recognized brand names. In one such study, a widely advertised brand of aspirin was shown to be more effective at relieving tension headaches than the same aspirin without a brand name, and, placebo with the same brand name was more effective than placebo without the brand name.[22] (Aspirin was more effective than placebo, brand or not.) The source of what people know, understand, or believe, is irrelevant: knowledge shapes the effectiveness of medical treatment.

Adherence

It is not unusual for drug trials to show that patients who take all the prescribed medicine do better than those who are less compliant with their medical instructions. In a recent re-analysis of a study called Beta Blocker Evaluation of Survival Trial (BEST), the author confirmed what had been noticed previously: that those who took more than 75% of their prescribed *placebos* did significantly better than those who were less compliant with their placebos. This was clearly true in the BEST study where 28% of highly adherent placebo patients died while 42% of less adherent placebo patients died.[23]

Similar results have been found in a number of trials dealing with a broad range of illnesses, among them infections incidental to cancer, heart attack in both women and men, and HIV infection in Zambia. A meta-analysis of 21 trials had similar conclusions.[24] An array of studies have attempted to find what life-style factors or other issues might account for this, but none have been found.

This is very hard to understand and even harder to study because one cannot randomize people to being adherent and not (just as one can't randomize people to being tall or short, witty or dull, male or female). There is a fairly widespread speculation that the high adherers are representative of an overall 'healthy behavior group' which includes good adherers. But one might imagine that the healthy behavior folks would eat better, smoke less, etc., and no one has been able to show that (or anything like it) for the placebo-treated patients in all those trials; nor for the adherent drug-treated patients (who also routinely do better in trials than the less adherent). Because this is such a mysterious—even if well known—phenomenon, I have no particular opinion if or how meaning plays a role in it.

History

In 2002, Walsh and colleagues showed that over a 19-year period from January 1981 through December 2000, a series of 75 controlled trials of treatment for depression had been done. Over that period, the average proportion of patients who responded to medication was 50.1%; the proportion of patients who responded to placebo was 29.7%. However, there was a strong correlation of both placebo improvement and drug improvement with the date of publication of the study. Over the two decades, the 'proportion responding to placebo has clearly increased, at a rate of approximately 7% per decade, and a similar increase has occurred in the fraction of patients responding to active medication'[25] (p. 1844). They concluded that the placebo healing rate for depression was 'variable, substantial, and growing.' They add: 'Some factor or factors associated with the level of placebo response must therefore have changed significantly during this period. Unfortunately, we were not able to determine the identity of these factors.' This seems a reasonably straightforward issue: over the period studied, there were major changes in how physicians, patients, families, indeed everyone, thought about depression. In 1970, one of the standard pharmacology textbooks stated that no drug was more effective than electroconvulsive therapy (ECT) for depression.[26] Subsequently, following on 'One Flew Over the Cuckoo's Nest' (1975), 'Listening to Prozac' (1994), plus innumerable other discussions, TV shows, Oprah, etc., 'everyone' agreed that depression could be treated medically with pills. The legalization of direct-to-consumer drug advertising in the USA in 1997 probably contributed. Medication becomes meaningful in historical fashion.

This situation, drug and placebo effectiveness varying through time, has been shown several other times, to one degree or another. The same authors who did the depression study did a similar study for bipolar disorder; they found only a few studies using a variety of outcome measures and report a weak trend of increased placebo effectiveness (from 1991 to 2005).[27] A comparative analysis of

20 trials between 1997 and 2008 showed that remission rate of Crohn's disease increased with year of publication, about 8% per year; however, the authors argue that the length of time till evaluation increased over the years, and they attribute the increased response to this. That is, people responded to increased amounts of inert treatment. A study of neuropathic pain found that placebo treatment increased over time (1996 to 2006), but that the difference was due to increased length of the trials over that period.[28] Again, people responded to increased amounts of inert treatment.

A series of acupuncture trials shows an increase in placebo effectiveness from 1988 to 2010 while treatment effect remained more or less constant, resulting in a lowering of overall specific effect (treatment minus placebo effect).[29]

A number of recent studies, reviewed here,[30] raise particularly troubling issues regarding temporal changes in treatments of schizophrenia. In effect, it appears that over the past 20 years, placebo-treated patients in trials of drugs for this pernicious disease have shown significant improvement from only a few points (on the PANSS scale) to as much at 20 points (a clinically significant amount). At the same time, the effectiveness of a succession of new drugs has declined by a few points. For a clearer presentation of these changes (if a less convincing discussion, see Alphs et al.[31]). The drugs considered here are often very toxic, with debilitating effects trailing along with their positive one. Interestingly, the control group patients often experience the same debilitating side effects that the drug patients experience, but almost always in smaller frequencies. We may wish to interpret this such that the control group patients, with a smaller proportion of the side effects, get a smaller portion of the drug effects, but enough that they can stick longer with the treatment than those with the full load of negative effects (in the studies compared, nearly half the patients dropped out[30]). It seems odd, but it's often the case that control group patients get the same side effects as the drug group gets. Note that this, too, is an odd use of language. Drugs have effects, some desirable, some not. When meaning creates a biologic response, it shouldn't seem surprising that all the responses of drug-treated patients occur, the good along with the bad. This may be what's happening in the trials of these lamented patients.

Surgery

All the cases considered so far have dealt with internal medicine. These often seem familiar and unsurprising. Placebo responses seem more plausible in such cases; they seem much more surprising in surgical care. Indeed, controlled trials of surgical procedures are very rare; since drugs are used largely for empirical reasons (we have learned from experience that willow leaf tea is effective in treating headaches), while surgical procedures are done for logical reasons (the skin is torn, he is bleeding, sew it up), it seems more reasonable to test the former than the latter. But when such tests of compellingly logical procedures are done, the results are often quite surprising. One famous case from the 1950s involved ligation of the internal mammary artery to ease angina pectoris, a procedure that was becoming quite popular; presumably the ligation increased the supply of blood to the coronary arteries by restricting it elsewhere (think what happens to the water going into the sink when you flush the toilet).[32] Two surgeons who had doubts organized small studies with patients randomized to receive the ligation, or only the skin incision. Patients reported significantly less pain, longer exercise times and reduced use of nitroglycerine, regardless of which group they were in.[33,34] A measure of the impact of these studies is that Google Scholar reports 420 citations of Cobb's paper (now up to 421).

More recently, orthopedic surgeon Bruce Moseley carried out a controlled trial of treatments for osteoarthritis of the knee; patients received arthroscopic debridement, arthroscopic lavage, or a simulated debridement, with only three small stab wounds in the knee. After 2 years, all patients had experienced the same modest improvement of about 5 points on a 100 point scale measuring walking and bending.[35]

It is common for older women to suffer painful stress fractures of vertebrae due to osteoporosis. In a recently developed procedure, clinicians, guided by fluoroscopy, inject medical cement (polymethylmethacrylate) into the fractures. Two recent studies compared this procedure with a sham procedure where no glue was injected into the fracture. Sham and verum groups had very nearly identical responses, with modest improvements in pain scores, physical functioning, quality of life, and a mix of other measures.[36,37] These trials elicited a good bit of comment, pro and con.[38–40]

Surgeons are the kings of medicine with the grandest reputations, and, often enough, the swagger to match. 'Self-effacing surgeon' seems almost an oxymoron. There is a substantial literature (which I reviewed here[41] (Chapter 4)) indicating that one of the most significant elements in enhancing the meaning response is the physician him/herself, encouraging us to think (logically) that much of the effectiveness of surgery is akin to the meaning response to drugs.

Genetics

For many years researchers have tried to determine before the fact what characteristics of patients led to meaning responses, testing every conceivable psychological, developmental, physiological characters that they could think of; all these studies failed to show effective results, or they were quickly falsified by other studies.

In a recent study, T. Kaptchuk's group at Harvard showed that reduction of symptoms of irritable bowel syndrome was maximized in patients with met/met homozygotes for the COMT gene; this gene plays an important role in dopamine catabolism which in turn is associated with reward, pain, memory and learning.[39]

All of those processes, of course, are involved in one way or another in interpreting meaning. By contrast, it seems unlikely that 'nothing' (i.e., a 'placebo') would have a significant effect on dopamine manufacture.

CONCLUSIONS

All (or, perhaps, almost all) of these cases, and many more I could cite, encourage us to banish from our tongues the terms 'placebo response' and 'placebo effect'. Placebos are inert; inert things don't do anything. But things do happen after placebos are administered. Among them are extraneous things like regression to the mean (some people selected for a study because of extreme conditions on some measure—high blood pressure, substantial back or bowel or head pain, serious allergies—will likely revert toward normal in time) or natural history (colds go away by themselves, as do many other conditions). These things are not due to placebos or meaning; they are due to study selection criteria, or to the nature of various illnesses which either wax and wane, or which, for some lucky few, simply wane.

But the most interesting things that happen after people are given placebos are the responses to the meaning of the treatment, the drugs, the surgery, the shape of the room, the paintings on the walls, the colors of the funny clothes they all wear (especially those funny hair nets! They must really really care about me if they are so concerned about the dust in their hair!). The charming (or nasty) parking lot attendant. The enthusiasm of the doctor, of the nurse, of the aide who sweeps up and who talks with me kindly and quietly. The doctor on TV (she's really an actor, but wow, can she figure out nasty diseases! and she's really cute, too.) And the heliports! We have two at our hospital! The plastic spine hanging on the door knob in the chiropractor's office; and the beautiful and antique-looking paintings of meridians at the acupuncturists office, along with all those fat books on Chinese herbs. The smell of rubbing alcohol and the stuff they use to clean the bathrooms which always smells like someone's idea of the out-of-doors. Number, color, and form are more easily quantified than may be plastic spines. But all these things work together to shape and form the dynamic elements of the medical encounter, what the encounter means to me. Nothing changes human beings more than meaning.[42] And, as one doctor put it 'Sometimes simply being silently present with a patient may be the most meaningful kind of care.'[43]

References

1. Moerman DE. Anthropology of symbolic healing. *Curr Anthropol.* 1979;20(1):59-80.
2. Moerman DE. *Meaning, Medicine and the 'Placebo Effect'.* Cambridge: Cambridge University Press; 2002.
3. Moerman DE, Jonas WB. Deconstructing the placebo effect and finding the meaning response. *Ann Intern Med.* 2002;136(6):471-476.
4. Moerman DE. Consciousness, 'symbolic healing', and the meaning response. *Anthropol Consc.* 2012;23(2):192-210.
5. Di Blasi Z, Kleijnen J. Context effects. *Eval Health Profess.* 2003;26(2):166-179.
6. Kirsch I, ed. *How Expectancies Shape Experience.* 1st ed. Washington, DC: American Psychological Association; 1999.
7. Benedetti F, Maggi G, Lopiano L, et al. Open versus hidden medical treatments: the patient's knowledge about a therapy affects the therapy outcome. *Prevent Treat.* 2003;6(1):1a.
8. Brody H, Waters DB. Diagnosis is treatment. *J Fam Practice.* 1980;10(3):445-449.
9. Brody H, Brody D. Three perspectives on the placebo response: expectancy, conditioning, and meaning. *Adv Mind-Body Med.* 2000;16(3):216.
10. Levi-Strauss C. *The Effectiveness of Symbols. Structural Anthropology.* Garden City, NY: Anchor Books; 1967:180-201.
11. Levi-Strauss C. *The Sorcerer and his Magic. Structural Anthropology.* Garden City, NY: Anchor Books; 1967:160-180.
12. de Craen AJ, Roos PJ, Leonard de Vries A, Kleijnen J. Effect of colour of drugs: systematic review of perceived effect of drugs and of their effectiveness. *BMJ.* 1996;313(7072):1624-1626.
13. Cattaneo AD, Lucchilli PE, Filippucci G. Sedative effects of placebo treatment. *Eur J Clin Pharmacol.* 1970;3:43-45.
14. de Craen AJ, Moerman DE, Heisterkamp SH, et al. Placebo effect in the treatment of duodenal ulcer. *Br J Clin Pharmacol.* 1999;48(6):853-860.
15. de Craen AJ, Tijssen JG, de Gans J, Kleijnen J. Placebo effect in the acute treatment of migraine: subcutaneous placebos are better than oral placebos. *J Neurol.* 2000;247(3):183-188.
16. Van der Geest S. The illegal distribution of western medicines in developing countries: pharmacists, drug pedlars, injection doctors and others. *Med Anthropol.* 1982;6(4):197-219.
17. Payer L. *Medicine and Culture: Varieties of Treatment in the United States, England, West Germany, and France.* First Owl Book ed. New York: Henry Holt and Company; 1996.
18. Moerman DE. General medical effectiveness and human biology: placebo effects in the treatment of ulcer disease. *Med Anthropol Quart.* 1982;14(4):13-15.
19. Katz D, Caplan AL, Merz JF. All gifts large and small: toward an understanding of the ethics of pharmaceutical industry gift-giving. *Am J Bioeth.* 2010;10(10):11-17.
20. Steinman MA, Harper GM, Chren MM, et al. Characteristics and impact of drug detailing for gabapentin. *PLoS Med.* 2007;4(4):e134.
21. Atherly A, Rubin PH. The cost-effectiveness of direct-to-consumer advertising for prescription drugs. *Med Care Res Rev: MCRR.* 2009;66(6):639-657.
22. Branthwaite A, Cooper P. Analgesic effects of branding in treatment of headaches. *Br Med J.* 1981;282(6276):1576-1578.
23. Pressman A, Avins AL, Neuhaus J, et al. Adherence to placebo and mortality in the Beta Blocker Evaluation of Survival Trial (BEST). *Contemp Clin Trials.* 2012;33:492-498.
24. Simpson SH, Eurich DT, Majumdar SR, et al. A meta-analysis of the association between adherence to drug therapy and mortality. *BMJ.* 2006;333(7557):15.
25. Walsh BT, Seidman SN, Sysko R, Gould M. Placebo response in studies of major depression: variable, substantial, and growing. *JAMA.* 2002;287(14):1840-1847.

26. Goodman LS, Gilman A. *The Pharmacological Basis of Therapeutics; a Textbook of Pharmacology, Toxicology, and Therapeutics for Physicians and Medical Students*. 4th ed. New York: Macmillan; 1970.
27. Sysko R, Walsh BT. A systematic review of placebo response in studies of bipolar mania. *J Clin Psychiatr*. 2007;68(8):1213-1217.
28. Quessy SN, Rowbotham MC. Placebo response in neuropathic pain trials. *Pain*. 2008;138(3):479-483.
29. We SR, Koog YH, Park MS, Min BI. Placebo effect was influenced by publication year in three-armed acupuncture trials. *Complement Ther Med*. 2012;20(1-2):83-92.
30. Khin NAMY-FC, Yang Yang, Yang Peiling, et al. Exploratory analyses of efficacy data from schizophrenia trials in support of new drug applications submitted to the US Food and Drug Administration. *J Clin Psychiatr*. 2012;73(6):856-864.
31. Alphs L, Benedetti F, Fleischhacker WW, Kane JM. Placebo-related effects in clinical trials in schizophrenia: what is driving this phenomenon and what can be done to minimize it? *Int J Neuropsychopharmacol*. 2012;1(1):1-12.
32. Kitchell JR, Glover RP, Kyle RH. Bilateral internal mammary artery ligation for angina pectoris: preliminary clinical considerations. *Am J Cardiol*. 1958;1:46-50.
33. Cobb L, Thomas GI, Dillard DH, et al. An evaluation of internal-mammary artery ligation by a double blind technic. *N Engl J Med*. 1959;260(22):1115-1118.
34. Dimond EG, Kittle CF, Crockett JE. Comparison of internal mammary ligation and sham operation for angina pectoris. *Am J Cardiol*. 1960;5:483-486.
35. Moseley JBJ, Wray NP, Kuykendall D, et al. Arthroscopic treatment of osteoarthritis of the knee: a prospective, randomized, placebo-controlled trial. Results of a pilot study. *Am J Sports Med*. 1996;24(1):28-34.
36. Buchbinder R, Osborne RH, Ebeling PR, et al. A randomized trial of vertebroplasty for painful osteoporotic vertebral fractures. *N Engl J Med*. 2009;M361(6):557-568.
37. Kallmes DF, Comstock BA, Heagerty PJ, et al. A randomized trial of vertebroplasty for osteoporotic spinal fractures. *N Engl J Med*. 2009;361(6):569-579.
38. Miller FG, Kallmes DF, Buchbinder R. Vertebroplasty and the placebo response. *Radiology*. 2011;259(3):621-625.
39. Noonan P. Randomized vertebroplasty trials: bad news or sham news? *AJNR Am J Neuroradiol*. 2009;30(10):1808-1809.
40. Weinstein JN. Balancing science and informed choice in decisions about vertebroplasty. *N Engl J Med*. 2009;361(6):619-621.
41. Moerman DE. *Meaning, Medicine and the 'Placebo Effect'*. Cambridge: Cambridge University Press; 2002.
42. Polanyi M, Prosch H. *Meaning*. Chicago: University of Chicago Press; 1977.
43. Bazari H. Gratitude, memories, and meaning in medicine. *N Engl J Med*. 2010;363(23):2187-2189.

CHAPTER 19

Placebo Effects in Complementary and Alternative Medicine: The Self-Healing Response

Harald Walach

Institute of Transcultural Health Studies, European University Viadrina, Frankfurt (Oder), Germany

BACKGROUND

A Note on Terminology and History

Complementary and alternative medicine (CAM) has moved from the fringe of scientific and popular interest to a more central place over the last two decades.[1] Whole systems of medicine (WSM),[2–4] such as homeopathy, Ayurveda, and Traditional Chinese medicine (TCM) have been around in their respective countries of origin for centuries as a kind of parallel medical culture. Since Virchow's seminal insights in the mid-19th century, Western biomedicine (BM) became the dominant way of doing medicine, and 'traditional' or 'natural' ways of medicine have moved to the background.[5,6] This is not so in many other countries, for example, in Africa or Asia, where those traditional systems still provide medical care in the majority of cases.[7] For instance, in India, more people are using Ayurveda, homeopathy, Siddhi or Unani medicine, the traditional systems, than can afford to use BM.[8–10] But far from being only 'poor people's medicine', those traditional systems also have something to offer. They have different conceptual approaches to health and disease that translate into different practical applications. Ayurvedic medicine, for instance, is very much a lifestyle and preventative system which uses phytopharmacologic substances only if someone is taken ill; otherwise, it emphasizes dietetic advice.[11,12] If someone is ill, there is the full arsenal of traditional medicine, from purgation to phytopharmacology, to surgery. Indeed, the Sushruta Samhita is probably the world's oldest textbook on surgery, dating back to roughly 600 BC.[13] The same is true for TCM.[14] Often people associate only acupuncture with this type of traditional medicine (TM); on the contrary, it has at its disposal a whole arsenal of phytopharmacologic, dietary and additional techniques such as specific diagnostic systems. While, in their countries of origin, BM is becoming the fashion, exports of these TM systems into the West have become fashionable for our culture. This results in a transcultural transformation of medical systems: what is being transported into the West and applied here is different from what it has been in the country of origin. It has transformed, and will continue to transform, the way BM is approaching health and disease, and our Western patients alike. Out of this amalgamation, new ways of providing medical care are born. Often such integrations are called 'integrative medicine' (IM),[15] which very often means that methods found to be useful or evidence-based in other cultures are being 'integrated' into treating patients from a BM point of view. However, sometimes it also means that different ways of thinking are being applied, involving more holistic concepts of treatment.

This fits with a counterculture, which has likely contributed to the rise of CAM in the West: the diversification, the analytical separation of the body into compartments—and the separation of our lives into disparate units—has led to a desire for holism. TM and CAM systems answer to that. Homeopathy, for instance, needs, in order to be effective, symptoms from all aspects of a patient's life. Not only, say, the headache symptoms that are bothering her—but also information about food preferences, menstruation problems, if there are any, mental and psychologic symptoms and other physical symptoms that might be present. Such a holistic approach, out of one hand, is appealing to patients during times when a patient has to see various specialists who do completely different things and often prescribe incompatible medicines and treatments for the same patient.

Thus, CAM can be seen as a medical counterculture serving the needs of a rather well-off, better educated, ecologically minded and probably politically more progressive group of patients.[16] The rise of CAM from a subculture to a counterculture, and then to a cultural co-player, officially started with the foundation of the Office of Alternative Medicine (OAM) at the NIH in 1991 through a political move.[17] It was bolstered by the biggest poll to date showing that US citizens spend more money out of pocket on CAM than on conventional medicine.[18,19] This piece of information, probably among the most frequently quoted, drove the message home: this is not marginal, this is real, and it is real money. As a sequel, OAM became the National Center for Complementary and Alternative Medicine (NCCAM), with a small but comparatively serious budget of 120–150 million US dollars to be spent on research. European countries, where various natural approaches such as traditional naturopathy, homeopathy, massage and spa treatments are very much part of the medical culture, have not matched this. But also here we can see a move towards institutionalization. Norway has its own national center. In the UK there was a move towards structure building through funding from the government at the beginning of the new millenium. And Europe, as a whole, has finally started to map out CAM provision and treatment and to formulate a research agenda in its first Pan-European research network CAMbrella.[20]

Comparatively early on, CAM researchers saw that the traditional BM model cannot be applied to WSM. The BM model is foremost a pharmacologic model that answers to the implicit research strategy of finding a cause, and then a target for treatment, and then a specific molecule to be targeted by a specific treatment. Its dominant rationale is to uncover such specific effects. What is rarely reflected, though, is this tacit presupposition and whether this is indeed the best approach to treatment. It has frequently been observed that the human organism is a complex system, and that disease can be viewed as a disturbance in the system's homeostatic self-organizing capacity.[3,21–23] While, in acute situations, the organism can get back to its natural balance with a little medical help, or just some time to recalibrate, this is not so easily possible in chronic disturbances. Here, it turns out that the disease itself becomes an active adaptation of a complex network to a new challenge, and disease can be reconceptualized as a network disturbance. Such a disturbed network can be brought back into its original balance by various methods that trigger, goad or even force the system out of its disease trough across and against an energetic barrier of disturbance, to use systems theoretical language, such that it can again activate its self-structuring capacity to re-establish its original or a new, healthier balance. In such a view, the comparatively static way of looking at disease as a disturbance with just one target, stemming from pharmacology, to be treated by a specific molecular agent, is not very useful. It was also very soon realized by CAM researchers, that the so-called nonspecific effects of treatments may well be the major aspects of many, if not all CAM treatments,[24,25] and that, hence, some more scientific study of the placebo effect, that has hitherto led a very secluded life in the litter basket of pharmacologic research, might be in order.

Hence, it was at the OAM that one of the major breakthrough conferences on the placebo topic was convened by its then director, Dr Wayne Jonas in 1995.[26] Lead researchers in the field, such as Dan Moerman, Howard Brody and Irving Kirsch, to name but a few, came together to discuss the scope and propose a potential new terminology. Out came the new definition of placebo as 'meaning-response.'[27] Placebo effects are those effects that happen, because, in an individual therapeutic situation, patients attribute particular types of meaning to a very specific therapeutic action as a whole. Thus, in this reconceptualization, placebo effects are individual. They are only partly predictable, since one would have to know the whole learning and cultural history of an individual to be able to understand how he or she might make meaning out of a situation. They emphasize the semiotic aspect of a treatment, i.e. the meaning that a patient sees in a therapeutic encounter, or generates from it, not necessarily the one it objectively has. And placebo effects are thus, by definition, complex psychologic processes that will involve conscious attribution processes and cognitive aspects such as forming expectations, imbuing actions with meaning, or forming plans for future actions such as health behavior changes. But they will also contain many well understood unconscious processes such as conditioning and learning, and, in addition, also less understood unconscious aspects such as emotional priming, or affect regulation.

In the wake of this meeting, OAM, and later NCCAM, started a research initiative on placebo effects which triggered a hitherto unseen interest in the subject area. Thus, CAM research is, in a way, the midwife to modern day placebo research. While this is surely not true for all research everywhere, a lot of the seminal studies in the field were funded or inspired by this original effort. Since those early days in the 1990s we have begun to understand that placebo effects are 'real.' They happen not only in patients' imagination: their effect can be traced in brain systems that are affected by various processes such as expectancy, and learning, which are part of our cultural and semiotic constitution as human beings existing in relationship with others, interacting through language, and forming mental maps of our environment and events likely to happen[28–35] (see also Ch 16 in this book). These effects may be small in clinical trials, which are explicitly constructed to control for the placebo effect and thus keep it as small as possible.[36] But they can be sizeable—up to a standard deviation—in experimental contexts which more clearly mimic the practical routine of patient–doctor encounters, which is geared towards higher expectations and trust.[37,38] Psychoneuroimmunologic research has uncovered some paths that also make mechanistically plausible how such nonspecific effects might be causal for a whole series of varied therapeutic effects.[39] Practically all major brain-transmitter systems have been shown to be involved in placebo effects. Thus pain reduction, as well as influences on the affective systems such as elevated mood, change of affective tone, can be plausibly understood as consequences of placebo effects. Indeed, the networks that produce placebo analgesia are largely overlapping with the brain networks that are responsible for regulation of emotion, as a recent meta-analysis has shown.[40] Change of focus affects the dopaminergic system and thus has direct impact on the endogenous opioid system, influencing pain perception.[41,42] Relaxation can act directly via downregulation of the HPNA axis or on the principal stress axis.[43] But it can also work indirectly via the inflammation reflex. All activated macrophages express acetylcholinergic receptors.[44] Activation of the parasympathetic pathway, such as in relaxation responses in general terms, will lead to a release of acetylcholine, which will thus lead to downregulation of inflammatory activity. This might happen systemically, or very topically in loci of subclinical or clinical inflammation. Those mechanisms have direct impact on inflammatory as well as noninflammatory aspects of pain and pain perception.

Thus we have a full arsenal of causal pathways at hand that make plausible how nonspecific treatments can have diverse therapeutic effects. This allows us now to reconceptualize many aspects of CAM treatments as activation of such self-healing capacities of our body using the meaning response. What I have just said is a more benevolent and more meaningful description of the often heard adage that 'all CAM is placebo'. This may be true—at least for some aspects of CAM—if understood in the above way.

IS CAM 'ALL PLACEBO'? A NOTE ON SPECIFICITY AND THE EFFICACY PARADOX

But is it true that 'all CAM is placebo'? Are there no specific effects at all? While my focus in this chapter is not on unraveling the potential specificity of CAM, it is important to state two things:

First, it is highly unlikely that therapeutic systems can build up the ritualistic power to muster all nonspecific effects possible, if not at least sometimes, and reliably enough, specific effects also play a role. I have used the metaphor of 'dwarfs on the shoulders of giants' to describe this situation.[25,45] During the middle ages this was a way of describing the power of the then present-day writers. They saw themselves as dwarfs sitting on the shoulder of giants, the authorities of the tradition.[46] I would like to reinterpret this image: specific effects are dwarfs sitting on the shoulder of the giants, the nonspecific effects. That is why therapies work well—in general—no matter which ones. In other words, it is only the combination of specific with nonspecific effects that creates the therapeutic potential of any therapy, and more likely than not they are not additive but synergistic in complex ways.[47,48] Thus, it likely makes no sense to separate them out. If you have a blue door you can separate the color from its function. You may contemplate painting it red or yellow, and this would not change its function as a door. But the interaction between specific and nonspecific effects is more likely similar to a blue work of art of Ives Klein, for instance, who painted beautiful deep blue paintings. You could not possibly contemplate changing the color and keeping the piece of art untouched, or understanding the effect of the piece of art by analyzing the different shades of blue that have been used. Another example of how we cannot separately analyze a highly synergistic system is that of horse and rider: Understanding how a rider on a horse can move, how fast, how far, how elegantly, or how he can jump, implies studying horse and rider together. Taking the rider off the horse and measuring how high he can jump or how fast he can run, and then observing the horse separately, and measuring the same variables, would not work, as both rider and horse will do completely different things when left on their own. It is only the complex synergistic system of a rider on a horse that will give us the full insight into the reach and possibilities of both together and the whole system. The horse enhances the outreach and speed of the rider, but so does the rider enhance the normal and natural range and movement of the horse. It is only the mindset of a particular view of science as analytical, and the world of relevance being completely analyzable into additive components, that has fostered such a view. I beg to differ and invite the reader to—at least for the moment—adopt the view with me that specific and nonspecific effects together are different in scope and magnitude than each one alone. And more importantly, it is likely that one cannot be had without the other.

The second point is this. It is, in general, difficult to prove specific efficacy of CAM interventions against a placebo. However, this is difficult also with conventional drugs, as the facts show that less than half of all individual trials of psychotropic drugs against anxiety or depression submitted to the FDA for approval are independently significant, because placebo effects are so high.[49,50] In that sense, CAM is not alone with its difficulties. But some CAM interventions have indeed shown at least some superiority against placebo. This has recently been shown in a large individual patient data meta-analysis of acupuncture in pain.[51] There is some indication that some phytotherapeutic drugs work better than placebo.[52,53] With homeopathy there is a longstanding dispute. While, by pure vote count procedures, the database of clinical trials in homeopathy has more significantly positive trials of homeopathic remedies against placebo than negative or indecisive ones,[54,55] a meta-analysis of a small selection of larger high-quality trials concluded that homeopathy is not different from placebo.[56] This analyis, however, has been criticized for not having conducted a sensitivity analysis and for not having produced a clear rationale for the cut-off point of the number of trials included.[57] For if more or fewer trials are included, the result is again different from placebo.[58]

But the point is this. Perhaps this question is a marginal one anyway. And the really interesting question is: how do CAM interventions produce such large therapeutic effects, no matter whether nonspecific or specific ones? And why do conventional procedures obviously not produce effects of the same size? Why, given that we have arguably the best medicine of all times, do people bother to seek out the help of purported 'ineffective' or 'disproven' therapies, to use a vocabulary that the critics of CAM often use, and do case reports and anecdotes thrive that report stunning effects, at least from hearsay? How does the magic work? Far from being able to answer the question, I would like to discuss some potential issues in what follows. In order to understand how therapies with little potential specific effect can still be very effective, and even more effective than therapies with a strong proven specific effect, it might be useful to discuss what I have termed the efficacy paradox (Fig. 19.1).

Imagine that we have two types of treatment, each tested against its appropriate placebo for specificity in the same condition, say chronic headache, and the same type of patients. Treatment x, we find, tested against its placebo, produces only marginally superior specific effects and is found to be not significantly different from control; its effect size vis-a-vis placebo is small. Treatment y, however, shows a strong specific effect; hence, in a trial it is significantly superior to placebo and thus we call it efficacious. But it can still be the case that the gross effect of treatment x, let's say homeopathy, is larger than the gross effect of treatment y, say pharmacologic treatment, although treatment y is efficacious against placebo, while treatment x is not. This can be because the nonspecific effect is a variable quantity. The placebo–control groups of trials control for various effects: measurement artifacts, regression to the mean, spontaneous improvements and fluctuations, plus the nonspecific effects of therapies. Assuming that the methodologic

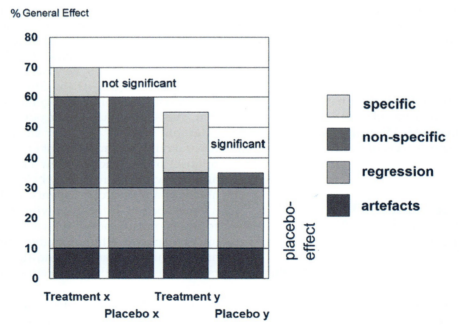

FIGURE 19.1 Efficacy paradox: an intervention with small and nonsignificant effects (treatment x) may be more effective overall than an efficacious one (treatment y); adapted from Walach.[59] *This figure is reproduced in color in the color section.*

artifacts across trials are roughly the same, it is highly implausible and factually wrong to assume that the nonspecific effects are constant across trials. Following statistical reasoning, such nonspecific effects have been conceptualized as error variance that can be neglected. However, it is more likely that it is not error variance but very systematic variance that is part and parcel of those nonspecific effects. Depending on the treatment, depending on patients' expectations, depending on the information they receive, these effects might vary tremendously.

This can be illustrated by the classical German Acupuncture (GERAC) Trials. These were among the largest acupuncture studies ever done and were conducted in three conditions: prevention of migraine,[60] treatment of chronic low back pain,[61] and treatment of osteoarthritis of the knee.[62] All three studies were randomized and placebo-controlled. All three studies also had a conventional control arm. This third conventional arm consisted of best-practice-guideline-oriented therapy and was thus deemed to be the best available conventional treatment option and thoroughly evidence-based in itself, i.e. based on placebo-controlled trials in the case of pharmacologic treatments. But not only pharmacologic treatments were applied in that best-practice arm; other treatments, such as physiotherapy or massage, were used if indicated. The placebo arms of these studies consisted of standardized superficial needling—only 1 mm deep insertion of the needles—in points that were agreed in advance by experienced practitioners to be non-therapeutic. Acupuncture was by formula, with a modest range of individualization representative of what an average patient could expect to receive from any well trained GP who had received an acupuncture diploma from one of Germany's acupuncture societies. Altogether, these were well planned trials independently sponsored by German statutory reimbursement insurance companies. The companies also recruited the patients, informing them that 'two types of acupuncture' were being tested against each other and against standard treatment. The point, which not many people realize, is that the patients of the insurance companies were promised acupuncture but only if they participated in the study. Thus an expectation had been built up in patients that they would receive some acupuncture. They did not know that one of these types of acupuncture would be sham, as it had been dubbed 'some novel kind of acupuncture,' which, descriptively speaking, was certainly true.

It is important to hold this piece of information in mind when looking at the results: in none of the studies was true acupuncture significantly better than sham acupuncture. In two of the studies, back pain and osteoarthritis of the knee, both sham acupuncture and true acupuncture were not only statistically but also clinically superior to best-practice-based conventional treatment. And in the third study, migraine prevention, all three treatment arms were equally effective. This is indirect evidence of the efficacy paradox come true: treatments that are clearly efficacious and evidence-based are only half as effective as sham treatments in two cases and equally effective with sham treatments in the last case (see also Ch 26 in this book). Why was this so? Two points stand out. First, patients expected to receive acupuncture. When they were randomized into one of the acupuncture arms, their expectation was met. We can hypothesize that they were then relieved, were more likely to anticipate improvements and more likely to be compliant. They were surely also more likely to rate improvements more positively. As the studies were not blinded, and the outcome assessment was based on patient-reported measurements, such a bias is likely. But it is certainly not enough to explain such a large effect. Second, patients in the conventional arm were probably disappointed. The insurance company had advertized acupuncture and now they did not receive it. Rather they had to undergo conventional treatment, which most patients had experienced already.

What happened in a study might happen in everyday practice. Patients form certain expectations about what they are about to experience. They read, they talk to other patients and friends. They go by word of mouth. They come with a certain mindset of preparedness or not, and thus nonspecific effects are more likely.

The same can be said about other therapies: homeopathy, for example, or spiritual healing. By and large, the indication that homeopathy is more than placebo is weak at best.[55,63] Especially in chronic headaches, a series of studies has been conducted which have not been able to replicate initial strong effects of homeopathy over placebo.[64–67] Thus, at least in headaches, homeopathy has not been shown to be superior to placebo. Yet, patients seek out homeopaths, especially patients with chronic headaches, and apparently with success. A careful prospective documentation study of classic homeopathy in chronic headaches in 230 adult patients shows a large effects size of 2.4 standard deviations of improvement against baseline after 2 years.[68] At the same time, the consumption of conventional medications was drastically reduced. Thus, independently and indirectly, the efficacy paradox can be seen here as well. Why is a therapy, without proven superiority over placebo, still highly effective in a practical sense? I am not aware that conventional pharmacologic therapies can actually produce effect sizes of this magnitude at such long follow-up periods. We were unable to detect a significantly different effect of spiritual distant healing over no-treatment in a blinded condition.[69] Nevertheless, in a waiting-list controlled study without sham-control, spiritual healing produced a sizable effect of d = 0.6 in patients who were deemed to be without any further treatment option by conventional standards.[70] Here again, we see the same phenomenon: what is arguably not different from placebo can have strong and clinically relevant effects, even in a situation where no proven treatment is available.

Thus, the efficacy paradox is alive and well, and many CAM treatments are a good example of how interventions that appear to lack strong specific efficacy—they may have some, but it is difficult to pinpoint it, or it is quite small—can nevertheless produce strong, lasting and clinically relevant effects. This is similar to a skillful child-rider steering a strong horse. Seen from a distance this might look like a horse running all on its own. So how are we to make sense of this situation? I suggest by redefining 'placebo effects' as self-healing response, and by seeing CAM interventions as a set of different methods to stimulate a complex system into recalibrating its set point that defines the normal state of the system, and/or to then reach this point of optimal balance by various means of self-regulation that are far from understood in their entirety and complexity.

JEROME D FRANK'S MODEL OF GENERAL HEALING EFFECTS OR COMMON FACTORS IN THERAPY

In 1961, Jerome D Frank had already produced a general model of healing in his classic study *Persuasion and Healing*.[71] In this model, four elements join together to promote a generic healing response:

- A therapeutic myth or narrative, shared by both patient and practitioner
- A persuasive therapeutic ritual
- A strong affective bond between patient and practitioner that conveys security, understanding and thereby promotes relaxation and a reduction of anxiety
- Convincing insignia of therapeutic power of the practitioner

Stemming from psychotherapy research, one could add a fifth element:[72]

- Mobilizing resources in the patient and empowering him or her

One can find different names for these aspects in the literature, but by and large the content will remain the same.[73] I suggest that this is also a good model for understanding a large part of what is done by any medical system, and especially by CAM methods. My assumption is that CAM practitioners are implicitly very good at using these factors, perhaps without even knowing that this is what they are doing.

THE COMMON MYTH

Many patients turn to CAM doctors or practitioners out of disappointment with the conventional system.[74] They may suffer from side-effects, do not want to continue taking drugs on a continuous basis, or do not improve with conventional treatments. Often, and this is especially true for complementary cancer treatments, they want to engage themselves to 'help the immune system,'[75] when conventional wisdom tells them that there is nothing to do, that they have been treated. So they turn to someone who shares their implicit view of the world and of themselves. They seek out the 'natural,' the 'soft' or more wholesome treatments and especially treatments that offer holistic approaches.[76] That implies a philosophical or even ideologic affinity to the therapeutic myth underlying some WSM and CAM interventions. And CAM practitioners have what many conventional doctors either lack or do not take seriously: time to talk to patients.[77,78] They explain what they do and why they do it. They explain their world view or the medical view of the system they represent. So a homeopath may say things like 'homeopathy is a system where we treat the whole person'—very pleasant, exactly what one is looking for; 'the medicines we use are actually devoid of any substances, it is only the information that we use'—now that is clever, is it not, sounds fascinating; 'in order to treat you properly I will have to take some time, an hour maybe, perhaps more, to listen to all you have to tell me, and I will perhaps also have to ask you about other diseases you might also have, or past ones'—now, finally, a doctor who wants to hear what is bothering me as a whole. And on it goes, until a good fit between homeopath and patient is established. We have little research on this process, but what we have documents: both for homeopaths and patients, the establishment of a good relationship is the most important focus and takes a lot of time.[79]

Most CAM therapists will take time to explain to patients what they do and why they do it, if patients are not already primed by popular media or the internet. And here patients find mainly soft-science versions that will speak of 'regulating life energies' or 'chi' or 'prana,' or the use of more sophisticated language about 'informational fields' and 'quantum approaches' to healing. All this can be very effective in establishing a common language and a common myth. And, indeed, this might have a strong placebo effect. We once studied the effect of, what we suppose, is a clear pseudo-machine, a so-called radionic treatment system.[80] In such a system, a random number generator chooses from various databases of names of homeopathic remedies, sentences of wisdom, healing colors, and so forth, the ones that 'conform' to a patient, and then 'send' it via an interface to a kind of crystal that is attached to the computer which is then said to 'radiate it out' into the universal 'energetic field', 'biasing quantum probabilities' that can then foster the healing of the person in question. This whole myth is being bolstered by the usage of some quantum terminology and some, in fact quite interesting and valid, data out of the Princeton Engineering Anomalies Research Lab that conducted many studies over decades on the influence of human intention on random systems.[81,82] Thus, a post-modern, but largely pseudo-scientific, myth is created that can be discerned as lacking a true scientific basis only by specialists or people knowledgeable enough to understand the lingo and its ramifications. To normal consumers it sounds very convincing and scientific, and most people don't care that much anyway, as long as some plausible story is in the background. And most of the time the complex myth is used to convince the practitioner who then buys the machine and uses a downgraded and simplified myth of his own to explain to patients.

In our study of this system we conducted a partially blinded experiment in our former department. People were free to formulate problems, health problems, personal problems, or whatever they wished changed in their lives. The problems were forwarded to the system operator who entered them into the database. Everything being remote, there was no interaction between system operator and participant, much as in a distant healing study. Participants knew about the system under study and received a kind of simplified version of the therapeutic myth. Half of the participants were randomized to wait, and half received their first treatment, and everyone was blind as to group assignment. In the second half of the study, people were told that they were being treated. They had to rate goal attainment after each treatment period. By that method we could indirectly estimate the 'placebo' component. When people knew that they were on 'radiation' they rated their goal attainment significantly higher and were actually quite surprised about how many of their problems were being helped by the treatment.

This is a nice illustration of how a common myth, and knowing about the treatment being applied, may even make a seemingly silly apparatus a powerful therapeutic catalyst, or, put differently: in our modern era electronic equipment is a prime trigger of faith in some power of healing, as people have ample experience with the miracles of our electronic age. The important thing, after all, is not what is being *really* done in a treatment, but what patients *think or believe* is being enacted.[83] One of our students, a medical doctor, uses this equipment together with some other elements, such as homeopathy.

In a systematic retrolective study of all allergic rhinitis patients in a season, 43 patients altogether—all with various pretreatment histories and dependent on topical antihistaminics or systemic cortisone during the pollen season—she saw good results in 93% of her patients, and even if one reduces this figure by those 8 patients who did not respond, or who could not be reached, this is still a stunning result. The patients had much fewer symptoms, and most of them could stop their topic or systemic medication—which, again, generates the myth of 'the only doctor far and wide that can help you with these symptoms'; and on it goes (these data are from an unpublished thesis of Dr Daniela Jobst to be prepared for publication).

CAM systems are under pressure from mainstream medicine and critical media. Thus they generate their apologetic narratives. These are mainly founded on stunning therapeutic case histories, e.g.[84] The histories themselves are very likely true in most cases. However, it is unclear how representative they are. Doctors and patients alike are interested only in the success stories, and they forget the rest. Thus, myths are created by word of mouth on the part of the patients and by selective memory on the part of the doctors.

CAM is not alone in using therapeutic myths. Conventional medicine does the same, and to good effect. One of the rare studies of surgery that was placebo-controlled compared true arthroscopy in osteoarthritis against a treatment arm where only half of the procedure was carried out, and against a sham operation, where only a small incision was made.[85] After 2 years, all three treatment arms were equally effective and could reduce pain considerably. But obviously there was nothing specific about the operation. But clearly, it speaks to an understandable narrative: debris in the joint has to be mechanically removed, and if it is, the knee will again function well. And so it does. Only it is not the removal of the debris, but the power of the therapeutic myth that works the miracle.

THE RITUAL

All therapy is ritualistic, even if the ritual is only the handing over of the prescription and the fetching of the medicine in the pharmacy, all combined with a few words of hope and suggestions of improvement. Frank has clearly established the ritualistic elements of conventional medicine. And it is not difficult to spot them in CAM treatments. Here, very different rituals are being enacted. TCM is full of rituals, starting with foreign ways of diagnosing, through looking at the tongue and palpating the pulse, inserting needles and burning moxa (a mixture of herbs with strong odor), handing out quite complex prescriptions for very strange herbal drugs, or even mixing the herbs in the clinic in front of the patient, all combined with individualized suggestions for diet. Homeopathy has established its own rituals, starting with a long-winding and complex case-taking process, handing out tiny sugar globules, taken out of the doctor's cabinet containing hundreds of other, similar looking ones, thereby suggesting individual importance. The prescription is given together with complex advice: the globules should not be taken in temporal vicinity to food, slowly sucked, not swallowed. Many homeopaths prohibit drinking coffee, at least on the day or days immediately following the intake of the remedies; some also advise using specific tooth paste, and a well reputed Swiss company even sells 'toothpaste compatible with homeopathy,' i.e. without peppermint. Next, patients are instructed to carefully observe themselves in order to be able to report back any changes. This is clever, for self-observation, if done with a stance of curiosity and mindfulness, will surely reveal the fluctuating nature of every type of symptom and will allow the patient to focus on the improvements, which will likely trigger a circle of positive expectation. But also negative experiences are used: if the symptoms get worse, the narrative defines this as a 'therapeutic aggravation,' which is 'a good prognostic sign' for future improvement. Although, from the point of view of homeopathy, there are also clear indications of situations when the remedy has been wrongly chosen, on the part of the patient the self-observation is likely to promote a positive circle.

Other therapeutic systems or doctors use their own rituals. Spiritual healers, depending on their training and model, will use certain movements of the hand in the 'aura field' of the patient. Indeed, a study of spiritual healing in chronic-pain patients, where a sham condition was created by an actor who had previously studied the movements of healers carefully, showed no difference between the true and the sham healing condition.[86] But both patient groups showed a strong improvement of $d = 1.2$ standard deviations from the baseline value, and 13 of 60 patients had an improvement of more than 50% after 8 weeks. This is considerable, if we keep in mind that the patients had been chronic-pain patients with many years of unsuccessful previous treatment history.

The myth of bioenergy has been mentioned. Building on it—and perhaps also on some efficacious principle we do not know enough about—bioresonance therapy can be very effective in practice. Indeed, in a prospective documentation study of 935 patients who suffered from untreatable chronic pain, 800 or 85.6% had good or sufficient pain relief. Some 18% of those can be considered healed, and 45% of those improved had considerable improvement. Half of the patients were treated only five times.[87] This is quite puzzling, given that all the patients had been pre-treated without success by various methods.

This myth of CAM methods as the last resort of patients close to desperation is something that is nearly a constant across modalities. Perhaps in such a situation patients have no clear expectation, or whatever they used to hold true about themselves, about medicine, about the world, is likely quite shaky by the time they see a CAM practitioner. This is the time when they are confronted with a new ritual that may capture their imagination, rekindle their hope and thus set in motion changes that ultimately lead to therapeutic improvements.

Rituals, it has been observed, stimulate our capacity to be attentive, to be present and to engage emotionally with a novel situation, perhaps even in a slightly hypnoid situation.[88] This change of attentional focus will, by default, affect the dopaminergic system.[89] The dopaminergic system itself has strong interconnections with the endogenous opioid system via connections in the periaqueductal gray of the midbrain.[90–92] Thus, it is immediately obvious that strong rituals will always, to some extent, activate this system. Although there is no direct evidence as yet, it can be plausibly expected from what we already know that this is the case. Rituals, then, can be a strong trigger for self-healing responses, and we see quite a few of those in CAM practice.

RELATIONSHIP AND THE ALLEVIATION OF ANXIETY

We know from the classic study of Thomas[93] that being positive is more decisive than giving a medicine. In this study he randomized 200 patients who had no serious medical problems in a 2×2 factorial design into four conditions. They were either treated or not; half of them received a positive message and half of them a neutral to negative one. Those who received a positive message were told that nothing was wrong with them and all will be well soon. Those who received the neutral message were told that some further testing was necessary because the findings were unclear and the symptoms might be signs of a more serious underlying condition. These are two quite ritualistic trajectories of patients with unclear symptoms meeting with either a clinically experienced or insecure GP. Some 64% of the patients with a positive consultation were improved after 2 weeks, but only 39% of the negative ones. Receiving treatment was irrelevant. This study is a tale of secure relationship and attachment, really. Attachment experiences from our early childhood are the condensation nuclei of templates for later experiences.[94–96] People who have a rather secure attachment history are less prone to psychologic problems and thus, as a consequence, also to pain and other psychosomatic conditions. By proxy and analogy, a strong attachment experience with a surrogate attachment figure may be able to heal and improve some of the deficits in earlier attachment history, we may assume. Therefore, therapeutic relationships that are always both a repetition and a corrective of earlier attachment experiences will have to be strong and reliable, if they are designed to alter and improve a dysfunctional attachment style. This is how I think CAM and WSM make use of this modality:

Most CAM therapies emphasize good patient–practitioner relationships. In some modalities, such as in homeopathy, this is functionally necessary for optimal treatment, and will be the result of frequent, intense and psychologically quite intimate interactions.[79] In other modalities it will be the style of practitioners in private practice in order to convince patients of their money's worth. Again some other modalities, such as massage, shiatsu, reflexology, cranio-sacral work all use very gently touching techniques, physical rather than verbal kinds of communication that tap into those early registers of attachment experiences, potentially correcting or improving them, on top of what else they may therapeutically do.

There is only little empirical evidence as of now, because most researchers have focused predominantly on the specific components of treatments. But an interesting first study of homeopathy in rheumatoid arthritis has shown that the consultation is much more important than the remedies being supplied.[97] A careful study that used only placebo acupuncture standardized the contact to various levels of intensity in irritable bowel syndrome patients.[98] One group had no contact and had to wait. One group had minimal contact and received practically no oral communication, only sham needling, while the third group received also conversation and intensive verbal contact. It was clearly visible that patient-reported outcome, as well as clinician-rated outcome by a blind assessor, in the third group was significantly superior and clinically twice as large as the results in the restricted communication group, which, again, had superior results to the waiting-list control group. This is the first direct experimental evidence that contact and communication itself is a therapeutic element.

Conversely, qualitative research on the conventional behavior of GPs has shown that their communication behavior is not really very therapeutic and professional. They often signal courtesy, but disinterest, and thus very likely neglect the strongest therapeutic factor they could possibly use,[99] and we know that implicit verbal signals, e.g. using words such as 'not' frequently or 'pain,' will yield a more negative outcome.[100,101] Although there is no comparative research, it can be safely assumed that one reason why patients seek out alternatives is that they feel better understood or listened to by CAM practitioners.[79]

Frank observed that a strong therapeutic relationship may work via relaxing the patient, alleviating his or her anxiety, giving hope and some security in all the turmoil. Psychotherapy research has since unconvered that the variance that is due to differences in methods is likely less important than the variance due to differences in therapists, and hence in personal style, personality and likely also attachments.[102,103] While the difference in method variance in a meta-analysis was close to zero, the variance attributable to the therapist was 8% in a remodeling of a famous comparative psychotherapy study.[104] In general, modern psychotherapy research has clearly pointed to the therapeutic alliance as one of the central elements in therapy,[105,106] and it is obvious that a good therapeutic alliance can be formed only if patients' preferences and styles are respected by the therapist.[107] The therapeutic alliance is the strongest therapeutic factor, and mindfulness is one of the best methods to foster it.[108] An interesting study in this respect is the one by Grepmair.[109] Here, 18 cognitive-behavioral psychotherapists in training, who worked in an inpatient psychotherapy clinic, were randomized to participate in a regular meditation session in the morning with a Soto-Zen master for an hour each day, or to carry on as usual. Their patients did not know of the practice of their therapists. Patient outcome was measured after 9 weeks and was drastically and significantly improved in those therapists who had followed the meditation practice in the morning.[109] It is implausible to assume that the therapists changed their modality or their interventions as such, as they all had a behavioral cognitive therapeutic training and needed to conform to this for their graduation. Thus, what they changed must have been on the level of relationship and how they engaged with their patients. Whether they may have caught more subtle cues due to the meditation training, were more empathic, or had a more competent behavior in shaping the relationship, is not known. In another study with physicians, mindfulness training led to a moderate improvement in empathy.[110] We know by now, through a disturbing study, that medical students who enter medical school with a high level of enthusiasm and empathy lose their empathy and idealism over the course of their training,[111] and thus very likely their most important resource. Thus it is not surprising that people turn to doctors where they find empathy and time to be listened to. It is empathy and time that, in the case of the Glasgow Homeopathic Hospital, are the most important factors in the opinion of patients,[78] and empathy correlates moderately with outcome.[112] If empathy can already be observed to have an effect on pain perception in mice,[113] we can assume that there is an intimate relationship also in humans. And indeed, it appears to be one of the strongest elements shaping the doctor–patient relationship.[114]

We can re-conceptualize many CAM modalities as ways of improving patient–doctor relationships and thereby allow for a modification of attachment experiences through the means of ample time provision, empathy, puzzling and novel rituals and narratives. It is therefore not at all surprising if strong effects in the change of pain perception and other symptoms have been consistently observed. It is completely inadequate to neglect and brush them aside as 'placebo effects,' as they are genuine self-healing responses, triggered by empathic relationships quite as in psychotherapy.

INSIGNIA OF POWER

Although Frank insisted that this is an important element in how nonspecific effects are being brought about, there has been little research in this area. Apart from some scattered evidence that branding an analgesic placebo will make it more potent,[115] there is little knowledge on this issue. Ethnographic evidence has provided us with colorful descriptions of the insignia of power of shamans, like feathers, masks, rattles, antlers, and the like.[116] Frank has paralleled them to the insignia of power in modern medical practice, such as the white coat, the stethoscope, the framed and gilded diplomas hung from the wall, or the prizes on the book shelf, etc. However, I am not aware that this has been formally evaluated in its real power to create positive expectations.

Interestingly, though, alternative practitioners and doctors display the same kind of diligence with their certificates of healing prowess. TCM practitioners normally enjoy displaying their titles and certificates in foreign Chinese letters, even though no one will be able to understand them. Although, at least in Germany, every doctor can practice homeopathy freely, there is a dense network of certification courses run by the German medical boards that also organize the exams. It is very doubtful that GPs in their own practice would make the effort to take exams again, if it were not for the patients who will doubtlessly prefer an examined, board-certified and hence purportedly more powerful homeopathic doctor to one without this exam and the legal possibility of pinning that new certificate to his board. While one aspect of the current discussion around board certification, legislation and voluntary registration of lay practitioners of CAM methods in Europe, is certainly around issues of safety and patient protection,[117] another aspect is that of signification of power. Who has been examined by the state or a legal body is more powerful than others. Critics therefore often attack first along the line of education and certification.[118] In the UK, all the academic institutions running CAM courses have come under pressure and have been attacked by activists in that university directorates and deans have been approached for validation material on the basis of the freedom of information act. This makes perfect sense from the point of view of therapeutic effectiveness: if official insignia of therapeutic strength are being reduced, therapeutic credibility, and hence results, will drop.

EMPOWERING PATIENTS AND MOBILIZING RESOURCES

There is little doubt within the psychotherapy research literature that mobilizing patients, tapping into their resources and helping them to use these is one of the most therapeutic acts a therapist can perform and likely also one of the strong common factors.[119] The literature on increasing self-efficacy and symptom improvement is huge. And as a very general and generic result of the research findings, we can state that increasing self-efficacy and helping patients to access their resources is a major therapeutic step.[120] While this is also increasingly seen as important by conventional medicine, CAM methods are, by and large, intrinsically prone to do this. While the theoretical stance of BM is that the experts—diagnosticians, the doctor, the system—know what the problem is, and how to treat it, thus making patients passive recipients of therapeutic acts that are being brought forward by others to the patient, most CAM modalities, if they are more than just prescribing a herbal remedy instead of a conventional one, require the patient to become active. A simple first step is: patients have to choose to step out of the conventional mainstream system. This in itself is a decision and means that they have to undergo a completely new type of treatment, tell their story again, undergo new procedures, probably pay out of their pocket for at least some of the treatment, if not all. Then, depending on the modality, their new therapist will not only do something to them but will require patients to cooperate—either by volunteering information, perhaps even very personal and intimate information, or by following certain lifestyle and dietary advices, or by engaging in a new type of regular practice, whether this is some small exercise program or some other lifestyle ritual does not really matter. Self-induced activity is the principal element here. Most practitioners are eclectic. That is, even hard-nosed homeopaths will rarely just prescribe globules, but will always give some advice, for instance regarding diet, or offer some exercises, perhaps relaxation tapes. Lifestyle advice is probably among the broadest categories of what CAM practitioners do across the board. That will always mean: giving responsibility about their lives back to patients, supporting their choices and thus empowering patients. Often, there is an alliance between the CAM doctor and patient against treatments and their providers that have been unsuccessful, although increasingly this competitive stance is being replaced by a joint effort to help patients. This means that patients will feel supported in their view, and in taking responsibility for their health and their lives.

What little research we have shows: CAM is seen as empowering and as increasing self-efficacy in patients.[121–123] Whether this is also the therapeutic mechanism we do not know. But the hypothesis is plausible that CAM, by definition and default, is more easily perceived by patients, than is BM, as helping to empower them. Thereby, CAM is using a very strong component of nonspecific therapeutic effectiveness. And again: it would be completely wrong to attribute this merely to the 'placebo' nature of CAM. More precisely, and more usefully, it is seen as using one factor from a whole set of common factors that stimulate self-healing.

SUMMING UP: THE SPECIFICITY OF NONSPECIFIC EFFECTS AND THE ELEGANCE OF REDUCING SIDE-EFFECTS BY 'PLACEBO'

It needs to be stated that the derogatory connotation of the word 'nonspecific,' in an age that worships and offers to the god of specificity, which is at the same time the idol of pharmacology, is, in my view, completely inadequate and out of place. The purported 'specificity' of, say, a serotonin agonist to treat acute migraine is quite nonspecific because the serotonin agonist will increase serotonin bioavailability of serotonin not only at receptor sites, where it is really needed, such as at the endothelium of the cranial vessels in the case of migraine, but also in other places in the body. The morphine given to a pain sufferer is specific to opioid receptors in the brain's pain network, but not only to those: it will also have numerous 'specific' but unwanted effects in the periphery, thus producing the side effects such as constipation. This is the reason why all 'specific' drugs also produce more or less intense side-effects. It is because they are not specific enough and their specificity refers only to receptor categories, and not to specific receptor sites.

A case can be made that, contrary to this situation, the purported and so-called nonspecific effects are actually even more specific, as placebo effects or therapeutic responses to nonspecific interventions happen in places where they are really needed. Placebo-induced endogenous opioid receptor activation has a much smaller target area than a drug-induced one, but it is exactly the receptors that are important that are being activated, but no other ones.[124] Placebo effects can be very specific.[32,125–128] If we take the novel definition of placebo effects as effects of the meaning of an intervention seriously, then, by that very token, these effects are individualized and thus specific. They are specific in a way that conventional pharmacologic interventions can never be, because they will always only be able to target classes of receptors, even though very specific ones, across the whole organism. Contrary to that, nonspecific interventions, because they are actually stimulating self-healing responses, are only targeting those systems and receptors that are in need of recalibration and that are really dysregulated.[3] This is especially true for treating pain. We have enough evidence that, in acute cases, pain can be modulated by top-down inhibitory regulation through opioid pathways, and placebo-induced analgesia does exactly this.[40] Empirically, we know that CAM treatments can also be effective in chronic pain, as demonstrated by a few examples referred to above. This will not only involve top-down inhibition of pain signaling, but remodeling of neuronal pain memory and a change in long-term potentiation at the level of neuronal circuits. How this works, we do not know. But we have data to suggest that CAM modalities can, at least sometimes, contribute to this very specific relearning. The emotional–cognitive regulation processes involved will probably activate very specific networks that are yet to be uncovered.

So what should become clear from this discussion is that, paradoxically speaking, nonspecific effects, like those from many CAM interventions, are in truth much more specific than the so-called specific interventions pharmacology uses. And this is likely the reason why many CAM modalities are preferred by patients because their potential to create side-effects is less. Indeed, the most elegant therapy yet to be invented would be one which would be able to stimulate only those physical or psychologic systems that are in need of regulation, leaving all the others alone, without any potential for harm, using a very generic, simple and easy-to-apply modality. The placebo component of many CAM modalities has already come close to this ideal. Hence, far from being relics of a distant past, where people did not know what to do with patients, to me they sound rather like heralds of a potential new era of therapy where we have learned to stimulate the body into healing itself only in those respects that need healing, and leaving everything else untouched.

References

1. Eardly S, Bishop FL, Prescott P, et al. A systematic literature review of Complementary and Alternative Medicine prevalence in EU. *Forschende Komplementärmedizin*. 2012;19(suppl 2):18-28.
2. Verhoef MJ, Lewith G, Ritenbaugh C, et al. Whole systems research: moving forward. *Focus Altern Complem Ther*. 2004;9:87-90.
3. Koithan M, Bell IR, Niemeyer K, Pincus D. A complex systems science perspective for whole systems of complementary and alternative medicine research. *Forschende Komplementärmedizin*. 2012;19(suppl 1):7-14.
4. Bell IR, Koithan M, Pincus D. Methodological implications of nonlinear dynamical systems models for whole systems of complementary and alternative medicine. *Forschende Komplementärmedizin*. 2012;19(suppl 1):15-21.
5. Meyer-Abich KM. *Was es bedeutet, gesund zu sein. Philosophie der Medizin*. München: Hanser; 2010.

6. Uexküll Tv, Wesiack W. *Theorie der Humanmedizin. Grundlagen ärztlichen Denkens und Handelns.* München: Urban & Schwarzenberg; 1988.
7. Wiedersheim R, Albrecht NJ, Lüken BJ, eds. *Traditionelle Heilsysteme und Religionen. Ihre Bedeutung für die Gesundheitsversorgung in Asien, Afrika und Lateinamerika.* Saarbrücken: Verlag Rita Dadder; 1991.
8. Bhardwaj SM. Medical pluralism and homoeopathy: a geographic perspective. *Soc Sci Med.* 1980;14B:209-216.
9. Satyavati GV. Some traditional medical systems and practices of global importance. *Ind J Med Res.* 1982;76(suppl):1-26.
10. Dinges M. Entwicklungen der Homöopathie seit 30 Jahren. *Zeitschrift für Klassische Homöopathie.* 2012;56(3):137-148.
11. Patwardhan B, Vaidya ADB, Chorghade M. Ayurveda and natural products drug discovery. *Curr Sci.* 2004;86:789-799.
12. Govindarajan R, Vijayakumar M, Pushpangadan P. Antioxidant approach to disease management and the role of 'Rasayana' herbs of Ayurveda. *J Ethnopharmacol.* 2005;99:165-178.
13. Balachandran P, Govindarajan R. Cancer – an ayurvedic perspective. *Pharmacol Res.* 2005;51:19-30.
14. Unschuld PU. When health was freed from fate: some thoughts on the liberating potential of early Chinese Medicine. *East Asian Sci Technol Med.* 2010;31:11-24.
15. Bell IR, Caspi O, Schwartz GER, et al. Integrative medicine and systemic outcomes research: Issues in the emergence of a new model for primary health care. *Arch Intern Med.* 2002;162:133-140.
16. Nissen N, Schunder-Tatzber S, Weidenhammer W, Johannessen H. What attitudes and needs do citizens in Europe have in relation to CAM? *Forschende Komplementärmedizin.* 2012;19(suppl 2):(In press).
17. Jonas WB. Complementary and alternative medicine and the NIH. *Clin Dermatol.* 1999;17:99-103.
18. Eisenberg D, Kessler RC, Foster C, et al. Unconventional medicine in the United States. Prevalence, costs and patterns of use. *N Engl J Med.* 1993;328:246-252.
19. Eisenberg DM, Davis RB, Ettner SL, et al. Trends in alternative medicine use in the United States, 1990-1997. *J Am Med Assoc.* 1998;280:1569-1575.
20. Weidenhammer W, Walach H, eds. *Complementary Medicine in Europe – CAMbrella.* Freiburg: Karger; 2012. Forschende Komplementärmedizin – Research in Complementary Medicine.
21. Hyland ME. The intelligent body and its discontents. *J Health Psychol.* 2002;7:21-32.
22. Hyland ME. Extended network learning error: a new way of conceptualising chronic fatigue syndrome. *Psychol Health.* 2001;16:273-287.
23. Pincus D, Metten A. Nonlinear dynamics in biopsychosocial resilience. *Nonlin Dynam Psychol Life Sci.* 2010;14:353-380.
24. Lewith G. Every doctor a walking placebo. *Complem Med Res.* 1987;2:10-17.
25. Walach H, Jonas WB. Placebo research: the evidence base for harnessing self-healing capacities. *J Altern Complem Med.* 2004;10(suppl 1):S103-S112.
26. Moerman DE, Jonas WB, Bush PJ, et al. Placebo effects and research in alternative and conventional medicine. *Chinese J Integr Trad West Med.* 1996;2:141-148.
27. Moerman DE, Jonas WB. Deconstructing the placebo effect and finding the meaning response. *Ann Intern Med.* 2002;136:471-476.
28. Benedetti F. How the doctor's words affect the patient's brain. *Eval Health Profess.* 2002;25:369-386.
29. Finniss DG, Benedetti F. Mechanism of the placebo response and their impact on clinical trials and clinical practice. *Pain.* 2005;114:3-6.
30. Enck P, Benedetti F, Schedlowski M. New insights into the placeo and nocebo responses. *Neuron.* 2008;59:195-206.
31. Pollo A, Carlino E, Benedetti F. Placebo mechanisms across different conditions: from the cinical setting to physical performance. *Phil Trans R Soc Biol Sci.* 2011;366:1790-1798.
32. Meissner K. The placebo effect and the autonomic nervous system: evidence for an intimate relationship. *Phil Trans R Soc Biol Sci.* 2011;366:1808-1817.
33. Meissner K, Kohls N, Colloca L, eds. *Placebo Effects in Medicine: Mechanisms and Clinical Implications.* London: Royal Society; 2011. (*Philosophical Transactions of the Royal Society Biological Sciences*; No. 366.)
34. Colloca L, Benedetti F. Placebos and painkillers: is mind as real as matter? *Nature Rev Neurosci.* 2005;6:545-552.
35. Colloca L, Petrovic P, Wager TD, et al. How the number of learning trials affects placebo and nocebo responses. *Pain.* 2010;151:430-439.
36. Hróbjartsson A, Gøtzsche PC. Is the placebo powerless? An analysis of clinical trials comparing placebo with no treatment. *N Engl J Med.* 2001;344:1594-1602.
37. Vase L, Riley JL, Price DD. A comparison of placebo effects in clinical analgesic trials versus studies of placebo analgesia. *Pain.* 2002;99:443-452.
38. Vase L, Jonas WB, Walach H. Placebo, meaning, and context issues in research. In: Lewith GL, Jonas WB, Walach H, eds. *Clinical Research in Complementary Therapies: Principles, Problems and Solutions.* Edinburgh: Churchill Livingstone; 2010:313-332.
39. Pacheco-López G, Engler H, Niemi m-B, Schedlowski M. Expectations and associations that heal: immunomodulatory placebo effects and its neurobiology. *Brain Behav Immun.* 2006;20:430-446.
40. Amanzio M, Benedetti F, Porro CA, et al. Activation likelihood estimation meta-analysis of brain correlates of placebo analgesia in human experimental pain. *Hum Brain Mapp.* 2012. doi:10.1002/hbm.21471.
41. Scott DJ, Stohler CS, Egnatuk CM, et al. Individual differences in reward resonding expli an placebo-induced expectations and effects. *Neuron.* 2007;55:325-336.
42. Scott DJ, Stohler CS, Egnatuk CM, et al. Placebo and nocebo effects are defined by opposite opioid and domaminergic responses. *Arch Gen Psychiatr.* 2008;65:220-231.
43. Black PH. Stress and the inflammatory response: a review of neurogenic inflammation. *Brain Behav Immun.* 2002;16:622-653.
44. Tracey KJ. Physiology and immunology of the cholinergic antiinflammatory pathway. *J Clin Invest.* 2007;117:289-296.
45. Walach H. Non-specific effects, placebo effects, healing effects – all doctors are healers. *Eur J Integr Med.* 2008;1(suppl 1):S33.
46. Klibansky R. Standing on the shoulders of the giants. *Isis.* 1936;26:147-149.
47. Di Blasi Z, Harkness E, Ernst E, et al. Influence of context effects on health outcomes: a systematic review. *Lancet.* 2001;357:757-762.
48. Kleijnen J, de Craen AJM, Van Everdingen J, Krol L. Placebo effect in double-blind clinical trials: a review of interactions with medications. *Lancet.* 1994;344:1347-1349.
49. Khan A, Khan S. Placebo in mood disorders: the tail that wags the dog. *Curr Opin Psychiatr.* 2003;16:35-39.
50. Khan A, Khan S, Brown WA. Are placebo controls necessary to test new antidepressants and anxiolytics? *Int J Neuropsychopharmacol.* 2002;5:193-197.
51. Vickers AJ, Cronin AM, Maschino AC, et al. Acupuncture for chronic pain: Individual patient data meta-analysis. *Arch Intern Med.* 2012:online first.
52. Linde K, Berner MM, Kriston L. St John's wort for major depression. *Cochrane Database System Rev.* 2008:(issue 4):Art. No.: CD000448.
53. Kaschel R. Specific memory effects of *Ginkgo biloba* extract EGb 761 in middle-aged healthy volunteers. *Phytomed.* 2011. doi:10.1016/j.phymed.2011.06.021§.

54. Milgrom LR. Under pressure: homeopathy UK and its detractors. *Forschende Komplementärmedizin.* 2009;16:256-261.
55. Walach H, Jonas WB, Ives J, et al. Research on homeopathy: state of the art. *J Altern Complem Med.* 2005;11:813-829.
56. Shang A, Huwiler-Münteler K, Nartey L, et al. Are the clinical effects of homeopathy placebo effects? Comparative study of placebo-controlled trials of homoeopathy and allopathy. *Lancet.* 2005;366:726-732.
57. Walach H, Jonas W, Lewith G. Letter to the Editor: are the clinical effects of homeopathy placebo effects? Comparative study of placebo-controlled trials of homoeopathy and allopathy. *Lancet.* 2005;366:2081.
58. Lüdtke R, Rutten ALB. The conclusions on the effectiveness of homeopathy highly depend on the set of analyzed trials. *J Clin Epidemiol.* 2008;61:1197-1204.
59. Walach H. Das Wirksamkeitsparadox in der Komplementärmedizin. *Forschende Komplementärmedizin und Klassische Naturheilkunde.* 2001;8:193-195.
60. Diener HC, Kronfeld K, Boewing G, et al. Efficacy of acupuncture for the prophylaxis of migraine: a multicentre randomised controlled clinical trial. *Lancet Neurol.* 2006;5:310-316.
61. Haake M, Muller HH, Schade-Brittinger C, et al. German Acupuncture Trials (GERAC) for chronic low back pain: randomized, multicenter, blinded, parallel-group trial with 3 groups. *Arch Intern Med.* 2007;167(17):1892-1898.
62. Scharf HP, Mansmann U, Streitberger K, et al. Acupuncture and knee osteoarthritis. *Ann Intern Med.* 2006;145:12-20.
63. Mathie R. The research evidence base for homeopathy: a fresh assessment of the literature. *Homeopathy.* 2003;92:84-91.
64. Brigo B, Serpelloni G. Homoeopathic treatment of migraines: a randomized double-blind controlled study of sixty cases (homoeopathic remedy versus placebo). *Berlin J Res Homoeopathy.* 1991;1:98-106.
65. Straumsheim P, Borchgrevink C, Mowinkel P, et al. Homoepathic treatment of migraine: a double blind, placebo controlled trial of 68 patients. *Br Homeopath J.* 2000;89:4-7.
66. Walach H, Gaus W, Haeusler W, et al. Classical homoeopathic treatment of chronic headaches. A double-blind, randomized, placebo-controlled study. *Cephalalgia.* 1997;17:119-126.
67. Whitmarsh TE, Coleston-Shields DM, Steiner TJ. Double-blind randomized placeo-controlled study of homoeopathic prophylaxis of migraine. *Cephalalgia.* 1997;17:600-604.
68. Witt CM, Lüdtke R, Willich SN. Homeopathic treatment of chronic headache (ICD-9: 784.0) – a prospective observational study with 2 year follow-up. *Forschende Komplementärmedizin.* 2009;16:227-235.
69. Walach H, Bösch H, Lewith G, et al. Efficacy of distant healing in patients with chronic fatigue syndrome: a randomised controlled partially blinded trial (EUHEALS). *Psychother Psychosom.* 2008;77:158-166.
70. Wiesendanger H, Werthmüller L, Reuter K, Walach H. Chronically ill patients treated by spiritual healing improve in quality of life: results of a randomized waiting-list controlled study. *J Altern Complem Med.* 2001;7:45-51.
71. Frank JD. *Persuasion and Healing: A Comparative Study of Psychotherapy.* Baltimore: Johns Hopkins University Press; 1961.
72. Grawe K, Grawe-Gerber M. Ressourcenaktivierung: Ein primäres Wirkprinzip der Psychotherapie. *Psychotherapeut.* 1999;44:63-73.
73. Messer SB, Wampold BE. Let's face the facts: common factors are more potent than specific therapy ingredients. *Clin Psychol: Sci Pract.* 2002;9:21-25.
74. Güthlin C, Lange O, Walach H. Measuring the effects of acupuncture and homoeopathy in general practice: an uncontrolled prospective documentation approach. *BMC Pub Health.* 2004;4:6.
75. Smithson J, Paterson C, Britten N, et al. Cancer patients' experiences of using complementary therapies: polarization and integration. *J Health Serv Res Policy.* 2010;15(suppl 2):54-61.
76. Bishop FL, Yardley L, Lewith GT. Why do people use different forms of complementary medicine? Multivariate associations betweeen treatment and illness beliefs and complementary medicine use. *Psychol Health.* 2006;21:683-698.
77. Stock B. Minds, bodies, readers: Rosenbach lectures, University of Pennsylvania 1999. *New Literary History.* 2006;37(3):489-501; 503-513; 515-524; 643-654.
78. Mercer SW, Reilly D. A qualitative study of patient's views on the consultation at the Glasgow Homoeopathic Hospital, an NHS integrative complementary and orthodox medical care unit. *Patient Educ Couns.* 2004;53:13-18.
79. Eyles C, Leydon GM, Brien SB. Forming connections in the homeopathic consultation. *Patient Educ Couns.* 2012:(In press).
80. Schneider R, Walach H. Randomized double-blind pilot study on psychological effects of a treatment with 'Instrumental Biocommunication'. *Forschende Komplementärmedizin.* 2006;13:35-40.
81. Dunne BJ, Jahn RG. Consciousness, information, and living systems. *Cell Mol Biol.* 2005;51:703-714.
82. Jahn RG, Dunne BJ. *Margins of Reality. The Role of Consciousness in the Physical World.* San Diego: Harcourt Brace Jovanovich; 1987.
83. Horne R. Patient's beliefs about treatment: the hidden determinant of treatment outcome? *J Psychosom Res.* 1999;47:491-495.
84. Mutter J, Naumann J, Güthlin C. Xenobiotikaausleitung bei einer Patientin mit Fibromyalgie, chronischer Erschöpfung und stammbetonter Adipositas. *Forschende Komplementärmedizin.* 2007;14:39-44.
85. Moseley JB, O'Malley K, Petersen NJ, et al. A controlled trial of arthroscopic surgery for osteoarthritis of the knee. *N Engl J Med.* 2002;347:81-88.
86. Abbot NC, Harkness EF, Stevinson C, et al. Spiritual healing as a therapy for chronic pain: a randomized, clinical trial. *Pain.* 2001;91:79-89.
87. Herrmann E, Galle M. Retrospective surgery study of the therapeutic effectiveness of MORA bioresonance therapy with conventoinal therapy resistant patients suffering from allergies, pain and infection diseases. *Eur J Integr Med.* 2011;3(e)e237-e244.
88. Rossano MJ. Setting our own terms: how we used ritual to become human. In: Walach H, Schmidt S, Jonas WB, eds. *Neuroscience, Consciousness and Spirituality.* Dordrecht: Springer; 2011:39-55.
89. Davidson RJ, Irwin W. The functional neuroanatomoy of emotion and affective style. *Trends Cogn Sci.* 1999;3:11-21.
90. de la Fuente-Fernández R, Schulzer M, Stoessl AJ. The placebo effect in neurological disorders. *Lancet Neurol.* 2002;1:85-91.
91. Bodnar RJ, Hadjimarkou MM. Endogenous opiates and behavior: 2002. *Peptides.* 2003;24(8):1241-1302.
92. Esch T, Stefano GB. The neurobiology of stress management. *Neuroendocrinol Lett.* 2010;31:19-39.
93. Thomas KB. General practice consultations: is there any point in being positive? *Br Med J.* 1987;294:1200-1202.
94. Meredith P, Ownsworth T, Strong J. A review of the evidence linking adult attachment theory and chronic pain: presenting a conceptual model. *Clin Psychol Rev.* 2008;28(3):407-429.
95. Main M, Goldwyn R, eds. *Assessing Attachment through Discourse, Drawings and Reunion Situations.* New York: Cambridge University Press; 1996.
96. Fonagy P, Leigh T, Steele M, et al. The relation of attachment status, psychiatric classification, and response. *J Consult Clin Psychol.* 1996;64:22-31.
97. Brien S, Lachance L, Prescott P, et al. Homeopathy has clinical benefits in rheumatoid arthritis patients that are attributable to the consultation process but not the homeopathic remedy: a randomized controlled clinical trial. *Rheumatol.* 2010. doi:10.1093/rheumatology/keq234.

98. Kaptchuk TJ, Kelley JM, Conboy LA, et al. Components of placebo effect: randomised controlled trial in patients with irritable bowel syndrome. *Br Med J*. 2008;336:999-1003.
99. Agledahl KM, Gulbrandsen P, Forde R, Wifstand A. Courteous but not curious: how doctors' politeness masks their existential neglect. A qualitative study of video-recorded patient consultations. *J Med Ethics*. 2011;37:650-654.
100. Lang EV, Hatsiopoulou O, Koch T, et al. Can words hurt? Patient–provider interactions during invasive procedures. *Pain*. 2005;114:303-309.
101. Arora NK. Interacting with cancer patients: the significance of physicians' communication behavior. *Soc Sci Med*. 2003;57:791-806.
102. Baskin TW, Tierney SC, Minami T, Wampold BE. Establishing specificity in psychotherapy: a meta-analysis of structural equivalence of placebo controls. *J Consult Clin Psychol*. 2003;71:973-979.
103. Wampold BE. *The Great Psychotherapy Debate: Models, Methods, and Findings*. Mahwaw: Lawrence Erlbaum; 2001.
104. Kim DM, Wampold BE, Bolt DM. Therapist effects in psychotherapy: a random-effects modeling of the National Institute of Mental Health Treatment of Depression Collaborative Research Program data. *Psychother Res*. 2006;16:161-172.
105. Grawe K, Donati R, Bernauer F. *Psychotherapie im Wandel. Von der Konfession zur Profession*. Göttingen: Hogrefe; 1994.
106. Ackerman SJ, Hilsenroth MJ. A review of therapist characteristics and techniques positively impacting the therapeutic alliance. *Clin Psychol Rev*. 2003;23:1-33.
107. Iacoviello BM, McCarthy KS, Barrett MS, et al. Treatment preferences affect the therapeutic alliance: implications for randomized controlled trials. *J Consult Clin Psychol*. 2007;75:194-198.
108. Lambert MJ, Simon W. The therapeutic relationship: central and essential in psychotherapy outcome. In: Hick SF, Bien T, eds. *Mindfulness and the Therapeutic Relationship*. New York, London: Guilford Press; 2008:19-33.
109. Grepmair L, Mitterlehner F, Loew T, et al. Promoting mindfulness in psychotherapists in training influences the treatment results of their patients: a randomized, double-blind, controlled study. *Psychother Psychosom*. 2007;76:332-338.
110. Krasner MS, Epstein RM, Beckman H, et al. Association of an educational program in mindfulness communication with burnout, empathy, and attitudes among primary care physicians. *JAMA*. 2009;302:1284-1293.
111. Newton BW, Barber L, Cleveland E, O'Sullivan P. Is there hardening of the heart during medical school? *Academ Med*. 2008;83:244-249.
112. MacPherson H, Mercer SW, Scullion T, Thomas KJ. Empathy, enablement, and outcome: an exploratory study on acupuncture patients' perceptions. *J Altern Complem Med*. 2003;9:869-876.
113. Langford DJ, Crager SE, Shehzad Z, et al. Social modulation of pain as evidence for empathy in mice. *Science*. 2006;312:1967-1970.
114. Jani BD, Blane DN, Mercer SW. The role of empathy in therapy and the physician-patient relationship. *Forschende Komplementärmedizin*. 2012;19:252-257.
115. Branthwaite A, Cooper P. Analgesic effects of branding in treatment of headaches. *Br Med J*. 1981;282:1576-1578.
116. Lévi-Strauss C. *Strukturale Anthropologie I*. Frankfurt: Suhrkamp (Französische Originalausgabe 1958); 1967.
117. Wiesener S, Falkenberg T, Hegyi G, et al. Legal status and regulation of complementary and alternative medicine in Europe. *Forschende Komplementärmedizin*. 2012;19(suppl 2):29-36.
118. Penston J. Why is alternative medicine alone under censure? *Br Med J*. 2012;344:e1632.
119. Caspar F. Psychotherapy research and neurobiology: challenge, chance, or enrichment? *Psychother Res*. 2003;13:1-23.
120. Leganger A, Kraft P, Roysamb E. Perceived self-efficacy in health behavriour research: conceptualisation, measurement and correlates. *Psychol Health*. 2000;15:51-69.
121. Segar J. Complementary and alternative medicine: exploring the gap between evidence and usage. *Health*. 2012;16:366-381.
122. Pincus D. Self-organizing biopsychosocial dynamics and the patient-healer relationship. *Forschende Komplementärmedizin*. 2012;10(suppl 1):22-29.
123. Flatt J. Decontextualized versus lived worlds: critical thoughts on the intersection of evidence, lifeworld, and values. *J Altern Complem Med*. 2012;18(5):1-9.
124. Petrovic P, Kalso E, Petersson KM, Ingvar M. Placebo and opioid analgesia – imaging a shared neuronal network. *Science*. 2002;295:1737-1740.
125. Meissner K, Ziep D. Organ-specificity of placebo effects on blood pressure. *Autonom Neurosci Basic Clin*. 2011;164:62-66.
126. Meissner K. Effects of placebo intervention on gastric motility and general autonomic activity. *J Psychosom Res*. 2009;66:391-398.
127. Montgomery GH, Kirsch I. Mechanisms of placebo pain reduction: an empirical investigation. *Psychol Sci*. 1996;7:174-176.
128. Benedetti F, Arduino C, Amanzio M. Somatotopic activation of opioid systems by target-directed expectations of analgesia. *J Neurosci*. 1999;19:3639-3648.

Conceptualizations and Magnitudes of Placebo Analgesia Effects Across Meta-Analyses and Experimental Studies

Lene Vase[1], Gitte Laue Petersen[2]

[1]Department of Psychology and Behavioural Sciences, School of Business and Social Sciences, Aarhus University, Aarhus, Denmark, [2]Danish Pain Research Center, Aarhus University Hospital, Aarhus, Denmark

INTRODUCTION

Traditionally, the placebo phenomenon has been closely related to medical practice, and through history the conceptualization of placebo effects has evolved with new ways of practicing, testing, and understanding medicine.[1-3] As placebo became a research topic in its own right, studies began to reveal small, medium, and large placebo effects, and researchers began to synthesize these findings via meta-analysis. A number of meta-analyses have been performed, however, with quite different results. The heterogeneous findings are related not only to the various methodologies used, but also to different conceptualizations of placebo analgesia effects.[4-7] A consistent finding across the meta-analyses is a highly variable magnitude of placebo analgesia effects. It is therefore of great interest to specify the factors that influence this magnitude. Meta-analyses may be of some help in this endeavor, but as experimental studies allow direct manipulation of potentially contributing factors, further study of the interplay between meta-analyses and experimental studies may be the best way to advance the understanding of placebo analgesia phenomena.

In this chapter, we briefly outline the development in the conceptualizations and definitions of placebo effects. The most important meta-analyses of placebo analgesia effects are presented, and we show how the various conceptualizations of placebo effects influence the findings of the meta-analyses. This leads to a review of experimental studies that corroborate and specify the results from the meta-analyses. Finally, we discuss new approaches to the investigation of placebo analgesia, and we outline the current status and developments within the field.

DEVELOPMENTS IN THE CONCEPTUALIZATIONS AND DEFINITIONS OF PLACEBO EFFECTS

The word placebo derives from the Latin phrase 'placere' which means 'I want to please'.[1] Traditionally, placebo has been conceptualized as a common method or medicine prescribed in order to please the patient, i.e. not because of its efficiency.[1,2,8,9]

With the introduction of the double-blind, randomized clinical trial (RCT), placebos became conceptualized as inert agents such as sugar pills, saline injections, and sham treatments.[1,2] In order to demonstrate the efficacy of a new pain medication, it was necessary to test it in comparison with an inactive placebo agent.[1] Thus, within this paradigm, placebo agents were used as control conditions, and placebo conditions were not controlled for factors relating to the administration of the inert agent (e.g. doctor–patient relationship, the patient's expectations toward treatment efficacy) or for factors relating to the disease or the design of the study (e.g. spontaneous remission, regression to the mean). The reduction in pain following administration of inert placebo agents is often termed a placebo response or a placebo effect.[10-13] This terminology is, however, misleading. Many diseases fluctuate over time (Fig. 20.1A). If an (inactive) pain treatment is given when pain levels are high, which is often the case when

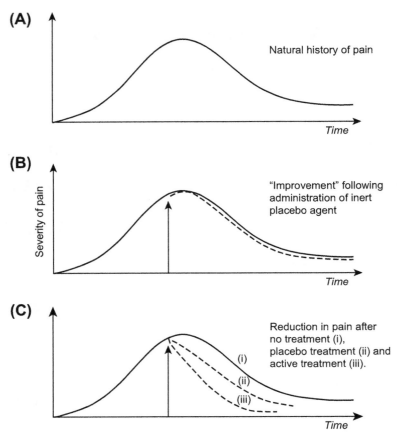

FIGURE 20.1 The development of pain is illustrated under three different conditions. (A) No-treatment control condition showing the natural history of the pain. (B) Treatment with an inert placebo agent showing that the pain reduction following administration of an inert treatment may be due to the natural history of the pain rather than to the treatment. (C) No treatment (i), placebo treatment, (ii) and active treatment (iii) allowing for a calculation of the placebo analgesia effect as the difference in pain levels between the no-treatment and the placebo treatment group or condition.

TABLE 20.1 Definitions of Placebo Agents and Placebo Effects

The placebo analgesic agent: 'an inert agent'[13 p. 35] and/or 'all external aspects of the therapeutic intervention that can be perceived by the patients or the experimental subject'[5 p. 207].

Empirical definition of placebo analgesia effects: 'The measured difference in pain across an untreated and a placebo-treated group or across an untreated and a placebo-treated condition within the same group (as in cross-over studies)'[5 p. 207]. It is important to distinguish the placebo analgesia effect from changes in pain following administration of an inert placebo agent. In the latter, changes in the natural history of the pain are not controlled for, and the pain-relieving effect following placebo administration can therefore not be distinguished from confounding factors such as spontaneous remission and regression to the mean[13 p. 20–23, 21].

Conceptual definition of placebo analgesia effects: '...meaning in the origins or treatment of illness'[18 p. 472] and/or '...the perception of the therapeutic intervention...'[4 p. 451].

Placebo-like effects: 'placebo effects without administration of placebo [inert agents]'[13 p. 35].

patients seek treatment or are included in a trial, it is not possible to deduce whether the subsequent pain reduction is due to the treatment or to the natural history of the pain disease (Fig. 20.1B). Hence, within such a design it is not possible to estimate the effect of the placebo treatment independently from confounding factors such as spontaneous remission and regression to the mean.[13]

As researchers became interested in investigating placebo phenomena in their own right, it was necessary to distinguish the effect of the placebo treatment from changes in the natural history.[14] This was done by including no-treatment control conditions, so the actual effect of the placebo treatment, *the placebo analgesia effect*, could be calculated as the difference in pain levels between the placebo-treated and the no-treatment group or condition (see Fig. 20.1C and Table 20.1).[5,13,15,16] Hence, the inclusion of a no-treatment group or condition enabled an estimation of the magnitude of placebo analgesia effects and at the same time made it possible to investigate factors contributing to it.

From an *empirical* standpoint, there is today agreement that the placebo analgesia effect is 'the measured difference in pain across an untreated and a placebo-treated group or across an untreated and a placebo-treated condition within the same group (as in crossover studies)[5 p. 207–208; 13]. At present, no single definition encapsulates the *conceptualization* of placebo analgesia effects; however, there appears to be agreement that placebo analgesia effects are related to 'the meaning in the origins or treatment of illness'[17–19] and/or to 'the perception of the therapeutic intervention'[4 p. 451]. The growing understanding of the contextual and psychosocial factors that influence placebo analgesia effects has made it possible to investigate the effects without administration of inert placebo agents. These effects have been termed *placebo analgesia-like* effects (see further description later) (see Table 20.1).[13,20]

META-ANALYSES OF PLACEBO ANALGESIA EFFECTS

Placebo Effects: No Control of the Natural History of Pain

One of the first meta-analyses of placebo effects was conducted by Henry Beecher and entitled the 'The Powerful Placebo'.[22] In this meta-analysis, the placebo effect was investigated in 15 clinical trials covering 9 different diseases including pain. Placebo was conceptualized as pharmacologically inert substances, and the effects were calculated as the difference in symptoms before and after treatment. Beecher[22] found that 35.2% (range: 15–58) of the patients reported satisfactory relief following placebo administration, and he argued that the placebo effects should be utilized to a higher extent in clinical practice.

This meta-analysis was conducted at a time when most clinical trials involved only an active and a placebo-treated group/condition. Thus, the pain relief following the administration of placebo was not adjusted or separated from confounding factors such as spontaneous remission and regression to the mean, which made it impossible to estimate the magnitude of an actual placebo analgesia effect. These and other methodologic problems have been pointed out,[8,23–26] but it is still a common assumption that approximately one third of patients respond positively to a placebo treatment.

Placebo Effects: Control of the Natural History of Pain

In 2001, Hrobjartsson and Gøtzsche published a meta-analysis entitled *Is the Placebo Powerless?*[27] which allowed for a calculation of the magnitude of placebo effects insofar as it included studies with a placebo treatment and a no-treatment group/condition (cf. Fig. 20.1C). The meta-analysis involved 114 randomized clinical trials covering 40 different diseases, including pain. No placebo effects were found across most of the conditions. Nevertheless, in the 27 studies with continuous outcome measures of pain, a significant effect of placebo was found as indicated by a pooled standardized mean difference of -0.27 (95% CI: -0.40 to -0.15), corresponding to a reduction in mean pain intensity of 6.5 mm on a 100 mm visual analog scale. Since pain is a subjective experience, and therefore subjectively reported, the authors suggested that the significant placebo analgesia effect might have been caused by response bias (see Table 20.2A).

The above meta-analysis aroused attention and was by some interpreted as if placebo effects did not exist.[28,29] However, the number and quality of the included studies,[30] the interpretation of the data,[31,32] and especially the conceptualization of placebo effects[33–37] were questioned and debated. Hrobjartsson and Gøtzsche conceptualized placebo in a practical sense as 'an intervention labeled as such'[27 p. 1595], primarily including inactive agents such as sugar pills, saline injections, and turned-off vibrators. Interestingly, the doctor–patient relationship[27 p. 1599] and the patient's expectations were not seen as part of the placebo effect.

Placebo Effects: Clinical Trials versus Placebo Mechanism Studies

In 2002, Vase and colleagues conducted two meta-analyses of placebo analgesia effects published in one study.[4] In these meta-analyses, all studies included placebo-treated and no-treatment groups/conditions in order for placebo analgesia effects to be calculated. Yet, the first meta-analysis included double-blind, randomized clinical trials in which placebo was used as a control condition, whereas the second meta-analysis included placebo mechanism studies in which placebo analgesia effects were the focus of the study. By looking at these two different types of meta-analysis, it was possible to examine the extent to which the study design influenced the magnitude of placebo analgesia effects. The first meta-analysis included 23 randomized clinical pain trials and, in line with Hrobjartsson and Gøtzsche, a small but significant placebo analgesia effect was found as indicated by an effect size of 0.15 (range: -0.95 to $+0.57$, Cohen's d). Interestingly, however, the second meta-analysis including 14 placebo analgesia mechanism studies found a large and quite significant placebo analgesia effect of 0.95 (range: -0.64 to $+2.29$, Cohen's d). Noteworthy, the magnitude was significantly different in the two types of meta-analysis ($p=0.003$)[4] (see Table 20.2A).

TABLE 20.2A Meta-Analyses of Placebo Analgesia Effects

Study	Type of study	Number of trials	Number of participants	Effect size	Range
Beecher 1955[22]	RCT[a] No natural history group	15	1082	35.2[b]	15–58
Hrobjartsson and Gøtzsche 2001[27]	RCT	27	1602	−0.27	−0.40 to 0.15 95% CI
Hrobjartsson and Gøtzsche 2004[40]	RCT	44	2833	−0.25	−0.35 to −0.16 95% CI
Hrobjartsson and Gøtzsche 2010[41]	RCT	60	4154	−0.28	−0.36 to −0.19 95% CI
Vase et al 2002[4]	RCT	23	1487	0.15	−0.95 to 0.57, Cohen's d
	Placebo mechanism studies	14	814	0.95	−0.64 to 2.29, Cohen's d
Vase et al 2009[6]	Placebo mechanism studies	24	602	1.00	0.12 to 2.51, Cohen's d

All the included studies are explicitly based on continuous measure of pain, except Beecher (1955).[22]
[a] RCT = randomized clinical trial.
[b] This is not an effect size, but a calculation stating that 35.2 ± 2.2% of patients report satisfactory relief following placebo administration. In this publication, the effect of placebo administration is not calculated separately for pain trials, so this measure includes conditions other than pain.

These meta-analyses indicate that when placebo is used as a control condition in clinical trials, the magnitude is generally low, whereas in placebo mechanism studies, the magnitude of placebo analgesia effect is generally high. This finding is not surprising. Clinical trials were developed to control for placebo effects and are not aimed at optimizing placebo effects. For example, participants are typically told via the informed consent that they may receive 'either an active pain medication or an inactive placebo agent', which may lead patients to have uncertain expectations about treatment effects. On the other hand, placebo mechanism studies often focus on the interaction with the health-care provider, and participants are typically given explicit verbal suggestions to the effect that 'this agent has been shown to powerfully reduce pain in some patients', thereby optimizing patients' potential for developing expectations of low pain levels.

The results of the two meta-analyses have been debated and recalculated,[6,7,38,39] but the finding that magnitudes of placebo analgesia effects are high in placebo mechanism studies remains.[4,6,7] When the mean effect size in the placebo mechanism studies was weighted according to the number of participants,[38] the effect size was 1.14, Cohen's d,[7] and when recalculated based on ambiguities stated by Hrobjartsson and Gøtzsche,[39] the effect size was 0.97, Cohen's d.[6] Moreover, the overall finding of small placebo effects in clinical trials and large placebo effects in placebo mechanism studies has been confirmed by subsequent meta-analyses. In 2004 and 2010, Hrobjartsson and Gøtzsche published updates of their meta-analysis from 2001, including separate analyses of pain trials using continuous outcome measures.[40,41] The update from 2004 included 44 clinical pain trials and found an effect size of −0.25 (95% CI: −0.35 to −0.16), whereas the update from 2010 included 60 clinical pain trials and found an effect size of −0.28 (95% CI: −0.36 to −0.19). Also, Vase and colleagues performed a new meta-analysis involving 24 newly identified placebo mechanism studies published in 21 studies[6] that yielded a large and significant placebo analgesia effect as indicated by an effect size of 1.00 (range: 0.12 to 2.51, Cohen's d) (see Table 20.2A).

Interestingly, meta-analyses of placebo mechanism studies have also started to specify how different types of placebo manipulation and pain induction may influence the magnitude of placebo analgesia effects.[4,6] In the first meta-analyses of placebo mechanism studies, the magnitude was larger in the three studies inducing placebo effects via verbal suggestions combined with a conditioning procedure as compared to the 14 studies using only verbal suggestions and one study using only conditioning. This is indicated by the effect sizes of 1.45 (range: 0.46 to 2.10, Cohen's d), 0.85 (range: −1.11 to 2.29, Cohen's d), and 0.83 (range: 0.82 to 0.84, Cohen's d), respectively, for the three types of study. These conditions add up to 18 because some of the 14 placebo mechanism studies included more than one placebo condition[4] (see Table 20.2B). In the second meta-analysis of placebo mechanism studies, the magnitude of placebo effects was larger in the nine studies using long duration pain stimuli > 20 seconds as compared to the 13 studies using short duration pain stimuli < 20 seconds as indicated by the effect sizes of 0.96 (range: 0.12 to 2.39, Cohen's d) and 0.81 (range: 0.26 to 1.30, Cohen's d), respectively. Two studies could not be classified

TABLE 20.2B Magnitude of Placebo Analgesia Effects Across Different Inductions of Placebo and Different Pain Types in Placebo Mechanism Studies

Study	Induction of placebo	Number of trials	Mean effect size	Range
Vase et al 2002[4]	Verbal suggestion + conditioning	3	1.45	0.46 to 2.10, Cohen's d
	Verbal suggestion	14	0.85	−1.11 to 2.29, Cohen's d
	Conditioning	1	0.83	0.82 to 0.84, Cohen's d
	Induction of pain			
Vase et al 2009[6]	Short duration (<20 s) pain stimuli	13	0.81	0.26 to 1.30, Cohen's d
	Long duration (>20 s) pain stimuli	9	0.96	0.12 to 2.39, Cohen's d
	– Hyperalgesic states	3	1.88	0.47 to 2.46 Cohen's d

as either using short or long duration stimuli in this meta-analysis. The largest placebo analgesia effect was found in the three long-duration stimuli studies that only tested placebo in relation to hyperalgesic states, as shown by an effect size of 1.88 (range: 0.47 to 2.46, Cohen's d)[6] (see Table 20.2B).

Although meta-analyses may specify some of the factors contributing to the magnitude of placebo analgesia effects, experimental studies are needed to directly manipulate and investigate the influence of the contributing factors.

EXPERIMENTAL STUDIES OF FACTORS INFLUENCING THE MAGNITUDE OF PLACEBO ANALGESIA EFFECTS

Verbal Suggestions given for Pain Relief

Pollo and colleagues[42] examined whether different types of verbal suggestion relating to uncertain and certain expectations of pain relief produced different placebo effects. Patients suffering from pain following thoracotomy were treated with an opioid analgesic and an intravenous saline infusion. The patients were allocated to three groups receiving different information about the saline infusion. One group served as a natural history condition and received no information about any analgesic effect of the infusion. A second group was told that 'the medication could be either a painkiller or a placebo'. This verbal suggestion is ambiguous and resembles the use of placebo in the randomized double-blind clinical trial, attempting to minimize the placebo effect. A third group was told that the basal infusion was 'a powerful painkiller', and this verbal suggestion is unambiguous and attempts to maximize the placebo effect. The three groups received identical treatments and only the verbal suggestions differed. Compared with the natural history group, the second (double-blind) group showed a reduced request for pain medication, and the third (deceived) group showed an even larger reduction in request for pain medication. During the study period, the course of pain was the same in the three groups, but the difference in verbal suggestions produced different magnitudes of placebo effects leading to changes in opioid intake.[42] This study suggests that verbal suggestions given for pain relief influence the magnitude of placebo analgesia effects.

Furthermore, Vase and colleagues examined whether the verbal suggestions typically embedded in clinical trials, or explicitly stated in placebo mechanism studies, influence the magnitude of placebo analgesia effects. Two almost similar experimental studies were conducted. In both studies, patients suffering from irritable bowel syndrome (IBS) were exposed to clinically relevant stimuli of rectal balloon distention under no-treatment, rectal placebo, and rectal lidocaine treatment conditions (Fig. 20.2). The first study was conducted as a double-blind, randomized clinical trial aiming to investigate the efficacy of lidocaine.[43] In this study, patients were informed via the informed consent that 'they would receive either lidocaine or inert saline jelly' in the treatment conditions[43] p. 225. The results were typical for a clinical trial: There was an effect of lidocaine, as indicated by a significant difference in pain levels between the lidocaine and the placebo condition, and there was a placebo effect as indicated by a significant difference in pain levels between the placebo and the no-treatment condition (Fig. 20.2, left panel). On the other hand, the second study was conducted as a placebo mechanism

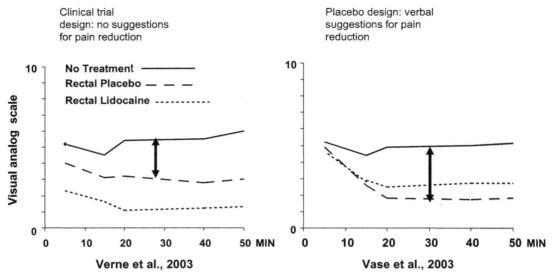

FIGURE 20.2 Comparison of natural history, rectal placebo and rectal lidocaine scores on pain intensity (VAS) during a 50-minute session in a clinical trial where no suggestions for pain relief are given, and in a placebo mechanism study where explicit verbal suggestions for pain relief are given. *(Based on results from references 43 and 44.)*

study aiming to study factors contributing to placebo analgesia effects.[44] In this study, patients were given explicit verbal suggestion to the effect that 'the agent you have just been given is known to powerfully reduce pain in some patients'[44] p. 19 which resulted in a significant placebo effect and no significant difference between pain levels in the placebo and the lidocaine condition (Fig. 20.2, right panel).

These results confirm the findings of the meta-analyses reviewed above[4,6,7,27,40,41] that the magnitude of placebo analgesia effects is low in clinical trials and high in placebo mechanism studies. These experimental studies further substantiate the meta-analyses by showing that verbal suggestions given for pain relief are likely to contribute to the difference in magnitude across the two types of study. In fact, the verbal suggestions given for pain relief in placebo mechanism studies appear to be so effective that they increase the magnitude of placebo analgesia to the level of active pain medication such as lidocaine. The finding that explicit verbal suggestions for pain relief are related to large magnitudes of placebo analgesia effects has been confirmed by subsequent experimental studies.[45–52]

Expectations and Emotions

It seems reasonable to assume that verbal suggestions for pain relief influence pain levels through the patients' expectations of pain relief and possibly by changes in emotional feelings. The verbal suggestion 'this agent is known to powerfully reduce pain in some patients' may lead patients to expect lower pain levels, and the prospect of low pain levels may reduce negative emotions such as anxiety.[5,47,53] In the experimental studies by Vase and colleagues,[44,46] the relationships between verbal suggestions for pain relief, expected pain levels, emotional feelings, and pain levels were directly investigated. In placebo mechanism studies similar to the one described above, IBS patients were asked about their expected pain levels, desire for pain relief, and anxiety levels.[44,46] These psychologic measures were obtained right after the (placebo or lidocaine) treatment had been administered, and before it had taken effect. Expected pain levels and emotional feelings accounted for up to 77% of the variance in pain levels following placebo administration, thereby indicating that expectations and reductions in negative emotions are central to placebo analgesia effects.[44,46] Strikingly, expected pain levels and emotional factors also contributed to the pain-relieving effect of lidocaine,[5,44] thereby demonstrating that placebo factors contribute not only to the efficacy of inert placebo agents, but also to the efficacy of active pain medication.[13]

In these and subsequent studies, the magnitude of placebo analgesia effects increased over time, whereas expected pain levels, desire for pain relief, and anxiety levels decreased over time.[5,47,49,54] These findings suggest that the placebo analgesia effect may be self-reinforcing. The verbal suggestion given for pain relief may lead patients to expect lower pain levels and to feel less anxious. These changes may in themselves contribute to the experience of low pain levels during the placebo analgesia effect, and the actual experience of low pain levels may further stimulate expectations of low pain levels and reduce negative emotions, thereby maintaining or increasing the effect. Several experimental studies have shown that expected pain levels and reductions in negative emotions appear to be pivotal to placebo analgesia effects.[53,55–62]

Brain Imaging Studies and Meta-Analyses of Placebo Analgesia Effects

Factors contributing to the magnitude of placebo analgesia effects in placebo mechanism studies have also been identified through brain-imaging studies. These studies yield important knowledge of the neurologic underpinnings of expectations and emotional feelings involved in placebo analgesia as well as the descending pain modulating structures involved in this effect.[47,49,51,63–71] Despite the heterogeneous design of the studies,[72] the increasing number of brain-imaging studies of placebo analgesia effects has enabled a meta-analysis of the brain areas consistently activated in placebo analgesia.[73]

Amanzio and colleagues[73] conducted a meta-analysis of nine fMRI (functional magnetic resonance imaging) and two PET (positron emission tomography) studies investigating cerebral hemodynamic changes to delineate consistent activation of brain regions associated with placebo analgesia. A total of 199 participants were involved and 162 activated foci were reported across the studies. The meta-analysis investigated two stages of the placebo analgesia effect: *expectation/anticipation*, before the pain starts where expectancies may be established, and *noxious stimuli*, during administration of painful stimuli where placebo effects may be obtained. During expectation/anticipation of analgesia, activated foci were found in the left anterior cingulate, right precentral and lateral prefrontal cortex, and in the left periaqueductal gray (PAG). During noxious stimulation, placebo-related activations were detected in the anterior cingulate and medial and lateral prefrontal cortices, in the left inferior parietal lobule and postcentral gyrus, anterior insula, thalamus, hypothalamus, PAG, and pons. These findings suggest a large overlap in the regions regulating emotional processes and placebo analgesia effects (e.g. the prefrontal cortices and the rostral anterior cingulate cortex), and they also show that nociceptive networks in the brain are down-regulated in parallel with subjectively measured placebo analgesia effects. Furthermore, this meta-analysis shows that meta-analyses may help systematize the findings of potential mechanisms involved in placebo analgesia effects. Nonetheless, in order to interpret these findings, it is essential to take the heterogeneity of the studies into account.[4,72,73]

New Ways of Investigating the Magnitude of Placebo Analgesia Effects

With the maturing understanding of the factors contributing to placebo analgesia effects, new ways of investigating the magnitude of the effects have emerged.[74] Verbal suggestions regarding the probability of receiving an active agent have been manipulated to a higher extent (from 0 to 100% probability) in order to test the influence on the magnitude of the effect.[70] Also, the balanced placebo design has been used where participants are randomized to receive either active or inactive placebo agents along with either correct or incorrect verbal suggestions, leaving four treatment groups.[75–77] The first group receives a placebo and is told that it is a placebo. The second group receives a placebo and is told that it is an active treatment. The third group receives an active treatment and is told that it is an active treatment. The fourth group receives an active treatment and is told that it is a placebo. In this manner it is possible not only to calculate the magnitude of the placebo analgesia effect (the difference in pain levels between the first two groups), but also to estimate the extent to which placebo factors, such as verbal suggestions for pain relief, influence the efficacy of active treatments (the difference in pain levels between the last two groups). Thus, this design allows for a more precise estimation of the contribution of pharmacologic effect versus placebo factors to the total efficacy of active pain treatments (for further information see Ch 16 in this book).

Lately, the magnitude of placebo analgesia effects and factors contributing to this effect have been investigated without administration of inactive agents as seen in the so-called open versus hidden administration of active agents.[78–82] These designs are in accordance with modern conceptualizations of placebo analgesia effects (cf. definitions and Table 20.1) in so far as placebo factors are investigated simply by manipulating whether or not patients *perceive* that they are receiving a treatment. In open administrations, a doctor comes to the bedside, injects an active medication along with a verbal suggestion for pain relief. This allows the patient to actively perceive that a treatment is being given and thus to develop expectations of low pain levels. In hidden administrations, on the other hand, the exact same active medication is given, but here it is administered without the patient's knowledge, for example through an infusion pump, so that the patient does not have the opportunity to develop expectations of treatment effect. The difference in pain levels between the open versus hidden administration of pain medication is conceptualized as a *placebo-like* effect.[13,20] Hence, the conceptualization and the investigation of placebo effects are undergoing a change from inactive agents in double-blind clinical trials towards a more explicit understanding and manipulation of factors influencing the magnitude of placebo analgesia effects.

CURRENT STATUS OF META-ANALYSES OF THE MAGNITUDE OF PLACEBO ANALGESIA EFFECTS

Lessons from Meta-Analyses

What do we know about the magnitude of placebo analgesia effects based on the meta-analyses outlined above? First, there is not *one* effect size that correctly or completely describes the magnitude. Different conceptualizations lead to different magnitudes, as seen in meta-analyses of placebo analgesia effects in clinical trials versus placebo mechanism studies. Second, there is a high degree of variability in the magnitude of placebo analgesia effects, not only across meta-analyses but also within meta-analyses. This may at least in part be due to different contextual and individual preconditions for obtaining,[83,84] as well as different ways of inducing placebo analgesia effects.[4–6] Third, it is possible to investigate the variability in placebo analgesia effects via meta-analyses to some extent, as seen in the meta-analyses of placebo mechanism studies in which different ways of inducing placebo effects and different types of pain lead to different magnitudes. However, the placebo phenomenon is complex, and studies investigating placebo mechanism are often heterogeneous, thereby limiting the validity of the results obtained from meta-analyses.[4,72,73] Fourth, the use of meta-analyses may be most successful when combined with experimental studies that verify, falsify, and/or specify the findings and interpretations. Having said that, the most prevalent finding based on the current meta-analyses of placebo analgesia effects is that the magnitude may range from small in clinical trials to large in placebo mechanism studies, depending on the way in which the placebo effect is induced. Apparently, the more the patient's perception of the treatment is optimized through interaction with the health-care provider, the experimental set-up, and the verbal suggestions for pain relief, the larger is the placebo analgesia effect.

Recent Developments in Meta-Analyses of Placebo Effect

Knowledge of factors influencing the magnitude of placebo effects can be utilized to increase or decrease the magnitude of placebo analgesia effects. Traditionally, there is an interest in increasing the magnitude of placebo effects in placebo mechanism studies as well as in clinical practice in order to optimize the investigation of placebo effects and the patients' treatment outcome. Conversely, in clinical trials there is an interest in minimizing placebo effects in order to prove efficacy of new active treatments.[1] Recently, increasing analgesic effects have been observed following administration of inactive placebo treatments in clinical trials that do not include a natural history control group/condition, and these large effects appear to be an obstacle to approval of supposedly new active medications and even approval of medications previously approved.[85–87] This has led to a renewed focus on the best ways to conduct clinical trials and to test assay sensitivity, i.e. the ability of a clinical trial to distinguish an effective treatment from a less effective or ineffective treatment.[88] Currently, it is discussed whether factors related to patient characteristics, study design, study sites, and outcome measures, may represent ways of improving assay sensitivity.[87] Accordingly, these factors are used as predictors in meta-analyses of effects in clinical pain trials with the aim of specifying factors that contribute to the increasing analgesic effect following placebo administration.[10,12,89] In these meta-analyses, the pain relief following administration of inactive placebo treatment is often denoted placebo analgesia effects[10,12,89]; however, it is important to be aware that these studies do not include a no-treatment control condition and therefore they do not, by definition, measure placebo analgesia effects (cf. conceptualizations and definitions of placebo effects). An alternative approach to understanding the increasing analgesic effects following placebo administration in clinical trials is to utilize the results from placebo mechanism studies. Patients participating in clinical trials are typically informed about possible adverse events of the active treatment under study. As different types of verbal suggestions for pain relief have been shown to lead to different magnitudes of placebo analgesia effects (cf. placebo mechanism studies outlined above), it is possible that different types of negative suggestions about potential adverse events could lead to different types of adverse events in the placebo arm of clinical trials. This hypothesis was tested by Amanzio and collagues[90] who looked at adverse events reported in the placebo arm of clinical trials using one of three classes of anti-migraine drugs: NSAIDs (nonsteroidal anti-inflammatory drugs), triptans, and anti-convulsants. Strikingly, the adverse events were frequent, and the type of adverse events in the placebo arms corresponded to those of the anti-migraine medicine in question.[89]

The approach of applying knowledge from the placebo mechanism literature to the investigation of the increasing analgesic effects following administration of inactive placebo agents in clinical trials without a natural history control group/condition could easily gain currency. For example, approximations of expectations and emotional feelings like randomization rate, strength of active medication, dosing regimen, as well as frequency and type of interaction with health-care professionals, could be used as predictors in clinical trials and meta-analyses, whereby it may be possible to find new ways of explaining the variability in analgesic effects in clinical trials. Such an approach could have far reaching implications for the way of testing pain medication and for the optimization of placebo factors in clinical practice (see Ch 21 in this book). Hence, in future studies it may be recommendable to use current knowledge from placebo mechanism studies to improve the conduction and interpretation of clinical pain trials.

Conflict of Interest

There are no conflicts of interest.

Acknowledgment

Work on this chapter is part of the Europain project and is funded by the Innovative Medicines Initiative Joint Undertaking (IMI JU), grant no. 115007. Further, it is supported by the MINDLab UNIK initiative at Aarhus University, which is funded by the Danish Ministry of Science, Technology and Innovation.

References

1. Andersen LO. Placebo – historisk og kulturelt. In: Andersen LO, Claësson MH, Hrobjartsson A, Nørbæk Sørensen A, eds. *Placebo. Historie, Biologi Og Effekt*: Akademisk Forlag A/S; 1997:69-118.
2. Harrington A, ed. *The Placebo Effect. an Interdisciplinary Exploration*. Cambridge, Massachusetts: Harvard University Press; 1997.
3. Andersen LO. *Før Placeboeffekten. Indbildningskraftens Virkning i 1800-Tallets Medicin*. København: Museum Tusculanums Forlag; 2011.
4. Vase L, Riley 3rd JL, Price DD. A comparison of placebo effects in clinical analgesic trials versus studies of placebo analgesia. *Pain*. 2002;99(3):443-452.
5. Vase L, Price DD, Verne GN, Robinson ME. The contribution of changes in expected pain levels and desire for pain relief to placebo analgesia. In: Price DD, Bushnell MC, eds. *Psychological Methods of Pain Control: Basic Science and Clinical Perspectives*. vol. 29. Seattle: IASP Press; 2004:207-232.
6. Vase L, Petersen GL, Riley 3rd JL, Price DD. Factors contributing to large analgesic effects in placebo mechanism studies conducted between 2002 and 2007. *Pain*. 2009;145(1-2):36-44.
7. Price DD, Riley 3rd JL, Vase L. Reliable differences in placebo effects between clinical analgesic trials and studies of placebo analgesia mechanisms. *Pain*. 2003;104:715-716.
8. Shapiro AK, Morris LA. Placebo effects in medical and psychological therapies. In: Garfield SL, Bergin AE, eds. *Handbook of Psychotherapy and Behaviour Change; An Empirical Analysis*. 2nd ed. New York: Wiley; 1978:369-409.
9. Ekeland TJ. *Meining som medisin: Ein analyse av placebofenomenet of implikasjonar for terapi og terapeutiske teoriar*. Psykologisk Fakultet, Universitetet i Bergen: Institut for samfunnspsykologi; 1999.
10. Pitz M, Cheang M, Bernstein CN. Defining the predictors of the placebo response in irritable bowel syndrome. *Clin Gastroenterol Hepatol*. 2005;3(3):237-247.
11. Dworkin RH, O'Connor AB, Audette J, et al. Recommendations for the pharmacological management of neuropathic pain: an overview and literature update. *Mayo Clin Proc*. 2010;85(suppl 3):S3-S14.
12. Hauser W, Bartram-Wunn E, Bartram C, et al. Systematic review: placebo response in drug trials of fibromyalgia syndrome and painful peripheral diabetic neuropathy – magnitude and patient-related predictors. *Pain*. 2011;152(8):h1709-h1717.
13. Benedetti F. *Placebo Effects. Understanding the Mechanisms of Health and Disease*. Oxford: Oxford University Press; 2009.
14. Fields HL, Price DD. Toward a neurobiology of placebo analgesia. In: Harrington A, ed. *The Placebo Effect. an Interdisciplinary Exploration*. Massachusetts, Cambridge: Harvard University Press; 1997:93-116.
15. Levine JD, Gordon NC, Fields HL. The mechanism of placebo analgesia. *Lancet*. 1978;2(8091):654-657.
16. Levine JD, Gordon NC, Bornstein JC, Fields HL. Role of pain in placebo analgesia. *Proc Natl Acad Sci USA*. 1979;76(7):3528-3531.
17. Moerman DE, Jonas WB. Toward a research agenda on placebo. *Adv Mind Body Med*. 2000;16(1):33-46.
18. Moerman DE, Jonas WB. Deconstructing the placebo effect and finding the meaning response. *Ann Intern Med*. 2002;136(6):471-476.
19. Moerman DE. The meaning response and the ethics of avoiding placebos. *Eval Health Prof*. 2002;25(4):399-409.
20. Benedetti F. Mechanisms of placebo and placebo-related effects across diseases and treatments. *Annu Rev Pharmacol Toxicol*. 2008;48:33-60.
21. Fields HL, Levine JD. Pain – mechanics and management. *West J Med*. 1984;141(3):347-357.
22. Beecher HK. The powerful placebo. *J Am Med Assoc*. 1955;159(17):1602-1606.
23. Wall P. The placebo effect: an unpopular topic. *Pain*. 1992;51:1-3.
24. Heeg MJ, Deutsch KF, Deutsch E. The placebo effect. *Eur J Nucl Med*. 1997;24(11):1433-1440.
25. Kienle GS, Kiene H. Placebo effect and placebo concept: A critical methodological and conceptual analysis of reports on the magnitude of the placebo effect. *Altern Ther Health Med*. 1996;2(6):39-54.
26. Kienle GS, Kiene H. The powerful placebo effect: fact or fiction? *J Clin Epidemiol*. 1997;50(12):1311-1318.
27. Hrobjartsson A, Gøtzsche PC. Is the placebo powerless? An analysis of clinical trials comparing placebo with no treatment. *N Engl J Med*. 2001;344(21):1594-1602.
28. Kolata G. Study casts doubt on the placebo effect. *The New York Times*. 24 May 2011.
29. Okie S. Analysis challenges 'placebo effect'. *The Washington Post*. 24 May 2011.
30. Einarson TE, Hemels M, Stolk P. Is the placebo powerless? *N Engl J Med*. 2001;345(17):1277; author reply 1278-1279.
31. Kaptchuk TJ. Is the placebo powerless? *N Engl J Med*. 2001;345(17):1277; author reply 1278-1279.
32. Shrier I. Is the placebo powerless? *N Engl J Med*. 2001;345(17):1278; author reply 1278-1279.
33. Miller FG. Is the placebo powerless? *N Engl J Med*. 2001;345(17):1277; author reply 1278-1279.
34. Lilford RJ, Braunholtz DA. Is the placebo powerless? *N Engl J Med*. 2001;345(17):1277-1278; author reply 1278-1279.
35. Kupers R. Is the placebo powerless? *N Engl J Med*. 2001;345(17):1278; author reply 1278-1279.
36. Wampold BE, Minami T, Tierney SC, et al. The placebo is powerful: estimating placebo effects in medicine and psychotherapy from randomized clinical trials. *J Clin Psychol*. 2005;61(7):835-854.
37. Miller FG, Rosenstein DL. The nature and power of the placebo effect. *J Clin Epidemiol*. 2006;59(4):331-335.
38. Hrobjartsson A, Gøtzsche PC. Unreliable analysis of placebo analgesia in trials of placebo pain mechanisms. *Pain*. 2003;104(3):714-715; author reply 715-716.
39. Hrobjartsson A, Gøtzsche PC. Unsubstantiated claims of large effects of placebo on pain: serious errors in meta-analysis of placebo analgesia mechanism studies *J Clin Epidemiol*. 2006;59(4):336-338; discussion 339-341.
40. Hrobjartsson A, Gøtzsche PC. Placebo interventions for all clinical conditions. *Cochrane Database Syst Rev*. 2004(3):CD003974.

41. Hrobjartsson A, Gøtzsche PC. Placebo interventions for all clinical conditions. *Cochrane Database Syst Rev.* 2010(1):CD003974.
42. Pollo A, Amanzio M, Arslanian A, et al. Response expectancies in placebo analgesia and their clinical relevance. *Pain.* 2001;93(1):77-84.
43. Verne GN, Robinson ME, Vase L, Price DD. Reversal of visceral and cutaneous hyperalgesia by local rectal anesthesia in irritable bowel syndrome (IBS) patients. *Pain.* 2003;105(1-2):223-230.
44. Vase L, Robinson ME, Verne GN, Price DD. The contributions of suggestion, desire, and expectation to placebo effects in irritable bowel syndrome patients. an empirical investigation. *Pain.* 2003;105(1-2):17-25.
45. Pollo A, Vighetti S, Rainero I, Benedetti F. Placebo analgesia and the heart. *Pain.* 2003;102(1-2):125-133.
46. Vase L, Robinson ME, Verne GN, Price DD. Increased placebo analgesia over time in irritable bowel syndrome (IBS) patients is associated with desire and expectation but not endogenous opioid mechanisms. *Pain.* 2005;115(3):338-347.
47. Price DD, Craggs J, Verne GN, et al. Placebo analgesia is accompanied by large reductions in pain-related brain activity in irritable bowel syndrome patients. *Pain.* 2007;127(1-2):63-72.
48. Chung SK, Price DD, Verne GN, Robinson ME. Revelation of a personal placebo response: its effects on mood, attitudes and future placebo responding. *Pain.* 2007;132(3):281-288.
49. Craggs JG, Price DD, Perlstein WM, et al. The dynamic mechanisms of placebo induced analgesia: evidence of sustained and transient regional involvement. *Pain.* 2008;139(3):660-669.
50. Kaptchuk TJ, Kelley JM, Conboy LA, et al. Components of placebo effect: randomized controlled trial in patients with irritable bowel syndrome. *BMJ.* 2008;336(7651):999-1003.
51. Elsenbruch S, Kotsis V, Benson S, et al. Neural mechanisms mediating the effects of expectation in visceral placebo analgesia: an fMRI study in healthy placebo responders and nonresponders. *Pain.* 2012;153(2):382-390.
52. Petersen GL, Finnerup NB, Norskov KN, et al. Placebo manipulations reduce hyperalgesia in neuropathic pain. *Pain.* 2012;153(6):1292-1300.
53. Flaten MA, Aslaksen PM, Lyby PS, Bjorkedal E. The relation of emotions to placebo responses. *Philos Trans R Soc Lond B Biol Sci.* 2011;366(1572):1818-1827.
54. Vase L, Norskov KN, Petersen GL, Price DD. Patients' direct experiences as central elements of placebo analgesia. *Philos Trans R Soc Lond B Biol Sci.* 2011;366(1572):1913-1921.
55. Montgomery GH, Kirsch I. Classical conditioning and the placebo effect. *Pain.* 1997;72(1-2):107-113.
56. Price DD, Milling LS, Kirsch I, et al. An analysis of factors that contribute to the magnitude of placebo analgesia in an experimental paradigm. *Pain.* 1999;83:147-156.
57. De Pascalis V, Chiaradia C, Carotenuto E. The contribution of suggestibility and expectation to placebo analgesia phenomenon in an experimental setting. *Pain.* 2002;96(3):393-402.
58. Petrovic P, Dietrich T, Fransson P, et al. Placebo in emotional processing – induced expectations of anxiety relief activate a generalized modulatory network. *Neuron.* 2005;46(6):957-969.
59. Charron J, Rainville P, Marchand S. Direct comparison of placebo effects on clinical and experimental pain. *Clin J Pain.* 2006;22(2):204-211.
60. Goffaux P, Redmond WJ, Rainville P, Marchand S. Descending analgesia – when the spine echoes what the brain expects. *Pain.* 2007;130(1-2):137-143.
61. Geers AL, Wellman JA, Fowler SL, et al. Dispositional optimism predicts placebo analgesia. *J Pain.* 2010;11(11):1165-1171.
62. Aslaksen PM, Flaten MA. The roles of physiological and subjective stress in the effectiveness of a placebo on experimentally induced pain. *Psychosom Med.* 2008;70(7):811-818.
63. Petrovic P, Kalso E, Petersson KM, Ingvar M. Placebo and opioid analgesia – imaging a shared neuronal network. *Science.* 2002;295(5560):1737-1740.
64. Wager TD, Rilling JK, Smith EE, et al. Placebo-induced changes in FMRI in the anticipation and experience of pain. *Science.* 2004;303(5661):1162-1167.
65. Bingel U, Lorenz J, Schoell E, et al. Mechanisms of placebo analgesia: RACC recruitment of a subcortical antinociceptive network. *Pain.* 2006;120(1-2):8-15. doi:10.1016/j.pain.2005.08.027.
66. Kong J, Gollub RL, Rosman IS, et al. Brain activity associated with expectancy-enhanced placebo analgesia as measured by functional magnetic resonance imaging. *J Neurosci.* 2006;26(2):381-388.
67. Nemoto H, Nemoto Y, Toda H, et al. Placebo analgesia: a PET study. *Exp Brain Res.* 2007;179(4):655-664.
68. Eippert F, Bingel U, Schoell ED, et al. Activation of the opioidergic descending pain control system underlies placebo analgesia. *Neuron.* 2009;63(4):533-543.
69. Eippert F, Finsterbusch J, Bingel U, Buchel C. Direct evidence for spinal cord involvement in placebo analgesia. *Science.* 2009;326(5951):404.
70. Kotsis V, Benson S, Bingel U, et al. Perceived treatment group affects behavioral and neural responses to visceral pain in a deceptive placebo study *Neurogastroenterol Motil.* 2012;24(10):935-e462.
71. Watson A, El-Deredy W, Iannetti GD, et al. Placebo conditioning and placebo analgesia modulate a common brain network during pain anticipation and perception. *Pain.* 2009;145(1-2):24-30.
72. Rainville P, Duncan GH. Functional brain imaging of placebo analgesia: methodological challenges and recommendations. *Pain.* 2006;121(3):177-180.
73. Amanzio M, Benedetti F, Porro CA, et al. Activation likelihood estimation meta-analysis of brain correlates of placebo analgesia in human experimental pain. *Hum Brain Mapp.* 2013;34(3):738-752.
74. Colloca L, Lopiano L, Lanotte M, Benedetti F. Overt versus covert treatment for pain, anxiety, and parkinson's disease. *Lancet Neurol.* 2004;3(11):679-684.
75. Ross S, Krugman AD, Lyerly SB, Clyde DJD. Drugs and placebos: a model design. *Psychol Rep.* 1962;10:383-392.
76. Flaten MA. Drug effects: agonistics and antagonistic processes. *Scand J Psychol.* 2009;50(6):652-659.
77. Rief W, Glombiewski JA. The hidden effects of blinded, placebo-controlled randomized trials: an experimental investigation. *Pain.* 2012;153:2473-2477.
78. Levine JD, Gordon NC, Smith R, Fields HL. Analgesic responses to morphine and placebo in individuals with postoperative pain. *Pain.* 1981;10(3):379-289.
79. Gracely RH, Dubner R, Wolskee PJ, Deeter WR. Placebo and naloxone can alter post-surgical pain by separate mechanisms. *Nature.* 1983;306(5940):264-265.
80. Levine JD, Gordon NC. Influence of the method of drug administration on analgesic response. *Nature.* 1984;312(5996):755-756.
81. Amanzio M, Pollo A, Maggi G, Benedetti F. Response variability to analgesics: a role for non-specific activation of endogenous opioids. *Pain.* 2001;90(3):205-215.
82. Bingel U, Wanigasekera V, Wiech K, et al. The effect of expectation on drug efficacy: imaging the analgesic benefit of the opioid remifentanil. *Sci Trans Med.* 2011;16(3):70ra14.
83. Geers AL, Helfer SG, Kosbab K, et al. Reconsidering the role of personality in placebo effects: dispositional optimism, situational expectations, and the placebo response. *J Psychosom Res.* 2005;58(2):121-127.
84. Zubieta JK, Yau WY, Scott DJ, Stohler CS. Belief or need? accounting for individual variations in the neurochemistry of the placebo effect. *Brain Behav Immun.* 2006;20(1):15-26.
85. Silberman S. Placebos are getting more effective. Drugmakers are desperate to know why. *Wired Magazine.* 2009:08.24.09.
86. Usdin S. Product discovery and development: shaking the cup for pain. *BioCentury.* 2011;19:13-14.

87. Dworkin RH, Turk DC, Peirce-Sandner S, et al. Considerations for improving assay sensitivity in chronic pain clinical trials: IMMPACT recommendations. *Pain*. 2012;153(6):1148-1158.
88. International Conference on Harmonisation. International conference on harmonisation. E10. choice of control groups and related issues in clinical trials. <http://www.fda.gov/downloads/RegulatoryInformation/Guidances/ucm125912.pdf> Accessed 01.08.2012.
89. Dworkin RH, Turk DC, Katz NP, et al. Evidence-based clinical trial design for chronic pain pharmacotherapy: a blueprint for ACTION. *Pain*. 2011;152(suppl 3):S107-S115.
90. Amanzio M, Corazzini LL, Vase L, Benedetti F. A systematic review of adverse events in placebo groups of anti-migraine clinical trials. *Pain*. 2009;146(3):261-269.

CHAPTER 21

The Contribution of Desire, Expectation, and Reduced Negative Emotions to Placebo Anti-Hyperalgesia in Irritable Bowel Syndrome

Donald D. Price[1], Lene Vase[2]

[1]Division of Neuroscience, Department of Oral and Maxillofacial Surgery, University of Florida, Florida, USA,
[2]Department of Psychology and Behavioural Sciences, School of Business and Social Sciences, Aarhus University, Aarhus, Denmark

INTRODUCTION

Many chronic pain conditions have lowered thresholds for pain (allodynia) and/or enhanced pain sensitivity (hyperalgesia). The overall pain intensity of chronic pain patients often reflects a combination of allodynia, hyperalgesia, and sometimes nociceptive stimulation (i.e. stimulation that produces tissue damage, or would produce tissue damage if maintained). These components of chronic pain are often partly sustained by peripheral pathologic conditions, such as inflammation, dysfunction of peripheral nerves and nerve endings, as well as central nervous system pathology. Chronic pain conditions are also greatly influenced by psychologic factors and hence central nervous system inhibitory and facilitatory mechanisms. Studies of patients with irritable bowel syndrome (IBS) have provided an interesting model that can be used to assess peripheral nervous system, central nervous system, and psychologic factors that induce and maintain this painful condition. It is an interesting model because reducing peripheral impulse input powerfully attenuates both ongoing and evoked rectal pain in IBS patients, yet so does placebo suggestion. In this chapter, we first review evidence that ongoing and evoked pain and associated secondary hyperalgesia are dynamically maintained by peripheral impulse input. This discussion is then followed by evidence that this type of pain can also be potently attenuated by psychologic factors such as placebo suggestions. Then we provide an explanation that accommodates both sets of observations and shows how it may apply to other forms of persistent pain, including neuropathic pain. The overarching aim of this chapter is to provide evidence for a synergistic interaction between peripheral and central contributions to pain and pain reduction (e.g. placebo factors) in IBS (Fig. 21.1). A main point of this explanation is that placebos provide powerful anti-hyperalgesic effects that have only recently been recognized. Such effects are likely to have important implications for treatment and understanding of IBS and other chronic pain conditions.

EVIDENCE FOR VISCERAL AND SOMATIC HYPERALGESIA IN IBS PATIENTS

The overall ideas outlined above begin with evidence that at least some IBS patients, and other patients with persistent neuropathic pain conditions, have both primary and secondary hyperalgesia. Several studies compared results of both clinically relevant painful rectal distention and painful cutaneous thermal stimulation (20 seconds at 45–47°C temperatures to hand and foot) in IBS patients with age/sex-matched normal control subjects.[1–3] Large magnitudes of visceral and somatic hyperalgesia were found in the first study of female IBS patients.[2] These patients gave higher pain ratings to phasic rectal distention pressures of 35 and 55 mmHg in comparison to normal control subjects, as in previous studies[4,5] (Fig. 21.2). Heat-pain sensitivity was tested in

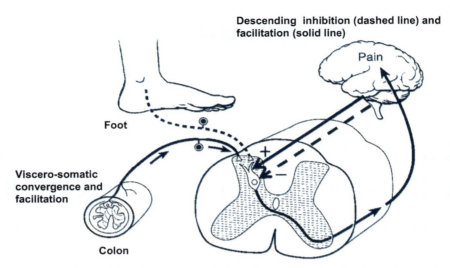

FIGURE 21.1 Schematic of descending and dorsal horn circuitry that mediates central sensitization and hyperalgesia as well as descending inhibition and anti-hyperalgesia.

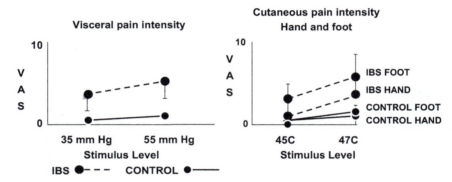

FIGURE 21.2 IBS patients' and normal control subjects' VAS pain intensity ratings of rectal distention pressures of 35 and 55 mmHg (left panel) and of thermal stimulation of the foot (right panel). Note that patients with IBS rate pain intensity from rectal distension and cutaneous heat stimuli ($p < 0.001$). Values are represented as means ± standard deviation, n = 12 IBS patients, 17 controls. *Based on data from reference 2.*

the same study by asking each subject to immerse his/her right hand (up to the level of the wrist) or right foot (up to the level of the right malleolus) in a circulating, heated, water bath at temperatures of 45°C and 47°C for 20 seconds. IBS patients rated cutaneous thermal pain in the hand and the foot as much more intense and unpleasant in comparison with control subjects, thereby demonstrating widespread secondary hyperalgesia (Fig. 21.2). Moreover, heat hyperalgesia was greater for the foot than for the hand among IBS patients. Very similar results were found in a study of male IBS patients.[1]

A third study of female IBS patients included both pain ratings and measures of pain-related brain activity using functional magnetic resonance imaging (fMRI).[6] In comparison with age- and sex-matched control subjects, IBS patients had both visceral and cutaneous thermal hyperalgesia, as measured by pain intensity and unpleasantness ratings, that was accompanied by corresponding increased activation of brain regions involved in pain processing, including ventroposterior lateral thalamus, somatosensory areas I and II, insular cortex, anterior cingulate cortex (ACC), and prefrontal cortical areas.[6] Thus, in comparison to age/sex-matched control subjects, IBS patients had increased pain-related activation within an entire network of brain areas, including those involved in early levels of somatosensory processing within the brain, such as the thalamus.[7] This widespread pattern suggests that widespread secondary hyperalgesia is the result of central sensitization that occurs at early levels of processing, such as the spinal dorsal horn (Fig. 21.1)

VISCERAL AND SOMATIC HYPERALGESIA IS DYNAMICALLY MAINTAINED BY TONIC PERIPHERAL IMPULSE INPUT

This pattern of heat hyperalgesia in some IBS patients, including a larger magnitude of hyperalgesia within the foot as compared to the hand, might

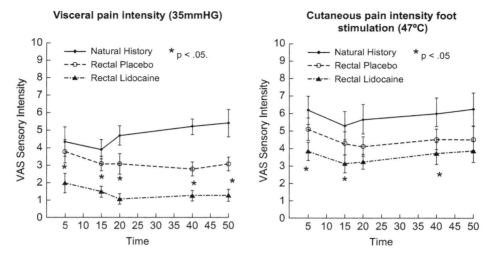

FIGURE 21.3 Left panel: VAS pain intensity ratings during rectal distention (35 mmHg). Three separate sessions were conducted: baseline or natural history (NH), rectal placebo (RP), and rectal lidocaine (RL). These sessions were conducted within a standard clinical trial design without suggestions for analgesia (10 IBS patients). Right panel: VAS pain intensity ratings of cutaneous heat pain in the same IBS patients, showing a similar reversal of heat hyperalgesia by lidocaine. *(Based on a comparison of results of references 25, 26 and 27.)*

be at least partially explained by a mechanism that partly relies on well established evidence that visceral (rectum/colon) and cutaneous (foot) nociceptive afferents converge onto common spinal lumbosacral neurons within the dorsal horn (Fig. 21.1)[8] and references therein. Tonic activity in rectal primary afferents could then sensitize the responses of these neurons to *both* rectal and thermal skin stimuli. Since dorsal horn neurons with both somatic and visceral input project to pain-related brain regions, the consequences of central sensitization and somatovisceral convergence would be reflected in primary and secondary hyperalgesia, and in some cases, allodynia. This possible mechanism is consistent with previous clinical observations showing that IBS patients often exhibit a number of extraintestinal pain symptoms such as back pain, migraine headaches, heartburn, dyspareunia, and muscle pain.[9] These symptoms may reflect widespread central hyperalgesic mechanisms. Similar to other pain conditions that likely depend on peripheral impulse input, such as complex regional pain syndrome (CRPS), postherpetic neuralgia, and fibromyalgia (FM), IBS patients seem to develop widely distributed hyperalgesia, possibly related to chronic peripheral impulse input from the rectum and colon.

Previously, hypersensitivity in IBS was thought to be limited to the gut.[9,10] However, the studies just discussed show that a subset of IBS patients demonstrate hyperalgesia to nociceptive stimuli applied to the extremities.[1-3,11,12] It has been proposed that processing of visceral and somatic stimuli may interact as a result of viscerosomatic convergence-facilitation in the lumbosacral spinal cord.[9,13] Tonic afferent impulse, especially from primary nociceptive afferents from the gut, may sensitize spinal cord neurons that exhibit somatotopic overlap with the gut, such as those associated with the foot and other parts of lower extremities. Pain in IBS patients is likely to be at least partly maintained by peripheral impulse input from the colon/rectum because several studies show that local rectal–colonic anesthesia normalizes visceral and somatic hyperalgesia in IBS patients and in rat models of IBS.[3,14,15] Given the presence of widespread zones of hyperalgesia in neuropathic pain, fibromyalgia (FM), and IBS patients, it is possible that hyperalgesia of these patients is at least partly maintained by tonic impulse input from nociceptive and/or non-nociceptive primary afferent neurons.

The role of tonic impulse input from the rectum of IBS patients was tested by administering controlled rectal distention and cutaneous heat stimuli before and after rectal administration of lidocaine gel or saline gel in a double-blind cross-over basis.[3] The comparison was ideal because it has been demonstrated that subjects cannot subjectively distinguish the two agents when applied rectally.[3] In comparison to saline placebo, lidocaine jelly completely normalized not only rectal hyperalgesia, as shown in Figure 21.3 (left), but also hyperalgesia to thermal stimuli applied to the foot (Fig. 21.3, right). These results were not caused by systemic absorption of lidocaine because (1) the lidocaine gel was directly applied to wall of the rectum, (2) blood levels of lidocaine remained below the lower limit of detection for 50 minutes, and (3) most of the effects were present well before (5 minutes after treatment) maximum systemic absorption would taken place even with liquid lidocaine (1–2 hours after treatment).[3] A similarly designed cross-over study found that intrarectal lidocaine produced 4 to 6 hours of large reductions in ongoing ('spontaneous') pain of IBS patients.[16] Thus, consistent with the role of peripheral impulse activity in some neuropathic pain conditions, tonic impulse input from a peripheral source dynamically maintains not only primary hyperalgesia from the rectum/colon but also the secondary hyperalgesia that is spatially remote (e.g. foot, hand) from the peripheral source of impulse input (i.e. rectum/colon). It also appears to maintain ongoing pain.[16]

ANIMAL MODELS OF HYPERALGESIA IN IBS

Animal models have produced very similar behavioral results[8,15,17] and some of these studies have extended understanding of neural mechanisms by providing recordings from primary afferent and dorsal horn neurons.[7,8] One animal model shows that 24% of rats initially treated with intracolonic trinitrobenzene sulfonic acid (TNBS) continue to display hypersensitivity to both rectal distension and somatic stimuli (e.g. heat stimulation of foot and tail) long after they healed from TNBS-induced colitis.[15] Like IBS patients, the somatic hypersensitivity was largest in lumboscacral dermatomes. This combination of hypersensitivity and histologically normal colons/rectums resembles a main characteristic of IBS, a condition that develops in 25% of persons who have had infectious diarrhea.[2] Similar to the study of IBS patients described above, intracolonic lidocaine administration in hypersensitive rats normalized both rectal and somatic hypersensitivity without producing detectable blood levels of lidocaine.[15] Another model of IBS used mustard oil injections in neonatal rats and produced delayed visceral and somatic hypersensitivity, similar to TNBS-treated rats.[8,17] Rats were tested in this model during adulthood and during a time of absence of colorectal histologic pathology. Compared with adult control rats treated neonatally with saline, adult rats treated neonatally with mustard oil enemas retained their condition of chronic visceral hypersensitivity manifested by increased contractility of abdominal muscles. These same rats also displayed hypersensitivity to cutaneous nociceptive stimulation within widespread regions but most predominantly in lumbosacral dermatomes, again similar to IBS patients and to the TNBS model.[8] Furthermore, their colorectal primary afferent neurons had higher spontaneous impulse activity and much higher impulse response to graded levels of rectal distension in comparison to rats treated only with rectal saline.[8,17] The same authors found that dorsal horn neurons receiving both rectal and somatic input had much higher levels of evoked impulse activity from both rectal distension and somatic stimuli as well as higher spontaneous activity, all in comparison to control rats.[8] The somatic hypersensitivity was greatest in lumbosacral dermatomes. All of these results strongly parallel those found for IBS patients and provide further evidence that both visceral and somatic hyperalgesia are dynamically maintained by increased activity in primary colorectal afferent neurons and by consequent sensitization of the dorsal horn neurons on which they synapse. These dorsal horn neurons are hyper-responsive to *both* visceral and somatic stimuli because tonic visceral impulse input sensitizes dorsal horn neurons that also receive input from other tissues. It thereby makes them hypersensitive to stimulation of other tissues such as skin (Fig. 21.1).

The Convergence-Facilitation Theory of Mackenzie (1909)[13]

Finally, a recent experiment involving IBS patients showed that repetitive rectal stimulation resulted in heat hyperalgesia of the foot, whereas repetitive painful heat stimulation of the foot resulted in rectal hyperalgesia.[18] Both of these forms of secondary hyperalgesia were blocked by dextromethorphan, an N-methyl-D-aspartate (NMDA receptor antagonist). Both forms of facilitation occurred only in a subset of IBS patients that had both somatic and visceral hyperalgesia. Results of this study, along with those described above, further support a convergence-facilitation theory originally proposed by MacKenzie in 1909.[13] It extends this theory by showing that the facilitation occurs in both directions (visceral → somatic and somatic → visceral) and that it occurs in pathophysiologic conditions. This study, along with those described above, used quantitative sensory tests and pharmacologic manipulations (e.g. local lidocaine gel, dextromethorphan) to show that secondary hyperalgesia is dynamically induced and maintained by tonic impulse activity from the rectum and colon, in the case of IBS, and that it is at least partly mediated by NMDA receptor mechanisms and central sensitization. Animal studies described above, including those which uniquely rely on neural recordings from primary colonospinal afferents and dorsal horn neurons on which they synapse, further support this overall explanation.[8,17]

PSYCHOLOGIC CONTRIBUTIONS TO HYPERALGESIA AND ANTI-HYPERALGESIA IN IBS

The primary afferent and dorsal horn mechanisms just described seem to be at odds with a prevailing and alternative viewpoint that numerous psychologic factors contribute to and may even be etiologic factors in 'functional' bowel disorders such as IBS.[9] These factors are integrally related to hypervigilance, level of somatic focus, or other factors related to emotional regulation. The relationship of negative affect to pain conditions such as IBS is well-documented in the literature.[19] In the large majority of published reports, the presence of negative emotions is associated with higher levels of pain. Induction of these emotions has also been shown to be related to pain report and pain behavior with some specificity to the type of emotion induced.[20-22] Interventions or instructions that reduce negative emotion also reduce pain report.[23,24] IBS patients have been shown to have a propensity for hypervigilance and somatic focus, and their painful symptoms may be at least partly maintained by these factors.[9] If so, then it should be possible to systematically enhance or reduce IBS pain by psychologic manipulations. One way of testing this possibility has been that of modulating evoked rectal and somatic pain in IBS patients by various types of placebo/nocebo suggestions.[25-27]

Placebo Effects Across Two Types of Study of IBS Patients

Our studies of placebo effects on evoked and spontaneous pain in IBS have used two general conditions, one in which the study was conducted as a clinical trial[3,16] and the other in which suggestions were given to enhance the placebo effect.[25-27] IBS patients in two clinical studies were given an informed consent form which stated that they 'may receive an active pain-reducing mediation or an inert placebo agent.' In one study there was a significant pain-relieving effect of rectal lidocaine as compared to saline placebo on pain evoked by 35 mmHg rectal distension pressure and there was a significant pain-relieving effect of rectal placebo as compared to the untreated baseline condition,[3] as has already been discussed in relation to Figure 21.3. The second study found no placebo effect on ongoing abdominal pain in IBS patients and a large effect of intra-rectal lidocaine.[16] Thus, when given standard and accurate instructions about possible agents they may receive, patients show highly variable placebo effects ranging from moderate to none and these effects usually (but not always) do not obscure the demonstration of effects of active agents. Results of these two studies can be compared to three studies in which IBS patients were told 'the agent you have just been given is known to significantly reduce pain in some patients' at the onset of each treatment condition.[25,27,28] In comparison to studies with the standard clinical trial instructions (Fig. 21.4, left), a much larger placebo analgesic effect was found in a study with enhanced placebo suggestions,[25] as shown in Figure 21.4 (Fig. 21.4, right, shows a greater placebo effect than is shown in Fig. 21.4, left). In fact, the magnitude of placebo analgesia was so high that there were no longer significant differences between the magnitude of rectal lidocaine and rectal placebo.[25,27,28] The comparison between outcomes using standard versus enhanced placebo suggestions indicate that, by adding an overt suggestion for pain relief, it is possible to increase the magnitude of placebo analgesia to a level that matches that of a known active agent. It is important to recognize that these effects reflect anti-hyperalgesic effects, because both rectal lidocaine and placebo suggestions normalized rectal hyperalgesia and did not eliminate all pain from balloon distention. Thus, pain ratings of IBS patients after either enhanced placebo or lidocaine were similar in magnitude to those of normal control subjects.[3,25]

Patient Experiences of Expectation and Desire for Relief as Proximate Causes of Placebo Responses

The reductions in evoked visceral pain were strongly predicted by ratings of expected pain associated with visceral stimulation and desire for relief, a prediction that follows from at least one current explanation of placebo analgesia.[26,29] However, placebo manipulations not only produced large reductions in visceral hyperalgesia but also in cutaneous heat hyperalgesia (foot), an effect not predicted by expectations.[25] This latter effect may be a result of a mechanistic link between somatic and visceral sensitization at the level of the dorsal horn as discussed above.

Three separate analyses from three studies provide converging lines of evidence that patient experiences of expectation and desire for relief serve as proximate mediators of placebo responses. Since desire and expectation constitute dimensions of emotions related to receiving and expecting results from medical treatments, emotions and emotional regulation may well represent the pivotal mediating factor in placebo analgesia. In the first study, hierarchical regression analysis established that ratings of expected pain coupled with desire for relief in the treatment condition accounted for 77% of the variability in pain ratings during the placebo condition[25] (Table 21.1). These ratings were obtained just after patients received placebo treatments (i.e. intrarectal saline gel). Approximately the same amount of variability (81%) was accounted for in the active treatment condition wherein IBS patients received intrarectal lidocaine, consistent with the view that placebo effects are embedded in active medical treatments. In a second analysis, data from two studies were pooled in order to determine whether *changes* in desire/expectancy ratings predicted *changes* in pain ratings across natural history and placebo conditions

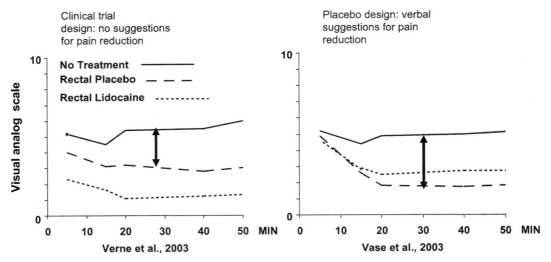

FIGURE 21.4 Comparisons of natural history (NH), rectal placebo and rectal lidocaine scores on visceral pain intensity ratings (VAS) during a 50-minute session within a clinical trial design (left panel), where no suggestions for pain relief are given and within a placebo design with verbal suggestions for pain relief (right panel). The X-axis refers to time in minutes. *(Based on results from references 3 and 25.)*

TABLE 21.1 The Contributions of Expectancy (E), Desire (D), and Anxiety (A) to Rectal Placebo Analgesia

Model	R^2 change	p
Vase et al[25]: D and E on pain ratings in placebo condition	0.77	0.003
Vase et al[26]: $\Delta E + \Delta D + \Delta DE$ on ΔPain (pain ratings in natural history – pain ratings in placebo)	0.38	0.02
Vase et al[27]: $\Delta E + \Delta D + \Delta A$ ΔPain, near onset of placebo ΔPain in 'late' period of placebo	0.13 0.52	>0.05 0.001

(i.e. placebo responses).[26] The two studies included one in which standard instructions for the possibility of receiving an active or inactive agent were included in the consent form and another in which enhanced placebo suggestions were given, as described above. The placebo response, which was calculated as natural history (NH) pain intensity minus pain intensity during the placebo (PL) condition, was regressed against change in expected pain across NH and PL conditions as well as change in desire across these conditions for 23 IBS patients. In other words, changes in expectation and desire were entered into a statistical analysis in which these factors and the multiplicative interaction between them (i.e. desire × expectation) served to predict placebo responses. As shown in Table 21.1, this entire desire–expectation model accounted for 38% of the variability in placebo responses (corresponding to a correlation coefficient of r=0.62). This analysis suggests that both desire for pain relief and expected pain intensity contribute to placebo analgesia, and a main factor is a multiplicative interaction between desire for pain reduction and expected pain intensity. This interaction is consistent with the desire–expectation model of emotions that shows that ratings of negative and positive emotional feelings are predicted by multiplicative interactions between ratings of desire and expectation[30] (Table 21.1). Desire to reduce pain would be related to an avoidance goal, according to a desire–expectation model of human emotions, and these results suggest that analgesia would be related to a reduction in desire for relief, a reduction in expected pain, and a consequent reduction in negative emotional feeling. This prediction was supported by a third study in which expectations of pain intensity, desire for relief, and anxiety were rated at two time points during intrarectal treatment with saline gel (placebo). The design of this study was similar to that of the two studies just described[25,26] and the analysis relied on change scores. As shown in Table 21.1 (bottom row), changes in desire and expectation did not significantly predict placebo responses near the beginning of placebo treatment (early), but accounted for a large amount of variability in placebo responses in the late phase of the placebo treatment.[27] However, the most interesting result of this study was that ratings of desire for pain reduction, expected pain, and anxiety all *decreased* during the same time that the placebo effect *increased*[25-27] and these variables came to explain larger amounts of the variance in the placebo effect over time (Table 21.1, bottom row). This result shows that the mediating variables of placebo analgesia, including expectation, desire, and emotions, are not static but change dynamically over time, possibly as a result of feedback from early results and perception of the dynamic aspects of the treatment procedure itself. Thus, perception of the process of being treated may be experientially distinguished from perceiving the results of treatment.

Somatic Focus Moderates Effects of Goals and Expectancy

Based on several interrelated experiments, Geers and his colleagues argue that the placebo effect is most likely to occur when individuals have a goal that can be fulfilled by confirmation of the placebo expectation, consistent with the desire–emotion model and the explanation just given.[31] The results of Geers and colleagues demonstrated a role for desire for an effect across a variety of symptom domains, including those related to positive approach goals or negative avoidance goals. In addition to the roles of goals, desires, and expectations in placebo responding, there is evidence that the degree of somatic focus has a moderating influence on these psychologic factors.[32] Somatic focus reflects the disposition to focus on body functions and to be vigilant to changes in them. In an experiment that induced expectations of unpleasant symptoms, individuals who expected to take a drug but who were given placebo tablets reported more placebo symptoms when they closely focused on their symptoms.[32] This type of interaction also has been proposed for approach goals.[30] Focusing more closely on symptoms can enhance their significance and implications and thereby increase or decrease the desire for their reduction. If catastrophic meanings are enhanced as a result of somatic focusing, nocebo effects could be developed. In contrast, positive placebo suggestions for analgesia or improvements in physical health lead individuals to selectively attend to signs of improvement, thereby enhancing the placebo effect. These types of influence can occur both during explicit placebo treatments, but more importantly, even when active treatments are given. When patients closely notice positive signs of pain reduction, this perception provides 'evidence' that the treatment has been effective regardless of whether the treatment is placebo or an active medication. Placebo effects, and for that matter nocebo effects, are potentially embedded in active treatments depending on what patients are told, behaviors and appearance of the caregivers, and numerous psychosocial contextual factors that occur during treatment.

If focusing on bodily symptoms or cues operates as a kind of feedback that supports factors underlying placebo responding, increasing the degree or frequency of somatic focusing could increase the magnitudes of placebo responses over time. This possibility is supported by observations showing that the growth of the placebo effect over time at least partly depends on the frequency of test stimuli. As discussed above, ratings of desire, expectation, and anxiety decrease over time along with the increase in placebo effect.[27] As shown in Figures 21.3 and 21.4, it took about 20 minutes for the placebo effect to increase to its maximum level in conditions wherein stimuli were applied at 7 times per 50 minutes.[3,25] This same pattern of increase was found in a subsequent experiment that applied stimuli more rapidly, 7 stimuli in 10 minutes.[28] The placebo effect increased to its maximum level during the first three stimuli and over 3–4 minutes with this more rapid stimulus frequency, unlike the experiment that used less frequent stimulation. Thus, feedback from the test stimuli serve as cues that signal increasing pain relief. More frequent test stimuli lead to more rapid pain reduction. These findings and interpretations are in agreement with an interview study showing that many of the patients who participated in the study had an inner focus on bodily symptoms at the beginning of the treatment.[33] Taken together, the studies of Geers, Vase, Price and their colleagues support a placebo mechanism wherein goals, desire, expectation, and consequent emotional feelings co-determine the placebo response. Somatic focus provides a self-confirming feedback that facilitates these factors over time, leading to less negative emotional feelings and higher expectations of avoiding aversive experiences. In other contexts, somatic focus could lead to positive feelings about obtaining pleasant consequences, such as feeling healthy, invigorated, or energized. In any case, the stimuli of the experiment can help confirm that the treatment is working if someone is expecting it to work. Since the stimuli self-reinforce expectations of pain reduction, reduce desire for relief, and consequently reduce negative emotions, the placebo effect increases more rapidly over time with higher frequencies of test stimulation. A similar dynamic may work for 'nocebo' responses, such as the increase in symptoms over time as a result of catastrophic thinking and expectations related to threatening cues.

Placebo Responses and Emotional Regulation

If the desire–expectation model is accurate, then placebo phenomena occur within the context of emotional regulation, and symptoms should be influenced by desire, expectation, and emotional feeling intensity across different settings and even regardless of whether or not these factors are evoked by placebo manipulations. Studies of healthy volunteers have shown that reductions in negative emotions, like stress, are related to placebo analgesia effects[34] whereas induction of fear may reduce or completely abolish the placebo analgesia effect.[35,36] These studies are in line with the understanding that expectations of treatment effects reduce negative emotions and contribute to pain relief.[24] A separate line of evidence for the role of expectancy in placebo analgesia includes studies that manipulate expectancy in non-placebo contexts. Two studies found large reductions in pain from expectancy manipulations[37,38] and one of these studies found corresponding reductions in pain-related brain activity.[37] Desire and emotions also influence pain in non-placebo contexts. Rainville and colleagues[38] have shown that hypnotic inductions of changes in desire for relief, as well as inductions of positive and negative emotional states, modulate pain in directions they claim are consistent with the desire–expectation model.

Thus, to put it simply, placebo responses seem to relate to feeling good (or less bad) about prospects of relief (avoidance goal) or pleasure (approach goal) that are associated with treatments or medications. These feelings can be separately influenced by desire and expectation or by the combination of both variables. These variables change dynamically, leading to enhanced placebo response over time. What are needed are explanations that incorporate knowledge obtained from neuroscience. For example, do placebo-induced changes in expectations, desires, and emotions simply lead to subjective biases about symptoms/effects or do they affect their biologic causes? These two alternatives would have somewhat different neurobiologic explanations.

Before turning to the neurobiologic underpinnings of these phenomena, it is worth mentioning that emotional factors have been shown to contribute to large anti-hyperalgesic placebo effects not only in IBS but also in neuropathic pain conditions where identifiable nerve damage has taken place.[39] Patients who had developed chronic neuropathic pain following thoracotomy rated positive and negative emotional feelings in sessions wherein they were exposed to placebo manipulations. Quantitative sensory testing conducted just after placebo manipulation showed significantly reduced areas of pinprick hyperalgesia, and magnitudes of placebo responses were significantly correlated with low levels of negative affect but not with positive affect. These findings are in accordance with the desire–expectancy model, and they suggest that emotional factors may be central to pain modulation across different types of chronic pain conditions.

CENTRAL NERVOUS SYSTEM MODULATION OF PAIN IN IBS

Neuroimaging of Placebo-Induced Anti-Hyperalgesia

These results, showing placebo effects on pain evoked by visceral and somatic stimuli, provide indirect support for a brain-to-spinal cord mechanism that reverses the sensitization of dorsal horn neurons that have both visceral and somatic primary afferent input. Reversal of hypersensitivity at the level of the dorsal horn should result in decreased pain-related activation at all subsequent supraspinal levels. A study tested this prediction by using the same methods of rectal distension and pain ratings scales as described above, in combination with fMRI brain imaging.[28] Similar to earlier studies described above, a large placebo effect was produced in IBS patients by suggestions, and this effect was accompanied by large reductions in visceral-evoked neural activity (as measured by BOLD) in the thalamus, first and second somatosensory cortices (i.e. S–1 and S–2), anterior, mid-, and posterior insular cortex, and anterior cingulate cortex, all areas that are part of the pain matrix. The widespread reduction in these areas, including those at early levels of processing (e.g. thalamus, S–1), is consistent with a descending brain-to-spinal cord mechanism. Clearly, more direct measures of spinal cord processing, such as neuroimaging of spinal cord activity, are needed to further test this hypothesis. In this regard, it is interesting that the placebo effect has been shown to be accompanied by reduced nociceptive activity in the human spinal cord.[40] In any case, widespread reduction of neural activity throughout the pain matrix tends to rule out a mechanism of modulation that involves only *selective* effects on forebrain areas involved in cognitive processing of pain without effects at earlier stages of processing, as proposed earlier.[41,42]

This neuroimaging study of IBS patients also provided the opportunity to analyze brain regions that were activated more during placebo analgesia than during the untreated baseline natural history condition.[43] Some of the activated regions are those known to be involved in the classic brain–spinal cord modulatory system, including the rostral ACC and bilateral amygdala, thereby providing neuroimaging results that support this classic mechanism.[44,45] These same placebo-activated regions are also known to be involved in emotions and emotional regulation, functions that are presently considered to be a part of endogenous pain modulation.[29,46–49] However, increased activation also occurred in relation to other phenomena likely to be involved in maintaining memory for the placebo suggestion and in developing expectations of pain reduction. Thus, some placebo-activated regions were those involved in neurolinguistic processes and memory, such as the parahippocampal gyrus, medial aspects of the left temporal lobe and left lentiform nucleus.[43] Other regions activated by placebo are known to comprise a network involved in associative thinking and included the left pre-cuneus, posterior cingulate, and aspects of the temporal lobe.[50] These regions were most active during the early part of the placebo condition, presumably at a time wherein subjects were attending to the memory of the placebo suggestions and to somatic feedback, as described earlier. The temporal profile of placebo-induced brain activations is consistent with the idea that the placebo effect increases early in its development as a result of a self-reinforcing confirmation of the efficacy of the treatment. Recall that we found that the placebo effect increases to its maximum magnitude over the first 2–3 of the 7 test stimuli,[3,25,27,28] 15 to 20 minutes, as shown in Figures 21.3 and 21.4, and over the first 3–4 minutes during the brain-imaging study. We have interpreted this progressive increase to be a consequence of a kind of feedback that is sustained by somatic focus and associative thinking that is characterized by loose and rapid associations between thoughts and images. As a result of feedback, the placebo analgesic effect increases over time and is accompanied by decreased desire for pain relief and decreased levels of expected pain[27,51] (Table 21.1). In patient interviews, for example, we have found that most IBS patients are thinking about the impending treatment and making associations between being given the treatment and likely outcomes.[33] This kind of thinking and attending was found to be much more common near the beginning of treatment than toward the end.[33]

These feedback mechanisms need not be strictly associated with deliberate administration of a placebo or nocebo agent. Thus, increases in catastrophizing, hypervigilance and somatic focusing that are known to be prevalent among IBS patients and other types of chronic pain patients could increase descending facilitation and hyperalgesia.[5,9,19,29] Searching for visceral cues that signal impending relief or worsening of symptoms could lead to a self-confirming feedback mechanism whereby increased or decreased pain could lead to still further changes in symptoms. Although this proposed mechanism has received some support from studies wherein placebo or nocebo agents are given, including neuroimaging studies, the role of feedback needs to be explored in contexts where no agent is given. Some lines of evidence for these types of mechanism are given in studies of neuropathic pain patients where high levels of pain catastrophizing has been related to increased levels of wind-up like pain and enhanced cortical activation.[51,52] Given the presence of ongoing widespread somatic and visceral hyperalgesia in IBS, it is reasonable to hypothesize that induction (e.g. nocebo), maintenance, and reversal of hyperalgesia (e.g. placebo) rely on dynamic regulation by networks of brain regions discussed above. This regulation is associated with somatic focus, hypervigilance, and emotional regulation that contribute to both inhibition and facilitation, depending on the ongoing expectations and other aspects of the psychologic state of these patients (Fig. 21.1).

NEUROCHEMICAL BASIS OF ANTI-HYPERALGESIA IN PLACEBO ANTI-HYPERALGESIC MECHANISMS

Placebo analgesic effects have been found to be reversed or diminished by opioid antagonism.[44,45,53–56] The potential involvement of endogenous opioids has led many to reason that placebo effects are fundamentally similar in their neurobiology to other expectancy effects.[8,29] Indeed, recent human imaging studies show that the expectancy of reward is accompanied by increased endogenous opioid activity in limbic and cortical areas.[56] Opiate antagonists, such as naloxone and naltrexone, block various conditioned effects on pain, including placebo analgesia.[53–56]

However, an involvement of endogenous opioids does not appear universal to all placebo analgesic effects. There is accumulating evidence that placebo anti-hyperalgesia is not directly mediated by endogenous opioid mechanisms. Within the context of this chapter, it is noteworthy that the large placebo anti-hyperalgesic effects discussed in relation to Figures 21.3 and 21.4 were replicated in a similarly designed study that tested whether such effects could be prevented by intravenous administration of 10 mg naloxone.[27] The two groups receiving naloxone and saline had no difference in the amount of pain reduction produced by placebo administration. Both groups showed large reductions in evoked rectal pain. Other human studies provide evidence that anti-hyperalgesic mechanisms are nonopioid. For instance, Kupers and colleagues tested a chronic low-back-pain patient who received saline infusion through an epidural catheter over a period of 50 days.[57] The patients experienced a large placebo analgesia effect, but this effect was not blocked by administration of naloxone. Also, Amanzio and Benedetti found that placebo effects produced by opioid conditioning but not nonopioid conditioning were related to endogenous opioids.[54] Unlike conditioning with morphine, they found that placebo analgesia (on tourniquet tolerance test) generated by conditioning with ketarolac was not prevented by naloxone[54] but was prevented by a specific cannabinoid antagonist, CB1.[58] The distinction also has been found in animal models of placebo analgesia. Among the few animal studies of placebo-induced analgesia, Guo and colleagues reported that placebo-induced analgesia was sensitive to naloxone, whereas analgesia induced by aspirin was insensitive.[53] This raises the possibility, as suggested by human imaging studies discussed above, that placebo-induced analgesia stems from the brain activity simulating the original (unconditioned) effect of the drug, an idea also put forward by Guo and colleagues.[53] In the case of morphine, this would be increased endogenous opioid release, thus being susceptible to naloxone. A study by Nolan and his colleagues introduced a novel preclinical model for studying placebo-induced analgesia that shows canonical features of placebo effects observed in humans.[59] They confirmed naloxone reversibility in an operant facial analgesia model, revealing the involvement of endogenous opioids. Due to its operant nature, and use of mild noxious stimuli only, this paradigm may better address the emotional aspects of placebo-induced analgesia and executive cortical control placed over them. In doing so, these experiments provide the foundation for methodologies that would be extremely difficult in human studies, such as single-neuron recordings, neurotransmitter depletion, focal lesions, and molecular/genetic manipulations.

A SYNERGISTIC INTERACTION BETWEEN PERIPHERAL IMPULSE INPUT AND CENTRAL FACILITATION?

It is surprising that primary and secondary hyperalgesia can be nearly completely reversed in some IBS patients either by removing the source of tonic peripheral impulse input (i.e. local rectal anesthesia) or by providing a form of central modulation through placebo suggestions (Fig. 21.1). One possible mechanism that could account for this combination of observations is that a synergistic interaction might occur between peripheral and central sources of facilitation of dorsal horn neuron responsiveness. That is, hypersensitivity of dorsal horn neurons might be dynamically maintained by impulses from *both* visceral structures and from brain–spinal cord descending facilitation. The latter mechanism has been strongly implicated in persistent pain conditions.[60,61] Attenuation of either peripherally induced dorsal horn sensitization or attenuation of descending facilitation may be sufficient to reverse most of the secondary hyperalgesia. A mechanistic model that might explain this paradox is that of a synergistic interaction between these two sources of facilitation. Many examples of synergy occur in pharmacology and central nervous system integration. In simple terms, synergy between two phenomena is demonstrated when their combined influence produces an effect (e.g. hyperalgesia or analgesia) that is distinctly greater than that produced by either influence by itself. To use a hypothetical example from pharmacology, suppose a small dose of drug A produces a 10% reduction in pain and a small dose of drug B produces a 15% reduction in pain but when these doses of both drugs are given together they produce a 75% reduction, not 25% (of course true pharmacologic synergy requires testing wide ranges of doses of each drug as well as their combinations). However, a synergistic relationship between the two influences also predicts that removing either influence alone would have a large effect, especially if both influences were normally present. If secondary hyperalgesia results from a synergistic influence from peripheral impulse input and descending facilitation, then local rectal anesthesia or placebo administration would each produce large anti-hyperalgesic effects (Fig. 21.1). This idea is also consistent with findings that indicate that psychologic factors (e.g. distress) strongly influence IBS pain. Such factors are associated with a main source of central facilitation. Of course, this hypothesis needs to be further tested. If this kind of synergistic interaction generalizes to other hyperalgesic states, it might have large implications for understanding how to assess the relative contribution of peripheral and central factors in other persistent pain conditions such as fibromyalgia and various forms of neuropathic pain. Central modulation may well interact with tonic peripheral impulse input to the spinal cord, so that widespread hyperalgesia reflects a confluence of both mechanisms. Thus, to improve our understanding of these phenomena it will be important to design studies that address both peripheral and central modulations that are likely to occur simultaneously and even synergistically. In the case of placebo analgesia, factors can serve as predictors of clinical outcomes and could be very useful in managing placebo responses and effects in clinical trials, for which there is a large amount of variability.[62] Thus, meta-analyses show effect sizes ranging from −0.95 to 0.57, Cohen's d in clinical trials of placebo effects (see Ch 20 in this book). This approach may have far reaching implications not only for explaining anti-hyperalgesic placebo effects but potentially for understanding the processing of a range of persistent pain conditions.

Acknowledgment

Work on this chapter is partly funded by the Innovative Medicines Initiative Joint Undertaking (IMI JU) Grant No. 115007.

References

1. Dunphy RC, Bridgewater L, Price DD, Robinson ME, et al. Visceral and cutaneous hypersensitivity in Persian Gulf war veterans with chronic gastrointestinal symptoms. *Pain.* 2003;102:79-85.
2. Verne GN, Robinson ME, Price DD. Hypersensitivity to visceral and cutaneous pain in the irritable bowel syndrome. *Pain.* 2001;93:7-14.
3. Verne GN, Robinson ME, Vase L, Price DD. Reversal of visceral and cutaneous hyperalgesia by local rectal anesthesia in irritable bowel syndrome (IBS) patients. *Pain.* 2003;105:223-230.
4. Mayer EA, Raybould H. Role of visceral afferent mechanisms in functional bowel disorders. *Gastroenterol.* 1990:1688-1704.
5. Naliboff BD, Munakata J, Fullerton S, et al. Evidence for two distinct perceptual alterations in irritable bowel syndrome. *Gut.* 1997;41:505-512.
6. Verne GN, Himes NC, Robinson ME, et al. Central representation of visceral and cutaneous hyperalgesia in irritable bowel syndrome. *Pain.* 2003;103:99-110.
7. Price DD. Psychological and neural mechanisms of the affective dimension of pain. *Science.* 2000;288(5472):1769-1772.
8. Al Chaer ED, Kawasaki M, Pasricha PJ. A new model of chronic visceral hypersensitivity in adult rats induced by colon irritation during postnatal development. *Gastroenterol.* 2000;119(5):1276-1285.
9. Mayer EA, Gebhart GF. Basic and clinical aspects of visceral hyperalgesia. *Gastroenterol.* 1994;107:271-293.
10. Zighelboim J, Talley NJ, Phillips SF, et al. Visceral perception in irritable bowel syndrome. Rectal and gastric responses to distension and serotonin type 3 antagonism. *Dig Dis Sci.* 1995;40:819-827.
11. Bouin M, Meunier P, Riberdy-Poitras M, Poitras P. Pain hypersensitivity in patients with functional gastrointestinal disorders: a gastrointestinal-specific defect or a general systemic condition? *Dig Dis Sci.* 2001;46:2542-2548.
12. Zhou Q, Fillingim RB, Riley 3rd JL, et al. Central and peripheral hypersensitivity in the irritable bowel syndrome. *Pain.* 2010;148:454-461.
13. Mackenzie J. *Symptoms and their Interpretation.* London: Shaw; 1909.
14. Price D, Mao J, Mayer D. *Central Consequences of Persistent Pain States.* Seattle: IASP Press; 1997.
15. Zhou Q, Price DD, Verne GN. Reversal of visceral and somatic hypersensitivity in a subset of hypersensitive rats by intracolonic lidocaine. *Pain.* 2008;139:218-224.
16. Verne GN, Sen A, Price DD. Intrarectal lidocaine is an effective treatment for abdominal pain associated with diarrhea – predominant irritable bowel syndrome. *J Pain.* 2005;68:493-496.
17. Lin C, Al-Chaer ED. Long-term sensitization of primary afferents in adult rats exposed to neonatal colon pain. *Brain Res.* 2003;971:73-82.
18. Verne GN, Price DD, Callam CS, et al. Viscerosomatic facilitation in a subset of IBS patients, an effect mediated by N-Methyl-D-aspartate receptors. *J Pain.* 2012 (in press).
19. Robinson ME, Riley JL. Psychosocial factors in pain. In: Gatchel R, Turk D, eds. *Negative Emotion in Pain.* New York: Guilford Press; 1990:333-345.
20. Rhudy JL, Meagher MW. Fear and anxiety: divergent effects on human pain thresholds. *Pain.* 2000;84:65-75.
21. Rhudy JL, Meagher MW. Negative affect: effects on an evaluative measure of human pain. *Pain.* 2003;104:617-626.
22. Zelman DC, Howland EW, Nichols SN, Cleeland CS. The effects of induced mood on laboratory pain. *Pain.* 1991;46:105-111.
23. McCracken LM, Gross RT. The role of pain-related anxiety reduction in the outcome of multidisciplinary treatment for chronic low back pain: preliminary results. *J Occup Rehabil.* 1998;8:179-189.
24. Flaten MA, Aslaksen PM, Lyby PS, Bjørkedal E. The relation of emotions to placebo responses. *Phil Trans R Soc.* 2011;366:1818-1827.
25. Vase L, Price DD, Verne GN, Robinson ME. 2004. The contributions of changes in expected pain levels and desire for pain relief to placebo analgesia. In: Price DD, Bushnell MC, eds. *Psychological Methods of Pain Control: Basic Science and Clinical Perspectives.* Seattle, WA: IASP Press; 2004:207-232.
26. Vase L, Robinson ME, Verne GN, Price DD. The contribution of suggestion, desire, and expectation to placebo effects in irritable bowel syndrome patients. *Pain.* 2003;105(1-2):17-25.
27. Vase L, Robinson ME, Verne GN, Price DD. Increased placebo analgesia over time in irritable bowel syndrome (IBS) patients is associated with desire and expectation but not endogenous opioid mechanisms. *Pain.* 2005;115:338-347.
28. Price DD, Craggs JG, Verne GN, et al. Placebo analgesia is accompanied by large reductions in pain-related brain activity in irritable bowel syndrome patients. *Pain.* 2007;127:63-72.
29. Price DD, Finnis DG, Benedetti F. A comprehensive review of the placebo effect: recent advances and current thought. *Annu Rev Psychol.* 2008;59:565-590.
30. Price DD, Barrell JJ. *Inner Experience and Neuroscience: Merging Both Perspectives.* Cambridge MA: MIT Press; 2012.
31. Geers AL, Weiland PE, Kosbab K, et al. Goal activation, expectations, and the placebo effect. *J Personal Social Psych.* 2005;89(2):143-159.
32. Geers AL, Helfer SG, Weiland PE, Kosbab K. Expectations and placebo response: a laboratory investigation into the role of somatic focus. *J Behav Med.* 2006;29(2):171-178.
33. Vase L, Nørskov KN, Petersen GL, Price DD. Patients' direct experiences as central elements of placebo analgesia. *Philos Trans R Soc Lond B Biol Sci.* 2011;366:1913-1921.
34. Aslaksen PM, Flaten MA. The roles of physiological and subjective stress in the effectiveness of a placebo on experimentally induced pain. *Psychosom Med.* 2008;70:811-818.
35. Lyby PS, Aslaksen PM, Flaten MA. Variability in placebo analgesia and the role of fear of pain – an ERP study. *Pain.* 2011;152:2405-2412.
36. Lyby PS, Forsberg JT, Åsli O, Flaten MA. Induced fear reduces the effectiveness of a placebo intervention on pain. *Pain.* 2012;153:114-1121.
37. Koyama T, McHaffie JG, Laurienti PJ, Coghill RC. The subjective experience of pain: where expectations become reality. *Proc Natl Acad Sci USA.* 2005;102(36):12950-12955.
38. Rainville P, Bao QV, Chetrien P. Pain-related emotions modulate experimental pain perception and autonomic responses. *Pain.* 2005;118(3):306-318.
39. Petersen GL, Finnerup NB, Nørskov KN, et al. Placebo manipulations reduce hyperalgesia in neuropathic pain. *Pain.* 2012;153:1292-1300.
40. Eippert F, Finsterbusch J, Bingel U. Direct evidence for spinal cord involvement in placebo analgesia. *Science.* 2009;326(5951):404-404.
41. Mayer EA, Berman S, Suyenobu B, et al. Differences in brain responses to visceral pain between patients with irritable bowel syndrome and ulcerative colitis. *Pain.* 2005;115:398-409.
42. Mertz H, Morgan V, Tanner G, et al. Regional cerebra activation in irritable bowel syndrome and control subjects with painful and nonpainful rectal distention. *Gastroenterol.* 2000;118:842-848.
43. Craggs JG, Price DD, Perlstein WM, et al. The dynamic mechanisms of placebo induced analgesia: evidence of sustained and transient involvement. *Pain.* 2008;139:660-669.
44. Fields HL. The placebo effect: An interdisciplinary exploration. In: Harrington A, ed. *Toward a Neurobiology of Placebo Analgesia.* Cambridge, MA: Harvard University Press; 1997:93-116.
45. Mayer DJ, Price DD. Central nervous system mechanisms of analgesia. *Pain.* 1976;2:379-404.
46. Petrovic P, Dietrich T, Fransson P, et al. Placebo in emotional processing – induced expectations of anxiety relief activate a generalized modulatory network. *Neuron.* 2005;46:957-969.
47. Petrovic P, Ingvar M. Imaging cognitive modulation of pain. *Pain.* 2002;95:1-5.

48. Eippert F, Bingel U, Schoell ED, et al. Activation of the opioidergic descending pain control system underlies placebo analgesia. *Neuron*. 2009;63:533-543.
49. Elsenbruch S, Rosenberger C, Bingel U, et al. Patients with irritable bowel syndrome have altered emotional modulation of neural responses to visceral stimuli. *Gastroenterol*. 2010;139:130-139.
50. Bar ME, Aminoff M, Mason J, Fenske M. The units of thought. *Hippocampus*. 2007;17(6):420-428.
51. Vase L, Nikolajsen L, Christensen B, et al. Cognitive and emotional sensitization contributes to wind-up-like pain in phantom limb pain patients. *Pain*. 2011;152:157-162.
52. Vase L, Egsgaard L, Nikolajsen L, et al. Pain catastrophizing and cortical responses in amputees with varying levels of phantom limb pain: a high-density EEG brain-mapping study. *Exp Brain Res*. 2012;218:407-417.
53. Guo JY, Wang JY, Luo F. Dissection of placebo analgesia in mice: the conditions for activation of opioid and non-opioid systems. *J Psychopharmacol*. 2010;24(10):1561-1567.
54. Amanzio M, Benedetti F. Neuropharmacological dissection of placebo analgesia: expectation activated opioid systems versus conditioning-activated specific subsystems. *J Neurosci*. 1999;19(1):484-494.
55. Benedetti F. The opposite effects of the opiate antagonist naloxone and the cholecystokinin antagonist proglumide on placebo analgesia. *Pain*. 1996;64(3):535-543.
56. Gardner EL. Addiction and brain reward and antireward pathways. *Adv Psychosom Med*. 2011;30:22-60.
57. Kupers R, Maeyaert J, Mélanie B, et al. Naloxone-insensitive epidural placebo analgesia in a chronic pain patient. *Anesthesiol*. 2007;106:1239-1242.
58. Benedetti F, Amanzio M, Rosato R, Blanchard C. Nonopioid placebo analgesia is mediated by CB1 cannabinoid receptors. *Nature Med*. 2011;17(10):1228-1230.
59. Nolan TA, Price DD, Caudle R, et al. Placebo-induced analgesia in an operant pain model in rats. *Pain*. 2012 (in press).
60. Gebhart GF. Descending modulation of pain. *Neurosci Biobehav Rev*. 2004;27:729-737.
61. Porreca F, Ossipov MH, Gebhart GF. Chronic pain and medullary descending facilitation. *Trends Neurosci*. 2002;25(6):319-325.
62. Vase L, Riley III JL, Price DD. A comparison of placebo effects in clinical analgesic trial versus studies of placebo analgesia. *Pain*. 2002;99:443-452.

CHAPTER 22

The Wound that Heals: Placebo, Pain and Surgery

Wayne B. Jonas[1], Cindy Crawford[1], Karin Meissner[2], Luana Colloca[3,4,5]

[1]Samueli Institute, Alexandria, VA, USA, [2]Institute of Medical Psychology, Ludwig-Maximilians-University Munich, Munich, Germany, [3]National Center for Complementary and Alternative Medicine (NCCAM), Bethesda, MD, USA, [4]National Institute of Mental Health (NIMH), National Institutes of Health, Bethesda, MD, USA, [5]Department of Bioethics, Clinical Center, National Institutes of Health, Bethesda, MD, USA

BACKGROUND

Pain

The treatment of pain is one of the more prevalent and persistent challenges to modern medicine. More than 25% of adults in the USA report significant daily pain, and 1 in 10 have pain that has lasted 1 year or more. The most difficult pain syndromes to manage are chronic low back pain, recurrent headache and persistent joint pain.[1]

Current research attempting to understand pain mechanisms has revealed a complex picture in which the biologic mechanisms of pain both interact in positive feedback loops and reach beyond the nervous system to other bio-psycho-social domains, creating depression, anxiety, sleep difficulties and other problems. A recent survey found that 1 in 5 Americans have pain that causes a major impact on employment, residence, or quality of life, personal function and mobility. The medical community is only partially successful in treating or helping patients to manage their pain[2]; a more holistic and process-oriented approach is needed.

When not functioning as it should—for prevention and repair—the pain system can be overwhelmed and become overly sensitive to any stimuli (allodynia) or create increased sensitivity to stimuli (hyperesthesia) or produce pain in non-damaged tissue (hyperalgesia). When this occurs, the pain is maladaptive. A patient's psychologic state influences their pain, and the perception of pain is modulated by factors beyond direct injury. These factors contribute to the persistence of pain after injury and healing, or where there is no injury. Pain is a bio-psycho-social phenomenon and not a strictly physiologic one.[3] (See Grahek[4] for a more complete summary of the changing theories of pain.) Thus, social and psychologic context and expectations are important in all approaches to pain. This is where the 'art' of medicine is as important as the science. Interestingly, research on placebo is now beginning to shed scientific light on this 'art' of medicine by exploring the neuronal bases of clinical benefits related to social and psychologic context and expectations in pain patients.[5] An example is provided by a study in chronic back pain (CBP) patients who participated in a double-blind brain-imaging clinical trial with a 2-week treatment with either lidocaine or placebo. Patients experienced a significant clinical benefit under placebo treatment. These placebo responses were predictable by using functional connectivity and high-frequency oscillations in different areas of the brain such as the medial prefrontal cortex and bilateral insula.[6] The fact that the neuronal population in the prefrontal cognitive and pain-processing regions predetermines the probability of placebo response in the clinical setting paves the way for integrated models of placebo and pain research in patients. Implementing new methodologic approaches for studying pain in patients suffering chronically, and treated with surgical treatments, opens new avenues for understanding the complexity of pain and guiding stratification of patients in surgical clinical trials.

The Complexity of Pain

The rationale for a holistic (bio-psycho-social) approach comes from the fact that mind–brain injuries and stresses (MBIS) share many common pathophysiologic and salutogenic mechanisms. The development, expression and durability of chronic pain and psychopathologies involve genotypic factors that are latent or code for phenotypes (e.g. of ion channels, neurotransmitters, receptors and synaptic elements) that differentially express themselves in response to both internal and external signals and contexts. Thus, predisposed individuals respond to environmental or psychosocial injury in ways that induce a core set of symptoms and dysfunctions. This set includes: (1) psychologic and emotional distress (e.g. depression, anxiety, anger), (2) cognitive impairment, (3) chronic and refractory pain, (4) drug/opioid desensitization (and abuse), and (5) somatic (sleep, appetite, sexual and energy) problems.

The interplay of psychosocial factors and biology is strikingly on display in poly-trauma injuries. The persistent and chronic nature of pain associated with traumatic injuries has been well documented.[7] Additionally, there is a high incidence of concurrent post-traumatic stress disorder (PTSD), depression, and anxiety with these injuries. The overlapping and multi-component nature of traumatic injuries, both psychologic and physical, has been recently characterized by Jonas et al as the 'trauma spectrum response' (TSR). They propose that mind–brain/body injuries, such as traumatic stress or traumatic brain injury (TBI), are more appropriately addressed by a constellation (whole systems) view of the impact of those injuries:[8] addressing this constellation in ways that include the process of care provides a more logical and likely efficient and effective framework than isolating them as 'co-morbidities,' each with their own specific treatments.

To match this complex situation, multi-disciplinary approaches to pain management have been developed that combine drugs, behavior, surgery, manipulation, and complementary medicine approaches. However, the dominant treatment continues to be medications, both prescription and over-the-counter. While medications are often necessary, their excessive use often results in overreliance, misuse, and abuse, effects that increase the more they are used.[9] Thus, alternatives are increasingly applied, including a rapid growth of interventional and surgical approaches, especially for the common conditions of back, head and joint pain. But are surgical and interventional approaches to pain specific? Or, given the major psychosocial overlay on pain, do they produce their effects more from the social context and psychologic impact of surgical rituals?

The Surgical Approach to Pain

Patients with chronic pain who do not respond to conventional pharmacologic treatments are recommended to receive therapeutic interventions such as surgery, invasive procedures and neurostimulation. Surgical procedures are widely used in medicine for pain, especially back and joint pain. With the development of minimally invasive procedures, the number of interventions for pain and related conditions has increased, including for low-back pain,[10] arthritis,[11] endometriosis,[12] and headache.[13]

However, invasive and classical surgical procedures are rarely evaluated with the same rigor as drugs. They are usually used without rigorous study designs involving randomization, allocation concealment and blinding. Blinding is challenging as a complex, invasive procedure requires an elaborate sham set up. Double-blinding is not possible as the physician and surgical team usually know if the true or sham intervention is being done. In addition, there is controversy about the ethical use of sham procedures, even with good informed consent.[14,15] Thus, it is often unknown to what extent the placebo response is contributing to pain improvement from these types of procedures.

A placebo response is defined as the outcome difference between a sham-surgery group and a no-treatment group. The specific effect of a surgery is the difference between the true surgery and a sham procedure. To separate what is called the placebo effects from the placebo response, a no-treatment control condition is needed in addition to a sham control. However, this design is almost never used to test surgery or invasive procedures in pain. Placebo responses come from complex factors that are imbedded in the surgical ritual and can be influenced by numerous factors, including the preparation for, and type of, procedure, the hospital-like setting, special costumes and preparation of the 'healer', involvement of multiple authoritative providers, collective, repeated suggestions to expect a positive outcome, a physical invasion of the body, and an elaborate ritual of delivery and recovery. In the West (and likely in the world), surgery, with its seeming miraculous ability to alter the body, and the assumption that it can permanently 'fix' a person's problem, has a power that takes on special significance not afforded to any other treatment approaches.[16] Thus, one would expect a significant contribution from placebo responses, especially for pain from invasive and surgical procedures. Several studies seem to support this hypothesis and are described below.

Osteoarthritis

Moseley et al,[17] for example, reported equal improvement of pain in osteoarthritis of the knee from arthroscopic surgery compared to a sham surgery. The study patients were randomly assigned to receive arthroscopic debridement, arthroscopic lavage, or placebo surgery. Patients in the placebo group received skin incisions and underwent a simulated debridement without insertion of the arthroscope. There was not a no-treatment comparison group. The arthroscopic debridement patients had three stab wounds made and an arthroscope inserted in the inferolateral portal; an inflow cannula was inserted in the superomedial portal, and the various operating instruments were inserted from the inferomedial portal. The study was carefully powered (approximately 100 per group) after a pilot study was done, and procedures for careful blinding of patients and those conducting outcomes was developed. The outcomes after arthroscopic lavage or arthroscopic debridement were no better than those after a placebo procedure. The reduced pain from any of the procedures persisted for 2 years post-operatively, questioning an often repeated myth about placebo effects—that they are short lived.

In a somewhat similar study of arthroscopic knee irrigation for osteoarthritis conducted by Bradley et al,[18] 180 patients were randomized to normal knee irrigation or sham irrigation. In the sham group, the needle was advanced to, but not through, the joint capsule. As in the study by Moseley et al, pain relief was statistically equivalent in both groups and the authors concluded that '[m]ost, if not all, of the effect of TI (tidal irrigation) appears to be attributable to a 'placebo response.'

The combined effect size of these two studies (as derived from the most important continuous outcome measure reported in the studies), involving a total of 340 patients (across both studies), was −0.15, which was statistically not significant but trended toward favoring the sham groups.[17,18] At this point, the evidence does not support doing invasive procedures for the pain of osteoarthritis.

Low Back Pain

A number of sham controlled studies have been done examining invasive procedures for the treatment of low back pain (LBP). Two large recent studies examined the use of vertebroplasty for pain from vertebral fractures due to lumbar discogenic disease.[19,20] Buchbinder et al[19] randomized 78 patients with vertebral fracture to vertebroplasty, or same procedures as vertebroplasty, up to the insertion of the 13-gauge needle to rest on the lamina. The central sharp stylet was then replaced with a blunt stylet. To simulate vertebroplasty, the vertebral body was gently tapped, and a cement-like chemical (PMMA), usually injected into the patient, was simply opened so that its smell permeated the room. Equal improvement in pain scores occurred in both groups measured at 1 week and at 1, 3, or 6 months after treatment.

In a similar study, reported in the same issue of *New England Journal of Medicine*, Kallmes et al[20] randomized 131 patients to either vertibroplasty, or a similar procedure, but with no use of cement. Their primary outcome measure was function using the RDQ (Roland–Morris disability score) at 1 month and a pain score. They reported that 'Improvements in pain and pain-related disability associated with osteoporotic compression fractures in patients treated with vertebroplasty were similar to the improvements in a control group.'

The combined effects of six sham controlled studies involving 396 patients examining invasive procedures for LBP show a non-significant effect size of 0.21.[19–24] At this point, the evidence does not support conducting invasive procedures for LBP. Despite this, the rate of vertebroplasty for LBP from vertebral fractures has not diminished.[25]

Headache

Two studies of medium size have examined invasive procedures for migraine headache compared to sham procedure. Dowson et al[13] randomized 139 patients with refractory migraine with aura to receive a PFO (patent foramen ovale) closure with a STARFlex septal repair implant or a sham procedure in which an incision was made with no implant. There were differences in headache severity or frequency between the real and sham repair groups.

In the second study, Guyuron et al[26] randomized 75 patients with moderate to severe migraine to either surgery in the predominant trigger sites (frontal, temporal and occipital)—with endoscopic removal of the glabellar muscles encasing the supraorbital and supratrochlear nerves and removal of a segment of the zygomaticotemporal branch of the trigeminal nerve—or removal of the greater occipital nerve. Sham surgery consisted of making superficial cuts in the appropriate location but with no surgery. They reported that '…surgical deactivation of peripheral migraine headache trigger sites is an effective alternative treatment for patients who suffer from frequent moderate to severe migraine headaches that are difficult to manage with standard protocols.'

The effect size of this study was moderate to large (0.69) and of the combined studies was 0.43, which is statistically significant (p < 0.03).[13,26] While it appears that invasive procedures can be effective beyond sham, one should note that the two studies were not consistent (one was effective and one was not) and the procedures and populations were markedly different from each other. More research is needed before such a procedure can be recommended.

Angina

Finally, there has been some research on surgery and invasive procedures for angina—the pain of coronary artery disease (CAD). During the 1950s, before the invention of coronary artery bypass graphting (CABG) or stenting, ligation of the internal mammary artery (LIMA) was a popular and reportedly successful procedure for angina. Two studies were conducted at that time randomizing patients to either real LIMA or to a sham procedure in which the artery was exposed but not ligated, all other surgical procedures being the same. Pain improvement (usually 40–50% improved) was identical in both groups in both studies, and exercise electrocardiograms were not altered by either procedure in the one study that measured that. The procedure was abandoned after these two studies were published and replaced with CABG. No sham controlled studies of CABG (or stenting) have been done.[27,28]

More recently, a number of invasive procedures for angina have been developed using the introduction of lasers and electrode manipulation of the myocardium. Leon et al[29] randomized 298 patients with percutaneous manipulation. The sham procedure consisted of having the laser in the room and turned on but no further procedure was performed. A comparison of placebo (sham) with two treatment groups (differing in the numbers of laser channels) using the 'Biosense DMR system' showed no differences in exercise duration (the primary end point), exercise time to the onset of chest pain, and exercise time to the appearance of ST-segment changes at 6 and 12 months.

Another study by Salem et al[30,31] randomized 82 patients to percutaneous myocardial laser revascularization (PMLR) plus optimal medical therapy or a sham procedure involving the laser catheter being inserted but connected to a hidden lead box plus optimal medical therapy. The results of this study suggested that PMLR therapy 'is reasonably safe and effective as symptomatic improvement in patients refractory to medical therapy, and that the clinical benefit is not attributable to placebo effect or investigator bias.' The combined effects size for these two studies was small and non-significant (0.35; $p = 0.27$). At this point, there is a weak recommendation against the use of these types of procedure for pain associated with angina.

There are a handful of other studies of invasive interventions for pain conditions such as endometriosis and cholia. These studies report variable effects, and there are insufficient attempts at replication to summarize their effects on these conditions.

Table 22.1 summarizes the current findings from invasive and surgical interventions for pain in studies that could be combined with summary estimates.

PLACEBO AND BRAIN STIMULATION FOR THE TREATMENT OF PAIN

Neurostimulation techniques have been used to treat chronic pain refractory to pharmacologic treatments. These techniques include invasive and noninvasive procedures. The invasive procedures include dorsal spinal cord stimulation (SCS), deep-brain stimulation (DBS), and motor cortex stimulation (MCS), while transcutaneous electrical nerve stimulation (TENS) and repetitive transcranial magnetic stimulation (rTMS) are considered noninvasive.[32] The efficacy of these procedures, the surgical targets and the kind of pain for which these interventions are indicated are subjects of investigation. In this regard, it is interesting to mention two studies indicating that expectations may at least contribute partially to the efficacy of these procedures. The first study evaluated the placebo effect of rTMS in a cross-over study design with neuropathic pain patients resistant to pharmacologic therapy. Patients were randomized to one of two possible arms: true rTMS followed by sham rTMS or vice versa. All the patients were informed that they would receive two sessions of rTMS which differed only for imperceptible changes in the stimulation parameters. The two rTMS sessions were performed 2 weeks apart in order to avoid carry-over effects of the first session of rTMS. The authors found that the participants experienced a substantial analgesic effect when they received the sham intervention after a session of real rTMS. However, patients experienced greater pain when they received the sham session or an rTMS intervention experienced as ineffective.[33] This study provides evidence that the placebo effect contributes to the effectiveness of rTMS in patients suffering from chronic neuropathic pain and that the magnitude of this effect depends upon prior successful or unsuccessful rTMS sessions.[34]

TABLE 22.1 Summary of Effect Sizes of Invasive Procedures and Surgery for Common Pain Conditions

Condition	Number of patients (number of studies)	Combined effect size[a]	p value
Low back pain	396 (6)	0.21	0.15
Osteoarthritis	340 (2)	−0.15	0.20
Angina	380 (2)	0.35	0.27
Headache	214 (2)	0.43	0.03

[a] Note the combined effect sizes are derived from the most important continuous outcome measure as reported in the studies. Data reported as collected and assessed for a systematic review and meta-analysis of 'Are surgery and invasive procedures effective beyond a placebo response?' (unpublished data from co-authors). According to Cohen's d, we report as no effect (<0.2), small (0.2–0.5), moderate (0.5–0.8), or large (>0.8).

Interestingly, it has been shown, in conditions other than pain,[35,36] that placebo effects are an important component of DBS. One study has investigated the effect of DBS under ON/OFF stimulation in patients with stimulation of the thalamus for treating chronic pain. The authors used a placebo manipulation of the intensity of stimulation—the stimulator was turned off surreptitiously—in order to study the contribution of expectation on pain reduction. The patients were evaluated for clinical pain and experimental pain under ON and OFF conditions. The thalamic stimulation reduced significantly clinical and experimental pain, and the reduction was in a range of 16 to 4% and 8 to 0.4% respectively.[37] These findings suggest that both DBS and placebo effects contribute to the overall analgesic effect and outline the importance of considering placebo controls in surgical procedures.

CONCLUSIONS

It appears that, except for migraine headache, the impact of invasive and surgical procedures on pain is not due to the specific nature of the surgical interventions. Even in the case of migraine the positive outcome from surgery came from one study, of moderate size. The other procedure in migraine was negative. In other words, it seems that improvement in pain from interventional procedures does not come from what the surgeon does with the knife, scope or needle, but from the social ritual of doing the intervention. Pain does improve, but not for the reasons that we do the procedure.

What are the clinical and policy implications of the current evidence for invasive procedures for pain? In order to better understand these implications we conducted a Grading of Recommendations, Assessment, Development and Evaluation (GRADE) analysis, which we report on in full detail in our full systematic review and meta-analysis (unpublished data).[38] GRADE is a standardized and widely used system for exploring recommendations. It includes: an estimate of the magnitude of effect; the confidence in the estimate of the effect, taking into consideration the power and sample size of the studies; and, a safety grade as reflected in reporting on adverse events. To conduct the GRADE we applied these criteria for each pain dataset. GRADE options range from 'strong recommendation against the treatment' through 'no recommendation' to 'strong recommendation for the treatment.'[38] Safety of the procedure comes into play here. Based on this analysis, the current evidence does not support a recommendation for instrumental procedures for common pain conditions, including low back pain, arthritis, and migraine. Although the procedures appeared relatively safe for these conditions, effect sizes were small and the authors feel that further research is very likely to change the confidence in the estimate of effect. For example, if there were to be one or more RCT studies, even with severe pain, these estimates might change. Invasive limitations are given a weak recommendation against use for angina due to CAD. Of the four randomized controlled trials mentioned above, it appears that there are safety concerns that include infrequent but serious adverse events and/or interactions, and further research is very likely to have an important impact on confidence in the estimate of effect and is likely to change the estimate. Of concern is that we found no studies comparing real to sham studies on the currently popular invasive interventions using stent insertion or CABG procedures. It is very possible that a large part of the reported effects from these procedures are due to placebo responses and ritual. The ethical implications of continuing to do these procedures without knowing whether or not they provide any specific benefit is in urgent need of national policy discussion.

Similar concerns exist regarding the use of invasive and non-invasive brain stimulations. In 2007, Cruccu et al[39] evaluated systematically the efficacy of these techniques with the scope to produce relevant recommendations. They searched the literature, from 1968 to 2006, looking for neurostimulation in neuropathic pain conditions. This Task Force concluded that the spinal cord stimulation (SCS) is efficacious in failed back surgery syndrome (FBSS) and complex regional pain syndrome (CRPS) type I (level B recommendation). High-frequency transcutaneous electrical nerve stimulation (TENS) may be better than placebo (level C), although worse than electro-acupuncture (level B). rTMS has transient efficacy in central and peripheral neuropathic pains (level B). TENS and r-TMS are considered suitable as short-treatment and add-on therapies. Motor cortex stimulation (MCS) is efficacious in central post-stroke and facial pain (level C). DBS has shown promise for phantom limbic pain and neuropathic pain.[32]

Despite these recommendations, the mechanisms of action are poorly known; placebo-controlled trials with an adequate number of patients are lacking.[32] Further research, including systematic review and meta-analyses, is also necessary to evaluate to what extend invasive and non-invasive brain stimulation help to manage drug-resistant pain.

Should we be using these procedures for pain? Do we need to test invasive interventions for pain with sham comparisons before using them? These are complicated questions. Certainly patients seem to benefit from being subjected to invasive procedures. However, the risks of surgical and invasive procedures are not minor; they include risks from anesthesia, permanent injury to the body, and psychologic stress as well as the time and productivity losses for all those involved. If most of the pain relief is due to placebo responses, certainly there are less risky and costly procedures that can be developed to produce benefit. Without more rigorous examination of the placebo factors, large numbers of patients will be exposed to risky and possibly unnecessary procedures for years. New interventional procedures will be invented and applied with certainty that they are specific or even necessary without knowing whether this is true.[40] It is currently felt to be unethical to deliver new drug treatments without testing them for their specific effects with placebo comparison arms. Why should it be different with invasive procedures?

On the other hand, what are the ethical implications of doing sham surgical procedures? Placebo controls are already controversial.[41] Recommending that all invasive procedures and surgery be tested against sham will certainly be even more controversial. Certainly, replacing a knee or removing a necrotic gall bladder involves structural changes that result in pain relief, and one would not recommend a sham knee replacement study. However, anatomic reasoning is not always correct, and determining this is not always easy or straightforward. Vertebroplasty procedures are done purportedly to correct the nerve root pressure from the anatomic collapse of a vertebra. While this certainly occurs from this procedure, it appears that the pain relief is not due to this structural change.

Do sham surgical studies change practice? The answer seems to be 'sometimes'. When the sham internal mammary studies of angina were published, the use of the procedure rapidly dropped off.[27,28] However, only marginal changes have occurred in the use of vertebroplasty for low back pain.[19,20] When these studies were published, the accompanying editorial rationalized their continued use under the guise of 'patient-centered' care, stating that[42]:

> 'Patients who are given the results of these two studies will probably not choose vertebroplasty. Informed choice matters. Patient empowerment is the best — if not the only — way to change the use of ineffective treatments short of refusing to pay for these procedures.'

With all the money behind an industry that delivers these procedures, certainly the industry itself cannot be trusted to make objective recommendations on which ones should be adopted. More rigorous and balanced judgment processes for setting standards of care and guidelines for interventional approaches to pain are needed.

The emerging body of research on placebo effects and its mechanisms in pain production and mitigation offers hope for improved and evidence-based tools for the physician and patient. Multiple pathways for enhancement of belief, reinforcement of physiologic response and establishment of optimal healing environments and processes can now be evidence based.[43] Without rigorous research on how placebo components can exacerbate or mitigate pain, we may be inadvertently making patients worse. For example, a study published in *Pain* showed that the normal procedure of telling patients that a procedure may hurt, and then expressing sympathy when it does, may increase both pain and anxiety during common interventional procedures.[44]

Moerman has called for a redefinition of placebo effects as the 'meaning response'.[45] Jonas and colleagues have developed a model and framework for ethically maximizing these responses in practice by creating an 'optimal healing environment' (OHE).[46] This model examines the components of the inner environment (the mind) such as expectancy and belief; the interpersonal environment such as empathy and communication processes; the behavioral environment including ritual and culturally appropriate care; and the external environment to enhance meaning and learning.[47] It is with such a redefinition of placebo that we will be able to put a science behind what has for centuries been relegated to the 'art' of medicine. With such a 'science of healing' we will be able to maximally improve our patients' lives and better relieve their pain and suffering. This, after all, is the core goal of medicine.[48]

References

1. CDC National Center for Health Statistics Press Office. New report finds pain affects millions of Americans. <http://www.cdc.gov/nchs/pressroom/06facts/hus06.htm>; 2006 Accessed 08.08.2012.
2. Peter D. Hart Research Associates. Americans talk about pain. <http://www.researchamerica.org/uploads/poll2003pain.pdf>; 2003 Accessed 08.08.2012.
3. Gatchel RJ. Perspectives on pain: a historical overview. In: Gachtel RJ, Turk DC, eds. *Psychosocial Factors in Pain: Critical Perspectives*. New York: Guilford Press; 1999:3-17.
4. Grahek N. *Feeling Pain and Being in Pain*. 2nd ed. Cambridge: MIT Press; 2007.
5. Tracey I. Getting the pain you expect: mechanisms of placebo, nocebo and reappraisal effects in humans. *Nature Med*. 2010;16(11): 1277-1283.
6. Hashmi JA, Baria AT, Baliki MN, et al. Brain networks predicting placebo analgesia in a clinical trial for chronic back pain. *Pain*. 2012; 153(12):2393-2402.
7. Turk D, Okifuji A. Psychological factors in chronic pain: evolution and revolution. *J Consult Clin Psychol*. 2002;70:678-690.
8. Jonas WB, Walter JAG, Fritts M, Niemtzow RC. Acupuncture for the trauma spectrum response: scientific foundations, challenges to implementation. *Med Acupunct*. 2011;23(4):249-262.
9. Sehgal N, Manchikanti L, Smith HS. Prescription opioid abuse in chronic pain: a review of opioid abuse predictors and strategies to curb opioid abuse. *Pain Physician*. 2012;15(suppl 3):ES67-ES92.
10. Friedly J, Standaert C, Chan L. Epidemiology of spine care: the back pain dilemma. *Phys Med Rehab Clin North Am*. 2010;21(4):659-677.
11. Khanna A, Gougoulias N, Longo UG, Maffulli N. Minimally invasive total knee arthroplasty: a systematic review. *Orthop Clin North Am*. 2009;40(4):479-489; viii.
12. Donnez J, Squifflet J, Donnez O. Minimally invasive gynecologic procedures. *Curr Opin Obstet Gynecol*. 2011;23(4):289-295.
13. Dowson A, Mullen MJ, Peatfield R, et al. Migraine Intervention With STARFlex Technology (MIST) trial: a prospective, multicenter, double-blind, sham-controlled trial to evaluate the effectiveness of patent foramen ovale closure with STARFlex septal repair implant to resolve refractory migraine headache. [Erratum appears in Circulation 2009;120(9):e71–72.] *Circulation*. 2008;117(11):1397-1404.

14. Miller FG, Kaptchuk TJ. Sham procedures and the ethics of clinical trials. *J Roy Soc Med*. 2004;97(12):576-578.
15. Miller FG, Wendler D. The ethics of sham invasive intervention trials. *Clin Trials*. 2009;6(5):401-402.
16. Johnson A. Surgery as placebo. *Lancet*. 1994;344(8930):1140-1142.
17. Moseley JB, O'Malley K, Petersen NJ, et al. A controlled trial of arthroscopic surgery for osteoarthritis of the knee. [Summary for patients in J Fam Pract 2002;51(10):813; PMID: 12401143.] *N Engl J Med*. 2002;347(2):81-88.
18. Bradley JD, Heilman DK, Katz BP, et al. Tidal irrigation as treatment for knee osteoarthritis: a sham-controlled, randomized, double-blinded evaluation. *Arthritis Rheum*. 2002;46(1):100-108.
19. Buchbinder R, Osborne RH, Ebeling PR, et al. A randomized trial of vertebroplasty for painful osteoporotic vertebral fractures. *N Engl J Med*. 2009;361(6):557-568.
20. Kallmes DF, Comstock BA, Heagerty PJ, et al. A randomized trial of vertebroplasty for osteoporotic spinal fractures. *N Engl J Med*. 2009;361(6):569-579.
21. Freeman BJ, Fraser RD, Cain CM, et al. A randomized, double-blind, controlled trial: intradiscal electrothermal therapy versus placebo for the treatment of chronic discogenic low back pain. *Spine*. 2005;30(21):2369-2377; discussion 2378.
22. Leclaire R, Fortin L, Lambert R, et al. Radiofrequency facet joint denervation in the treatment of low back pain: a placebo-controlled clinical trial to assess efficacy. *Spine*. 2001;26(13):1411-1416; discussion 1417.
23. Nath S, Nath CA, Pettersson K. Percutaneous lumbar zygapophysial (facet) joint neurotomy using radiofrequency current, in the management of chronic low back pain: a randomized double-blind trial. *Spine*. 2008;33(12):1291-1297.
24. van Kleef M, Barendse GA, Kessels A, et al. Randomized trial of radiofrequency lumbar facet denervation for chronic low back pain. *Spine*. 1999;24(18):1937-1942.
25. Lin CC, Shen WC, Lo YC, et al. Recurrent pain after percutaneous vertebroplasty. *Am J Roentgenol*. 2010;194(5):1323-1329.
26. Guyuron B, Reed D, Kriegler JS, et al. A placebo-controlled surgical trial of the treatment of migraine headaches. *Plastic Reconstr Surg*. 2009;124(2):461-468.
27. Cobb LA, Thomas GI, Dillard DH, et al. An evaluation of internal-mammary-artery ligation by a double-blind technic. *N Engl J Med*. 1959;260(22):1115-1118.
28. Dimond EG, Kittle CF, Crockett JE. Comparison of internal mammary artery ligation and sham operation for angina pectoris. *Am J Cardiol*. 1960;5:483-486.
29. Leon MB, Koronowski R, Downey WE, et al. A blinded, randomized, placebo-controlled trial of percutaneous laser myocardial revascularization to improve angina symptoms in patients with severe coronary disease. *J Am Coll Cardiol*. 2005;46(10):1812-1819.
30. Salem M, Rotevatn S, Stavnes S, et al. Release of cardiac biochemical markers after percutaneous myocardial laser or sham procedures. *Int J Cardiol*. 2005;104(2):144-151.
31. Salem M, Rotevatn S, Stavnes S, et al. Usefulness and safety of percutaneous myocardial laser revascularization for refractory angina pectoris. *Am J Cardiol*. 2004;93(9):1086-1091.
32. Nguyen JP, Nizard J, Keravel Y, Lefaucheur JP. Invasive brain stimulation for the treatment of neuropathic pain. *Nature Rev Neurol*. 2011;7(12):699-709.
33. Andre-Obadia N, Magnin M, Garcia-Larrea L. On the importance of placebo timing in rTMS studies for pain relief. *Pain*. 2011;152(6):1233-1237.
34. Colloca L. Learned placebo analgesia in sequential trials: what are the Pros and Cons? *Pain*. 2011;152(6):1215-1216.
35. Benedetti F, Colloca L, Lanotte M, et al. Autonomic and emotional responses to open and hidden stimulations of the human subthalamic region. *Brain Res Bull*. 2004;63(3):203-211.
36. Pollo A, Torre E, Lopiano L, et al. Expectation modulates the response to subthalamic nucleus stimulation in Parkinsonian patients. *Neurorep*. 2002;13(11):1383-1386.
37. Marchand S, Kupers RC, Bushnell MC, Duncan GH. Analgesic and placebo effects of thalamic stimulation. *Pain*. 2003;105(3):481-488.
38. GRADE Working Group. Grading of Recommendations Assessment, Development and Evaluation (GRADE). <http://www.gradeworkinggroup.org/>; 2012 Accessed 08.08.2012.
39. Cruccu G, Aziz TZ, Garcia-Larrea L, et al. EFNS guidelines on neurostimulation therapy for neuropathic pain. *Eur J Neurol*. 2007;14(9):952-970.
40. Tilburt J, Emmanuel EJ, Kaptchuk TJ, et al. Prescribing 'placebo treatments': results of national survey of US internists and rheumatologists. *BMJ*. 2008;337:a1938.
41. *World Medical Association Declaration of Helsinki: ethical principles for medical research involving human subjects*; 2008.
42. Trials of vertebroplasty for vertebral fractures. *N Engl J Med*. 2009;361(21):2097-2100.
43. Meissner K, Kohls N, Colloca L, eds. *Placebo effects in medicine: mechanisms and clinical implications. Phil Trans Roy Soc B*. 2011;366:1781-1930 (1572).
44. Lang EV, Hatsiopoulou O, Koch T, et al. Can words hurt? Patient–provider interactions during invasive procedures. *Pain*. 2005;114(1-2):303-309.
45. Moerman D, Jonas WB. Deconstructing the placebo effect and finding the meaning response. *Ann Intern Med*. 2002;136(6):471-476.
46. Jonas WB. Reframing placebo in research and practice. *Phil Trans Roy Soc B*. 2011;366(1572):1896-1904.
47. Jonas WB, Chez R. Toward optimal healing environments in health care. *J Alternat Complem Med*. 2004;1(suppl 1):S1-S6.
48. Cassell E. *The Nature of Suffering and the Goals of Medicine*. Oxford: Oxford University Press; 1994.

CHAPTER 23

What are the Best Placebo Interventions for the Treatment of Pain?

Karin Meissner[1], Klaus Linde[2]

[1]Institute of Medical Psychology, Ludwig-Maximilians-University Munich, Munich, Germany,
[2]Institute of General Practice, Klinikum rechts der Isar, Technische Universität München, Munich, Germany

INTRODUCTION

The question posed by this chapter—whether some placebo interventions are more powerful than others—is not new. As early as 1955 Louis Lasagna, one of the first placebo researchers, expected sham injections to be more effective than placebo pills.[1] In general, more invasive and impressive therapeutic rituals are believed to be more powerful interventions than less spectacular ones.[2]

However, how can one placebo intervention be more effective than another, if placebos themselves are inert? In an attempt to overcome the shortcomings of the traditional placebo concept some authors have suggested conceptualizations that did not focus on the placebo vehicle itself but on the context in which the intervention is performed. For example, Moerman and Jonas[3] emphasized that most elements of medicine convey some meaning for the patient and can thereby contribute to the success of a treatment—the physician's attire, manner, style and language, and also the diagnosis and prognosis (see also Ch 18 in this book). In a similar approach, other authors have emphasized that placebos are effective via the psychosocial context in which they are embedded.[4,5] Such contextual factors include characteristics of the patient (experience, expectation), of the provider (personality, beliefs), of the provider–patient relationship (style, and content of the communication between doctor and patient), of the setting in which the treatment is being performed, and of the treatment procedure itself (e.g. pill or injection, device or surgery).[4]

The two contextual factors on which patients traditionally set their greatest hope are the doctor's skills and the kind of treatment provided. In terms of hope for recovery, the treatment is a very important factor, and it becomes even more significant when the treatment itself is impressive and potentially harmful. The characteristics of treatment procedures may thus have direct consequences for their potential to induce placebo effects.

In this chapter, we will try to answer the question as to whether some placebo interventions are systematically more effective than others, and which factors may account for this. We will focus on the field of pain for two reasons. First, this is the area of placebo research which is growing the fastest and which is best understood. Second, the biggest meta-analysis on the placebo effect so far has confirmed a small, but consistent, placebo effect on pain for clinical populations.[6] However, we will also include results on other conditions to see whether the results for the field of pain can be generalized to other fields.

THE EFFICACY PARADOX

The question of whether some placebos are more effective than others in the treatment of pain is important, as the gold standard for evaluating the efficacy of interventions is still the placebo-controlled, randomized trial. Placebo effects, together with other factors (such as the natural history of the disease, regression to the mean, experimenter or subject bias, and error in measurement or reporting) can lead to substantial improvement in the placebo groups of clinical trials. One of the underlying assumptions of such trials is that the placebo effect is a constant background noise, which is more or less stable in all trials of a given condition. If placebo effects are not constant, however, it could happen that placebo-controlled trials do not provide a correct estimate of the

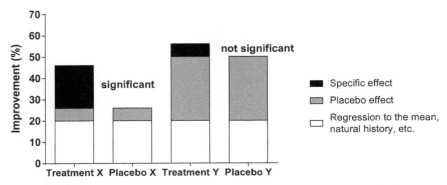

FIGURE 23.1 The efficacy paradox – hypothetical four-arm study to compare both specific and placebo (nonspecific) effects of different types of treatment. Patients are randomized to one of four arms, each two of which represent a double-blind sub-study with an active and a corresponding placebo arm. It is assumed that regression to the mean and spontaneous changes are equal in all groups. Treatment Y has a smaller specific effect than treatment X, but its total (specific + placebo) effect is larger as it is associated with a larger placebo effect. *(Figure adapted from Walach.[61])*

overall clinical effects of a treatment. For example, if in a trial of a treatment 'X', the placebo intervention induces a small placebo effect, 'X' may have a greater chance to show superiority above placebo, and thus may be more readily regarded as effective compared to a treatment 'Y', in which the placebo intervention produces a more powerful placebo effect. This could be true even if the absolute treatment effect of 'Y' is substantially larger than that of 'X'. This paradoxical situation which can arise when placebo effects are not constant across treatment modalities has been called 'the efficacy paradox' (Fig. 23.1).[7]

Is there evidence for a differential effectiveness of various placebo control procedures in the field of pain? The most straightforward way to investigate this question is to compare the effects of different types of placebo treatment directly in a randomized controlled trial (RCT). Alternatively, differential placebo effects may be compared indirectly between trials that include different placebo control groups. This is usually done by meta-analytic approaches. Before we focus on evidence available from direct and indirect approaches, we will briefly summarize ideas put forward in the literature on why certain types of placebo treatment should be more potent than others.

HYPOTHESES FROM THE LITERATURE

As mentioned above, in the 1950s Louis Lasagna had already hypothesized that sham injections are more effective than placebo pills.[1] Leslie[8] assumed that the size of tablets and capsules was important: '…the tiny ones suggesting great strength and the jumbo ones impressing by its heroic size,' but: 'Nothing can approach the psychotherapeutic impressiveness of puncture of the integument by a hollow needle.'[8 pp. 860–861]. The underlying assumption is that more impressive placebo interventions are more potent than others.

It also has been claimed that the best placebos are interventions with some mystical or magical connotation.[9–13] 'Patients may be led to believe that a treatment sets in motion a supernatural mechanism, which is able to stop the disease and restore health to the usual.'[10 p. 326, authors' translation]. The more a doctor is known for a specialty, the more power the patient expects from his knowledge and potency.[11] Not only primitive or ancient healing ceremonials bear magical elements. "The white coat, medicinal smells, apparatus and instruments of modern medicine can likewise create a magical atmosphere in which pain suddenly subsides, or by which the prescribed treatment receives its actual effectiveness."[11 p. 373, authors' translation]. Recently, Kaptchuk[12 p. 1854] compared the magic of healing rituals to theatre, especially tragedy, in that the audience—like patients—experiences emotional agitation (e.g. fear and dread) and physical arousal (e.g. tears, coldness and shuddering). 'Through dramatic enactment, both theatre and healing would entice the patient into an "as if" or "could be" world of open possibilities.' Thus, the more dramatic a healing ritual, and the more it appeals to mysterious powers, or in modern times technology,[12,13] the greater should be the patient's expectation and thus the potential of the treatment procedure to generate placebo effects.

Furthermore, it has been claimed that the occurrence of subjective sensations, such as heat or a special taste, may increase expectations and hence the placebo effect.[8,10] For example, Leslie[8 pp. 859–860] argued that medications have to taste or smell like 'medicine,' and bitterness rather than ordinary nastiness would carry a strong placebo effect, because patients have by association come to expect this. Similarly, placebo treatments that leave visible signs such as pain, erythema, or scars would be expected to trigger positive expectations—as is the case with saline injections, sham acupuncture, and sham surgery (see also Ch 16 in this book).

Beecher focused on negative emotions that may enhance the placebo effect.[14] He argued that placebos could affect only psychologic processes; therefore, the greater the importance of the psychologic component in a situation, the greater the opportunity for the placebo to produce an effect. Beecher referred to the example of surgical procedures, which are not only costly of time and money, but also dangerous to health or life, rendering the circumstances surrounding surgery full of stress and anxiety.

Another reason why sham acupuncture and sham surgery could be especially effective placebo interventions for the treatment of pain has been proposed by Liu and Yu.[15] These authors hypothesized that placebo interventions which—in addition to triggering expectations—also comprise sensory (e.g. painful) stimulation and emotional factors (e.g. fear and anxiety induced by the painful stimulus) would induce the strongest placebo effect on pain. They assumed that all three factors would stimulate the endogenous opioid system and would thereby independently contribute to the magnitude of placebo analgesia.[15]

In conclusion, there is some consensus in the literature to expect stronger effects from more impressive placebo interventions, such as sham acupuncture or sham surgery, as these evoke larger expectations and/or go along with conscious sensations and/or strong emotions.

EVIDENCE FROM DIRECT COMPARISONS

In 2002, Kaptchuk et al[16] were the first to review the evidence for a differential placebo efficacy by focusing on trials that included two different kinds of placebo intervention, namely, sham devices and inert pills. They collected seven relevant trials from the literature: four trials on pain in chronic arthritis,[17-20] and one trial each on hypertension,[21] schizophrenia,[22] and varicose veins.[23] The four trials on arthritic pain provided evidence for a larger placebo effect of parenteral placebos[17,19,20] and sham acupuncture[18] as compared to oral placebo pills. From the three studies on conditions other than pain, two studies showed positive results: hypertensive patients reacted more strongly to parenteral placebos than to oral placebos,[21] and patients with varicose veins found greater relief from a topical vs. an oral placebo.[23] Only the study in schizophrenic patients did not find an effect of any placebo intervention, and hence no difference in-between.[22] Even though most of these trials suffered from methodologicl weaknesses, such as lack of randomization and small sample size, the results warranted further studies on this issue.[16]

In 2006, Kaptchuk and his team published an RCT in which 270 patients with chronic arm pain were randomized to either sham acupuncture or inert pills.[24] After 2 weeks of single-blind treatment with one of the different placebo treatments, patients were re-randomized to either remain on their treatment or to receive active treatment, namely, acupuncture or amitriptyline. Contrary to the hypothesis, patients in both placebo groups showed similar pain levels after the 2 weeks of single-blind placebo treatment. During the 6 weeks after re-randomization, however, the patients who had continued on sham acupuncture showed greater reduction in pain over time than did those who had remained on placebo pills. A further result points to the complexity of the research question: one of the secondary outcomes, namely, an arm-function score, turned out to be significantly more improved in the oral placebo group after 2 weeks of single-blind treatment than in the sham acupuncture group, and this difference vanished during the subsequent double-blind period. The initial superiority of the oral placebo with respect to arm function was mainly due to an improvement of the ability-to-sleep subscale—possibly a consequence of informing the patients that sleepiness is a side effect of amitriptyline.[24] Thus, both the type of placebo intervention and the process of informed consent obviously had an effect on the outcome, showing that placebo responses are heterogeneous and influenced by many context factors.

Based on these results, we have recently performed a systematic review that aimed to retrieve all RCTs from the literature; this allowed us to compare different types of placebo control (unpublished data). Even though an extensive literature search was performed, we could identify only 11 RCTs with two or more placebo control groups. These did not include the pain trials of Kaptchuk's review,[16] which, due to methodologic shortcomings, did not fulfill our inclusion criteria. We found four other trials that focused on pain; two of them were on acute pain conditions (acute musculoskeletal pain[2] and postoperative pain[25]) and two investigated chronic arm pain.[24,26] In none of the pain studies did we see clear differences in favor of the more impressive placebo intervention. Only the study by Kaptchuk et al[24] provided evidence that sham acupuncture reduced symptoms more than inert pills after 6 weeks of treatment, whereas the primary outcome (pain after 2 weeks) did not show the expected result (see above). A better efficacy of sham acupuncture, compared to placebo pill, could not be verified in two non-pain trials—one on male sexual dysfunction and another on asthma.[27,28] Yet, two of the seven trials on conditions other than pain suggested greater benefit by the more intense placebo, but again only for a secondary outcome: topical placebo cream turned out to be better than oral placebo in the treatment of varicose veins with regard to venous refill time, while foot volume and minimal ankle circumference only showed a tendency for improvement,[23] and patients with Raynaud's phenomenon improved on a clinical rating scale more with EMG biofeedback (that served as a control for temperature biofeedback) than with oral placebos, but the main outcome, namely the frequency of Raynaud attacks, did not change differentially.[29]

In summary, available RCTs that directly compared the effects of different placebo interventions did not provide clear evidence that more impressive placebo interventions are generally better than less impressive ones, either in the treatment of pain or in other conditions. However, RCTs that allow such a direct comparison are extremely rare, and the existing ones vary widely with regard to indications, interventions, and outcomes. Thus, evidence from direct comparisons is yet too sparse to get a satisfactory answer to the question posed above.

EVIDENCE FROM INDIRECT COMPARISONS

The second approach to test the hypothesis of a differential placebo efficacy is to compare the effect sizes of different placebo interventions across trials by meta-analytic approaches.

An interesting result for a differential effectiveness of different kinds of placebo intervention emerged from a Cochrane meta-analysis by Hróbjartsson and Gøtzsche on the effects of placebo versus no treatment in all clinical conditions.[6,30] The meta-analysis was based on RCTs that included both a placebo and a no-treatment control group, thus controlling for factors such as regression to the mean and natural history of disease, which—in addition to placebo effects—are known to contribute to the improvement in placebo groups in clinical trials. The latest update of this meta-analysis included 202 trials on 60 clinical conditions.[6] In this dataset, a small but significant placebo effect above no-treatment was confirmed for pain. Furthermore, in one of the subgroup analyses the effects of 'physical', 'pharmacologic,' and 'psychologic' placebo interventions were compared. Results demonstrated that the average placebo effect was larger in trials using 'physical' placebos (typically sham acupuncture) than in typical drug trials that used 'pharmacologic' placebos as controls. By performing a re-analysis of the same dataset, we complemented these results by showing that sham acupuncture was the most effective 'physical' placebo.[31]

Since these results were derived from indirect comparison across trials with high heterogeneity in-between, the findings must be interpreted with caution as differences may be due to confounding factors, such as differing study populations, study designs, study sites and approaches. This caveat applies also to a couple of further meta-analyses that aimed to compare the effects of different placebo interventions across RCTs for a given condition. For example, in 2002, de Craen et al performed a systematic review of RCTs on acute migraine and found that the subcutaneous placebo interventions were, on average, associated with greater headache relief than oral placebos.[32] In fields other than pain, a similar approach has been used to look for possible differences between drug placebos and a de-tuned transcranial electromagnetic stimulation device in RCTs on depression; however, no difference between these placebo interventions could be shown.[33] Two further systematic reviews investigated whether the improvement on placebo may be larger in complementary and alternative medicine (CAM) trials compared to conventional ones.[34,35] One of the reviews focused on patients with visceral pain (irritable bowel syndrome, IBS) and reported a similar responder ratio under CAM placebos (typically Chinese herbs, probiotics and dietary supplements) as compared to conventional drug placebos.[34] Interestingly, the placebo responder ratio increased with longer trial duration and more office visits in CAM trials, while in conventional drug trials the placebo responder ratios decreased as office visits increased, suggesting an enhanced placebo effect of CAM office visits.[34] The second systematic review tested the hypothesis that RCTs of classical homeopathy often fail because placebo effects are substantially higher than in conventional medicine. This hypothesis, however, was not supported by the study results: there was no difference between the rate of improvement in the placebo groups of classical homeopathic trials compared with the conventional medicine trials.[35]

Our group has recently performed a meta-analysis that aimed to compare the effectiveness of different kinds of placebo intervention in the prophylaxis of migraine attacks.[36] We focused on migraine prophylaxis because a variety of prophylactic treatments for this condition exist, including drugs, acupuncture, biofeedback, and surgery, all of which have been tested in RCTs. Our meta-analysis was based on 79 placebo-controlled RCTs. These trials used seven different kinds of placebo control, including orally administered placebos for a pharmacologic drug, orally administered placebos for a CAM intervention, injected placebos for a pharmacologic drug, sham acupuncture, sham surgery, a de-tuned electromagnetic stimulation device, and cognitive-behavioral sham treatments (such as sham biofeedback or pseudo-meditation). Results showed that more patients improved on sham acupuncture and on sham surgery than on oral pharmacologic placebos. The remaining placebo interventions were associated with improvements similar to those achieved with oral placebo pills. However, between-study heterogeneity was higher than what was expected by chance, implying a risk of confounding. An explorative multivariable analysis that controlled for 15 different factors showed 'sham acupuncture' and 'sham surgery' to be the only independent predictors for the magnitude of improvement. Furthermore, 40 trials were subjected to a network meta-analysis, which is an extension of traditional meta-analysis and allows multiple pairwise comparisons across a range of interventions while preserving randomization.[37–40] The network we could form from the dataset allowed us to compare sham acupuncture with oral pharmacologic placebos, cognitive-behavioral sham treatment, and no treatment. Results confirmed that sham acupuncture was more effective than oral pharmacologic placebos, and that both placebo interventions were more effective than no treatment. However, these results cannot verify a causal relationship between the type of placebo intervention and related improvements. Nonetheless, our data lend support to the hypothesis that sham acupuncture, and possibly also sham surgery, may reduce the number of migraine attacks more effectively than other placebos, including injections and machines turned off.[36]

DISCUSSION

The results of our two systematic reviews, which compared different placebo interventions either directly, in randomized trials, or indirectly in trials on migraine prophylaxis, appear a bit contradictory. This could be due to confounding in the analysis of migraine trials; however, it could also be due to a lack of statistical power in the direct comparisons. The effects of active pain treatments over placebo are often small to moderate in size, and it would be unrealistic to expect large differences between different placebos. Even if relatively small in size, differential placebo effects could have important implications for the methodology and interpretation of clinical trials (see below).

While the enhanced effect of sham acupuncture and sham surgery might be due partly to the physiologic effects of skin injury,[41,42] it is likely to be related also to the special combination of contextual factors that these interventions entail. Kerr et al[43] described sham acupuncture as follows: "The placebo acupuncture ritual typically includes an initial introductory conversation between the patient and the practitioner, a brief use of touch palpation (usually on the wrist or the abdomen) to carry out a 'diagnosis,' followed by a touch treatment usually consisting of non-penetrating placebo needles that sit on top of the skin and cause patients to feel a tactile sensation that they believe (…) is a real acupuncture treatment. The sham needles are identical in appearance to regular acupuncture needles except that the shaft of the sham needle retracts into the sheath instead of penetrating the skin (in a manner analogous to a theatrical sword (…)). Patients usually lie still for 20–30 min or more, at which point, the practitioner returns to remove the 'needles'." (p. 785). Thus, sham acupuncture involves intense communication with the therapist, tactile sensations by the needles placed at different body sites, and visual stimuli that appear 'as if' real needles had been used. In the case of penetrating sham acupuncture, painful sensations may additionally arise.

Sham surgery is a likewise complex ritual. The following description of a sham surgery procedure is taken from the publication of an RCT on arthroscopic surgery for osteoarthritis of the knee. 'To preserve blinding in the event that patients in the placebo group did not have total amnesia, a standard arthroscopic débridement procedure was simulated. After the knee was prepped and draped, three 1-cm incisions were made in the skin. The surgeon asked for all instruments and manipulated the knee as if arthroscopy were being performed. Saline was splashed to simulate the sounds of lavage. (…) The patient was kept in the operating room for the amount of time required for a débridement. Patients spent the night after the procedure in the hospital and were cared for by nurses who were unaware of the treatment-group assignment.'[44] (p. 82). Thus, also in sham surgery procedures, an array of contextual factors is involved that may enhance patient expectations towards the treatment and thereby increase the size of placebo effects. These factors include tactile, auditory and visual stimuli as well as the high amount of time and attention provided by the surgeons and nurses, plus the many little interventions that surgery procedures usually entail (e.g. drugs, injections, infusions, skin preparation, etc.).

Support for the importance of perceptual characteristics for the size of placebo effects is provided by a recent study on experimental pain. When a pepper spray was used as the purported analgesic, the pain threshold of healthy individuals was significantly higher as compared to that following the administration of an inert placebo spray, which did not produce any side effect.[45] Even though these results have to be replicated in patients, the study confirms earlier notions that subjective sensations may increase expectations, and thus the placebo effect.[8]

The fact that sham acupuncture and sham surgery are composed of many sub-interventions may also contribute to their enhanced effectiveness. Support for the augmentation of placebo effects by higher dosing is provided by an early study that showed increased placebo effects on alertness with a higher number of placebo pills,[46] as well as by a meta-analysis of clinical trials on duodenal ulcer that found a regimen of four placebo pills per day to be associated with a larger improvement than achieved with a regimen of two placebo pills per day.[47]

Furthermore, contextual factors related to the interaction between patients and practitioners have been shown to constitute important determinants for the placebo effect. For example, a systematic review of RCTs that examined the impact of different levels of expectancy and emotional support on health outcomes found most robust effects when both factors were combined.[4] Some years later, Kaptchuk et al randomized patients with IBS to 3 weeks on a waiting list (observation), placebo acupuncture alone ('limited'), or to placebo acupuncture with a patient–practitioner relationship augmented by warmth, attention, and confidence ('augmented').[48] It was found that sham acupuncture was significantly more effective when the patient–practitioner relationship was enhanced, suggesting that the supportive interaction with a practitioner is a potent component of the sham procedure. Further underscoring the importance of the patient–provider interaction, several meta-analyses found a positive association between the number of office visits and improvements on placebo in trials of Crohn's disease,[49] ulcerative colitis,[50] and complementary and alternative medicine (CAM) trials of IBS.[34] However, in conventional drug trials of IBS, an opposite relationship was found.[51] This may reflect a heightened therapeutic effect, especially of CAM office visits for IBS patients, when compared to conventional medical visits.[34]

A further issue that may explain the high placebo effects, especially in acupuncture trials, may be related to the selection of patients. Patients who enter acupuncture trials usually have high expectations of this type of treatment, especially since acupuncture is not part of mainstream medicine in western countries. The importance of *a priori* expectations is underscored by the finding that the patients' expectancies when entering an acupuncture trial predicted the subsequent improvement in chronic pain by sham acupuncture.[52] Furthermore, and due to high costs in terms of time, the motivational concordance of patients participating in acupuncture and surgery trials may, on average, be higher than in a trial where the treatment necessitates simply the intake of a pill three times a day. Motivational concordance with a given treatment is known to correlate positively with the response to placebo interventions.[53–56]

Finally, a special characteristic of sham acupuncture trials is that patients are frequently not told that they might receive a 'placebo,' but that 'two different types of acupuncture' are being compared, or that they might receive a type of 'acupuncture that is not considered fully adequate according to Chinese medicine.'[57] Compared to typical drug-trial information such as 'you have a 50% chance of receiving placebo' such a wording might induce higher expectations and thus foster placebo responses.

In conclusion, different factors may contribute to the enhanced success of complex sham interventions, such as sham acupuncture and sham surgery. It seems reasonable to assume that a combination of several of these contextual factors increases the placebo effect in a manner analogous to a dose response.[12]

IMPLICATIONS FOR CLINICAL TRIAL METHODOLOGY AND DECISION-MAKING

Overall, the available evidence for differential placebo effects is still relatively limited. But particularly the findings in the area of acupuncture suggest that the efficacy paradox could be a reality in some situations in the treatment of pain. For the methodology of clinical pain trials this would have two implications. First, a bigger improvement in the placebo group can considerably enhance the sample size necessary to prove superiority of active treatment. A meta-analysis suggested that in fact most sham-controlled acupuncture trials are probably largely underpowered to detect the probably relatively small 'specific' effect.[58] Second, there is a need for more trials that compare different types of treatment directly, thus bypassing the problem of differential 'placebo baselines' in different types of treatment. However, such trials cannot be blinded, and researchers should be aware of the potential bias when including participants who have a strong belief in favor of one type of intervention.

The problems the efficacy paradox could create for decision-making are well demonstrated by experiences from Germany. Between 2001 and 2006 large research programs were launched to provide an evidence base for deciding whether acupuncture treatment for chronic low-back pain, chronic pain due to osteoarthritis, and chronic headache should be reimbursed in the statutory sickness system which covers about 90% of the German population. In a whole series of large randomized trials and observational studies, a puzzling picture came up (see Cummings[59] for a summary). Compared to no-treatment or usual-care-only, acupuncture was consistently associated with statistically significant and clinically relevant benefit for all conditions investigated. Compared to sham acupuncture, only one of eight trials found a significant difference in the primary analysis. However, both acupuncture and sham acupuncture were found to be superior to standard treatment based on German guidelines available at that time for low-back pain and osteoarthritis of the knee. Based on this evidence, the German health authorities decided to reimburse acupuncture for chronic low-back pain and chronic pain due to osteoarthritis. Otherwise, they would have withheld a treatment shown to be superior to (reimbursed) standard treatment. Critics argued that the trials were biased and that German statutory health insurance now pays for placebo treatment. And why then not pay for sham acupuncture, too?

We believe that there needs to be a systematic discussion among all relevant stakeholders (including patients, healthcare professionals, researchers and political decision-makers) when and under what circumstances a treatment should be considered effective and reimbursed when it has no, or only small, effects over placebo or sham treatment.

CONCLUSIONS AND FUTURE DIRECTIONS

In this chapter, we have tried to answer the question of whether and why some placebo interventions may be more effective than others. By collecting evidence from different approaches our results suggest that, in the field of pain, such differences appear to exist, with more impressive placebo interventions, especially sham acupuncture and sham surgery, being more powerful interventions than less impressive ones. In conditions other than pain, evidence is still too scarce to draw any valid conclusion. It is most probably the special composition of context factors that renders the more impressive placebo interventions more effective than others. Future studies should try to identify these factors as well as their mutual interplay. This would be of importance not only for improving clinical trial methodology in order to find the best treatment option for patients; identifying and understanding the context factors that push treatment expectations would also help to improve pain management. We know today that the placebo effect can enhance not only the effectiveness of inert interventions but also of active pain treatments.[60] More knowledge about how to optimize the treatment expectations of patients could, therefore, improve the success of any pain treatment.

References

1. Lasagna L. Placebos. *Sci Am*. 1955;193(2):68-71.
2. Schwartz NA, Turturro MA, Istvan DJ, Larkin GL. Patients' perceptions of route of nonsteroidal anti-inflammatory drug administration and its effect on analgesia. *Acad Emerg Med*. 2000;7(8):857-861.
3. Moerman DE, Jonas WB. Deconstructing the placebo effect and finding the meaning response. *Ann Intern Med*. 2002;136:471-476.
4. Di Blasi Z, Harkness E, Ernst E, et al. Influence of context effects on health outcomes: a systematic review. *Lancet*. 2001;357(9258):757-762.
5. Benedetti F. How the doctor's words affect the patient's brain. *Eval Health Prof*. 2002;25(4):369-386.
6. Hróbjartsson A, Gøtzsche PC. Placebo interventions for all clinical conditions. *Cochrane Database Syst Rev*. 2010(1):CD003974.
7. Walach H. The efficacy paradox in randomized controlled trials of CAM and elsewhere: beware of the placebo trap. *J Altern Complement Med*. 2001;7(3):213-218.
8. Leslie A. Ethics and practice of placebo therapy. *Am J Med*. 1954;16(6):854-862.
9. Shapiro AK. The placebo effect in medical and psychological therapies. In: Garfield SL, Bergin AE, eds. *Handbook of Psychotherapy and Behaviour Change*. New York: John Wiley; 1978:369-410.
10. Haas H, Fink H, Härtefelder G. Das Placeboproblem. In: Beckett AH, Büchi J, Chen KK, et al, eds. *Fortschritte der Arzneimittelforschung*. Vol. 1. Stuttgart: Birkhäuser Verlag; 1959:279-454.
11. Clauser. Ü. ber seelische Wirkungen der Arznei. *Deutsche Medizinische Wochenschrift*. 1956;81:370-376.
12. Kaptchuk TJ. Placebo studies and ritual theory: a comparative analysis of Navajo, acupuncture and biomedical healing. *Philos Trans R Soc Lond B Biol Sci*. 2011;366(1572):1849-1858.
13. Kaptchuk TJ. The placebo effect in alternative medicine: can the performance of a healing ritual have clinical significance? *Ann Intern Med*. 2002;136(11):817-825.
14. Beecher HK. Surgery as placebo. A quantitative study of bias. *JAMA*. 1961;176:1102-1107.
15. Liu T, Yu CP. Placebo analgesia, acupuncture and sham surgery. *Evid Based Complement Alternat Med*. 2011;2011:943147.
16. Kaptchuk TJ, Goldman P, Stone DA, Stason WB. Do medical devices have enhanced placebo effects? *J Clin Epidemiol*. 2000;53:786-792.
17. Morison RA, Woodmansey A, Young AJ. Placebo responses in an arthritis trial. *Ann Rheum Dis*. 1961;20:179-185.
18. Thomas M, Eriksson SV, Lundeberg T. A comparative study of diazepam and acupuncture in patients with osteoarthritis pain: a placebo controlled study. *Am J Chin Med*. 1991;19(2):95-100.
19. Traut E, Passarelli EW. Placebos in the treatment of rheumatoid arthritis and other rheumatic conditions. *Ann Rheum. Dis*. 1957;16:18-22.
20. Passarelli EW, Traut EF. Study in the controlled therapy of degenerative arthritis. *AMA Arch Intern Med*. 1956;98(2):181-186.
21. Grenfell RF, Briggs AH, Holland WC. A double-blind study of the treatment of hypertension. *JAMA*. 1961;176:124-128.
22. Goldman AR, Witton K, Scherer JM. The drug-giving ritual, verbal instructions and schizophrenics ward activity levels. *J Nerv Ment Dis*. 1965;140:272-279.
23. Saradeth T, Resch KL, Ernst E. Placebo-treatment for varicosity – dont eat it, rub it. *Phlebology*. 1994;9(2):63-66.
24. Kaptchuk TJ, Stason WB, Davis RB, et al. Sham device vs. inert pill: randomised controlled trial of two placebo treatments. *BMJ*. 2006;332:391-397.
25. Thipphawong JB, Babul N, Morishige RJ, et al. Analgesic efficacy of inhaled morphine in patients after bunionectomy surgery. *Anesthesiology*. 2003;99(3):693-700; discussion 696A.
26. Pain in the neck and arm: a multicentre trial of the effects of physiotherapy, arranged by the British Association of Physical Medicine. *Br Med J*. 1966;1(5482):253-258.
27. Aydin S, Ercan M, Caskurlu T, et al. Acupuncture and hypnotic suggestions in the treatment of non-organic male sexual dysfunction. *Scand J Urol Nephrol*. 1997;31(3):271-274.
28. Wechsler ME, Kelley JM, Boyd IO, et al. Active albuterol or placebo, sham acupuncture, or no intervention in asthma. *N Engl J Med*. 2011;365(2):119-126.
29. Wigley FM, Wise R, Haythornthwaite J, et al. Comparison of sustained-release nifedipine and temperature biofeedback for treatment of primary Raynaud phenomenon – results from a randomized clinical trial with 1-year follow-up. *Arch Intern Med*. 2000;160(8):1101-1108.
30. Hrobjartsson A, Gotzsche PC. Placebo interventions for all clinical conditions. *Cochrane Database Syst Rev*. 2004(3):CD003974.
31. Linde K, Niemann K, Meissner K. Are sham acupuncture interventions more effective than (other) placebos? A re-analysis of data from the Cochrane review on placebo effects. *Forsch Komplementmed*. 2010;17(5):259-264.
32. de Craen AJ, Tijssen JG, de GJ, Kleijnen J. Placebo effect in the acute treatment of migraine: subcutaneous placebos are better than oral placebos. *J Neurol*. 2000;247(3):183-188.
33. Brunoni AR, Lopes M, Kaptchuk TJ, Fregni F. Placebo response of non-pharmacological and pharmacological trials in major depression: a systematic review and meta-analysis. *PLoS One*. 2009;4(3):e4824.
34. Dorn SD, Kaptchuk TJ, Park JB, et al. A meta-analysis of the placebo response in complementary and alternative medicine trials of irritable bowel syndrome. *Neurogastroenterol Motil*. 2007;19(8):630-637.

35. Nuhn T, Ludtke R, Geraedts M. Placebo effect sizes in homeopathic compared to conventional drugs – a systematic review of randomised controlled trials. *Homeopathy*. 2010;99(1):76-82.
36. Meissner K, Fässler M, Rücker G, et al. *Are some placebo interventions more effective than others? Results from a systematic review on migraine prophylaxis*:submitted; 2012.
37. Li T, Puhan MA, Vedula SS, Singh S, Dickersin K, Ad The. Hoc Network Meta-analysis Methods Meeting Working group. Network meta-analysis – highly attractive but more methodological research is needed. *BMC Med*. 2011;9(1):79.
38. Mills EJ, Bansback N, Ghement I, et al. Multiple treatment comparison meta-analyses: a step forward into complexity. *Clin Epidemiol*. 2011;3:193-202.
39. Puhan MA, Bachmann LM, Kleijnen J, Ter Riet G, Kessels AG. Inhaled drugs to reduce exacerbations in patients with chronic obstructive pulmonary disease: a network meta-analysis. *BMC Med*. 2009;7:2.
40. Salanti G, Higgins JP, Ades AE, Ioannidis JP. Evaluation of networks of randomized trials. *Stat Methods Med Res*. 2008;17(3):279-301.
41. Birch S. A review and analysis of placebo treatments, placebo effects, and placebo controls in trials of medical procedures when sham is not inert. *J Altern Complement Med*. 2006;12(3):303-310.
42. Lund I, Lundeberg T. Are minimal, superficial or sham acupuncture procedures acceptable as inert placebo controls? *Acupunct. Med*. 2006;24:13-15.
43. Kerr CE, Shaw JR, Conboy LA, et al. Placebo acupuncture as a form of ritual touch healing: a neurophenomenological model. *Conscious Cogn*. 2011;20(3):784-791.
44. Moseley JB, O'Malley K, Petersen NJ, et al. A controlled trial of arthroscopic surgery for osteoarthritis of the knee. *N Engl J Med*. 2002;347(2):81-88.
45. Rief W, Glombiewski JA. The hidden effects of blinded, placebo-controlled randomized trials: an experimental investigation. *Pain*. 2012;153(12):2473-2477.
46. Blackwell B, Bloomfield SS, Buncher CR. Demonstration to medical students of placebo responses and non-drug factors. *Lancet*. 1972;1(7763):1279-1282.
47. de Craen AJ, Moerman DE, Heisterkamp SH, et al. Placebo effect in the treatment of duodenal ulcer. *Br J Clin Pharmacol*. 1999;48(6):853-860.
48. Kaptchuk TJ, Kelley JM, Conboy LA, et al. Components of placebo effect: randomised controlled trial in patients with irritable bowel syndrome. *BMJ*. 2008;336(7651):999-1003.
49. Su C, Lichtenstein GR, Krok K, et al. A meta-analysis of the placebo rates of remission and response in clinical trials of active Crohn's disease. *Gastroenterology*. 2004;126(5):1257-1269.
50. Ilnyckyj A, Shanahan F, Anton PA, et al. Quantification of the placebo response in ulcerative colitis. *Gastroenterology*. 1997;112(6):1854-1858.
51. MacPherson H, Thorpe L, Thomas K. Beyond needling – therapeutic processes in acupuncture care: a qualitative study nested within a low-back pain trial. *J Altern Complement Med*. 2006;12(9):873-880.
52. Linde K, Witt CM, Streng A, et al. The impact of patient expectations on outcomes in four randomized controlled trials of acupuncture in patients with chronic pain. *Pain*. 2007;128(3):264-271.
53. Hyland ME. Motivation and placebos: do different mechanisms occur in different contexts? *Philos Trans R Soc Lond B Biol Sci*. 2011;366(1572):1828-1837.
54. Hyland ME, Whalley B. Motivational concordance: an important mechanism in self-help therapeutic rituals involving inert (placebo) substances. *J Psychosom Res*. 2008;65(5):405-413.
55. Hyland ME, Whalley B, Geraghty AW. Dispositional predictors of placebo responding: a motivational interpretation of flower essence and gratitude therapy. *J Psychosom Res*. 2007;62(3):331-340.
56. Whalley B, Hyland ME. One size does not fit all: motivational predictors of contextual benefits of therapy. *Psychol Psychother*. 2009;82(Pt 3):291-303.
57. Linde K, Dincer F. How informed is consent in sham-controlled trials of acupuncture? *J Altern Complement Med*. 2004;10(2):379-385.
58. Linde K, Niemann K, Schneider A, Meissner K. How large are the nonspecific effects of acupuncture? A meta-analysis of randomized controlled trials. *BMC Med*. 2010;8:75.
59. Cummings M. Modellvorhaben Akupunktur – a summary of the ART, ARC and GERAC trials. *Acupunct Med*. 2009;27(1):26-30.
60. Colloca L, Lopiano L, Lanotte M, Benedetti F. Overt versus covert treatment for pain, anxiety, and Parkinson's disease. *Lancet Neurol*. 2004;3(11):679-684.
61. Walach H. The efficacy paradox in randomized controlled trials of CAM and elsewhere: beware of the placebo trap. *J Altern Complement Med*. 2001;7(3):213-218.

CHAPTER 24

How Communication between Clinicians and Patients may Impact Pain Perception

Arnstein Finset

Department of Behavioral Sciences in Medicine, Institute of Basic Medical Sciences,
Faculty of Medicine, University of Oslo, Oslo, Norway

INTRODUCTION

The Patient–Clinician Relationship: A Contextual Factor in Pain Management

The idea that the patient–clinician relationship may affect pain and other symptoms is not new. Hippocrates actually suggested that '… the patient … may recover his health simply through the contentment with the goodness of the physician.'[1] But exactly *how* the 'goodness of the clinician'—or the patient–clinician relationship in a more modern language—promotes recovery has most often not been explicitly investigated in clinical trials. By remaining an unspecified contextual component of treatment, the relationship and communication patterns between clinicians and patients have been understood as elements in the placebo response by many authors.[2,3]

In recent years, more attention has been given to contextual effects of treatment, including the patient–clinician relationship, and their effect on health outcome.[3-7] In most research on patient–clinician relations different elements in the interaction between patient and clinician have been operationalized as communication behavior, either studied in natural settings by different observational methods[8] or experimentally, by training or assigning clinicians to communicate in specific ways. Only few of the studies on the effect of communication behavior on outcome have applied a placebo design, but the findings from many of these studies may be interpreted within a placebo model.[9,10]

In a comprehensive review of the role of expectations in the placebo effect and their use in the delivery of health care, Crow et al have presented a model of the placebo effect (Fig. 24.1).[2] The model points to four sets of determinants for the placebo response: patient factors, practitioner factors, factors related to the interaction between practitioner and patient, and finally other aspects of the treatment and the setting (shape and color of the pill, cost of treatment etc.). The model of Crow et al includes only one mechanism, as the role of expectancy was the topic of that particular review. The model specifies three types of outcome: objectively measures health status, patient report and health-care utilization.

The present chapter considers the impact on pain perception of one of the determinants in Crow's model, the interaction between clinician and patient. It obviously takes two to tango, but most studies on the effect of the practitioner–patient interaction on health outcome referred to in the chapter focus on the practitioner's part of the equation: how do communication behaviors of the clinician impact pain perception? I will attempt to specify the different elements of the ongoing communication in clinical encounters and examine their potential impact on the patient's experience of pain.

How Communication may Affect Pain

In Crow et al's model of the placebo response, expectancy is the only mechanism investigated,[2] and expectancies represent a major mechanism of the placebo effect.[3] In many of the papers discussed in the present chapter, and summarized in Table 24.1, the researchers test the effect of the clinician's communication behavior, aiming to create positive patient expectations about pain relief. But in a number of the studies the researchers go beyond a mere formulation of a positive expectation, such as 'This drug is effective and will decrease the pain quickly after taking it'[11] or 'You will be better within a week.'[12] We will suggest three main categories of communication behavior that may add significant qualities

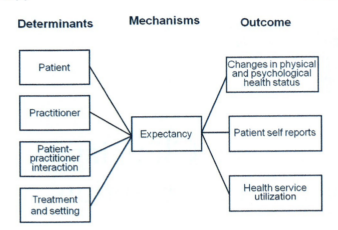

FIGURE 24.1 A model of determinants for expectancy effects on outcome *(from Crow et al[2])*.

to the formulation of an expectation: a persuasive and enthusiastic way to convey the message, a warm and empathic communication style, and an active involvement of the patient in the consultation.

There are a few examples among the papers of a suggestive or persuasive communication style. In an early study, the doctor is instructed to present information about pain and treatment 'in a manner of enthusiasm and confidence.'[13] In another early study the effect of acupuncture is 'strongly suggested.'[14] In early discussions of placebo effects, in particular in the context of psychotherapy, the importance of a persuasive delivery was often mentioned and emphasized,[15] but the degree of persuasiveness in the way messages are communicated are sparsely specified in many studies.

Other studies emphasize emotional aspects of the relationship. Clinicians are in some studies instructed to display a warm and friendly attitude,[14] 'emphasize patient comfort and well-being'[16] or be empathic by attempting to 'understand and share the participants' pain.'[17] This emphasis on emotional aspects of the patient–clinician relationship is elaborated both in clinical communication research in general[18,19] and in the placebo literature.[3,20] For instance, Shapiro and Shapiro point to the relevance, for the placebo response, of the therapist's variables underlying Rogerian nondirective psychotherapeutic approach, such as genuineness, empathy and unconditional positive regard for the patient.[21]

Finally, in a number of the studies discussed in the chapter, communication behavior of the clinician is supposed to involve the patient actively in communication. For instance, in one study the clinician is instructed to 'explore patients' own thoughts about the illness, the personal meaning of the illness and own attributions of causes.'[22,23] The relevance of patient experiences has recently been discussed in the literature on placebo analgesia.[24]

In most studies presented in this chapter, a communication behavior intended to obtain positive health effect, such as pain relief, was contrasted to a neutral behavior with little presumed effect on pain. But the clinician's way of communication may also increase the pain, creating a nocebo effect.[3] For instance, in a small study on the effect of information about headache after lumbar puncture, almost half of those who were told to expect a headache afterwards reported one, while all but one of those not warned about a potential headache remained headache-free ($p < 0.05$).[25] (See Chs 25 and 27 in this volume.)

Figure 24.2 presents a revised model, drawn for the purpose of this chapter. The model specifies communication behavior as the independent variable, and expectancy, degree of persuasiveness, emotional quality of the interaction and involvement of the patient as the potentially effective aspects of the communication behavior, with pain perception as the main outcome.

Selection of Studies

The selection of papers for this chapter has been based on a number of sources, such as numerous searches of databases with different sets of keywords (Medline, SCOPUS), from reference lists of the papers identified in the searches and other relevant papers, as well as on personal knowledge of the relevant research literature.

In selecting papers, I attempted to include only papers investigating the impact of a more or less specified element of clinician–patient communication behavior and pain perception. Most studies on the impact of communication on health outcomes *other than* in the context of pain, and studies of the impact on pain of independent variables *other than* communication, are therefore not included. In some of the studies, other relevant outcome measures are investigated, such as pain behavior, quality of life and health-care utilization in a broad sense (for instance, the use of analgesics). A few examples of experimental studies which elucidate relevant aspects of the main topic will also be referred to. Papers applying a placebo design and other studies have been included.

THE IMPACT OF EXPECTANCY IN CLINICAL STUDIES

One of the first studies which attempted to investigate the potential effects of physician–patient communication on health outcome was a trial by Egbert et al as early as 1964; it was designed to 'determine the effect of instruction, suggestion and encouragement upon the severity of postoperative pain.'[13] Ninety-seven patients who were to undergo elective intra-abdominal operations were randomized to a 'special care group' or a control group. An anesthetist saw all patients on the

TABLE 24.1 Overview of Studies

| Author | Setting, Sample, Design | How communication is operationalized ||||| Outcome |
		Information	Expectancy	Persuasion, Confidence	Warm Atmosphere, Empathy	Patient Involvement, Agreement	
Egbert et al 1964[13]	Clinical setting, surgery patients, N=97 Intervention group: Surgeon gives information Control group: No information	Information about pain	Partly: pain is common and normal	Information given 'in a manner of enthusiasm and confidence'	Not specified	Not specified	Less use of analgesics, but not less pain
Reading et al 1982[26]	Clinical setting, elective laparoscopy, N=59 Intervention group I: specific information II: un specific reassurance	Information about pain	Group I: yes. Group II: No specific information about surgery	Information given in a reassuring and supportive way (group I and II)	Not specified	Not specified	Less use of analgesics ($p<0.05$), but not less pain
Thomas 1987[30]	General practice, N=200 Two groups: positive and negative consultations	Information about diagnosis	Positive cons.: Recovery to be expected	Positive cons.: Patients told confidently that they would be better	Not specified	Not specified	Better symptom recover in intervention group ($p<0.001$)
Knipschild & Arntz 2005[12]	General practice, N=128 Two groups: Explanation of illness. Positive expectation.2. No explanation	I will tell you precisely what the matter is with you (followed by a clear explanation)	You will be better within a week or so.	Not specified	Not specified	No	No effect on pain of intervention
Gryll & Katahn 1978[32]	Clincal setting, dentistry, N=160 2×2×2×4 design (Status, Attitude ×2, Message)	Yes	Oversell and undersell condition	Not specified	Warm and neutral condition (attitude)	Encourage interaction with patient	Most effect of message (oversell, $p<0.001$) and interaction oversell and attitude ($p<0.05$)
De Craen et al 2001[11]	Clinical setting, Pain clinic, N=112 2×2 design Expectancy (positive vs. negative)×pill vs. placebo	Yes	Intervention: Pill effective, will decrease the pain quickly Ctrl: Medication is limited and ... not ... beneficial in all patients	Not specified	Not specified	No	No effect
Street et al 2012[31]	Acupuncture clinic N=311 2×2 design High vs. neutral expectations, real vs. sham acupuncture	Yes	Yes I expect this will work, you should see improvement etc	Not specified	Not specified	Not specified	High expectation condition predicted lower pain indirectly through effect on patient satisfaction ($p<0.05$)

(Continued)

TABLE 24.1 Overview of Studies — cont'd

| Author | Setting, Sample, Design | How communication is operationalized |||||| Outcome |
		Information	Expectancy	Persuasion, Confidence	Warm Atmosphere, Empathy	Patient Involvement, Agreement	
Berk et al.[14] 1977	Clinical setting, Acupuncture, N=42, 2×2 design	About treatment	Acupuncture is effective therapy	Effect of acupuncture 'strongly suggested'	Warm and friendly	Calling patient attention to needle sensation	Borderline effect of positive milieu ($p=0.053$)
Kaptchuk et al 2008[22]	Clinical setting, Acupuncture, N=289 Three conditions: Augmented Neutral Waiting list ctrl	About treatment	(d) to communicate positive treatment expectations and	Not specified	Be warm and empathic in the interaction with the patient	Explore thoughts about the illness, the meaning, attributions, (e) to use active listening skills	Significant better symptom relief in augmented condition ($p<0.001$)
White et al 2012[16]	Clinical setting, Acupuncture, N=221 Empathic condition vs. control	Not specified	Not specified	Not specified	(a) Greet patients in a friendly, warm manner, (d) emphasize patient comfort and well-being	(b) Permit patients to enter conversation with their practitioner, (c) comply with participants' wishes, providing detailed answers to questions	No effect on pain of empathic condition. One provider achieved better pain reduction ($p<0.002$)
Sambo et al 2010[17]	Experimental setting, N=30, 3 conditions: High empathy Low empathy Alone	Not specified	Not specified	Not specified	High vs. low empathy	Not specified	No main effect of empathy Low empathy × hi attachment anxiety> more pain ($p<0.01$)
Bass et al 1986[41]	Survey, N=272 Correlational design	No	Not specified	Not specified	Included in the patient-centered concept	Included in the patient-centered concept	Patient discussed headache fully predicted less headache ($p<0.001$)
Shaw et al 2011[42]	Clinical settings, Back pain patients, N=97 Correlational design	Included in RIAS	Not specified	Not specified	Included in RIAS	Included in RIAS	Questions on treatment, facilitation and rapport building associated with more pain

Study	Setting	Column 3	Column 4	Column 5	Outcome
Staiger et al 2005[44]	Clinical setting, physiotherapy, N=380 Correlational design	Not specified	Not specified	Not specified	Clinician-patient agreement	Agreement predicted higher quality of life, not pain
Bieber et al 2006[48]	Clinical setting Effect of training N=111	Not specified	Not specified	Not specified	Physicians in intervention group trained in shared decision making	No effect on pain
Chassany et al 2006[49]	Clinical setting Effect of training GPs: N=180 GP (96 trained) in Patients: N=842	Information about pain	(Describe likely evolution of pain)	Not specified	Propose the idea of therapeutic partnership	Make sure that patient has said all he/she wants to	Patients in trained-GP group had better overall pain relief (p<0.0001)
Aiarzaguena et al 2007[50]	Clinical setting Effect of training GPs: N=39 Patients: N=156	Information about illness	Not specified	Not specified	Empathic response to patient concerns	Patients encouraged to Describe feelings	SF-36 scores in intervention group better in intervention group, including bodily pain (p<0.003)
Flaten et al 2006[53]	Experimental setting, N=84 2×2 design	Information about drug vs. Information about pain	Positive expectancy	Not specified	Supportive interaction encouraged vs. minimal interaction	Attention to patient vs. minimal attention	Both intervention>less pain, only in males
Sarinopoulos et al (in press)	fMRI experiment after arranged consultation, N=9	Not specified	Not specified	Not specified	Included in the patient-centered concept	Included in the patient-centered concept	Reduced pain related neural activation in anterior insula in patient-centered group

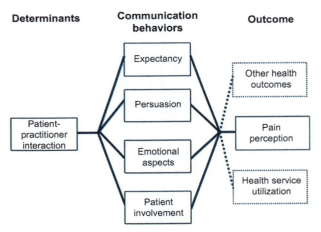

FIGURE 24.2 A model of the effects of patient–practitioner interaction on pain perception, based on an original model by Crow et al;[2] see Figure 24.1.

day before surgery to describe procedures. The anesthetist did not give the patients in the control group any information about postoperative pain. The special care group, however, was given detailed information about where they would feel pain, how severe it would be, and how long it would last. They were told that having pain after abdominal operations was common and normal, and that pain could be relieved by relaxation. Simple relaxation instructions were also given during the conversation between the patient and the anesthetist. The authors pointed out that 'the presentation was given in a manner of enthusiasm and confidence'. The paper does not report data on pain perception, but patients in the intervention group requested significantly less analgesics ($p<0.01$), appeared more comfortable and in better condition physically and emotionally, and were discharged from hospital at an average of 2–3 days earlier than patients in the control group ($p<0.01$).

Fifty years after that study was performed it is easy to criticize the researchers for applying a design which makes it difficult to decide what element of the intervention that actually had the effect on the patients' reduced demand for analgesics. What was most important of the content of the information given, the relaxation instructions or the 'enthusiasm and confidence' that characterized the doctor's communication style? However, whatever shortcomings the study has, it was the first systematic attempt to specify the effects of communication on pain and pain behavior.

Subsequent studies have further investigated different strategies to prepare patients for surgery, aiming to reduce postoperative pain and promote a speedy recovery. In one study 59 women who were undergoing elective laparoscopy were randomly assigned to three experimental conditions: a pre-operative interview providing information about surgery in a reassuring and supportive way, an interview in which the clinician provided reassurance in general terms, with no specific information about surgery, and finally a control condition, in which no contact was made prior to surgery.[26] As with the findings of Egbert et al, no difference was found in pain perception, but the group who received specific and reassuring information about surgery was found to use significantly less analgesics than the two other groups ($p<0.05$). The findings from this study indicated that a generally reassuring attitude on the part of the clinician has no effect on outcome if it is not related specifically to information about the procedure. However, from that particular study we do not know whether a persuasive communication style characterized by reassurance and support was a prerequisite to make information produce the changes in pain behavior (lower use of analgesics).

In a number of subsequent studies of preparation for medical procedures, the emphasis was more on teaching patients relaxation and other techniques and the use of different media for presentation of information (tapes, booklets etc.) than on communication style in face-to-face interaction. Reduction in postoperative pain has been reported in studies where the intervention involved cognitive coping techniques,[27] stress inoculation techniques[28] and relaxation training.[29] Taken together, these early studies of preparation for surgery showed that preparing patients psychologically for the operation might pay off. Gradually, interventions became more targeted to promote coping and self efficacy, as cognitive behavior therapies developed in the field of pain management.

In 1987 the British physician K. Bruce Thomas published a paper which has become a much quoted classic in the early literature; this paper deals with the effects of communication style on health outcome.[30] Two hundred patients in general practice, who had reported a number of different symptoms but with no objective findings of disease and without a specific diagnosis, were given arranged consultations by general practitioners in a 2×2 factorial design. The patients were given either a consultation conducted in a 'positive manner,' with and without treatment, or a consultation conducted in a 'non-positive manner,' called a negative consultation, with and without treatment. A main element in the positive consultations was to give the patient a firm diagnosis and to tell him or her confidently that he or she would be better in a few days. When patient satisfaction was assessed 2 weeks after the consultation, there were significant differences between positive and negative groups, but not between the treated and untreated groups. There were also significant differences in subjective symptoms. Of the patients in the positive conditions, 64% reported to have become better, compared with 39% in the negative consultation groups ($p=0.001$). There were no significant differences between treated and untreated groups. In the paper there is no specification on the effect on pain, but pain conditions such as abdominal pain, back pain, leg pain and headache were among the most common complaints in the data set. In his conclusion, Thomas posed the following rhetorical question: 'For a thousand years the action of the placebo has made vast numbers of patients feel better; have we today produced a consultation in which the placebo does not act?'[30]

Even if Thomas' study has been frequently quoted, it took almost 20 years before the Dutch researchers Knipschild and Arntz performed a replication of Thomas' study.[12] A total of 128 patients were randomized into two groups. The instructions to the GPs on how to treat them were similar to those in Thomas' study. The patients in the 'positive group' were, according to the authors '… given a clear diagnosis and were told that they would soon be better'. The patients in the other group were told that they '… probably had no serious underlying disease but that the GP did not know exactly what was wrong.' These latter patients were advised to come back to the GP later if necessary. Two weeks after the consultation no significant differences were found between the two groups, neither in patient satisfaction nor in subjective symptoms. Thomas's study and the Dutch replication are quite similar. There is, however, a nuance in the description of the intervention. In Thomas's study it is explicitly stated that the doctor *confidently* told the patient that he or she would be better in a few days.[30] The formulations in the articles on the Dutch group do not specify the degree of persuasion in how the messages with the positive outcome expectations were communicated.[12]

De Craen et al performed a similar study with a a 2 × 2 factorial, randomized, placebo-controlled, double-blind design.[11] Chronic pain patients attending a chronic pain outpatient clinic were randomized to receive a single oral dose of 50 mg tramadol or placebo. Moreover, patients were also randomized to two different conditions regarding the content of the information given about the drug. In the positive condition, the patient was told that '… this is a medication that recently became available in the Netherlands. This drug, according to my experience, is very effective and will decrease the pain quickly after taking it.' In the neutral condition, the instruction was as follows: 'My own experience with this medication is limited and my impression is that it will not be beneficial in all patients. The pill becomes effective almost immediately, if it is going to have an effect.' Clinicians were instructed to use their own wording when presenting the message to the patient. There is no other information in the paper on how the message was communicated. This is one of the very few studies that combine an active drug vs. placebo design with a systematic variation of communication variables. In this study all findings were negative. There was no significant effect of tramadol vs. placebo, no significant main effect of the positive vs. neutral instruction, and no interaction effects.[11]

Recently, Street et al conducted a randomized controlled trial (RCT) to test the effect of an expectancy instruction (high vs. neutral expectations) on pain in an acupuncture study (real vs. sham acupuncture) with patients with knee osteoarthritis.[31] Acupuncturists in the High Expectations group were trained to convey hope and optimism for treatment using positive statements such as 'I expect this will work,' 'you should see improvement,' 'I have had excellent results with this treatment,' 'I have had a lot of success with these kinds of symptoms,' and 'I am very optimistic that this will work for you.' The positive expectation intervention had no main effect on pain. However, the clinician's communication about treatment effectiveness during an initial visit significantly predicted patients' satisfaction with acupuncture 4 weeks into treatment. Satisfaction scores, in turn, predicted patient-reported pain 6 weeks following treatment ($p < 0.05$).[31]

THE IMPACT OF EMOTIONAL COMMUNICATION

While the studies described so far emphasize expectancy, with varying degrees of persuasion, a number of studies have investigated the emotional component of the placebo responses, with instructions to provide a warm atmosphere and empathic communication style.[22,23,32]

A key construct in emotional communication is empathy. Although different definitions of empathy exist, recent conceptualizations have converged on the idea that therapeutic empathy is comprised of at least three primary components: (1) cognitive (accurately recognizing the client's experience); (2) affective (sharing the client's feelings); and (3) behavioral (expressing empathy to the client).[33-35]

In an early study specifying the potential effect of an emotionally warm atmosphere, Gryll and Kathan investigated context effects when giving dental patients mandibular-block injections required for their dental treatment.[32] Before the injection was administered the patient received a placebo, which was a 100-mg, light-green spansule capsule, supposed to relieve tension, anxiety, and sensitivity to pain in association with the injection. The study did not include an active drug condition; all patients received a placebo or no capsule. The design was a 2 × 2 × 2 × 4 factorial design with four independent variables: (1) status of the individual delivering the communication of tile effects of the pill administration to the patient (i.e. dentist or dental technician); (2) the attitude of the dentist toward the patient (i.e. warm or neutral); (3) the attitude of the dental technician toward the patient (i.e. warm or neutral); and (4) the type of message given to the patient concerning the anticipated effects of the pill (i.e. oversell, undersell, saliva, no pill). Dependent variables were the patients' report of pain experienced from mandibular-block injection, as well the effects of the placebo on anxiety and fear of the injection.

The results were complex. The most powerful variable was type of message, with lowest pain in the oversell condition ('This is a recently developed pill that I've found to be very effective in reducing tension, anxiety, and sensitivity to pain. It cannot harm you in any way. The pill becomes effective almost immediately'). The main effect of oversell was statistically significant ($p < 0.001$) with reasonably large effect sizes. The effects of the attitudes of the dentist ($p < 0.05$) and dental assistants ($p < 0.01$) were also significant, as well as the interaction of message content and attitude ($p < 0.001$). The most effective condition in terms of pain reduction was a positive expectation presented by a warm practitioner in a positive atmosphere.[32]

A number of studies have used acupuncture vs. sham acupuncture as part of the intervention. Acupuncture is a treatment which lends itself to placebo trials. It is easy to perform a placebo condition by placing needles in a way which provides sham acupuncture. A number of studies have applied sham acupuncture to investigate placebo effects.

In an early study applying real and placebo acupuncture Berk et al investigated the effects on pain in a sample of 42 volunteers with bursitis and/or tendonitis of the shoulder.[14] The design was a 2×2 factorial design with acupuncture vs. placebo acupuncture and a so-called positive vs. negative milieu as the independent variables. The positive milieu was defined in terms of three different elements: (a) A prepared statement read to the patients before treatment, strongly suggesting that acupuncture was an effective therapy. (b) During the treatment itself, the acupuncturist actively engaged the patients in the therapeutic process by calling their attention to the needle sensations and rewarding them for feeling them. (c) The doctor–patient relationship was supposed be 'warm and friendly'. The negative milieu condition was characterized by (a) a statement which strongly emphasized the shortcomings, doubts, and inconsistencies that surround acupuncture, (b) no engagement of the patient in the therapeutic process, and (c) a minimum of doctor–patient interaction. On average, patients in general reported a reduction in discomfort after treatment. There was no significant effect of active vs. placebo acupuncture, and a borderline significant effect of a positive milieu ($p = 0.053$).

More recently, Kaptchuk et al investigated the impact of clinician communication behavior on symptoms of patients with irritable bowel syndrome (IBS).[22,23] The researchers conducted a clinical trial of 289 patients with IBS. There were three conditions, the so-called augmented and neutral interaction conditions, both receiving placebo acupuncture, and a waiting-list control group. In the augmented condition, the practitioner was given rather specific instructions about communication style. They were instructed (a) to be warm and empathic in the interaction with the patient, (b) to explore psychosocial stressors, (c) to explore patients' own thoughts about the illness, the personal meaning of the illness, and own attributions of causes, (d) to communicate positive treatment expectations, and (e) to use active listening skills. However, the practitioners were explicitly instructed not to apply specific cognitive or behavioral techniques, known from clinical studies of IBS treatment. The consultations were videotaped, and the consultations in the augmented group were found to be characterized by a communication style that may be characterized as patient-centered. The practitioners in the Limited group, on the other hand, were instructed and trained to minimize their interaction with the patient in a neutral and businesslike manner, without being outright negative.

The most pain-relevant dependent variable was a symptom severity scale, a questionnaire that measures the sum of the participant's evaluation on a 100 point scale of each of five items: severity of abdominal pain, frequency of abdominal pain, severity of abdominal distension, dissatisfaction with bowel habits, and interference with quality of life.

There were significant effects of condition, in relation to all dependent measures (global improvement, percent change in symptoms, change in symptom severity and quality of life); the augmented condition was most effective at both 3 and 6 weeks after treatment (levels of significance ranging from <0.005 to <0.001). The effect sizes were medium to large.[22] A number of patient characteristics, such as patient extraversion, agreeableness, openness to experience, and female gender, were associated with placebo response, but these effects held only in the augmented group.[23]

In a recent study, White et al examined the placebo effect of sham acupuncture on pain in patients with osteoarthritis.[16] As in the study of Kaptchuk et al[22] they sought to use an empathic communication style. The empathic condition is defined slightly differently from the one in the study of Kaptchuk et al. Practitioners should (a) greet patients in a friendly, warm manner, (b) permit patients to freely enter into conversation with their practitioner, (c) comply with participants' wishes, providing detailed answers to questions, and (d) emphasize patient comfort and well-being.

In the non-empathic condition, the consultation was supposed to be more 'clinical' in nature. Practitioners should (a) greet patients in an efficient manner, and quietly show them to the treatment cubicle, (b) discuss only matters directly relating to the treatment, (c) keep explanations as short as possible. If patients attempted to enter into any discussion, the practitioner should say: 'I'm sorry, but because this is a trial I am not allowed to discuss this with you.'

The primary outcome was pain (VAS) at 1 week post-treatment. On average, the patients improved, with no difference between real and sham acupuncture. In contrast to the study of Kaptchuk et al, there was no significant effect of the empathic condition on pain, but one specific practitioner achieved a significantly greater pain reduction than one of the other practitioners in the trial (p=0.002). The researchers also conducted interviews with patients. Patients' beliefs in treatment and confidence in outcome predicted treatment effect independent of experimental condition.[16]

A recent non-clinical experiment indicates that there may be individual differences regarding the effect of empathy on pain.[17] Thirty healthy subjects participated in a study in which thermal pain was inflicted in a laboratory. They took part in a pain experiment either without the company of other individuals or in the presence of two observers of the opposite sex; the observers had been instructed to show high empathy or low empathy with the patient during the experiment. The effect on pain of the empathic behavior of the experimenter depended on individual differences in the attachment styles of the subjects. Subjects with higher attachment anxiety responded with less pain in the empathy condition than in the non-empathy condition. For subjects without high attachment anxiety scores, the empathy status of the observer had no significant effect on pain. Moreover, there was a main effect of social presence on autonomic responses to pain, with less pain with an observer present, independently of the empathy status of the observer and the attachment style of the subject.[17]

PROMOTING PATIENT INVOLVEMENT AND COMMON GROUND: THE PATIENT-CENTERED INTERVIEW

One of the main concepts of the literature on physician–patient relations is patient-centered communication.[36,37] Important elements of the patient-centered communication include how practitioners take the patient perspective, show empathy, and allow for a full discussion of the patients' concerns.[18,19] Patient-centered practitioners should also provide information and attempt to elicit positive expectations and emotions, with recognition of patient resources, self efficacy and coping.[36] A main concept in the patient-centered tradition is common ground, pointing to the importance of obtaining a shared understanding of the patient's conditions and concerns.[38]

Although few studies have investigated the impact of patient-centered interviewing on pain, a number of researchers have discussed how a patient-centered approach may be beneficial.[39,40]

Empirical evidence for the potential benefit of a patient-centered approach is a study from The Headache Study Group of the University of Western Ontario.[41] A total of 272 patients presenting to general practitioners with a new complaint of headache took part in a 1-year prospective study. Data from both patients and physicians covering a large number of variables were collected at the first visit. Follow-up data were collected by mailed questionnaires or telephone interviews. At 1-year follow-up only three variables from the first visit data set predicted resolution of symptoms independently in a logistic regression analysis. The strongest predictor was whether or not patients had reported after the first visit that they had discussed their headache fully with the physician. The two other predictors were an organic final diagnosis and absence of visual problems accompanying headache. Neither age, gender, medication, referral nor the presence of psychosocial problems contributed independently to symptom resolution.

Shaw et al investigated the effects of patient–provider communication on 3-month recovery from acute low back pain in 97 patients.[42] Consultations with a clinician (mostly a physician or a nurse) were analyzed according to the Roter's Interaction Analysis System (RIAS).[43] Associations between RIAS data and pain ratings at 3 months follow-up were investigated. The number of biomedical questions and questions regarding treatment was significantly associated with more pain after 3 months, but also typically patient-centered variables such as the providers' facilitation of patients' concerns and positive rapport building. It is difficult to interpret these findings. Hypothetically, one explanation could be that facilitation and rapport building actually caused more pain, for instance by legitimizing and reinforcing pain behaviors. However, the authors have a different explanation. They suggest that providers may distinguish high-risk patients at the first consultation and adapt their communication patterns when interacting with them. This interpretation is also supported by the findings that certain patient behaviors, such as emotional and negative rapport building, were associated with more pain at follow-up.[42]

An important element in the patient-centered approach is to establish common ground with the patient and reach agreement about symptoms and treatment.[38] Staiger et al investigated the effect of perceived patient–physician agreement on patient-reported outcome 1 month later.[44] High agreement was associated with patient satisfaction, and SF-36 domains of mental health, social functioning and vitality, but not on pain. A few studies on provider–patient agreement have focused on the other side of the coin, namely, on the disagreements which often occur between physicians and patients on how to understand the pain.[45] For instance, Allegretti et al found frequent mismatch between physicians and patients with low back pain regarding understanding of both the etiology and treatment.[46] However, we know little on the impact of such disagreement on pain.

A number of studies describe the effects of communication skills training programs, often designed to promote a patient-centered communication style.[47] Three studies have measured the effect of communication skills training on patients' pain perception.

Bieber et al trained physicians in shared decision making and investigated whether patients who saw physicians after training reported different outcomes than patients who saw untrained physicians. They found that training had an impact on communication patterns, but not on pain.[48] Chassany et al trained 84 GPs to improve their communication with patients with osteoarthritis, aimed at improving pain management. During training, the physicians discussed methods of pain evaluation and treatment, such as the effects of words during prescribing and how to manage a therapeutic contract with patients. They were trained to inform patients about pain mechanisms, describe the likely evolution of their pain, to make sure that patients discussed their concerns fully with the physician, and to promote a therapeutic partnership. The sum of daily pain intensity differences (SPID) between intervention group and control group in the direction of less pain was significantly larger in the intervention group ($p<0.0001$). The reduction in VAS (0–100 mm) in the intervention group from baseline to 2 weeks was also significant ($p=0.01$), but the difference between the groups was only 4.5 mm.[49]

Aiarzaguena et al trained 19 GPs in a communication skills intervention; different communication skills aimed to avoid stigma associated with psychosocial suffering, devlop a safe and warm atmosphere, and improve the patient's quality of life.[50] SF-36 scores were better in the intervention group, including bodily pain ($p<0.003$), but the threshold of clinical relevance suggested by Ferguson et al[51] was not reached.

In the studies which investigated the effect of different patient-centered communication strategies, the designs did not include behavior to explicitly promote positive expectancies in patients, which is considered a core element in the placebo response.[52] In an experiment with healthy volunteers. Flaten et al therefore attempted to differentiate between an expectancy induction and emotional aspects of the communication between the study administrators (providers) and subjects ('patients').[53] Eighty-four subjects took part in the study. Pain was induced by the submaximum tourniquet technique, and all patients were given an analgesic to relieve pain during the experiment. A 2×2 factorial design was applied. Expectancy was manipulated by giving positive and neutral information about the drug respectively to separate groups. Moreover, emotional aspects of interaction were introduced in the manipulation of information giving and interaction with the experimenter about pain and the subject's feelings during the experiment. Subjects in the Pain Information groups were informed by the experimenter about the pain-induction procedure and what pain to expect, emphasizing that, although the procedure would be painful, it is completely without risk. The experimenter repeated the information during the experiment and responded to questions from subjects about pain and the procedure in a positive and supportive manner. Subjects in the No Pain Information groups received minimal information about the effects of the induction procedure, and the experimenters were instructed to engage only minimally in communication with the volunteers during the procedure. Pain tolerance was significantly higher, and pain unpleasantness less, in the group that received information about both the drug and the pain, but only in male subjects, possibly reflecting the fact that the nurses who administered the experiment were all female.[53]

In evaluating the effect of the Pain Information variable it is difficult to differentiate between the effect of the content of the information about pain and the effect of a more supportive interpersonal context for the experiment. Interestingly, the information about the pain stimulus had a gradually stronger analgesic effect over the course of the experiment, with maximum analgesia seen at 40 minutes. Thus, it may have been the more supportive nature of the interaction with the nurses, and not the content of the information about the pain stimulus, that affected pain levels.[53]

In a recent laboratory study. Sarinopoulos et al investigated the effect of patient-centered and a more conventional clinician-centered interviewing style on subsequent processing of pain stimuli, applying fMRI to study neural activation patterns during painful stimulation.[54] After a clinical interview the investigators conducted an fMRI experiment in which painful stimuli were presented together with a picture of the interviewing doctor or an unknown doctor. During the presentation of pain stimuli anterior insula activation increased, as expected. However, in patients who had undergone the patient-centered interview the anterior insula activation was less when the stimulation was accompanied by a picture of the interviewing doctor than by a picture of an unknown doctor. This effect was not seen in patients who had been interviewed according to a conventional doctor-centered approach. The findings indicate that a patient-centered consultation may have an effect on pain processing after the consultation, and could represent a contribution to our understanding of the mechanisms underlying placebo analgesia. Moreover, the study confirms the potential applicability of fMRI in the study of analgesic procedures, even in clinical settings.[55]

PSYCHOSOCIAL INTERVENTIONS IN PAIN MANAGEMENT

Most of the studies summarized in Table 24.1 concern interaction between clinicians and patients in regular medical consultations. But over the last few years we have seen the development of a number of well defined therapeutic approaches to psychologic treatment of pain, which certainly apply a number of communication behaviors and techniques, but apply these techniques within the framework of specified psychologic treatment.[56]

The most well established psychologic treatment approach in contemporary pain management is cognitive behavioral therapy (CBT). Gatchel et al have provided a definition of CBT in the context of pain management: 'The term CBT varies widely and may include self instructions ... relaxation or biofeedback, developing coping strategies, changing maladaptive beliefs about pain and goal setting ... varying selection of these strategies ... embedded in a more comprehensive pain management program that includes functional restoration, pharmacotherapy, and general medical management.'[57] CBT has been extensively evaluated in studies of treatment,[58] and a number of different treatment approaches has developed within the general framework of CBT.[59]

In recent years so-called acceptance-based traditions, such as acceptance commitment therapy (ACT), have been applied in pain management.[60,61] A recent meta-analysis concludes that acceptance-based therapies in many studies are shown to provide small to medium effects on physical and mental health in chronic pain patients, with effect sizes quite comparable to those of cognitive behavioral therapy.[62]

Motivational interviewing (MI), a treatment approach with a strong and explicit emphasis on patient-centered interviewing techniques, such as open-ended questions and empathy, and with a specific emphasis on how to handle patient ambivalence to behavioral change, has also been applied to pain treatment and rehabilitation programs. However, few studies have investigated the specific effect of MI on pain. In a recent RCT, an MI approach was compared to usual-care to improve cancer pain management.[63] Patients randomized to the MI group reported significant improvement in their ratings of pain-related interference with function, as well as general health, vitality, and mental health, but not reduced pain intensity. The intervention included a video and a pamphlet on managing cancer pain, and the paper did not provide data on specific communication behavior. The researchers concluded that MI may be a useful strategy to help patients decrease attitudinal barriers toward cancer pain management and to better manage their cancer pain.

A relevant question in the context of this chapter is to what extent the effects of CBT, ACT and MI may be attributed to the therapist's communicative behavior or to other elements of the program, such as relaxation, mindfulness or biofeedback sessions, graded exercise, patient-education elements, such as videos and leaflets, home assignments etc. In the literature on these treatment approaches applied to pain management, there are hardly any studies which evaluate the effects of specific and distinct clinician communication behaviors on pain perception. Treatment outcome is most often described in terms of the effects of a more or less comprehensive program, not the effects of specified therapist communication style. However, therapist behaviors are obviously crucial elements in the treatment program. In psychotherapy research a distinction is made between specific and common factors. The specific factors include the therapeutic techniques specific to each therapeutic method, whereas common factors are elements seen in all or most therapies. The most important common factors include the therapeutic alliance between therapist and patient, the degree of empathic behavior displayed by the therapist, and more specific behaviors such as expectancies promoted by the therapist.[64,65] Such common factors may obviously also be observed in medical consultations, and have been exemplified in some of the studies reviewed in the present chapter.

DISCUSSION AND CONCLUSION; SUGGESTIONS FOR FUTURE RESEARCH

Taken together, the findings from the studies discussed in this chapter are mixed. Some studies indicated positive effects of the clinician's communication behavior on pain, while in other studies there were negative findings. Moreover, some of the studies indicated nuanced findings in terms of individual differences in the effect on pain perception, both regarding patient characteristics[17,53] and clinician effects.[16]

Of the 19 selected studies, an effort to elicit a positive expectancy is present in at least 10 studies. In two of these studies we see significant effects in the direction of pain reduction,[22,32] in one a borderline significant effect,[14] in one an effect in males, but not in females,[53] and in three studies positive effects in terms of 'getting better' of symptoms (which include pain)[30] or reducing the use of analgesics.[13,26] In one study communication to elicit a positive expectation had an indirect effect on pain, mediated by patient satisfaction.[31] In two studies an expectancy induction was sufficient to promote pain reduction,[32,53] but in both cases the studies also included a condition with an emotional component, which was associated with further positive effects on pain reduction. The other five studies in which pain reduction was reported included communication elements other than expectancy, such as persuasion (information given 'in a manner of enthusiasm and confidence'), empathic relationship to the patient, and/or an emphasis on patient involvement in the consultation.[13,14,22,23,26,30]

Only 2 of the 10 studies aiming to elicit a positive expectancy reported no effects on pain or other relevant variables.[11,12] There were no specifications of the degree of persuasion in how the message was delivered in these studies, and emotional aspects were not mentioned.

While the importance of persuasion is an obvious aspect in marketing,[66] it is relatively seldom explicitly discussed in the placebo literature. We have found explicit examples of the emphasis on persuasive communication in only four of the papers. The role of persuasion in the placebo response in clinical settings warrants further discussion.

Another quality of the interventions was the extent to which there was an emphasis on emotional aspects, such as a warm and friendly atmosphere and empathic clinician behavior. The role of emotions in placebo analgesia is important.[20] It is well established that negative emotions increase pain, and it has been hypothesized that one factor in placebo analgesia could be a reduction in negative emotions. Lyby et al have recently shown, in an experimental study, that fear was related to reduced placebo analgesia, in terms of both self reported fear of pain[67] and experimentally induced pain.[68] A potential explanation of why a warm and empathic communication style seems to augment the expectancy effect in some of the studies could be reduction of negative emotions.[22,32] Explicit measures of negative emotions, before and after the consultations, should therefore be added in future studies. In a review on the effect of empathy on pain management, Tait concludes that 'despite the relatively limited evidence regarding the role of empathy in pain medicine, the larger literature on clinical empathy yields results that are relevant to the management of chronic pain and headache, especially when considered in conjunction with the neurophysiology of empathy' (p. 110).[33]

Increased patient involvement, giving the patient an opportunity to fully discuss his or her concerns, and a patient-centered patient–clinician relationship with emphasis on a process towards finding common ground, seems, in some of the reviewed studies, to be associated with a patient experience of less pain[41,49,50] or better quality of life.[44,50] Most of these studies have a correlational design or are intervention studies designed to test the effect of communication skills training; this makes it difficult to pinpoint exactly which communication behaviors are responsible for an effect on pain perception. However, there is growing evidence that the way patients—after the actual interaction with a clinician (or an experimenter in an experiment)—experience and remember the situation may be associated with successful pain reduction.[24,69]

Statistically significant and positive effects on pain perception in clinical studies are reported in five studies.[22,32,41,49,50] But even if the effects on pain perception are statistically significant, the effect sizes and clinical significance vary. Gryll et al do not report effect sizes, but from the data presented it appears that the effect size of the oversell condition is rather large, and further reinforced by a warm attitude.[32] The effect sizes in the study of Kaptchuk et are also medium to large.[22] However, in the studies reporting effects of communication skills training, the effects are small, and whether or not they are clinically significant is a matter of discussion.[49,50] Moreover, it is difficult to measure the size of the effect in Bass' study on the outcome of communication on headache. The outcome criterion is dichotomized, whether symptom resolution is present or not. However, the statistical association between the communication variable, whether patients felt that they had discussed their headache fully with the physician, was strong and robust.[41]

To summarize, we have discussed four different elements of communication which may impact on pain perception (expectancy, persuasion, emotional aspects, patient involvement). We have found some evidence that each of these factors may have impact on pain perception. There is some evidence that a combination of the elements may have stronger effects than each of them separately. But to the extent that positive effects are found, they are—with a few excpetions—relatively small.

Future studies should develop designs making it possible to differentiate better between the effects of different elements of the interventions and to investigate the effects of communication behavior on proximal outcomes which may serve as potential mediators on an effect on pain perception.[31] Most research on doctor–patient relations so far has been in the form of clinical studies. More emphasis should be given to experimental studies which may better specify the different elements of communication behavior and elucidate the underlying mechanisms of how the doctor's words may affect the patient's brain.[6]

References

1. Hippocrates. *On Decorum and the Physician*. Vol. II. London: William Heinemann; 1923.
2. Crow R, Gage H, Hampson S, et al. The role of expectancies in the placebo effect and their use in the delivery of health care: a systematic review. *Health Technol Asses*. 1999;3:1-96.

3. Benedetti F. *The patient's Brain. The Neuroscience Nehind the Doctor–Patient Relationship.* Oxford: Oxford University Press; 2011.
4. Stewart MA. Effective physician–patient communication and health outcomes: a review. *Can Med Assoc J.* 1995;152:1423-1433.
5. Di Blasi Z, Harkness E, Ernst E, et al. Influence of context effects on health outcomes: a systematic review. *Lancet.* 2001;357:757-762.
6. Benedetti F. How the doctor's words affect the patients's brain. *Eval Health Prof.* 2002;25:369-386.
7. Street Jr RL, Makoul G, Arora NK, Epstein RM. How does communication heal? Pathways linking clinician–patient communication to health outcomes. *Patient Educ Couns.* 2009;74:295-301.
8. Eveleigh RM, Muskens E, Van Ravesteijn H, et al. An overview of 19 instruments assessing the doctor–patient relationship: different models or concepts are used. *J Clin Epid.* 2012;65:10-15.
9. Benedetti F, Amanzio W. The placebo response: how words and rituals change the patient's brain. *Pat Educ Couns.* 2011;84:413-419.
10. Bensing J, Verheul W. The silent healer: the role of communication in placebo effects. *Pat Educ Couns.* 2010;80:293-299.
11. De Craen AJM, Lampe-Schoenmaeckers AJEM, Kraal JW, et al. Impact of experimentally induced expectancy on the analgesic efficacy of tramadol in chronic pain patients: A 2 × 2 factorial, randomized, placebo-controlled, double-blind trial. *J Pain Sympt Manag.* 2001;21:210-217.
12. Knipschild P, Arntz A. Pain patients in a randomized trial did not show a significant effect of a positive consultation. *J Clin Epid.* 2005;58:708-713.
13. Egbert LD, Battit GE, Welch CE, Bartlett MK. Reduction of postoperative pain by encouragement and instruction of patients. A Study of patient rapport. *N Engl J Med.* 1964;270:825-827.
14. Berk SN, Moore ME, Resnick JH. Psychosocial factors as mediators of acupuncture therapy. *J Consult Clin Psychol.* 1977;45:612-619.
15. Frank JD. *Persuasion and Healing.* 2nd ed. Baltimore: Johns Hopkins University Press; 1975.
16. White P, Bishop FL, Prescott P, et al. Practice, practitioner, or placebo? A multifactorial, mixed-methods randomized controlled trial of acupuncture. *Pain.* 2012;153:455-462.
17. Sambo CF, Howard M, Kopelman M, et al. Knowing you care: effects of perceived empathy and attachment style on pain perception. *Pain.* 2010;151:687-693.
18. Suchman AL, Markakis K, Beckman HB, Frankel R. A model of empathic communication in the medical interview. *JAMA.* 1997;277:678-682.
19. Finset A. 'I am worried, doctor.' Emotions in the doctor–patient relationship. *Pat Educ Couns.* 2012;88:359-363.
20. Flaten MA, Aslaksen PM, Lyby PS, Bjørkedal W. The relation of emotions to placebo responses. *Phil Trans R Soc B.* 2011;366:1818-1827.
21. Shapiro AK, Shapiro E. *The powerful Placebo.* Baltimore: Johns Hopkins University Press; 1997.
22. Kaptchuk TJ, Kelley JM, Conboy LA, et al. Components of placebo effect: randomised controlled trial in patients with irritable bowel syndrome. *BMJ.* 2008;336:999-1003.
23. Kelley JM, Lembo AJ, Ablon JS, et al. Patient and practitioner influences on the placebo effect in irritable bowel syndrome. *Psychosom Med.* 2009;71:789-797.
24. Vase L, Nørsjolv KN, Petersen GL, Price DD. Patients' direct experience as central elements of placebo analgesia. *Ohil Trans R Soc B.* 2100;366:1913–1921.
25. Daniels AM, Sallie R. Headache, lumbar puncture, and expectation. *Lancet.* 1981;1(8227):1003.
26. Reading AE. The effects of psychological preparation on pain and recovery after minor gynaecological surgery: a preliminary report. *J Clin Psychol.* 1982;38:504-512.
27. Ridgeway V, Mathews A. Psychological preparation for surgery: a comparison of methods. *Br J Clin Psychol.* 1982;21:271-280.
28. Wells JK, Howard GS, Nowlin WF, Vargas MJ. Presurgicalanxiety and postsurgical pain and adjustment: effects of a stress inoculation procedure. *J Consult Clin Psychol.* 1986;54:831-835.
29. Scott LE, Clum GA. Examining the interaction effects of coping style and brief interventions in the treatment of postsurgical pain. *Pain.* 1984;20:279-291.
30. Thomas KB. A group of 200 patients who presented in general practice with symptoms but no abnormal. *BMJ.* 1987;294:200-1202.
31. Street RL, Cox V, Kallen MA, Suarez-Almazor ME. Exploring communication pathways to better health: Clinician communication of expectations for acupuncture effectiveness. *Pat Educ Couns.* 2012;89:245-251.
32. Gryll SL, Katahn M. Situational factors contributing to toe placebo effect. *Psychopharmacol.* 1978;57:253-261.
33. Tait RC. Empathy: necessary for effective pain management? *Curr Pain Headache Rep.* 2008;12:108-112.
34. Wynn R, Wynn M. Empathy as an interactionally achieved phenomenon in psychotherapy. Characteristics of some conversational resources. *J Pragmat.* 2006;38:1385-1397.
35. Thwaites R, Bennett-Levy J. Conceptualizing empathy in cognitive behaviour therapy: making the implicit explicit. *Behav Cog Psychother.* 2007;35:591-612.
36. Levenstein JH, McCracken EC, McWhinney IR. The patient centered clinical model.1.A model for the doctor–patient interaction in clinical medicine. *Fam Pract.* 1986;3:24-30.
37. Bensing J. Bridging the gap: the separate worlds of evidence-based medicine and patient centered medicine. *Pat Educ Couns.* 2000;39:17-25.
38. Stewart MA, Brown JB, Donner A, et al. The impact of patient centered care on outcomes. *J Fam Pract.* 2000;49:796-804.
39. Lærum E, Indahl A, Skouen JS. What is 'The good back consultation'? A combined qualitative and quantitative study of chronic low back pain patients. Interaction with and perceptions of consultation with specialists. *J Rehab Med.* 2006;38:255-262.
40. Gulbrandsen P, Madsen HB, Benth JS, Lærum E. Health care providers communicate less well with patients ith chronic low back pain – a study of encounters at a back pain clinic in Denmark. *Pain.* 2010;150:458-461.
41. Bass MJ, McWhinney IR, Dempsey JB. Predictors of outcome in headache patients presenting to family physicians – a one year prospective study. *Headache.* 1986;26:285-294.
42. Shaw WS, Pransky G, Roter DL, et al. The effects of patient-provider communication on 3-months recovery from acute low back pain. *JABFM.* 2011;24:16-25.
43. Roter D, Larson S. The Roter interaction analysis system (RIAS): utility and flexibility for analysis of medical interactions. *Pat Educ Couns.* 2002;46:243-251.
44. Staiger TO, Javik JG, Deyo RA, et al. Patient–physician agreement as a predictor of outcomes in patients with back pain. *J Gen Internal Med.* 2005;20:935-937.
45. Frantsve LME, Kerns RD. Patient-provider interactions in the management of chronic pain: current findings within the context of shared medical decision making. *Pain Med.* 2007;8:25-35.
46. Allegretti A, Borkan J, Reis S, Griffiths F. Paired interviews of shared experiences around chronic low back pain: classic mismatch between patients and their doctors. *Fam Pract.* 2010;27:678-683.
47. Rao JK, Anderson LA, Inui TS, Frankel RM. Communication interventions make a difference in conversations between physicians and patients: a systematic review of the evidence. *Med Care.* 2007;45:340-349.
48. Bieber C, Muller KG, Blumenstiel K, et al. Long-term effects of a shared decision-making intervention on physician–patient interaction and outcome in fibromyalgia. A qualitative and quantitative 1 year follow-up of a randomized controlled trial. *Patient Educ Couns.* 2006;63:357-366.
49. Chassany O, Boureau F, Liard F, et al. Effects of training on general practitioners' management of pain in osteoarthritis: a randomized multicenter study. *J Rheumatol.* 2006;33:1827-1834.

50. Aiarzaguena J, Grandes G, Gaminde I, et al. A randomized controlled clinical trial of a psychosocial and communication intervention carried out by GPs for patients with medically unexplained symptoms. *Psychol Med*. 2007;372:283-294.
51. Ferguson RJ, Robinson AB, Splaine M. Use of the reliable change index to evaluate clinical sigtnificance in SF-36 outcomes. *Qual Life Res*. 2002;11:509-516.
52. Finniss DG, Benedetti F. Mechanisms of the placebo response and their impact on clinical trials and clinical practice. *Pain*. 2005;114:3-6.
53. Flaten MA, Aslaksen PM, Finset A, et al. Cognitive and emotional factors in placebo analgesia. *J Psychosom Res*. 2006;61:81-89.
54. Sarinopoulos I, Hesson AM, Gordon C, et al. Patient-centered interviewing is associated with decreased responses to painful stimuli: an initial fMRI study. *Patient Educ Couns*. [in Press].
55. Borsook D, Becerra LR. Breaking down the barriers: fMRI applications in pain, analgesia and analgesics. *Mol Pain*. 2006;2:30.
56. Turk DC, Swanson KS, Tunks ER. Psychological approaches in the treatment of chronic pain patients – when pills, scalpels, and needles are not enough. *Can J Psychiatr*. 2008;53:213-223.
57. Gatchel RJ, Peng YP, Peters ML, et al. The biopsychosocial approach to chronic pain: scientific advances and future directions. *Psychol Bull*. 2007;133:581-624.
58. Morley S, Eccleston C, Williams A. Systematic review and meta-analysis of randomized controlled trials of cognitive behaviour therapy and behaviour therapy for chronic pain in adults, excluding headache. *Pain*. 1999;80:1-13.
59. Morley S. Efficacy and effectiveness of cognitive behaviour therapy for chronic pain: progress and some challenges. *Pain*. 2011;152(suppl 3):S99-S106.
60. McCracken LM, Vowles KE. A prospective analysis of acceptance of pain and values-based action in patients with chronic pain. *Health Psychol*. 2008;27:215-220.
61. Thompson M, McCracken LM. Acceptance and related processes in adjustment to chronic pain. *Curr Pain Headace Rep*. 2011;15:144-151.
62. Veehof MM, Oskam MJ, Schreurs KMG, Bohlmeijer ET. Acceptance-based interventions for the treatment of chronic pain: a systematic review and meta-analysis. *Pain*. 2011;152:533-542.
63. Thomas ML, Elliott JE, Rao SM, et al. A randomized, clinical trial of education or motivational-interviewing- based coaching compared to usual care to improve cancer pain management. *Oncol Nurs Forum*. 2012;39:39-49.
64. Imel ZE, Wampold BE. The importance of treatment and the science of common factors in psychotherapy. In: Brown SD, Lent RW, eds. *Handbook of Counseling Psychology*. 4th ed. Hoboken, NJ: Wiley; 2008.
65. Schnur JB, Montgomery GH. A systematic review of therapeutic alliance, group cohesion, empathy, and goal consensus/collaboration in psychotherapeutic interventions in cancer: uncommon factors? *Clin Psychol Rev*. 2010;30:238-247.
66. Garvin DA, Roberto MA. Change through persuasion. *Harvard Business Rev*. 2005;83:104-112:149.
67. Lyby PS, Aslaksen PM, Flaten MA. Is fear of pain related to placebo analgesia? *J Psychosom Res*. 2010;68:369-377.
68. Lyby PS, Forsberg JT, Åsli O, Flaten MA. Induced fear reduces the effectiveness of a placebo intervention on pain. *Pain*. 2012;153:1114-1121.
69. Vase L, Robinson ME, Verne GN, Price DD. Increased placebo analgesia over time in irritable bowel syndrome (IBS) patients is associated with desire and expectation, but not endogenous opiod mechanisms. *Pain*. 2005;115:338-347.

CHAPTER 25

Nocebos in Daily Clinical Practice
The Potential Side Effects of the Treatment Context and the Patient–Doctor Interaction on Pain in Clinical Populations

Bettina K. Doering, Winfried Rief

Department of Clinical Psychology & Psychotherapy, Philipps University Marburg, 35032 Marburg, Germany

INTRODUCTION

In clinical practice, many patients discontinue drug treatments because of side effects, albeit the reported symptoms cannot be fully attributed to the drug. Alternatively, these patients are prescribed additional drugs to alleviate the side effects, thus increasing unnecessarily the amount of medication intake and also the costs for the health-care system. Clinical practice also reports patients not responding to otherwise adequate therapeutic strategies with no obvious explanation for the treatment's inefficiency. These observations may, in many cases, be manifestations of the nocebo effect. The following sections will elaborate on this clinical relevance of the nocebo effect and propose strategies to minimize its occurrence.

The traditional definition of nocebo effects describes negative responses to inert treatments, such as the occurrence of side effects in the placebo-arm of a randomized clinical trial. Reeves and colleagues published a widely cited example of the nocebo effect:[1] a young man who was participating in a clinical trial of an antidepressant medication intoxicated himself with 29 pills in a suicide attempt. He was hospitalized with a significant hypotension requiring intravenous fluids to stabilize his blood pressure. An emergency unblinding of his trial randomization revealed that he had been assigned to the trial's control group and thus had taken 29 placebo pills. After this information had been disclosed to the patient, his physical symptoms subsided rapidly. This case report impressively illustrates how adverse events can develop in response to a patient's expectations about the putative properties of an (inert) substance.

However, the nocebo effect is not limited to clinical trials that involve the administration of an inert substance. Research has accepted a broader definition in the last years. According to this account, nocebo effects are induced by a patient's negative response expectancies about the treatment outcome.[2] They may therefore also arise in the context of the administration of a pharmacologically active substance and alter the treatment outcome, analogous to the treatment-modulating effects of placebo-related positive response expectancies.[3] While early accounts of placebo and nocebo effects may have been dismissed as indications of a reporting bias or subjective experiences without physiologic correlates, recent studies have convincingly demonstrated that nocebo effects entail physiologic changes and can be substantiated using brain-imaging and EEG methods.[4-8] In analogy to placebo effects, the processes of associative learning and expectancy formation are thought to contribute to the development of nocebo effects.

When considering the nocebo effect, it is important to distinguish its two typical phenomenologies. On the one hand, nocebo effects manifest themselves in the development of side effects during a treatment, i.e. symptoms that had not been perceived before the intake of the medication and that are attributed to the drug. This may be the case in the placebo arm of a clinical trial, but also in daily clinical practice, as the following sections will demonstrate. On the other hand, nocebo effects may become apparent in the worsening of the pre-existing medical condition that was the target of the therapeutic regimen. This manifestation is also highly important in daily clinical practice. Both types of nocebo effect have to be clearly distinguished from symptom fluctuations, the natural course of the disease and statistical artifacts in clinical trials. This distinction has become ingrained in placebo research, thus requiring the comparison with a no-treatment control group in order to determine 'true' placebo effects. Unfortunately, it has been less embraced in nocebo research. Many meta-analyses of nocebo effects in clinical trials report only the side effects of the placebo arm without considering a natural history control group.[9,10] This is certainly also due to the fact that clinical trials with a no-treatment control group are scarce.

It is increasingly accepted that nocebo effects occur in daily clinical practice and that they negatively influence the treatment outcomes across a wide range of medical conditions and therapeutic modalities. Barsky and colleagues discuss these consequences of nocebo-induced side effects in their groundbreaking review.[11] The nocebo phenomenon may lead to an increase in the symptom burden and may distress the patient. Additionally, nocebo-induced side effects significantly influence a patient's decision to adhere to a prescribed treatment, as is demonstrated by the number of patients that withdraw from the placebo arm of a clinical trial due to reported side effects.[12] Thus, in the clinical setting, the nocebo phenomenon may lead to non-adherence to an otherwise adequate treatment. Nocebo effects can lead to extra medical visits and the initiation of additional treatments in an attempt to alleviate the nocebo symptoms. Considering the societal cost of nocebo effects, the nocebo phenomenon can also significantly add to health-care costs by increasing healthcare utilization and by motivating the non-adherence to a medication or the prescription of additional drugs.

Thus, the identification and reduction of nocebo effects in daily clinical practice is of therapeutic as well as of economic relevance. In the analysis of nocebo effects, it is helpful to consider the components of the medical encounter that may inadvertently contribute to a nocebo effect. Previous research has identified the following crucial factors in the formation of nocebo expectations: the information about possible negative treatment outcomes, the patient's anticipation of these outcomes, and the direct as well as indirect experiences of negative therapeutic outcomes.[2] These factors are shaped by different aspects of a therapeutic encounter. The communication of the diagnosis, the information about the treatment and its risk and benefit, and the verbal interaction during a treatment, may all be a source of nocebo responses. The following sections will analyze these elements of the treatment with a special focus on the treatment and management of pain. Ultimately, the goal must be to delineate an approach to minimize the concomitant nocebo effects.

BELIEFS ABOUT ILLNESSES AND MEDICATIONS

Nocebo effects may arise in response to the information a patient receives about a treatment. While this information is certainly part of the therapeutic encounter (cf. sections 'Communicating a diagnosis and test results' and 'Initiating a treatment', below), it is also conveyed outside the therapeutic context. Important sources of such additional information are the media, peers and relatives, or the observation of another person's experiences with a certain drug or illness. If this information contributes to beliefs about illnesses and medications regarding their negative treatment outcome, it is conducive to the development of nocebo phenomena.

In the domain of pain, cognitive factors, such as the meaning that is ascribed to the pain by an individual, and the beliefs about pain, are of special importance.[13] Beliefs about pain include e.g. the causal attribution of pain and how to react to the pain. Beliefs about pain develop over the life span; such beliefs are based on the individual's experiences with pain, the observation of pain in others, and communication about pain experiences. Examples of maladaptive beliefs about chronic pain are the conviction that pain is always dangerous and that the appropriate reaction to chronic pain is to rest and reduce activity. The maladaptive function of such beliefs is expressed in fear-avoidance models: a catastrophizing interpretation of pain results in fear, avoidance and disuse, which, in turn, lead to a deterioration of the pain.[14] This effect can be interpreted as a nocebo phenomenon because the respective negative expectancies are associated with a worsening of the medical condition: fear-avoidance beliefs not only predict the persistence of pain in patients suffering from acute pain, they are also considered a risk factor for the development of pain in the general population.[13,15]

In addition, various studies demonstrate that external information can influence the development of nocebo effects, for example health warnings[16] or modern health worries.[17] A laboratory study investigated the occurrence of headache after mobile phone use, a phenomenon which has often been covered in the media.[18] The headache is presumed to be a result of radiofrequency fields (RF) of the mobile phone. In this sample of participants, who described themselves as 'electromagnetic hypersensitive,' the real exposure to RF was compared to sham RF exposure regarding the resulting discomfort, the occurrence of pain, and the location of the pain. The participants were blind to the randomized exposure conditions. The experiment elicited headache and discomfort; however, these measures did not differ in the RF and the sham trials. Thus, headache after mobile phone use may be regarded as a nocebo effect that may be strengthened by the media coverage.

These findings draw attention to information processes that happen outside the therapeutic context, and to the beliefs that a patient already holds when starting the treatment. On a societal level, it would be desirable to sensitize the representatives of professional journals and the daily press to the possible consequences of their reports about side effects of medical treatments and their influence on the development of health worries and associated nocebo effects.[18,19] On the level of the health-care providers, it stresses the importance of assessing the respective beliefs of the patients and of subsequently addressing these concerns. Importantly, the health-care practitioner needs to tailor the respective information to the individual patient and should try to identify patients that are at high risk for nocebo effects.[11] These *a priori* beliefs can be measured using standardized instruments, such as the Beliefs About Medicine Questionnaire (BMQ).[20] The BMQ has proven to be a valid predictor of the occurrence of nonspecific side effects in the treatment of rheumatoid arthritis.[21] Two of its subscales assess the respondent's orientation to 'modern medicines' in general. If this questionnaire was employed more frequently in clinical practice it could be evaluated for its use as a screening tool to identify patients at high risk for nocebo effects. However, standardized instruments cannot elicit specific concerns as successfully as an individual interview. It therefore seems advisable to explore the patient's attitudes toward a medication proactively, before he or she enters any treatment, and to encourage the patient to voice possible existing concerns regarding the treatment regimen and address them. This holds especially true for the treatment and management of pain, because even the drug-naive patient may hold an exaggeratedly negative attitude towards many pharmacologic treatments (e.g. cortisone), due to their alleged side effects. The goal of this strategy is to identify patients with a pre-existing negative treatment expectancy that has been shaped by external sources of information.

COMMUNICATING A DIAGNOSIS AND TEST RESULTS

When a patient starts treatment, the health-care provider will usually explore the patient's symptoms and run tests in order to diagnose the patient's pathology. In the domain of pain, especially in chronic pain, the diagnostic process merits special attention. Over-investigation, i.e. the implementation of repeated medical tests that do not lead to additional information poses a serious problem in this population and is even discussed as an iatrogenic risk factor for pain chronification.[22] This over-investigation may be caused by the health-care professional's anxiety not to overlook any medical evidence. Considering, however, that any diagnostic procedure may induce worry and anxiety in the patient, and that an invasive diagnostic procedure may even cause additional harm, a nocebogenic effect of over-investigation must be taken into consideration.

At the beginning of the diagnostic process, patients may be anxious about their condition and the test results. This holds especially true for patients with chronic pain who are often unsure about the meaning of their symptoms and the impact that the illness will have on their everyday lives.[13] Anxiety itself has been discussed as a factor in the occurrence of nocebo effects.[23] Therefore, some have argued that it may even benefit the patient not to know all the possibly serious diagnoses that a health-care practitioner may have in mind when running diagnostic tests.[24] Practitioners could instead explain that the tests are necessary to rule out other conditions, and that details would be disclosed only after the test results are known. This procedure is thought to reduce worry and unnecessary anxiety on the patient's part. Whether this approach is ethically acceptable, and whether it would benefit all patients equally, merits further debate. Anxiety, however, also limits patients' cognitive capacity to follow the verbal content of the therapeutic encounter and may thus bias their understanding of the communicated test results.

Effective strategies of communicating test results and diagnoses therefore merit special attention. The importance of providing information about normal test results for patients' reassurance, and the subsequent occurrence of symptoms, was investigated in a randomized controlled trial (RCT) in patients with chest pain.[25] The study demonstrated that providing patients with information about the test, and explaining the meaning of normal test results before testing, improved subjective reassurance and reduced the likelihood of future reports of chest pain. Importantly, this information was provided before the testing. Essential parts of the information concerned the interpretation of a normal test result, i.e. a risk for cardiovascular diseases as low as for anyone in the general population, and the mentioning of other, less serious reasons for chest pain. It is interesting to note that the statements were formulated positively and colloquially ('normal,' 'low risk,' etc.). On the contrary, the use of medical jargon to convey normal test results ('the scan is negative') may even contribute to nocebo effects.[26,27]

While the latter example concerns the communication of normal test results, less is known about the possible nocebogenic effect of the communication strategies used to inform the patient about a medical diagnosis. Such effects could arise with regard to the communication of the prognosis, the disease recurrence, or the spread of disease, and may foster a more negative expectancy than is necessary regarding the outcome of treatment. The nocebogenic effect of fear-avoidance beliefs in pain has already been mentioned (cf. section 'Beliefs about illnesses and medication', above). Unfortunately, these maladaptive beliefs may also be fostered in the therapeutic interaction. Health-care providers sometimes may even strengthen fear-avoidance beliefs through the recommendation to refrain from everyday activities.[28,29] The diagnostic label itself may influence the patient's illness beliefs and thus lead to a deterioration in the symptoms as a nocebo effect.[30] In a survey among general practitioners and physical therapists, more than 60% indicated that they would advise a patient to refrain from painful movements, more than 30% believed that a reduction in pain is necessary for return-to-work, and more than 25% reported that they considered sick leave a good treatment for back pain.[31] The communication of a diagnosis and its prognosis may therefore be accompanied by catastrophizing remarks of a fear-avoidant health-care provider and thus impact negatively on the patient's pain.

Due to the limited amount of empirical research, adjacent areas of research may be informative about how to avoid nocebo effects in communicating a diagnosis. For instance, research has accumulated a considerable amount of guidelines on 'how to break bad news' to the patient.[32-34] An important feature of these communication strategies aligns well with studies of the nocebo effect and concerns the provision of truthful but optimistic statements,[35] i.e. focusing on realistic positive outcomes while providing an honest view of the consequences associated with a diagnosis. This may entail a positive framing of the prognosis, i.e. 'Forty percent of the patients get better' instead of 'The medical condition deteriorates for 60% of the patients.' Additionally, establishing a trustful therapeutic relationship may be of great importance, even at this early stage of the therapeutic encounter, in order to facilitate the communication of a diagnosis and to avoid nocebo effects.[36,37] This can be accomplished by the use of the basic principles of patient-centered communication.[38]

INITIATING A TREATMENT

Before any treatment is initiated, the health-care provider must inform the patient about the costs and benefit of the various treatment options and obtain informed consent to the treatment. As it is in this phase that a patient's expectation about the outcome of treatment is shaped most directly, this process is highly relevant to the development of nocebo effects. The cost of a treatment may be seen in the occurrence of side effects, while the benefit pertains to the symptom alleviation that is to be expected. In the treatment of chronic pain, the selection of analgesic drugs often follows the algorithm of the 'pain ladder' as proposed by the World Health Organization.[39] This algorithm advocates a step-wise approach to pharmacologic pain management, starting with non-opioids and then continuing with mild opioids and strong opioids. Therefore, the analgesic benefit of the drugs needs to be weighed against common side effects, e.g. in opioids: respiratory depression, vomiting, somnolence, circulatory changes, pruritus, cognitive impairment, and constipation.[40] The fear of potential side effects and their actual occurrence are highly relevant to the analgesic treatment because they influence the patients' adherence to the prescribed medical regimen.[41]

Communication about the side effects of a treatment can produce these very side effects, as observational and experimental research has demonstrated. One experimental study compared the occurrence of nocebo effects in patients with persistent arm pain that received either sham acupuncture or a placebo pill.[42] The patients were falsely informed that they had a 50% chance of receiving an active treatment. The side effects that the patients experienced were completely dependent upon the information they received during the informed consent, i.e. pain during treatment, local redness or swelling after the sham acupuncture and drowsiness, headache, or dry mouth after the placebo pill (said to contain amitriptyline). In support of this specificity hypothesis of nocebo effects, meta-analyses of clinical trials demonstrate that the profile of side effects experienced in the placebo arms of the trials mirrors the side effect profile of the active drug under investigation.[9,10,43,44] In a systematic review of anti-migraine clinical trials, adverse event profiles of triptans, anticonvulsants and nonsteroidal anti-inflammatory steroids (NSAIDs) were compared in their respective placebo arms.[45] The adverse events in the placebo arms mimicked those of the anti-migraine medication against which the placebo was compared. For example, anorexia and memory difficulties are typical side effects of anticonvulsants and were observed only in the placebo arm of the respective trials. The occurrence of nocebo effects is not restricted to the administration of inert placebos. Inclusion of a specific side effect in the informed consent form of a verum increases its prevalence in the respective treatment group compared to another group receiving the same medication but without the respective side effect outlined in the informed consent.[46,47]

If the disclosure of the side effects or risks may lead to the very same side effects, this poses an ethical dilemma to the clinical practice of informed consent.[48] Research has advanced different strategies in order to minimize the possible nocebogenic effect of the informed consent process.[24,48,49] For once, a procedure of contextualized informed consent has been proposed[24] that encourages the health-care provider to individually adapt the amount of information disclosed to the specific situation. In a similar vein, 'authorized concealment' has been discussed as a strategy of modifying the informed consent procedure.[48] In this approach, patients are informed about the nocebo phenomenon and then asked whether they would prefer not to be informed about potential treatment side effects that are distressing but do not pose a serious threat to their health. However, both approaches merit further discussion under an ethical, or at least legal perspective, which at the moment hinders their integration into clinical practice.

Other strategies, however, can be implemented to minimize nocebo effects within the existing informed consent approach. Educating patients about the nocebo effect and the underlying mechanisms without changing the amount of information disclosed has been discussed as one such strategy.[11] However, as a case report illustrates,[50] it may be difficult to convince a patient that he is experiencing a nocebo effect once the side effect has occurred. This approach therefore needs further empirical research. Another option is to prepare patients for the occurrence of side effects, regardless of their drug-specific or nonspecific nature, and to help patients to cope with them.[11] This strategy assumes that patients will be less anxious about a side effect when they know that it is not medically dangerous and can even be reframed positively as an indicator that the drug is taking its effect. The reduction of anxiety may help to prevent nocebo effects. In the case of opioid medication for chronic pain, the health-care provider could thus inform the patient about frequent side effects such as nausea, pruritus, and somnolence, and stress at the same time that these side effects are likely to diminish within the first weeks of the treatment and can be reframed as 'the drug taking its effect'. Side effects which are likely to persist, such as constipation in opioid therapy, need to be addressed in a coping focused manner, i.e. by preparing the patient for the symptom and informing the patient about possibilities to counteract the diminished gastrointestinal motility.

Additionally, the form in which the information about the risks of a specific side effect is presented merits attention. A series of studies demonstrated that a qualitative description of the frequency of side effects (i.e. very rare, rare, uncommon) led to a gross overestimation of risk compared to quantitative descriptions.[51] The effect of the framing of information on nocebo effects was demonstrated in a study that prospectively examined the occurrence of side effects in relation to different types of side effect disclosure.[52] The authors demonstrated that when the information about the frequency of side effects was formulated in a positive way, i.e. focusing on the percentage of patients that do not experience the side effect, the patients reported fewer side effects than when the same information was formulated in a negative way, i.e. focusing on the percentage of patients that do experience the side effect. It is important to note that the information communicated is essentially equivalent in both cases. The framing manipulation did not impact on the choice of treatment option, thus demonstrating no biasing effect on the autonomy of informed consent. On the contrary, biasing effects of framing mode were found in another study that investigated the attractiveness of different treatments for lung cancer:[53] treatment options were rated more attractive if they were described in terms of the probability of living or the probability of dying. Accordingly, the frame of reference for the description of alternative therapies should be held constant, e.g. both described in the probability of living, in order to avoid biasing effects. The examples argue that the informed consent procedure has a potential to reduce nocebo effects if it describes possible side effects in a coping focused manner, if it uses quantitative descriptors, and if it frames risks focusing on the percentage of patients that do not suffer from the side effect.

Even though the possible costs of a treatment in terms of side effects have received greater attention in nocebo research, the framing of the benefits of a treatment is equally important, because it also directly contributes to the patient's treatment expectancies. The influence of experimentally induced negative treatment expectancies on the medical outcomes in patient populations has not often been researched, due to the ethical concerns about the induction of a nocebo effect. However, experimental research in healthy participants demonstrates that nocebo suggestions are capable of inducing allodynic or hyperalgesic effects, thus exacerbating pain or turning non-painful stimuli nociceptive.[54] In a brain-imaging study of healthy participants, a nocebo suggestion (exacerbation of pain) abolished the analgesic effect of remifentanil in a heat-pain paradigm.[55] Similar results have been obtained in other studies using different paradigms of experimentally induced pain with varying medications in healthy participants.[5,56,57] A recent review also implicates nocebo effects in the management of postoperative pain.[58] While most studies investigate the nocebo instruction concurrently with the treatment, one study investigated whether a nocebo suggestion that was given once before the treatment could still impact nocebo responses after longer pain experiences.[7] This approach mirrors closely the discussion of risk and benefit in the informed consent procedure at treatment initiation. The healthy participants underwent a repetitive nociceptive stimulation over 8 days, the effect of which should normally diminish over time due to processes of habituation. The experimental group, however, received a nocebo instruction that the pain would increase over time. The results demonstrate that the experimental group reported a constant and unchanging pain during the complete trial that was also associated with specific patterns of brain activation. Thus, a negative suggestion of pain exacerbation conveyed only once before a longer experimental timeframe was able to produce a lasting nocebo hyperalgesia.

Explicit nocebo instructions of pain exacerbation will hopefully seldom be found in clinical settings. However, different scenarios are conceivable where nocebo suggestions may be given inadvertently. On the one hand, a health-care practitioner may clearly favor one treatment option and depict the other options as ineffective or very risky. In this case, if a treatment option is presented negatively to the patient, and is later administered after all other options have failed, the patient's expectancies will have changed and he may demonstrate unfavorable physiologic responses to the treatment.[59] On the other hand, nocebo suggestions may be given while informing the patient about the disease (cf. section 'Communicating a diagnosis and test results', above). Careless nocebo information may impact the treatment outcome even with a time-delay. Concerning nocebo effects, the results encourage a truthful but optimistic discussion of possible treatment outcomes. If uncertainty about a treatment's effectiveness is communicated to the patient, this may negatively influence treatment outcomes.[60] Additionally, an exaggerated devaluation of other treatment options as risky or ineffective should be avoided to prevent nocebo effects in case the patient should choose to recur to these options. Naturally, the health-care practitioner should not raise false hopes. However, encouraging an optimistic view seems advisable. In the management of pain, there may be instances when a treatment option can be described as a powerful treatment unequivocally. In instances when this is not the case, e.g. in chronic pain conditions, the health-care practitioner may point out that even if one treatment option fails there still will be several other effective options available.

TREATMENT IMPLEMENTATION

Nocebo effects occur also during treatment implementation. Health-care practitioners often accompany their procedures with descriptions of the next treatment steps; they try to prepare the patient for oncoming pain by informing him about the forthcoming experience or try to create a more relaxed atmosphere through the use of a humorous description of the event. Even though this approach is often taken to ease the patient, these verbal or nonverbal interactions may inadvertently create nocebo effects.

A nocebogenic effect that may arise through an attempt to inform the patient about a procedure that has been discussed in the literature, namely, the interruption of an ongoing treatment. This phenomenon has been studied using the open–hidden paradigm in conditions of pain, anxiety and Parkinson's disease.[61,62] In the domain of pain, postoperative patients were administered morphine. Then, the administration of morphine was interrupted. In one group, this interruption was announced by the doctor. In the other group the treatment was discontinued without the patients' knowledge. Therefore, this latter group did not expect a treatment interruption or a worsening of their pain, while the former group was able to form negative expectancies about the discontinuation of the medicine. The increase in pain after the discontinuation of morphine reported to the open (nocebo) group was significantly greater than that reported in the hidden group. Additionally, the open group also requested more additional analgesic medication. This raises the question as to whether patients should always be informed when a treatment is discontinued, or a medication dose is reduced, because this may subsequently lead to anxiety, worry and nocebo effects. In the area of medically supervised drug withdrawal, especially with benzodiazepines, it is already current practice to reduce the amount of benzodiazepines surreptitiously while the patient does not know how much of the drug he is going to receive or when the next reduction in dose is to be expected. However, this practice requires the patient's informed consent beforehand. A similar consent could also be sought from patients undergoing a long-term analgesic medication in order to avoid nocebo effects at the discontinuation or when there is a reduction in the dose of the medication.

In addition to information about the facts of a given treatment, the importance of the therapeutic interaction, both verbal and non-verbal, has been emphasized for the occurrence of placebo and nocebo phenomena.[11,35,58,60,63] However, only a few studies have investigated the nocebo component of this interaction more thoroughly. One of these studies examined whether the communication that accompanied a local anesthetic injection influenced the patients' self-reported pain during the procedure.[64] In the nocebo condition, the injection was announced as 'You are going to feel a big bee sting; this is the worst part of the procedure.' In the placebo (i.e. reassuring) condition, the same procedure was announced by 'We will give you a local anesthetic that will numb the area and you will be comfortable during the procedure.' The nocebo suggestion resulted in reports of significantly higher pain levels than the placebo condition. This is noteworthy because the nocebo phrasing that was used in this study may also be employed in a realistic clinical setting without any nocebogenic intention.[65] In an even more ecologically valid study, the interactions between health-care providers and patients during an interventional radiologic procedure were videotaped, and all statements that warned the patient of a painful experience were analyzed.[66] In this study, all warnings of pain resulted in greater pain and higher anxiety levels than when no warning was given. Sympathizing with the patient after a painful experience did not have beneficial effects either, but led to higher anxiety levels. Thus, the attempt to make the situation more controllable to the patient, and to empathize, seemed to yield adverse effects, even though a patient's feelings of self-efficacy and an empathic therapeutic climate are thought to contribute to placebo effects. The seemingly contradictory findings can be reconciled by a closer examination of the respective wording of the warnings and commiserations. The warnings in the study were all negatively loaded, i.e. they referred to negative experiences such as 'bad,' 'sharp pain,' 'uncomfortable,' or 'hurting,' even when they intended to announce that 'this will hurt only a bit.' In accordance with negative affective priming effects,[67] these statements informed the patient that pain was to be expected and thus contributed to nocebo effects. In a similar vein, the sympathizing remarks were negatively loaded ('That hurts! This is the hardest part') and contributed to the patient's anxiety. Thus, nocebogenic remarks seem to have been delivered with the best intentions, even though a more internally valid investigation of this phenomenon seems warranted. It could therefore be advisable to focus in the communication on the desired outcome (e.g. numbness, comfortable) instead of on the outcome to be avoided (e.g. no pain, less hurt).

THE ROLE OF TREATMENT EXPERIENCE

Previous treatment experiences may influence later administered treatments in terms of nocebo effects through different mechanisms. One vivid example of how nocebo effects may be formed comes from an observational study of inadvertent pain conditioning in newborn children of diabetic mothers who were exposed to multiple heel lances for the monitoring of blood glucose.[68] The painful heel lancing was always preceded by a painless skin cleansing. The experimental group, and a group of control children who had not been exposed to heel lances, then underwent a venipuncture which also involved a skin cleaning. Pain responses during the venipuncture were scored by blind raters. The experimental group displayed more pain responses than the control group, thus showing hyperalgesia. Importantly, the experimental group also demonstrated anticipatory pain during the skin cleansing. This can be interpreted as a conditioned nocebo response: due to the learning experience that a (painless) skin cleansing precedes a painful experience, the skin cleansing suffices to elicit a pain response. According to Pavlovian conditioning, it acquires the properties of a conditioned stimulus that elicits the conditioned nocebo response. Thus, neutral cues that are associated with a painful or distressing experience may come to elicit the pain or distress even without the experience really taking place. Medical treatment is always accompanied by external stimuli that may become conditioned nocebo stimuli. However, this finding is most relevant for treatments that are invasive or painful and that are administered repeatedly, because learning processes develop over time. If such situations occur in the clinical setting, the health-care practitioner needs to be aware of the potential conditioning of hyperalgesic nocebo responses. In order to avoid the conditioning, the saliency of the conditioned stimulus needs to be reduced. Saliency refers to the properties of the originally neutral stimulus in the treatment context that make it more easily noticeable or distinct so that it can be associated with the unconditioned stimulus in the learning process. Additionally, the contingency of the conditioned stimulus (e.g. treatment room) and the unconditioned stimulus (i.e. pain) can be weakened. This could be achieved by varying the treatment surroundings in order to change salient cues or by providing alternative salient cues that change with each treatment trial.

Once a treatment has been administered the patient invariably forms an evaluation regarding its cost and benefit. In the case of chronic or recurrent diseases these evaluations are highly liable to shape the patient's expectancies about the next treatment sequence. It has therefore been discussed whether negative previous treatment experiences may contribute to a nocebo effect for the subsequent treatments. This discussion refers to the role of learning in nocebo effects. An experimental study in healthy participants using an electrical stimulation paradigm compared the effects of verbal nocebo suggestions and a nocebo conditioning procedure on pain.[54] In this study, no difference in the magnitude of the nocebo effect was found, whether it was induced by a verbal suggestion alone or whether the verbal suggestion was preceded by a nocebo conditioning, thus arguing against an effect of previous experiences on the magnitude of subsequent nocebo effects. However, another study using the same paradigm demonstrated that multiple repetitions of the conditioning procedure contributed to the stability of the nocebo effect.[69] Thus, conversely to placebo effects, the number of learning trials does not seem to increase the magnitude of the effect but to contribute to its stability. These results must be interpreted with caution because they refer to healthy participants. Yet this could mean that, in a clinical context, the experience of an ineffective treatment could lead to a nocebo response that will soon extinguish if only experienced once; however, it will be more stable and relevant for subsequent treatments if more negative experiences have occurred previously. Because pain is often recurring and chronic in nature, and requires repeated treatment, this illustrates the need to take into account the previous experiences a patient has with the treatment.[70] Recommendations as how to manage these negative treatment expectancies derived from previous experience are sparse due to a lack of empirical research. Possibly, a change in the mode of application of the drug, if feasible, may help to weaken the association. If alternative therapeutic options that are adequate for the treatment are available, these may prove more effective because they would not be hampered in their effectiveness by nocebo expectations. Explaining the reasons why a treatment may not have helped the last time, and why it could be effective now, or an explanation in which way the current treatment differs from the ineffective one, should surely be considered.

CONCLUSIONS

The empirical results convincingly demonstrate that nocebo effects occur not only in the laboratory but also in the clinical setting. Nocebogenic expectancies may even be formed before the actual treatment begins and can be engendered later by the various components of the therapeutic process. Even the outcome of a treatment itself may influence nocebo effects in subsequent treatments. Figure 25.1 shows suggestions for minimizing the nocebo effects that can be drawn from a review of the existing literature.

The therapeutic relationship lies at the heart of the prevention of nocebo effects. Because a climate of anxiety and worry, and an expectation of negative treatment outcomes, seem to be relevant in the occurrence of nocebo effects, the health-care practitioner should take all steps necessary to prevent or diminish these conditions. Practical recommendations to advance interpersonal healing, and to promote placebo effects as one form of interpersonal healing,[71] may also be helpful in minimizing nocebo effects: speak positively about treatments, provide encouragement, develop trust, provide reassurance, support relationships, respect uniqueness, explore values, and create ceremony.[72] The implementation of these principles is seen in the suggestions for the discrete treatment stages.

Specifically, in the context of pain, it seems important to explore a patient's fear-avoidance beliefs when the patient starts treatment. Additionally, the exploration must take into account previous experiences with ineffective analgesic treatments and side effects. In the diagnostic process, the nocebogenic effect of an over-investigation, and the possible fear-avoidance beliefs of the health-care professional, must be considered. When initiating a pharmacologic analgesic treatment, a clear distinction should be communicated between common side effects that will subside after the lead-in phase and drug-specific side effects that are likely to persist during the treatment. Persistent side effects should be addressed in the coping-focused manner outlined. During treatment, the verbal interaction

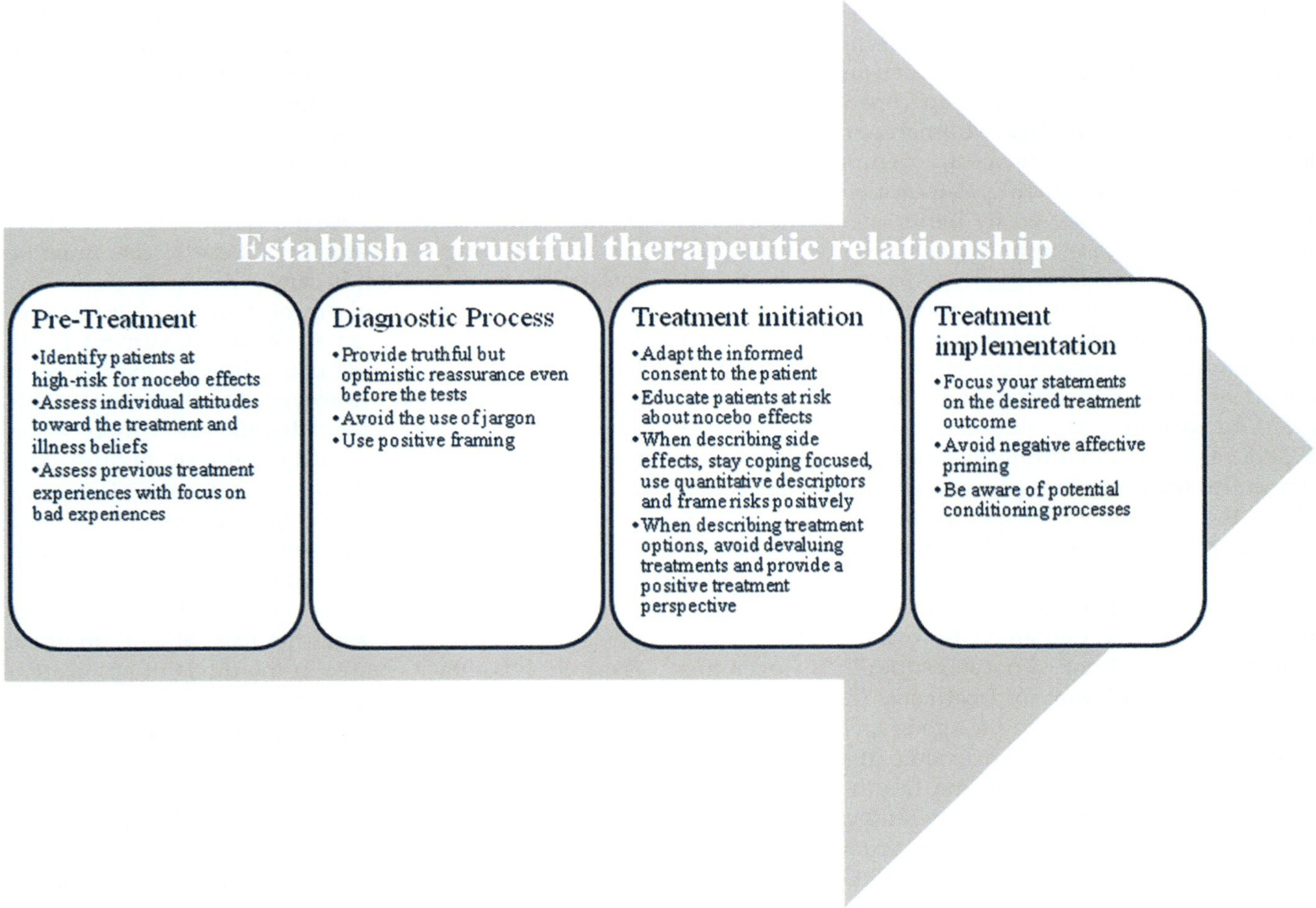

Establish a trustful therapeutic relationship

Pre-Treatment
- Identify patients at high-risk for nocebo effects
- Assess individual attitudes toward the treatment and illness beliefs
- Assess previous treatment experiences with focus on bad experiences

Diagnostic Process
- Provide truthful but optimistic reassurance even before the tests
- Avoid the use of jargon
- Use positive framing

Treatment initiation
- Adapt the informed consent to the patient
- Educate patients at risk about nocebo effects
- When describing side effects, stay coping focused, use quantitative descriptors and frame risks positively
- When describing treatment options, avoid devaluing treatments and provide a positive treatment perspective

Treatment implementation
- Focus your statements on the desired treatment outcome
- Avoid negative affective priming
- Be aware of potential conditioning processes

FIGURE 25.1 Suggested strategies for the minimization of nocebo effects at the different stages of treatment.

should be centered on the positive treatment outcome, avoiding negative affective priming of pain sensations. Measures should be taken to prevent an inadvertent pain conditioning.

These guidelines are based on the empirical evidence up to date. However, our knowledge about the prevention of nocebo effects is still far from complete. Future research will certainly help us to minimize nocebo effects and personalize the treatment so that we can provide optimal treatments for our patients. Yet even today, the increasing awareness of nocebo effects and the accumulating body of research have the potential to contribute to a better treatment outcome.

References

1. Reeves RR, Ladner ME, Hart RH, Burke RS. Nocebo effects with antidepressant clinical drug trial placebos. *Gen Hosp Psychiatr.* 2007;29(3):275-277.
2. Colloca L, Miller FG. The nocebo effect and its relevance for clinical practice. *Psychosom Med.* 2011;73:598-603.
3. Rief W, Bingel U, Schedlowski M, Enck P. Mechanisms involved in placebo and noceco responses and implications for drug trials. *Clin Pharm Ther.* 2011;90:722-726.
4. Lorenz J, Hauck M, Paur RC, et al. Cortical correlates of false expectations during pain intensity judgments–a possible manifestation of placebo/nocebo cognitions. *Brain Behav Immun.* 2005;19:283-295.
5. Benedetti F, Amanzio M, Vighetti S, Asteggiano G. The biochemical and neuroendocrine bases of the hyperalgesic nocebo effect. *J Neurosci.* 2006;26:12014-12022.
6. Scott DJ, Stohler CS, Egnatuk CM, et al. Placebo and nocebo effects are defined by opposite opioid and dopaminergic responses. *Arch Gen Psychiatry.* 2008;65:220-231.
7. Rodriguez-Raecke R, Doganci B, Breimhorst M, et al. Insular cortex activity is associated with effects of negative expectation an nociceptive long-term habituation. *J Neurosci.* 2010;30:11363-11368.
8. Kong J, Gollub RL, Polich G, et al. A functional magnetic resonance imaging study on the neural mechanisms of hyperalgesic nocebo effect. *J Neurosci.* 2008;28:13354-13362.
9. Mitsikostas DD, Mantonakis LI, Chalarakis NG. Nocebo is the enemy, not placebo. A meta-analysis of reported side effects after placebo treatment in headaches. *Cephalalgia.* 2011;31:550-561.
10. Mitsikostas DD, Chalarakis NG, Mantonakis LI, et al. Nocebo in fibromyalgia: meta-analysis of placebo-controlled clinical trials and implications for clinical practice. *Eur J Neurol.* 2012;19:672-680.
11. Barsky AJ, Saintfort R, Rogers MP, Borus JF. Nonspecific medication side effects and the nocebo phenomenon. *JAMA.* 2002;287:622-627.
12. Rief W, Avorn J, Barsky AJ. Medication-attributed adverse effects in placebo groups. *Arch Intern Med.* 2006;166:155-160.
13. Gatchel RJ, Peng YB, Peters ML, et al. The biopsychosocial approach to chronic pain: Scientific advances and future directions. *Psychol Bull.* 2007;133:581-624.
14. Leeuw M, Goosens MEJB, Linton SJ, et al. The fear-avoidance model of musculoskeletal pain: current state of scientific evidence. *J Behav Med.* 2007;30:77-94.
15. Chou R, Shekelle P. Will this patient develop persistent disabling low back pain? *JAMA.* 2010;303:1295-1302.
16. Winters W, Devriese S, van Diest I, et al. Media warnings about environmental pollution facilitate the acquisition of symptoms in response to chemical substances. *Psychosom Med.* 2003;65: 332-338.
17. Petrie KJ, Broadbent EA, Kley N, et al. Worries about modernity predict symptom complaints after environmental pesticide spraying. *Psychosom Med.* 2005;67:778-782.
18. Stovner LJ, Oftedal G, Straume A, Johnsson A. Nocebo as headache trigger: evidence from a sham-controlled provocation study with RF fields. *Acta Neurol Scand.* 2008;188(suppl):67-71.
19. Mortazavi SM, Ahmadi J, Shariati M. Prevalence of subjective poor health symptoms associated with exposure to electromagnetic fields among university students. *Bioelectromagn.* 2007;28:326-330.
20. Horne R, Weinman J, Hankins M. The beliefs about Medicines Questionnaire: the development and evaluation of a new method for assessing the cognitive representation of medication. *Psychol Health.* 1999;14:1-24.
21. Nestoriuc Y, Orav EJ, Liang MH, et al. Prediction of nonspecific side effects in rheumatoid arthritis patients by beliefs about medicines. *Arthritis Care Res.* 2010;62:791-799.
22. Kouyanou K, Pither CE, Wessely S. Iatrogenic factors and chronic pain. *Psychosom Med.* 1997;59:597-604.
23. Meissner K, Bingel U, Colloca L, et al. The placebo effect: advances from different methodological approaches. *J Neurosci.* 2011;31: 16117-16124.
24. Wells R, Kaptchuk TJ. To tell the truth, the whole truth, may do patients harm: the problem of the nocebo effect for informed consent. *Am J Bioeth.* 2012;12:22-29.
25. Petrie KJ, Müller JT, Schirmbeck F, et al. Effect of providing information about normal test results on patients' reassurance: randomised controlled trial. *BMJ.* 2007;334:352-355.
26. Brewin TB. Three ways of giving bad news. *Lancet.* 1991;337: 1207-1209.
27. Häuser W, Hansen E, Enck P. Nocebo phenomena in medicine: their relevance in everyday clinical practice. *Deusches Ärzteblatt.* 2012;109:459-465.
28. Houben RM, Ostelo RW, Vlaeyen JWS, et al. Health care providers' orientations towards common low back pain predict perceived harmfulness of physical activities and recommendations regarding return to normal activity. *Eur J Pain.* 2005;9:173-183.
29. Darlow B, Fullen BM, Dean S, et al. The association between health care professional attitudes and beliefs and the attitudes, beliefs, clinical management, and outcomes of patients with low back pain: a systematic review. *Eur J Pain.* 2012;16:3-17.
30. Buitenhuis J, de Jong PJ. Fear avoidance and illness beliefs in post-traumatic neck pain. *Spine.* 2011;36:S238-S243.
31. Linton SJ, Vlaeyen J, Ostelo R. The back pain beliefs of healthcare providers: are we fear-avoidant? *J Occup Rehabil.* 2002;12: 223-232.
32. Lamont EB, Christiakis NA. Complexities in prognostication in advanced cancer: "To help them live their lives the way they want to". *JAMA.* 2003;290:98-104.
33. Ptacek JT, Eberhardt TL. Breaking bad news: a review of the literature. *JAMA.* 1996;276:496-502.
34. Buckman R. *How to Break Bad News – a Guide for Health Care Professionals.* Baltimore: John Hopkins University Press; 1992.
35. Benedetti F, Amanzio M. The placebo response: how words and rituals change the patient's brain. *Patient Educ Couns.* 2011; 84:413-419.
36. Benedetti F, Lanotte M, Lopiano L, Colloca L. When words are painful: unraveling the mechanisms of the nocebo effect. *Neurosci.* 2007;147:260-271.
37. Olshansky B. Placebo and nocebo in cardiovascular health: implications for healthcare, research, and the doctor-patient relationship. *J Am Coll Cardiol.* 2007;49:415-421.
38. Ong LML, De Haes JCJM, Hoos AM, Lammes FB. Doctor-patient communication: a review of the literature. *Soc Sci Med.* 1995;40:903-918.
39. WHO (World Health Organization). *Adherence to Long-Term Therapies. Evidence for Action.* Geneva: WHO; 2003.

40. McCarberg BH, Barkin RL. Long-acting opioids for chronic pain: pharmacotherapeutic opportunities to enhance compliance, quality of life, and analgesia. *Amer J Ther*. 2001;8:181-186.
41. Graziottin A, Gardner-Nix J, Stumpf M, Berliner MN. Opioids: how to improve compliance and adherence. *Pain Practice*. 2011;11:574-581.
42. Kaptchuk TJ, Stason WB, Davis RB, et al. Sham device v inert pill: randomised controlled trial of two placebo treatments. *BMJ*. 2006;332:391-397.
43. Rief W, Nestoriuc Y, Weiss S, et al. Meta-analysis of the placebo response in antidepressant trials. *J Affective Disord*. 2009;118:1-8.
44. Papadopoulos D, Mitsikostas D. Nocebo effects in multiple sclerosis trials: a meta-analysis. *Mult Scler*. 2010;16:216-228.
45. Amanzio M, Corazzini LL, Vase L, Benedetti F. A systematic review of adverse events in placebo groups of anti-migraine clinical trials. *Pain*. 2009;146:261-269.
46. Mondaini N, Gontero P, Giubilei G, et al. Finasteride 5 mg and sexual side effects: how many of these are related to a nocebo phenomenon? *J Sex Med*. 2007;4:1708-1712.
47. Myers MG, Cairns JA, Singer J. The consent form as a possible cause of side effects. *Clin Pharmacol Ther*. 1987;42:250-253.
48. Miller F, Colloca L. The placebo phenomenon and medical ethics: rethinking the relationship between informed consent and risk-benefit assessment. *Theor Med Bioeth*. 2011;32:229-243.
49. Barsky AJ, Goodson JD, Lane RS, Cleary PD. The amplification of somatic symptoms. *Psychosom Med*. 1988;50:510-519.
50. Stern RH. Nocebo responses to antihypertensive medications. *J Clin Hypertens*. 2008;10:723-725.
51. Berry DC, Knapp P, Raynor DK. Provision of information about drug side effects to patients. *Lancet*. 2002;359:853-854.
52. O'Connor AM, Pennie RA, Dales RE. Framing effects on expectations, decisions, and side effects experienced: The case of influenza immunization. *J Clin Epidemiol*. 1996;49:1271-1276.
53. McNeil BJ, Pauker SG, Sox HJJ, Tversky A. On the elicitation of preferences for alternative therapies. *N Engl J Med*. 1982;306:1259-1262.
54. Colloca L, Sigaudo M, Benedetti F. The role of learning in nocebo and placebo effects. *Pain*. 2008;136:211-218.
55. Bingel U, Wanigasekera V, Wiech K, et al. The effect of treatment expectation on drug efficacy: imaging the analgesic benefit of the opioid remifentanil. *Sci Transl Med*. 2011;3:70ra14.
56. Benedetti F, Pollo A, Loopiano L, et al. Conscious expectation and unconscious conditioning in analgesic, motor, and hormonal placebo/nocebo responses. *J Neurosci*. 2003;23:4315-4323.
57. Dworkin SF, Chen AC, LeResche L, Clarke DW. Cognitive reversal of expected nitrious oxide analgesia for acute pain. *Anaesth Analg*. 1983;62:1073-1077.
58. Svedman P, Ingvar M, Gordh T. 'Anxiebo,' placebo, and postoperative pain. *BMC Anesthesiol*. 2005;5:9.
59. Chae Y, Kim SY, Park HS, et al. Experimentally manipulating perceptions regarding acupuncture elicits different responses to the identical acupuncture stimulation. *Physiol Behav*. 2008;95:515-520.
60. Benedetti F. How the doctor's words affect the patient's brain. *Eval Health Prof*. 2002;25:369-386.
61. Colloca L, Lopiano L, Lanotte M, Benedetti F. Overt versus covert treatment for pain, anxiety, and Parkinson's disease. *Lancet Neurol*. 2004;3:679-684.
62. Benedetti F, Carlino E, Pollo A. Hidden administration of drugs. *Clin Pharm Ther*. 2011;90:651-661.
63. Kelley JM, Lembo AJ, Ablon JS, et al. Patient and practitioner influences on the placebo effect in irritable bowel syndrome. *Psychosom Med*. 2009;71:789-797.
64. Varelmann D, Pancaro C, Cappiello EC, Camann WR. Nocebo-induced hyperalgesia during local anesthetic injection. *Anesth Analg*. 2009;110:868-870.
65. Sakala C. Letter from North America: understanding and minimizing nocebo effects in childbearing women. *Birth*. 2007;34:348-350.
66. Lang EV, Hatsiopoulou O, Koch T, et al. Can words hurt? Patient-provider interactions during invasive procedures. *Pain*. 2005;114:303-309.
67. Zajonc RB. Feeling and thinking: Preferences need no inferences. *Am Psychol*. 1980;35:151-157.
68. Taddio A, Shah V, Gilbert-MacLeod C, Katz J. Conditioning and hyperalgesia in newborns exposed to repeated heel lances. *JAMA*. 2002;288:857-861.
69. Colloca L, Petrovic P, Wager TD, et al. How the number of learning trials affects placebo and nocebo responses. *Pain*. 2010;151:430-439.
70. Colloca L, Finniss D. Nocebo effects, patient-clinician communication, and therapeutic outcome. *JAMA*. 2012;307:567-568.
71. Miller FG, Colloca L, Kaptchuk TJ. The placebo effect: illness and interpersonal healing. *Perspect Biol Med*. 2009;52:518-539.
72. Barrett B, Muller D, Rakel D, et al. Placebo, meaning and health. *Perspect Biol Med*. 2006;49:178-198.

CHAPTER 26

The Potential of the Analgesic Placebo Effect in Clinical Practice – Recommendations for Pain Management

Regine Klinger[1], Herta Flor[2]

[1]Outpatient Clinic of Behavior Therapy, Department of Psychology, University of Hamburg, Hamburg, Germany,
[2]Department of Cognitive and Clinical Neuroscience, Central Institute of Mental Health/Medical Faculty Mannheim, Heidelberg University, Mannheim, Germany

INTRODUCTION

Corticosteroid injection is a commonly used treatment for chronic back pain. In 1991 Carette et al[1] published a study that was recently included in a Cochrane review[2] where they examined the efficacy of methylprednisolone. The authors showed a significant reduction in chronic back pain as assessed with a numeric rating scale at both post-treatment and the 6-month follow-up; however, the reduction in the placebo group was comparable. These data suggest that a placebo can have clinically significant effects; these effects need to be compared with the effects in an untreated group of patients with back pain (natural history group) to show their superiority over spontaneous remission. Nevertheless, the fact that patients can reduce their pain via placebo effects is rarely explicitly used in the treatment of acute and chronic pain. The purpose of this chapter is to discuss whether research on placebo effects can be transferred to treatments with clinical patients, and whether the efficacy of pain treatment can be boosted by placebo effects. This perspective is based on specific knowledge about the underlying mechanisms of placebo effects.

PLACEBO RESPONSES IN PATIENTS

Most studies on placebo effects have been conducted on healthy participants. The studies that were conducted in patients are usually clinical trials where placebo effects are viewed as a control, and often as a nuisance, rather than an object of study, because they are confounded with the effects of the active treatment. However, if one examines the effects of placebos in clinical trials, and compares them to treatment as usual, an added effect can be determined. Examples are found in the German acupuncture trials (GERAC)[3,4] where verum acupuncture, following the principles of traditional Chinese medicine, was compared to (i) placebo acupuncture, which consisted of sham acupuncture with superficial needling at non-acupuncture points, and (ii) conventional treatment, consisting of a combination of drugs, physical therapy, and exercise based on guidelines for the treatment of back pain by the Committee on Drug Treatment of the German Medical Association.[5] A positive outcome was defined as at least 33% improvement, or better, on three pain-related items on the Von Korff Chronic Pain Grade Scale Questionnaire[6,7] or 12% improvement or better on the back-specific Hannover Functional Ability Questionnaire.[8] Patients who were unblinded, or who had recourse to other than permitted concomitant therapies during follow-up, were classified as nonresponders regardless of improvement in symptoms. In this study, verum and placebo acupuncture yielded positive results—but both were significantly better than the best available conventional treatment for back pain treatment; this indicated a therapeutic effect of placebo that was still effective 6 months after the completion of the treatment. In fact, the placebo acupuncture treatment was about 30% better than the conventional treatment for back pain. Patients in the conventional therapy group who received a treatment program according to German guidelines improved about 30% less than those who had placebo sham acupuncture. This difference was significant, suggesting a powerful placebo effect that was still present in the long-term follow-up, in this case 6 months.

Even more prolonged placebo effects were present in a study by Flor et al.[9,10] Here, electromyographic feedback of the back muscles was compared to placebo and standard medical treatment in a group of chronic back pain patients. The placebo treatment consisted of feedback of another patient's data to the patient and was called 'myotronics' to make it equally appealing as biofeedback. Both were termed new and promising treatments for chronic back pain. Whereas there was no significant long-term effect of standard medical treatment in these patients, both placebo and biofeedback were still effective 2 years post-treatment with placebo, leading to a 30% reduction in chronic low back pain compared to biofeedback, which attained a 60% reduction. This is, to our knowledge, the longest study of a clinical placebo effect; it suggests that these effects may have extended time courses that go well beyond the personal interaction they entail.

Another very interesting study was conducted by Kaptchuk, Kelley et al.[11,12] These authors investigated patient and practitioner influences on the placebo effect in irritable bowel syndrome. Two types of placebo acupuncture treatment were employed; one was placebo acupuncture augmented with personal interactions, the other was a placebo acupuncture limited to neutral, business-like interactions. Both were compared to a waiting-list group. Both treatments with placebo acupuncture led to symptom improvements that were beyond the normal course, which was tested by a waiting-list group. The authors found that a positive therapeutic relationship (which was realized in the augmented treatment group) can further increase the effects of placebo acupuncture in comparison to a treatment in which the practitioners minimized interaction with the patient (limited treatment group).

In a still unpublished study, we examined the effects of placebo in a group of 48 chronic back pain patients with longstanding chronic pain where we induced the placebo effect either via instruction or via instruction and additional conditioning. Conditioning was realized by manipulating the experience of the patients related to an experimental pain stimulus. As noted in Chapter 16, expectation refers to the induction of a positive drug effect by maximizing the patient's expectations by verbal instruction, whereas conditioning is provided by having the patient experience a real pain reduction. This may or may not change the patient's overt expectation. In daily clinical routine, both mechanisms are usually closely connected. The latter is based on the rules of classic conditioning, where a formerly neutral or conditioned stimulus (CS), such as a placebo pill, acquires the properties of a biologically significant or unconditioned stimulus (US), such as pain reduction, by pairing it with the US. This one-trial learning or several pairings of the CS and US lead to a conditioned response (CR, in this case pain reduction), which is usually similar to the unconditioned response (UR, here pain reduction, although in certain circumstances the CR can be opposite in direction to the UR). The main difference between the two conditions is thus verbal instructions about a placebo effect compared to the veridical experience of pain reduction in relation to the placebo. Specifically, in this study, one group of patients was told that they would receive a tincture that would be highly effective in reducing pain and physical impairment and would improve movement ability. The other group was instructed that it would receive a neutral, totally inactive tincture, a placebo. In reality, both groups received an identical neutral liquid without any active substance. The tincture was water, lightly colored red, with a bitter taste of quinine. The patients received a cotton wool swab filled with this tincture, which they were asked to put into their mouth. Both groups (one group with the instruction 'tincture is highly effective in pain reducing' and one group with the instruction 'tincture is neutral, a totally inactive tincture') were divided again, resulting in four sub-groups in total. One group from each category was additionally conditioned (resulting in two groups with conditioning) (see Fig. 26.1). The basis for the conditioning process was the application of a painful electrical stimulus, the strength of which was individually determined. Then, in the conditioning phase, the painful stimulation was reduced by 50% when the tincture was given. The patients were unaware of this procedure, so they could assume that the tincture was the reason for the reduction in pain. The patients had to rate their pain and execute standardized daily activities (according to items in the 'Daily Activity Questionaire' [8,13], see Table 26.1]) such as bending forward to pick up a small object from the floor, rising from a prone position, putting on socks, bending sideways from a sitting position to pick up a small object on the floor, just beside the chair, before and after the intake of the fluid. Pain behavior was defined as the time it took to perform those exercises.

We found a significant effect on both dependent variables: the pain ratings during the exercises were reduced after the intake of the tincture and it took the patients less time to perform the exercises. The mere instruction that an active substance, rather than a placebo was given, significantly reduced that amount of pain behavior in this group. In addition, the rating of the pain elicited by the exercises was also significantly reduced. When the effects

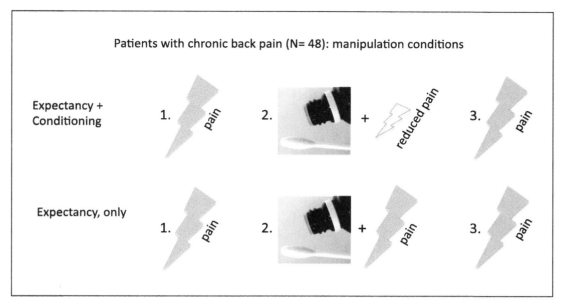

FIGURE 26.1 Diagram of the two manipulation conditions in the experiment 'Placebo intervention in patients with chronic back pain'. 1. Expectancy with conditioning: under the placebo intervention with the sham tincture; unbeknown to the patients, the individually determined electrical pain stimulus was reduced to a level of 50%. 2. Expectancy only: the patients experienced the same experimental pain level as in the individually determined condition.

TABLE 26.1 Exercises of Daily Activities

1. Bending forward to pick up a small object from the floor
2. Rising from a prone position
3. Putting on socks
4. Bending sideways from a sitting position to pick up a small object on the floor beside the chair

The patients were asked to perform these exercises before and after the intake of a sham-opioid tincture. The exercises were choosen from items in the 'Daily Activity Questionnaire.'[8,13] They were evaluated by the observation category system 'Tübingen Pain Behavior Scale' [TBS[39,40]].

of conditioning were examined, we found that the conditioning not only significantly boosted the expectancy effect in the condition where the patients believed that they had taken an opioid, but it also yielded a significant effect in the condition where the patients had been told that a placebo would be given.

These data suggest that both pain ratings and pain behaviors of chronic back pain patients are significantly reduced by expectation, and that this effect is enhanced by conditioning. This has important implications if one considers the patients' history with pain treatments. Conditioning was achieved by reducing pain perception of an experimental pain stimulus. Successful treatment trials in the patients' history should similarly boost treatment effects and unsuccessful treatment attempts might increase nocebo effects, i.e. they may induce negative expectations about the efficacy of a certain treatment ([14], see Ch 11). Likewise, expectation about outcomes should influence treatment effects. How prior experience and expectancy interact is completely unknown, but there could be multiplicative effects.

COMPARISON OF PLACEBO EFFECTS IN HEALTHY CONTROLS AND PATIENTS

There are only very few studies that compared the placebo responses of healthy people and patients, and it is not clear whether the placebo response in patients and in healthy people are the same. A clinical application of placebo effects requires specific knowledge about effects of placebos in different individuals. In comparison to healthy people, pain patients may have completely different learning histories pertaining to the experience with treatments, especially with medication. Thus, it is likely that this could change the action of placebo effects. For example, patients are more likely to have taken analgesic medication, thus the extent of experienced pre-conditioning in the sense of having learnt to associate medication with positive or negative effects, or no effects, is much higher and should lead to a more distinctive placebo response. Moreover, based on the long history of their disorder, their need for pain relief should be much higher. Against this background, it can be expected that the result of placebo effects is of higher importance for them than for healthy people. Whereas a positive result of a supposed painkiller can have a greater effect in patients, a failure of the drug could increase the risk of disappointment and lead to a later lack of effect of analgesics and even nocebo effects.

Klinger et al[15] compared the analgesic placebo effect in 48 patients with atopic dermatitis, a painful disorder of the skin, with that of 48 matched healthy controls. For these patients, ointments that reduce the irritation of the skin are of special relevance. Most of the patients had a long history of using diffent typs of ointment, always combined with a high expectation of healing. The aim of this study was to analyze the mechanisms of the placebo effect of ointments in patients with atopic dermatitis and a group with healthy skin. The patients with atopic dermatitis were expected to show a stronger placebo effect than the control group because of their previous experience as a natural form of preconditioning with ointments and the higher importance they would assign to the effectiveness of ointments. It was thus hypothesized that patients, compared to healthy controls, would have higher expectations of medication specifically designed for the treatment of their disorder. The central questions were: (1) is the placebo effect attained through a process of classic conditioning or through expectancy, and can a combination of both increase the effect? (2) can the effect be maintained over time, and how? (3) do patients with atopic dermatitis develop a stronger placebo effect than healthy controls when an ointment declared as pain-reducing (placebo) is applied?

In the experimental design we used a completely neutral ointment and coupled this ointment with two different instructions and conditioning. The participants were told that they were part of a clinical study to test the effectiveness of a new analgesic ointment. The ointment was being tested against a cream with no active substance. The participants would be randomly assigned to the group treated with the ointment or to the control group with the medically ineffective cream. All participants were informed prior to the experiment that pain stimuli would be applied before the application and that half of them would be given an effective pain-relieving ointment and half would receive an ineffective medication. The following instructions and pain experiences were systematically varied:

1. 'Instruction.' Half of the patients and half of the controls were given the information that they belonged to the group that had received the ointment with the analgesic effect ('the ointment reduces pain'). The other half were told they had received the 'ointment with no effect' ('ointment is neutral').
2. 'Classic conditioning.' All four groups were divided again, resulting in eight subgroups with 12 people in each. Four groups participated in the conditioning procedure and four groups did not. The conditioning was performed by reducing the intensity of the painful stimulus by 50% during phase 2 of the experiment, i.e. the participants were thus led to believe that the ointment had a pain-reducing effect.

In summary, there was a comparison of patients and controls and expectation and expectation plus conditioning. Both groups showed an expectancy-related placebo response. Interestingly, we found a significant difference between the patients and the healthy controls with respect to conditioning (see Fig. 26.2). Over time, only the healthy controls maintained their placebo response to expectation, whereas the patients only maintained the effect when they received additional conditioning. Without conditioning, i.e. the veridical experience of pain relief, the placebo effect was dramatically reduced in the patients.

It could be that patients are more tuned to their bodily sensations and expect more from medications and ointments. Thus, the lack of expected pain relief through the application of the ointment may have led to a sense of disappointment, which may have cancelled out their placebo effect. Thus, prior experience may interact with current experience and alter the response to placebo. This suggests that effects of medication have to be explained in a realistic manner because patients may otherwise be disappointed about small effects. In addition, in clinical practice, both expectations and prior experience of the patients should be assessed because experience seems to be so important, especially with long-term effects. In the explanation of the treatment, this information needs to be considered and perhaps also a trial experience with the treatment needs to be applied accordingly. This topic definitely needs more investigation.

USE OF PLACEBO EFFECTS IN CLINICAL PRACTICE

This leads to the very important clinical application of placebo effects.cf.[16] Several meta-analyses have concluded that placebo effects in clinical trials are much smaller than the placebo effects found in laboratory studies.[17,18] This is likely related to the different type of pain (acute versus chronic) and the different types of placebo instruction used in these studies, in addition to procedural differences.

Clinical Implications Based on the 'Open–Hidden Paradigm'

In this respect the 'open–hidden paradigm'[19–21] (see also Ch 3) yielded important information. This work showed that open application is significantly more

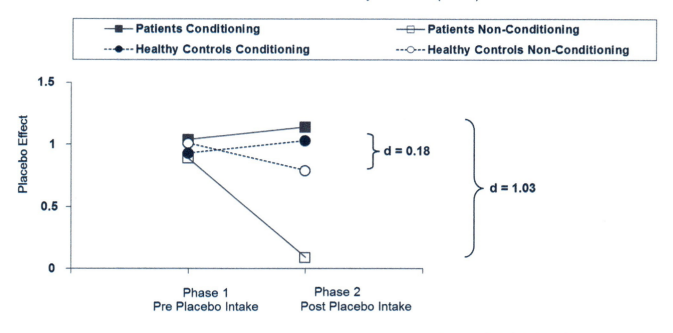

FIGURE 26.2 Results of the expectancy and conditioning manipulations in the patients and controls. Exploratory effect sizes d for the differences between the patients with atopic dermatitis (AD; N=48) and the healthy controls (HC; N=48): Factors 'Phase * Conditioning.'

effective in reducing pain than an application of an active pain-relieving substance about which the patient is unaware. The utilization of placebo effects in clinical settings should not be confused with substituting analgesic medication by placebo. Every effective analgesic has a pharmacologically active component and a psychologic (placebo) component. Thus, we suggest increasing the effectiveness of the analgesic substance by adding an inherent placebo component.

Based on this, the German guidelines for the treatment of acute perioperative and post-traumatic pain adopted a section on the therapeutic use of placebo effects ([22], see Table 26.2). Based on the current understanding of the analgesic placebo effect, the recommendation to exploit the additive placebo effect of an active, pharmacologically relevant analgesic drug can now be incorporated in everyday clinical practice.[23,24] It demonstrates an innovative approach to pain management.

This guideline opens possibilities for the practical clinical application of placebo effects in pain management. Taking into account the described mechanisms of the placebo effect (see Ch 16), a range of possibilities emerges for the deliberate induction of additive 'placebo effects' and for their application in the clinical field in an ethically acceptable way. This applies to any analgesic, but also to other medical and psychologic treatments used in pain management. The high effect size of this additive placebo component can thus be utilized in many settings ranging from the treatment of acute and postoperative pain to the treatment of chronic pain conditions. One of the future tasks of the public health system must be to educate professionals in the health-care service about the properties and underlying mechanisms of placebos so that they can optimize the placebo component of their active treatments. The knowledge of placebo effects and their application should also be emphasized more in the education of health-care professionals. Knowing the principles and mechanisms behind the effect permits a wide range of applications without violating ethical principles.[25–27] Again, it is worth mentioning that the use of placebos unknown to the patients is critical from an ethical point of view (see also Ch 27 in this book).

From this point of view the question is not if but rather how we can deliberately boost the efficacy of pain treatments by applying placebo effects. Known placebo mechanisms, such as expectancy or conditioning, or modeling of placebo effects by others, can be employed and it is to be expected that these interact in clinical practice.

TABLE 26.2 Recommendation in the Guideline 'Treatment of Acute Perioperative and Post-Traumatic Pain' [22]

- The placebo effect as an additional part of analgesic medication (open medication) in pain management should be used as much as possible by providing positive and realistic information.
- The nocebo effect should be reduced as far as possible by avoiding negative or anxiety-inducing information.

GRADE OF RECOMMENDATION A

- If active pain management is possible and provided, the administration of placebo drugs, about which the patient has not been informed and instructed, is not ethically acceptable. They should not be used for postoperative pain management outside of research studies.

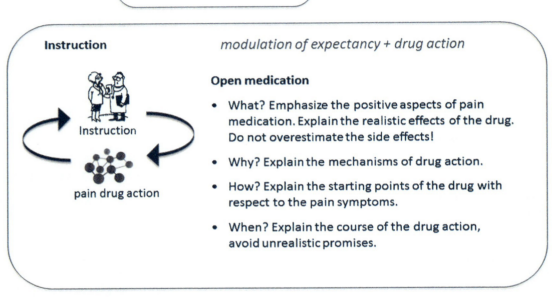

FIGURE 26.3 Applications of placebo effects with emphasis on shaping expectancies.

Shaping Placebo Effects: Boosting Positive Expectancy via Instruction

Expectancy is shaped by instructions and therefore in clinical practice it is very important to be aware what the patient is told and how the message is told. By contrast, conditioning and modeling rely on association; therefore, in clinical practice, pain-reducing medication can be associated with positive results and situations in the patient him- or herself or in other patients around him or her. Figure 26.3 summarizes how expectancy can be maximized in clinical practice. Maximizing expectancy involves a clear explanation of the analgesic medication that is prescribed, information on potential positive effects with a focus on being realistic about the expected changes, and a clear description of, but not an overemphasis on, side effects. This is a major problem in patient information sheets, which often focus exclusively on negative information and fail to discuss positive effects in an equally clear manner. Usually, indications of the drug are listed but they should be augmented by a description of positive effects that have been demonstrated, such as reduction in pain, improved mobility, or better mood. It is also important to explain the mechanisms of the action of the drug, the time frame in which the drug works, and what the course of the drug action is—and also to avoid unrealistic promises. This positive communication about drug effects not only involves verbal communication but it also entails nonverbal aspects such as a friendly manner and a positive attitude towards the patient as well as the conveyance of personal competence. Since the cost of a treatment seems to raise the outcome expectancy, and thus also efficacy related to it,[28] information about this aspect and the value of the medication in general might also be useful. Finally, it is important that the therapists themselves believe in the efficacy of a certain treatment. It was previously shown that therapists who were told that a drug would be less effective also achieved lower placebo effects[29] compared to those who believed that the drug is very effective.

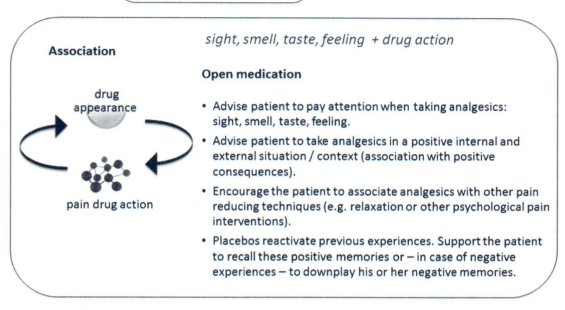

FIGURE 26.4 Applications of placebo effects with emphasis on context associations.

Shaping Placebo Effects: Boosting Positive Treatment Effects by Conditioning

The emphasis on conditioning mechanisms would mean turning the focus on associations related to analgesic medication (see Fig. 26.4). Specifically, this would entail advising patients to concentrate on sensory aspects of the medication such as sight, smell, taste, and texture. It would also be useful to take the medication in a positive situation. This precludes the patient waiting until the pain is maximal because this is not only an association of the intake of the medication with an unpleasant situation, which may reduce its efficacy, but may also negatively reinforce medication intake. As already noted by Fordyce[30] the reduction of pain induced by medication intake in a state of high pain sets a negative reinforcement learning process into action whereby medication intake is increased because it reduces a negative event (pain). Thus, both from a classic conditioning and an instrumental learning point of view, analgesics should not be taken when peak pain is present but rather independently of the pain in a time-contingent fashion. This also maximizes the pharmacologic effect of the drug, which works best with a steady level of the active substance present when chronic pain is the problem. It is also useful if patients associate the intake of analgesics with other pain-coping strategies, such as relaxation or increased activity. In the case of negative experiences, positive memories should be recalled to minimize the effects of pain peaks or other negative events. This experience can also be made via social learning by observing a positive effect of an analgesic in another patient either in the medical setting or via the use of instructional video presentations or testimonials.[31] These can also be used on the internet. It should be noted that positive experiences and associations of treatments with cues and contexts can exert effects by boosting positive expectations; however, these effects can also form implicit memories that influence the treatment outside of the patient's awareness. A major problem is the use of mainly hidden medications specifically in inpatient care or in nursing homes. Most patients in these settings are unable to tell which medication they are taking and what the purpose of the specific medication is. Even infusions are mainly 'hidden' on inpatient wards, i.e. usually the patients neither know what is in the infusion nor can they see the infusion bags. This most likely dramatically reduces the efficacy of the medications they receive and could easily be changed by making the administration of drugs more open. This involves not only the use of labels but also of colors, descriptions of the effects of the drugs that are given, and positive social interaction around the drug. This would also be very important in the context of nursing homes and specifically for patients with dementia or Alzheimer's disease. It was found that persons with Alzheimer's disease show a loss of the efficacy of placebo responses correlated with reduced connectivity of the frontal lobes with the rest of the brain. This altered connectivity is related to short attention span, poor working and short-term memory and therefore a reduced capacity to acquire and maintain explicit expectations in the form of declarative memory. By contrast, nondeclarative memory is intact in these patients, and therefore many aspects of conditioning related to placebo effects will be active in these patients and may be more effective than verbal instructions. In addition, this reduced placebo response also reduced the efficacy of analgesic medication, suggesting that alternative mechanisms of boosting placebo effects need to be considered.[32] For example, the medication requires clear and repeated instructions on its use and on its effects.

Placebo Analgesia: Interdependence of Responses and Reward Processing

An interesting finding relates to the interdependence of responses to reward and placebo analgesia. Scott et al[33] showed that there are person-related differences in activation in the dopaminergic mesolimbic reward pathway that predict not only the response to reward but also a large proportion of the variance (up to 28%) in the placebo response. Similarly, Schweinhardt et al[34] found that dopamine-related traits predicted placebo analgesia together with gray matter density in reward-related brain regions. Thus, reward processing and placebo analgesia may share common pathways. Maximizing the chance to activate the reward system may thus also improve placebo analgesia. This could alo entail using pain diaries that focus on being pain-free rather than on the amount of pain that is experienced. This might turn the patient's attention to indicators of the pleasant state of having less pain rather than the unpleasant state of being in pain. This is especially interesting with respect to chronic pain because chronic pain may be associated with a shift of attention to indicators of pain and with alterations in the processing of reward. Thus, in chronic pain, aversive processing may supersede appetitive responding and this could, in turn, negatively bias the response to reward and thus placebos. Unfortunately, these interactions have so far not yet been systematically evaluated.

PLACEBO ANALGESIA: INTERACTIONS WITH ATTITUDES TOWARDS MEDICATION AND PRIOR EXPERIENCE

The better the psychobiologic mechanisms behind placebo analgesia are understood, the more we will be able to deduce findings about its clinical application. The research conducted in recent years has focused mainly on demonstrating the neurobiologic correlates of the placebo effect and the active mechanisms of 'expectancy,' 'classic conditioning,' and 'social learning.' At the same time, little attention has been paid to possible specific interactions with existing attitudes towards medication in general (e.g. positive/negative attitudes) or prior experience (positive/negative) in clinical populations. In most research, the placebo manipulation itself (inducing specific expectations) is an independent variable, but not the already existing pattern of attitudes and the learning history. The few studies on healthy humans or patients[35,36] suggest that prior experience with pain and expectations about pain greatly alter pain processing and may also have an effect on the response to placebo analgesia.[35-37]

In an ongoing study we are examining how previous experience and expectations of pain relief affect pharmacologic and psychologic placebos in chronic pain patients. Previous experience with medication is being assessed, as are expectations about the efficacy of analgsics. Then the patients receive a placebo intervention where they expect an active analgesic. First results suggest that both experience and expectation modulate the placebo effect, which has long-lasting consequences as determined by a 1-week follow-up where both pain ratings and pain behaviors were affected.

SUMMARY

In this chapter we pointed out that little is known about the ability to transfer findings from research on placebo effects in healthy humans to patients. Initial evidence suggests that the experience of pain controls via conditioning may be especially powerful in pain patients. Clinical work with patients should use mechanisms such as expectancy, classic conditioning and social observational learning in boosting placebo effects in clinical practice and thus increase the efficacy of pain treatment. Placebos are not a substitute for pain medication. However, the efficacy of the pain treatment can be deliberately boosted by applying placebo mechanisms.[38] Placebo effectivness enhances the pharmacologic part of analgetic medication. From an ethical point of view, this potential should not be withheld from the patients in need of analgetics. Even though evidence for this position already exists, further translational research is needed to underpin the clinical implications of placebo analgesia and must focus on variables related to the therapist, the patient and the treatment that influence the efficacy of placebos.

Acknowledgment

This work was supported by grants to R.K (Kl 1350/3-1) and H.F. (Fl 156/33-1) funded by the Deutsche Forschungsgemeinschaft within the transregional research unit 'Expectation and conditioning as basic processes of the placebo and nocebo response' (FOR 1328/1).

References

1. Carette S, Marcoux S, Truchon R, et al. controlled trial of corticosteroid injections into facet joints for chronic low back pain. *N Engl J Med.* 1991;325:1002-1007.
2. Staal JB, de Bie R, de Vet HC, et al. Injection therapy for subacute and chronic low-back pain. *Cochrane Library.* 2010;2:1-50.
3. Haake M, Müller HH, Schade-Brittinger C, et al. German Acupuncture Trials (GERAC) for chronic low back pain: randomized, multicenter, blinded, parallel-group trial with 3 groups. *Arch Intern Med.* 2007;167:1892-1898.
4. Haake M, Müller HH, Schade-Brittinger C, et al. The German multicenter, randomized, partially blinded, prospective trial of acupuncture for chronic low-back pain: a preliminary report on the rationale and design of the trial. *J Altern Complement Med.* 2003;9: 763-770.
5. Arzneimittelkommission der Deutschen Ärzteschaft (AKDÄ). *Empfehlungen zur Therapie von Kreuzschmerzen [Treatment recommendations for low back pain].* 2nd ed Düssseldorf, Germany: Nexus GmbH; 2000.
6. Smith BH, Penny KI, Purves AM, et al. The Chronic Pain Grade questionnaire: validation and reliability in postal research. *Pain.* 1997;71:141-147.
7. Von Korff M, Ormel J, Keefe FJ, Dworkin SF. Grading the severity of chronic pain. *Pain.* 1992;50:133-149.
8. Kohlmann T, Raspe H. Der Funktionsfragebogen Hannover zur alltagsnahen Diagnostik der Funktionsbeeinträchtigung durch Rückenschmerz [Hannover Functional Questionnaire in ambulatory diagnosis of functional disability caused by backache]. *Rehabilitation.* 1996;35:I-VIII.
9. Flor H, Haag G, Turk DC, Köhler H. Efficacy of EMG biofeedback, pseudotherapy, and conventional medical treatment for chronic rheumatic back pain. *Pain.* 1983;17:21-31.
10. Flor H, Haag G, Turk DC. Long-term efficacy of EMG biofeedback for chronic rheumatic back pain. *Pain.* 1986;27:195-202.
11. Kelley JM, Lembo AJ, Ablon JS, et al. Patient and practitioner influences on the placebo effect in irritable bowel syndrome. *Psychosom Med.* 2009;71:789-797.
12. Kaptchuk TJ, Kelley JM, Conboy LA, et al. Components of placebo effect: randomised controlled trial in patients with irritable bowel syndrome. *BMJ.* 2008;336:999-1003.
13. Raspe H. Back pain. In: Silman A, Hochberg MC, eds. *Epidemiology of the Rheumatic Diseases.* Oxford: Oxford University Press; 2001: 309-338.
14. Colloca L, Finniss D. Nocebo effects, patient-clinician communication, and therapeutic outcomes. *JAMA.* 2012;307:567-568.
15. Klinger R, Soost S, Flor H, Worm M. Classical conditioning and expectancy in placebo hypoalgesia: a randomized controlled study in patients with atopic dermatitis and persons with healthy skin. *Pain.* 2007;128:31-39.
16. Price DD, Finniss DG, Benedetti F. A comprehensive review of the placebo effect: recent advances and current thought. *Annu Rev Psychol.* 2008;59:565-590.
17. Vase L, Petersen GL, Riley 3rd JL, Price DD. Factors contributing to large analgesic effects in placebo mechanism studies conducted between 2002 and 2007. *Pain.* 2009;145:36-44.
18. Vase L, Riley 3rd JL, Price DD. A comparison of placebo effects in clinical analgesic trials versus studies of placebo analgesia. *Pain.* 2002;99:443-452.
19. Benedetti F, Mayberg HS, Wager TD, et al. Neurobiological mechanisms of the placebo effect. *J Neurosci.* 2005;25:10390-10402.
20. Benedetti F, Amanzio M, Maggi G. Potentiation of placebo analgesia by proglumide. *Lancet.* 1995;346:1231.
21. Levine JD, Gordon NC. Influence of the method of drug administration on analgesic response. *Nature.* 1984;312:755-756.
22. Deutsche interdisziplinäre Vereinigung für Schmerztherapie (DIVS) [German Interdisciplinary Association of Pain Treatment]. In: Laubenthal H, Becker M, Sauerland S, Neugebauer E, eds. *S3-Leitlinie Behandlung akuter und perioperativer posttraumatischer Schmerzen [S3-Guideline, 'Treatment of acute perioperative and posttraumatic pain'].* Köln, Germany: Deutscher Ärzte-Verlag – AWMF-Reg.-Nr. 041/001, http://www.awmf.org; 2008.
23. Klinger R. Das Potenzial des analgetischen Plazeboeffektes: S3-Leitlinien-Empfehlung zur Behandlung akuter und perioperativer Schmerzen [The potential of the analgetic placebo effect – S3-guideline recommendation on the clinical use for acute and perioperative pain management]. *Anasthesiol Intensivmed Notfallmed Schmerzther.* 2010;45:22-29.
24. Klinger R, Thomm M, Bryant M, Becker M. Patienteninformation und -aufklärung [patient information and -education]. In: Laubenthal H, Becker M, Sauerland S, Neugebauer E, eds. *Deutsche interdisziplinäre Vereinigung für Schmerztherapie (DIVS) [German Interdisciplinary Association of Pain Treatment], S3-Leitlinie Behandlung akuter und perioperativer posttraumatischer Schmerzen [S3-Guideline, 'Treatment of acute perioperative and posttraumatic pain'].* Köln, Germany: Deutscher Ärzte-Verlag – AWMF-Reg.-Nr. 041/001, http://www.awmf.org; 2008:19-22.
25. Finniss DG, Kaptchuk TJ, Miller F, Benedetti F. Biological, clinical, and ethical advances of placebo effects. *Lancet.* 2010;375:686-695.
26. Brody H, Colloca L, Miller FG. The placebo phenomenon: implications for the ethics of shared decision-making. *J Gen Intern Med.* 2012;27:739-742.
27. Miller FG, Colloca L. The legitimacy of placebo treatments in clinical practice: evidence and ethics. *Am J Bioeth.* 2009;9:39-47.
28. Waber RL, Shiv B, Carmon Z, Ariely D. Commercial features of placebo and therapeutic efficacy. *JAMA.* 2008;299:1016-1017.
29. Gracely RH, Dubner R, Deeter WR, Wolskee PJ. Clinicians' expectations influence placebo analgesia. *Lancet.* 1985;1:43.
30. Fordyce WE. *Behavioral Methods for Chronic Pain and Illness.* St Louis: Mosby; 1976.
31. Colloca L, Benedetti F. Placebo analgesia induced by social observational learning. *Pain.* 2009;144:28-34.
32. Benedetti F, Arduino C, Costa S, et al. Loss of expectation-related mechanisms in Alzheimer's disease makes analgesic therapies less effective. *Pain.* 2006;121:133-144.
33. Scott DJ, Stohler CS, Egnatuk CM, et al. Individual differences in reward responding explain placebo-induced expectations and effects. *Neuron.* 2007;55:325-336.
34. Schweinhardt P, Seminowicz DA, Jaeger E, et al. The anatomy of the mesolimbic reward system: a link between personality and the placebo analgesic response. *J Neurosci.* 2009;29:4882-4887.
35. Colloca L, Benedetti F. How prior experience shapes placebo analgesia. *Pain.* 2006;124:126-133.
36. Andre-Obadia N, Magnin M, Garcia-Larrea L. On the importance of placebo timing in rTMS studies for pain relief. *Pain.* 2011;152: 1233-1237.
37. Colloca L, Petrovic P, Wager TD, et al. How the number of learning trials affects placebo and nocebo responses. *Pain.* 2010;151: 430-439.
38. Colloca L, Miller FG. Harnessing the placebo effect: the need for translational research. *Philos Trans R Soc Lond B Biol Sci.* 2011;366: 1922-1930.
39. Flor H. *Psychobiologie des Schmerzes [Psychobiology of pain].* Bern, Switzerland: Huber Verlag; 1991.
40. Flor H, Turk DC. *Chronic Pain: an Integrated Biobehavioral Approach.* Seattle, WA: IASP Press; 2011.

CHAPTER 27

Placebo and Nocebo: Ethical Challenges and Solutions

Luana Colloca[1,2,3], Franklin G. Miller[3]

[1]National Center for Complementary and Alternative Medicine (NCCAM), National Institutes of Health, Bethesda, MD, USA, [2]National Institute of Mental Health, National Institutes of Health, Bethesda, MD, USA, [3]Department of Bioethics, Clinical Center, National Institutes of Health, Bethesda, MD, USA

INTRODUCTION

Placebo research has undergone a boom in the past few years, particularly in the field of pain.[1] Driving that interest is how placebo effects shape neural circuits in the brain, perception of pain, and pain management. Indeed, there is increasing scientific attention to inner self-healing systems and how expectations and emotions influence pain perception and pain-related symptoms in patients. Research aimed at elucidating placebo effects started after World War II, with pioneering experiments by Stewart Wolf[2,3] and Henry Beecher[4-8] who first had the intuition that placebo effects influence pain and other symptoms and the response to active treatments such as painkillers.

In recent years, the mechanisms by which placebo effects take place from both psychologic and neurobiologic points of view have been investigated. The use of brain-imaging techniques has greatly accelerated such mechanistic research. Research on placebo effects has explored the involvement of opioids and nonopioids in placebo analgesia. Verbal, conditioned, and observational cues can create strong expectations which influence the brain placebo response and can lead to the release of endogenous opioids, cannabinoids, and dopamine—the same chemicals released in the brain in response to pain-relieving medications.[9]

One classic example of placebo analgesia is when a patient experiences pain relief from expecting an active drug. Furthermore, patients can form conditioned analgesic responses to characteristics like a pill's color and can experience pain relief when a placebo pill of the same color is given. Additionally, placebo analgesic responses can be elicited by observing pain relief in others.[10] The ability to form placebo analgesic effects through different kinds of learning represents an interesting perspective to understand placebo-induced healing processes, including release of pain-relieving endogenous pain inhibitors in the brain.[11]

However, whereas we know that pain can be altered by placebo mechanisms, the crucial question is whether placebo-induced modulation of pain may influence the time course and outcomes of pathophysiologic processes such as neuropathic and chronic pain. Investigation of such mechanisms of placebo has unraveled several important questions requiring more rigorous scientific investigation. For instance, why do some patients respond to expectations and placebos and some not at all? How can we predict and understand the duration of placebo effects over time? Under what circumstances can use of placebo treatments be endorsed as potentially effective interventions in clinical practice?

Although these and many other questions remain open for future laboratory investigation, interest in the impact of placebos appears to be increasing in the medical community. As is common in the realm of novel medical research, ethical dilemmas are unearthed and examined. Of paramount interest is the question of whether or not it is reasonable to recommend and prescribe placebo interventions in clinical practice and harness placebo effects for improving pain management. The ramifications and outcomes of these questions are likely to have a profound impact on patient care and the health-care field in general.

TOWARDS PLACEBOS IN CLINICAL PRACTICE

Promoting placebo responses within routine clinical care, using standard interventions, is generally considered desirable and without ethical controversy. Yet rigorous knowledge is not well developed regarding optimal strategies to promote placebo responses in communicating with patients and in prescribing and administering medically indicated treatments. It is controversial, however, whether and how physicians can recommend and administer treatments that lack any specific efficacy deriving from their inherent or characteristic properties but are likely to have placebo efficacy, based on evidence that placebos produce outcomes for patients that are superior to no treatment or usual medical care.

We have argued that there may be a legitimate place within contemporary medicine for using strategies and interventions to promote placebo analgesic effects, provided that there is consistent evidence from laboratory research and randomized, controlled trials (RCTs) that placebo treatments produce significantly improved outcomes.[12,13] Much research supports the concept that a placebo effect is powerful enough to influence pain perception and clinical symptoms. Whereas it is clear that placebo mechanisms influence pain pathways from the frontal area of the brain to the spinal cord,[1,14] most of the experimental placebo analgesia studies have involved healthy volunteers with outcomes monitored for a short time. Therefore, the clinical significance and generalizability of this research is unclear. Treatments that are likely to be effective solely or primarily by means of the placebo response should be evaluated rigorously in randomized trials comparing them with no-treatment or usual-care groups.

Accordingly, it is worth exploring the magnitude of placebo effects in research settings through systematic reviews of clinical trials. Indeed, some reviews have shed light on placebo effects in clinical trials by looking at patients who were randomly assigned to either placebo or no treatment.[15–17] By comparing outcomes in patients randomized to placebo under blind conditions with those randomized to no treatment, Hróbjartsson and Gøtzsche considered the effect of three types of placebo: 1. pharmacologic (e.g. a pill), 2. physical (e.g. a manipulation), and 3. psychologic (e.g. conversation), along with different types of outcome: binary (e.g. the proportion of alcohol abusers and non-alcohol abusers) versus continuous (e.g. the amount of alcohol consumed). The resulting placebo effect on subjective, continuous outcomes, most notably in pain relief, was statistically significant (SMD -0.28 (95% CI -0.36 to -0.19). Variations in the magnitude of placebo effects were partially explained by trial designs and whether patients were informed about the inclusion of inert substances. Larger placebo effects were present when patients were not informed that they would receive a placebo intervention. Meta-regression analyses showed a positive association between the magnitude of placebo effects and physical placebo interventions (e.g. sham acupuncture) and outcomes (larger effects in patient-reported outcomes than in observer-reporter outcomes).[17] The authors suggested that the observed significant effect of placebos on subjective outcomes like pain may have been due to biased reports of subjects: those receiving placebos likely believed that they were receiving an active treatment intervention, while those in the no-treatment groups knew that they were not receiving a treatment intervention.

Brain-imaging technology has been instrumental in elucidating some of these potentially biased reports. Subjective, 'biased' reports seem to be related to brain and body changes when studied in controlled environments with brain-imaging techniques.[18] Interestingly, Vase et al conducted a meta-analysis aimed at comparing the placebo analgesic effects observed in laboratory settings versus clinical trials.[19] These researchers included 23 clinical trials from the meta-analysis by Hrobjartsson and Goetzche[15] and 14 studies that investigated placebo analgesic mechanisms. The magnitudes of the placebo analgesic effects were dramatically higher in studies investigating placebo analgesic mechanisms compared with clinical trials in which placebos served as a control.

In a follow-up meta-analysis with many more laboratory studies (from 14 to 21 studies including control, placebo treatment, randomization and pain measures), the authors found that the magnitude of placebo analgesia in laboratory settings was fivefold larger than analgesia in placebo control studies.[20] Such a difference might be due to the context features of clinical trials versus experimental settings (see Ch 20 in this book). Participants' expectations may vary based on receiving different information about treatments and differing perceptions of the interest of investigators. Trialists typically avoid giving verbal suggestions of analgesia in favor of neutral instructions, whereas investigators looking at the placebo mechanisms tend to emphasize the analgesic properties of placebo treatments and procedures. Yet, it is important to stress that the laboratory effects of placebo analgesia were of short duration and they mostly involved healthy volunteers.

CLINICIANS' ATTITUDES TOWARDS PLACEBOS

Recent surveys of clinicians report widespread use of placebo therapy, along with physician behaviors and motivations in prescribing placebo interventions.[21–23] For example, in Denmark, 503 physicians were asked about the use of placebo described as 'an intervention not considered to have any "specific effect" on the condition treated, but with a possible "unspecific" effect.' Eighty-six percent of general practitioners, 54% of hospital-based clinicians and 41% of private specialists have prescribed placebos at least once in the last year with a trend to reach 10 times for half of general practitioners. Commonly as placebo treatments, physicians prescribed antibiotics (70% of general practitioners, 33% of hospital-based physicians, and 18% of private specialists), physiotherapy (59% of the general practitioners, 24% of hospital-based physicians, and 13% of private specialists), sedatives (45% of the general practitioners, 24% of hospital-based physicians, and 10% of private specialists), and vitamins (48% of the general practitioners, 10% of hospital-based physicians, and 9% of private specialists). 'Inert' placebo treatments (talc and sugar pills, saline solution, etc.) were rarely prescribed. The primary motivation for prescribing placebos was 'to follow the wish of the patient and avoid conflict.'[21,22]

Similar results have been found in a study in the USA where a random sample of 1200 internists and rheumatologists was surveyed for the use of placebos. Physicians were asked to indicate which of several placebo treatments they had used in the past year, defined as 'a treatment whose benefits derive from positive patient expectations and not from the physiologic mechanism of the treatment itself.' Fifty-five percent of the physicians reported having recommended at least one of a list of interventions as a placebo treatment during the past year: 41% recommended use of over-the-counter analgesics, 38% vitamins, 13% sedatives, and 13% antibiotics. Only 5% reported using pure placebos, such as sugar pills and saline injections. When asked about their frequency of recommending a therapy 'primarily to enhance patient expectation,' 46% reported doing so at least two or three times per month. Of those physicians who reported recommending one or more placebo treatments in the past year, 68% described this recommendation to their patients as 'a medicine not typically used for your condition but may benefit you.'[21,22]

A systematic review of empirical studies from 12 different countries indicated that the prevalence of placebo use in clinical practice varies between 17 and 80% among physicians and 51 and 100% among nurses. Motivation for the use of placebos among physicians and nurse also varies substantially.[24]

These surveys on the use of placebo treatments among physicians suggest that placebo treatments are prescribed in a less than transparent way and that physicians may not be clear themselves about what they are trying to accomplish. It appears that physicians with some frequency engage in behaviors conflicting with professional norms relating to medically indicated treatment by rationalizing the use of placebos as a way to comply with the pressure of patient demand.

The practice of using active treatments (e.g. antibiotics, sedatives) with the purpose to evoke placebo effects becomes problematic because of potential side effects, making this type of clinical approach a poor candidate for placebogenic treatments. The intentional use of placebos in clinical practice must be ultimately grounded in scientific rationale, professional integrity, and respect for patients' autonomy. Accordingly, health practitioners who intend to use placebos should determine the appropriateness of their decisions by addressing the following questions:

1. Is there evidence of benefit for using placebos as compared with no-treatment/usual care in this specific circumstance?
2. Are the risks associated with the procedure or intervention low?
3. Are the costs of intervention low or modest?
4. Can the intervention be presented without deception?
5. Is there any possibility to inform the patient about the placebo effect without instilling negative expectations and producing bad reactions?
6. Can such a placebo effect be evoked without placebos?

This set of questions may help health practitioners to evaluate the appropriateness of placebo on a case-by-case basis, the associated risks, the overall benefits, the respect for the patient's autonomy as well as the impact on the doctor–patient relationship.

PATIENTS' ATTITUDES TOWARDS PLACEBOS

In considering the pros and cons of using placebos in clinical practice during decision-making processes, the opinion of patients must be taken into account. However, little systematic research has been conducted on patient attitudes to placebos in clinical practice. A survey of Swedish patients asking for an evaluation of some clinical scenarios and general statements about placebos, revealed that 78% of 83 patients believe that physicians should follow the wishes of the patient to receive treatment 'even if the treatment is tantamount to placebos in the opinion of the physician.' Seventy six percent of them believe that a placebo would be acceptable in severe circumstances, such as terminally ill patients, because 'it preserves the patient's hope without making her final time unbearable.'[25]

Fassler et al recruited 414 patients in the Canton of Zurich to explore their attitudes towards the use of placebos in clinical practice. Seventy percent of patients wanted to be explicitly informed when receiving a placebo, and 54% of patients would be disappointed to learn that they had been treated with pure placebo ('sugar pill').[26] Moreover, patients' attitudes differ considerably based on education and cultural factors.

The majority of these surveys have been conducted in countries other than the USA. Therefore, we probed patients' attitudes towards placebos in clinical practice in the USA by using a combination of general questions and detailed scenarios, and including a large and demographically diverse sample. In general, the surveys revealed that the acceptability of placebos among patients is high. Seventy six percent of respondents judged that it was acceptable for a doctor to recommend a placebo treatment if she thought that it would benefit and not harm the patient, and 50% of responders considered it acceptable if the doctor is uncertain of the benefit. Approximately 70% stated that it is acceptable for a doctor to offer a safe placebo if it addresses a patient's need to feel like she is being given something to get better. Only 21% of respondents judged that it is *never* acceptable for doctors to recommend placebo treatments.

Overall, these findings reveal that patients are prone to use placebos to address a patient's need. Also, patients' opinions seem to be at least to some extent in line with the American Medical Association Code of Medical Ethics, Opinion 8.083, which recommends that '*a placebo must not be given merely to mollify a difficult patient, because doing so serves the convenience of the physician more than it promotes the patient's welfare.*'[27]

PLACEBOS AND THE DECLARATION OF HELSINKI

In the context of clinical trials, the Declaration of Helsinki states—in a note of clarification inserted in 2002[28]—that placebos are acceptable despite proven effective treatment under some conditions, and this was codified in the revised version of 2008[29]:

The benefits, risks, burdens and effectiveness of a new intervention must be tested against those of the best current proven intervention, except in the following circumstances:

✓ *The use of placebo, or no treatment, is acceptable in studies where no current proven intervention exists; or*

✓ *Where for compelling and scientifically sound methodological reasons the use of placebo is necessary to determine the efficacy or safety of an intervention and the patients who receive placebo or no treatment will not be subject to any risk of serious or irreversible harm.*

Extreme care must be taken to avoid abuse of this option.

Placebo controls also pose ethical issues when they are used to evaluate procedures which are invasive, such as sham surgery trials. Nevertheless, some surgical techniques are sufficiently low risk to legitimize the use of minimally invasive sham surgery as a rigorous way to evaluate subjective outcomes (e.g. pain relief). It is important to note that sham-controlled surgery trials can provide important evidence about the efficacy of specific treatments. For example, a sham-controlled surgery trial for osteoarthritis of the knee has reported comparable clinical benefits between sham (involving skin incision without manipulation of the knee) and active group (involving skin incision with manipulation of the knee).[30] Similar findings have been found for vertebroplasty, a common surgical treatment for painful osteoporotic vertebral fractures. When the surgical treatment was compared with a sham procedure in which needles were introduced into the back without injecting cement, the clinical improvement in pain did not differ at 1 week or at 1, 3, or 6 months after treatment.[31] Thus, at least in some circumstances, placebo-controlled surgical trials are both scientifically necessary and ethically appropriate.[32-34] It remains controversial as to whether surgical interventions with low-risk profiles and outcomes better than no treatment can be legitimate in clinical routine practice and be covered by health-care systems.[34]

THE DILEMMA OF DECEPTION

Deception in Laboratory Encounters

Very often, placebo research involves elements of deception. Typically, deception entails deliberate, misleading communication about the goal of the research, the nature of experimental procedures and the psychologic manipulations (e.g. surreptitiously reducing pain intensity to provide the experience of analgesia).[35]

The general guidelines of the American Psychological Association (APA) for the use of deception in research are listed as follows with a brief explanation in the context of placebo research:

1. Psychologists do not conduct a study involving deception unless they have determined that the use of deceptive techniques is justified by the study's significant prospective scientific, educational, or applied value and that effective nondeceptive alternative procedures are not feasible.[36]

Deception in placebo research is adopted to create expectations of benefit (or expectations of negative symptoms, as in nocebo research). Specifically, deception is often adopted to make participants believe that a certain procedure (e.g. a placebo cream, sham needle or electrode) is able to induce analgesia. For example, all the stimuli are set at the painful control level and participants are deceptively informed that they would receive a painkiller or an analgesic intervention. This use of deception is often necessary to test for placebo responsiveness.

2. Psychologists do not deceive prospective participants about research that is reasonably expected to cause physical pain or severe emotional distress.[36]

Strictly speaking, placebo analgesia research which involves the use of painful stimuli is inconsistent with this guideline. It is reasonable to presume that it should be qualified to rule out severe or lasting pain. Participants in placebo research should have the possibility to stop the delivery of painful stimuli and study participation at any time.

3. Psychologists explain any deception that is an integral feature of the design and conduct of an experiment to participants as early as is feasible, preferably at the conclusion of their participation, but no later than at the conclusion of the data collection, and permit participants to withdraw their data.[36]

Each participant should be informed about the true nature of research and the features of the protocol at the end of their individual participation in the study (during the debriefing process) by the investigators. When informed about the use of deception, participants should be also offered the opportunity to withdraw their data from the study.

In laboratory placebo research, it is possible to adopt an 'authorized deception' approach during the consent process when false and misleading descriptions of certain aspects of the study are implemented.[37,38] Study participants are informed that there will be elements of deception in the research but without indicating specifically what these are. This approach gives prospective subjects a fair opportunity to decide whether to participate in research that involves deception. A suggested statement to be included in the consent form is:

> 'You should be aware that the investigators have intentionally misdescribed certain aspects of the study. This use of deception is necessary to obtain valid results. However, an independent ethics committee has determined that this consent form accurately describes the risks and benefits of the study. The investigator will explain the misdescribed aspects of the study to you at the end of the experiment.'[38]

The use of authorized deception in the informed consent process has been tested by Martin and Katz,[39] who randomly assigned healthy participants to an authorized deception group or a deceptive group without authorized deception. Healthy volunteers received a deceptive placebo analgesic procedure. Interestingly, the authors found that authorized deception did not influence the size of placebo-induced placebo analgesia; nor did it influence recruitment and retention of participants. These findings suggest that the investigators may enhance the ethics of placebo research without jeopardizing the scientific validity of the protocol. Additionally, Martin and Katz found that informing participants about the nature of the placebo manipulation does not cause distress and lack of trust in research.[39] Therefore, these researchers did not expect any strong negative reactions or lasting negative consequences from debriefing participants about the details regarding the use of deception in their study. It seems that there are no risks associated with the 'authorized deception' relating to the purpose of the study and the procedures employed. If these findings find future confirmation, an authorized deception approach in which all subjects are informed about the use of deception may represent the ethically preferred approach to the study of placebo mechanisms. Providing prospective subjects who do not want to be deceived with the opportunity to decline to participate is consistent with the principle of autonomy.

Deception Associated with Placebo Interventions

The placebo effect poses challenges and questions for medical ethics. Can it be ethical for physicians to recommend placebo treatments in a deceptive way, or should placebos be recommended or administered only transparently?

Placebo research calls for rethinking the balance between beneficence (or, more precisely, beneficence plus nonmaleficence) and respect for autonomy. The benevolent use of deception to invoke a placebo effect in clinical practice has been the object of philosophical analysis.[40,41] Some argue that the use of placebo treatments in clinical practice must be consistent with the professional integrity of clinicians and respect for patients' values and preferences. Deceptive administration of placebos can violate these values.[12,13] In addition to violating respect for the autonomy of patients, deceptive efforts to promote placebo effects might encourage medicalization and drug dependence.[42,43] Others believe that these concerns need to be balanced against the potential for promoting clinically meaningful placebo responses without adverse treatment effects.[44-46]

One suggestion for bypassing deception is to prescribe placebos transparently. However, very little research has been conducted to understand whether placebo interventions can be prescribed overtly without deception.[47-49] Recently, Kaptchuk and colleagues studied the role of placebos given transparently, accompanied by information about the placebo effect, in patients with chronic pain due to irritable bowel syndrome. An RCT showed significant improvement in symptom severity after receiving open-label placebo as compared to a no-treatment control group with matched patient-provided interactions.[49] This was a small, pilot study. Additional randomized trials in various groups of patients are needed to assess whether open placebos can produce clinically significant symptomatic improvement. A basic question that deserves greater attention is whether discrete action, such as the action of taking pills and treatment rituals, are necessary to optimally evoke clinical benefit by virtue of placebo effects.

THE IMPACT OF THE CLINICIAN–PATIENT RELATIONSHIP

It is important to emphasize that placebo effects can be obtained without any placebos. In a study conducted by Thomas, the impact of a positive or negative consultation with a clinician on the patient's optimism was demonstrated along with the results of the treatment.[50] Thomas found that how a clinician presents a prognosis and treatment has a significant impact on the satisfaction and healing time of the patient. Patients who were positively consulted had a much higher 2-week recovery rate than those who were negatively consulted. In this experiment, all of the patients were suffering from minor illnesses, and patients would have healed with time, but the manner in which the treatment was administered made an impact on the amount of time it took for their illnesses to heal. Half of the patients in both types of consultation group were given treatment that was a placebo, and half were not given any treatment. There was no difference in recovery time between placebo treatment and no-treatment groups of both consultations, but there was a considerable difference in the recovery times between the two different forms of consultation group, demonstrating that the intervention style of the clinician accounted for the difference. Although these findings need to be replicated, they have important implications. Based on Thomas' findings, being more optimistic may have caused the patients to recover faster. Having a positive expectancy of the treatment may successfully relieve minor illnesses. This concept can be also applied to the role of the clinician in being able to alter the patient's mindset and expectancy to enhance the effectiveness of painkillers.

Some therapeutic effects may result from social interactions between clinicians and patients. The open–hidden approach for studying placebo effects clearly illustrates the impact of a verbal and non-verbal communication relating to expectations of symptomatic relief.[51] For example, in patients with postoperative pain, substantially higher doses of morphine are required when the drug is delivered by a computerized infusion device without the patient knowing when to expect medication, as compared to the routine clinical practice of open injection of morphine and assurance of pain relief. These contrasting open and hidden interventions permit isolation of the impact of placebo effects on therapeutic outcomes without the use of a placebo.

Kaptchuk and colleagues further demonstrated the role of the clinician–patient relationship in placebo and pain management.[52] Patients with irritable bowel syndrome treated with sham acupuncture in the context of a randomized, sham-controlled trial were assigned to three groups: no treatment, treatment with minimal patient–clinician contact, and augmented patient–clinician contact, as described below. The point of the experiment was to investigate how clinical outcome was affected by the time spent with the patient, and the manner in which clinicians presented themselves and their treatment to the patient. The group receiving sham acupuncture with minimal patient–clinician contact had only a 5-minute meeting with each other, whereas the augmented patient–clinician group met with each other for 45 minutes and had a meaningful session regarding the expectancy of the treatment to work and the clinician's concern for the wellbeing of the patient. The enhanced communication and attention made a significant impact on the effectiveness of the treatment. The augmented group had the best results in terms of relief from IBS symptoms, emphasizing the importance of empathetic clinical attention.[52] Those receiving the sham acupuncture and minimal attention had superior outcomes to the no-treatment group, suggesting the prospect of benefit from the treatment ritual.

Low-risk and low-cost interventions in the arena of complementary and alternative medicine (CAM) provide a promising avenue for evaluating the potential for promoting clinically significant benefit from placebo responses by means of treatment interventions without deception.[12,13]

The physician would explain to the patient that efforts to take advantage of the placebo response may provide beneficial symptomatic improvement. Such a way for promoting placebo-based healing processes represents a general strategy consistent with professional norms and ethics of decision-making processes. In guiding the patient's decision, Brody and colleagues suggested a specific framing with respect to treatments, which may work primarily by means of placebo analgesic effects.[53] In the case of low back pain, a statement by the physician such as the following seems ethically justified as well as likely to be therapeutically effective:

> 'As we've discussed before, medical science tells us that we all have built-in responses and chemicals that can help us get over symptoms more quickly, and that can add effectiveness to medicines and other treatments. So my job as your doctor is to try to work with you to turn on those powerful forces you have inside your body, to accompany whatever other treatments I think you could benefit from. Now, how can we apply this to your back pain? We know that one thing that turns on those powerful inner chemicals is your own expectation that a treatment is going to work. As I talk with you about acupuncture, I sense that you have a lot of confidence in that approach. So going with acupuncture could give you the best of both worlds, the physical effects of the needles plus that extra boost from your own confidence.'[53]

THE NOCEBO AND ITS IMPLICATIONS FOR HOW DOCTORS CONSULT WITH THEIR PATIENTS

In considering the ethical issues related to placebos and placebo effects, it is important to consider the nocebo effect—a phenomenon that negatively influences clinical outcomes. Indeed, nocebo effects are common in clinical trials and practice and can produce discontinuation of trial participation, alteration of treatment schedules, and lack of adherence to treatment.[54] Nocebo effects in clinical practice can elicit negative symptoms, which may be ascribed mistakenly as adverse treatment effects. Communicating to patients potential side effects of drugs may produce nocebo effects.[55]

Nocebo effects (or placebo adverse effects) have been observed in systematic reviews of randomized, double-blind, placebo-controlled studies for migraine treatments.[56,57] A systematic review of randomized placebo-controlled clinical trials, including 56 trials for triptans, 9 trials for anticonvulsants, and 8 trials for nonsteroidal anti-inflammatory drugs (NSAIDs), revealed a high rate of adverse events in the placebo arms of trials matching those described for real drugs. For example, anticonvulsant placebos produced anorexia, memory difficulties, paresthesia and upper respiratory tract infection—all adverse events reported in the side effect profile of this class of anti-migraine drugs.[56,57] The findings from these studies raise the clinically and ethically important issue of how physicians should frame information so that the truth relating to risks of treatments is preserved and the probability of producing harm is minimized.[54]

Wells and Kaptchuk proposed that physicians practice what they call 'contextualized informed consent,' which attempts to take into account the possible side effects, the patient being treated, and the disease involved by conveying information about side effects to provide the most complete picture of the treatment with the least potential to cause harm.[58] However, the ethical analysis of their approach needs to be specified more concretely in order to support their goal in a way that is consistent with the principle of respect for patient autonomy.[59] For example, adverse effects with the potential for serious or irreversible harm should not be concealed, as patients cannot make informed choices without disclosure of serious risks.

To minimize nocebo effects consistent with patient autonomy and disclosure of serious risks, a technique of 'authorized concealment' can be also considered in certain circumstances.[13,54] According to the authorized concealment approach, patients prescribed a particular drug would be asked if they are willing to agree not to receive information about certain types of side effect. The authorized concealment may be appropriate for relatively mild and transient side effects. Consistent with this approach, a physician who is recommending a given drug to a patient might communicate in the following way:

> 'A relatively small proportion of patients who take Drug X experience various side effects that they find bothersome but are not life-threatening or severely impairing. Based on research, we know that patients who are told about these sorts of side effect are more likely to experience them than those who are not told. Do you want me to inform you about these side effects or not?'[54]

Such an authorized concealment disclosure might be promising when the risk profile is moderate, e.g. headache, mild fatigue, nausea. However, many patients might be aware of the potential side effects of a treatment through the widespread access to internet websites containing information about medical treatments. In particular, learning about potential side effects is a process which can be guided by a trusted clinician. Telling patients about the nocebo phenomenon could help mitigate it. We have learned that nocebo adverse effects occur frequently in clinical trials and practice due to the impact of words and information. Sensitive verbal instructions can be a potent mechanism for the reduction of fear and anxiety, with general significance insofar as anxiety exacerbates a wide range of distressing symptoms. Thus, patients need to know that negative outcomes may be a phenomenon produced unconsciously in their brains and not necessarily due to the treatment they are receiving.

The manner in which information during a painful procedure is framed may influence pain perception and experience. Gentle and relief-oriented information, such as 'We are going to give you a local anesthetic that will numb the area and you will be comfortable during the procedure,' produced different pain outcomes in women at term gestation requesting epidural analgesia during the anesthetic procedure as compared to a typical description of the procedure: 'You are going to feel a big bee sting; this is the worst part of the procedure.' After the local anesthetic injection, a blinded observer came into the room to assess the patient's pain. Women in labor told to expect pain comparable to a bee sting during the local anesthetic injection (nocebo group) scored pain higher than those receiving the procedure along with gentle positive words.[60]

Nocebo studies demonstrate that merely knowing about potential adverse effects may lead to nocebo responses. In general, the clinical implications of nocebo effects illuminate the importance of cognitive appraisal in symptom worsening and the power of words in the context of clinician–patient interaction. It is important to appreciate that physicians cannot avoid framing the information they provide patients during clinical encounters, including the informed consent process.

Not only is information disclosure necessarily selective; truthful information can be provided in different ways. Alternative ways of framing information about benefits and risks of symptomatic treatments can influence placebo and nocebo responses. Because information framing is unavoidable, clinicians should become more self-conscious about this aspect in shaping conversations with patients regarding benefits and risks of symptomatic treatments, with the aim of promoting optimal outcomes while respecting patient autonomy.

WHAT TRANSLATIONAL RESEARCH IS BEING DONE, OR SHOULD BE DONE?

Bridging the gap between mechanistic research and clinical practice is one of the most important goals for future research in pain and placebo. There is a need to identify the circumstances and requirements for the transfer of knowledge from molecular and animal models to clinical practice.[61] In the field of placebo and nocebo effects, most research has been conducted in humans. However, pain and placebo mechanisms can be studied in animal models with the potential for dramatically increasing our scientific knowledge at the molecular level.

An emblematic model of the advantage of studying molecular mechanisms in animal findings came from the field of the immune system. Animal results in the immune system represent an elegant and critical example of translational placebo research, in which the evidence of pain modulation, via conditioning, has been transferred from bedside to bench. After the early observations by Cohen and Ader providing experimental evidence that immunologic placebo responses can be obtained in mice by a sodium saccharin solution administered after repetitively pairing the solution (conditioned stimulus) with the immunosuppressive drug cyclophosphamide (unconditioned stimulus),[62] recent studies have explored the biochemistry of placebo analgesia.[63,64]

This line of research suggests that conditioned placebo substitution can be understood as a specific technology for promoting placebo responses, as distinct from more informal expectation-related interventions in the context of clinical encounter. More research will be needed to evaluate the therapeutic potential and clinical feasibility of placebo conditioning in the field of pain conditions, including management of clinical acute and chronic pain. It is worthwhile exploring whether analgesic effects can be potentiated, extended, and manipulated, after repetitively pairing a conditioned stimulus with an effective painkiller (unconditioned stimulus). These approaches are supported by recent studies investigating the biochemistry of placebo analgesia in animals, demonstrating that placebos given after morphine or aspirin repetitively elicit placebo analgesic effects even if they were injected merely with saline solution.[64–66] Similarly, placebo substitution experiments have been performed in pediatric and adult populations, respectively in children with attention deficit hyperactivity disorders and adults with psoriasis.[67–69] Using placebos to extend the analgesic effects of active painkillers is in line with professional integrity and patients' autonomy. In fact, there is evidence that these effects can be triggered while placebos are delivered overtly.

Another promising circumstance to also harness placebo effects is when standard treatments are not accompanied by adequate improvement. For example, physicians might propose a series of interventions oriented around promoting placebo responses, reducing symptoms of illness, and improving coping ability. For example, a physician might recommend acupuncture treatments,[70] described as a treatment that may work either by the physical stimulus of the needling or by promoting a placebo response, for managing pain discomfort. This should be explained to patients that, although clinical trials have demonstrated that traditional acupuncture is not always better than sham acupuncture treatment, both have been shown to be considerably better than either no treatment or usual care.

Finally, physicians can teach patients to gain relief without any placebo interventions. An early study by Egbert et al demonstrated that encouragement and instructions reduced pain in post-intra-abdominal operation patients. The 'active placebo action' consisted in explaining to patients what to expect during the post-operative period, and in teaching how to relax, breathe, and move. Compared to a control group, patients who were encouraged and informed by a physician required half the dosage of narcotics to manage the post-operative pain.[71] Overall, much more clinically oriented research is needed to learn how to promote placebo responses optimally in clinical practice.

Acknowledgments

The opinions expressed are the views of the authors and do not necessarily reflect the policy of the National Institutes of Health, the Public Health Service, or the US Department of Health and Human Services.

This research was supported by the Intramural Research Program of the National Center for Complementary and Alternative Medicine (NCCAM) and the National Institute of Mental Health (NIMH).

References

1. Colloca L, Benedetti F. Placebos and painkillers: is mind as real as matter? *Nat Rev Neurosci.* 2005;6(7):545-552.
2. Wolf S, Doering CR, Clark ML, Hagans JA. Chance distribution and placebo reactor. *J Lab Clin Med.* 1957;49(6):837-841.

3. Wolf S, Pinsky RH. Effects of placebo administration and occurrence of toxic reactions. *J Am Med Assoc*. 1954;155(4):339-341.
4. Beecher HK. The placebo effect and sound planning in surgery. *Surg Gynecol Obstet*. 1962;114:507-509.
5. Beecher HK. Surgery as placebo. A quantitative study of bias. *JAMA*. 1961;176:1102-1107.
6. Beecher HK. The powerful placebo. *J Am Med Assoc*. 1955;159(17):1602-1606.
7. Lasagna L, Mosteller F, Von Felsinger JM, Beecher HK. A study of the placebo response. *Am J Med*. 1954;16(6):770-779.
8. Beecher HK, Keats AS, Mosteller F, Lasagna L. The effectiveness of oral analgesics (morphine, codeine, acetylsalicylic acid) and the problem of placebo 'reactors' and 'non-reactors'. *J Pharmacol Exp Ther*. 1953;109(4):393-400.
9. Benedetti F, Amanzio M, Rosato R, Blanchard C. Nonopioid placebo analgesia is mediated by CB1 cannabinoid receptors. *Nat Med*. 2011;17(10):1228-1230.
10. Colloca L, Benedetti F. Placebo analgesia induced by social observational learning. *Pain*. 2009;144(1-2):28-34.
11. Colloca L, Miller FG. How placebo responses are formed: a learning perspective. *Philos Trans R Soc Lond B Biol Sci*. 2011;366(1572):1859-1869.
12. Miller FG, Colloca L. The legitimacy of placebo treatments in clinical practice: evidence and ethics. *Am J Bioeth*. 2009;9(12):39-47.
13. Miller FG, Colloca L. The placebo phenomenon and medical ethics: rethinking the relationship between informed consent and risk-benefit assessment. *Theor Med Bioeth*. 2011;32(4):229-243.
14. Meissner K, Bingel U, Colloca L, et al. The placebo effect: advances from different methodological approaches. *J Neurosci*. 2011;31(45):16117-16124.
15. Hrobjartsson A, Gotzsche PC. Is the placebo powerless? An analysis of clinical trials comparing placebo with no treatment. *N Engl J Med*. 2001;344(21):1594-1602.
16. Hrobjartsson A, Gotzsche PC. Is the placebo powerless? Update of a systematic review with 52 new randomized trials comparing placebo with no treatment. *J Intern Med*. 2004;256(2):91-100.
17. Hrobjartsson A, Gotzsche PC. Placebo interventions for all clinical conditions. *Cochrane Database Syst Rev*. 2010(1):CD003974.
18. Tracey I. Getting the pain you expect: mechanisms of placebo, nocebo and reappraisal effects in humans. *Nat Med*. 2010;16(11):1277-1283.
19. Vase L, Riley 3rd JL, Price DD. A comparison of placebo effects in clinical analgesic trials versus studies of placebo analgesia. *Pain*. 2002;99(3):443-452.
20. Vase L, Petersen GL, Riley 3rd JL, Price DD. Factors contributing to large analgesic effects in placebo mechanism studies conducted between 2002 and 2007. *Pain*. 2009;145(1-2):36-44.
21. Hrobjartsson A, Norup M. The use of placebo interventions in medical practice–a national questionnaire survey of Danish clinicians. *Eval Health Prof*. 2003;26(2):153-165.
22. Tilburt JC, Emanuel EJ, Kaptchuk TJ, et al. Prescribing 'placebo treatments': results of national survey of US internists and rheumatologists. *BMJ*. 2008;337:a1938.
23. Meissner K, Hofner L, Fassler M, Linde K. Widespread use of pure and impure placebo interventions by GPs in Germany. *Fam Pract*. 2012;29(1):79-85.
24. Fassler M, Meissner K, Schneider A, Linde K. Frequency and circumstances of placebo use in clinical practice – a systematic review of empirical studies. *BMC Med*. 2010;8:15.
25. Lynoe N, Mattsson B, Sandlund M. The attitudes of patients and physicians towards placebo treatment – a comparative study. *Soc Sci Med*. 1993;36(6):767-774.
26. Fassler M, Gnadinger M, Rosemann T, Biller-Andorno N. Placebo interventions in practice: a questionnaire survey on the attitudes of patients and physicians. *Br J Gen Pract*. 2011;61(583):101-107.
27. American Medical Association. American Medical Association Code of Medical Ethics, Opinion 8.083: Placebo Use in Clinical Practice. 2007; Available at http://www.ama-assn.org/ama/pub/physician-resources/medical-ethics/code-medical-ethics/opinion8083.shtml.
28. Helsinki DO. World Medical Association 53rd WMA General Assembly: Washington. 2002 (Note of Clarification on paragraph 29 added) .http://www.wma.net/en/30publications/10policies/b3/.
29. Helsinki DO. World Medical Association 59th WMA General Assembly: Seoul, October 2008. http://www.wma.net/en/30publications/10policies/b3/.
30. Moseley JB, O'Malley K, Petersen NJ, et al. A controlled trial of arthroscopic surgery for osteoarthritis of the knee. *N Engl J Med*. 2002;347(2):81-88.
31. Buchbinder R, Osborne RH, Ebeling PR, et al. A randomized trial of vertebroplasty for painful osteoporotic vertebral fractures. *N Engl J Med*. 2009;361(6):557-568.
32. Miller FG, Kallmes DF. The case of vertebroplasty trials: promoting a culture of evidence-based procedural medicine. *Spine (Phila Pa 1976)*. 2010;35(23):2023-2026.
33. Miller FG, Kallmes DF, Buchbinder R. Vertebroplasty and the placebo response. *Radiology*. 2011;259(3):621-625.
34. Wulff KC, Miller FG, Pearson SD. Can coverage be rescinded when negative trial results threaten a popular procedure? The ongoing saga of vertebroplasty. *Health Aff (Millwood)*. 2011;30(12):2269-2276.
35. Colloca L, Benedetti F, Porro CA. Experimental designs and brain mapping approaches for studying the placebo analgesic effect. *Eur J Appl Physiol*. 2008;102(4):371-380.
36. American Psychological Association A. http://www.apa.org/ethics/code/index.aspx. 2010;par. 8.07 ref. 39.
37. Miller FG, Kaptchuk TJ. Deception of subjects in neuroscience: an ethical analysis. *J Neurosci*. 2008;28(19):4841-4843.
38. Miller FG, Wendler D, Swartzman LC. Deception in research on the placebo effect. *PLoS Med*. 2005;2(9):e262.
39. Martin AL, Katz J. Inclusion of authorized deception in the informed consent process does not affect the magnitude of the placebo effect for experimentally induced pain. *Pain*. 2010;149(2):208-215.
40. Foddy B. A duty to deceive: placebos in clinical practice. *Am J Bioeth*. 2009;9(12):4-12.
41. Lichtenberg P, Heresco-Levy U, Nitzan U. The ethics of the placebo in clinical practice. *J Med Ethics*. 2004;30(6):551-554.
42. Bok S. The ethics of giving placebos. *Sci Am*. 1974;231(5):17-23.
43. Asai A, Kadooka Y. Reexamination of the ethics of placebo use in clinical practice. *Bioethics*. 2013;27(4):186-193.
44. Touwen DP, Engberts DP. Those famous red pills-Deliberations and hesitations. Ethics of placebo use in therapeutic and research settings. *Eur Neuropsychopharmacol*. 2012;22(11):775-781.
45. Dobrila-Dintinjana R, Nacinovic-Duletic A. Placebo in the treatment of pain. *Coll Antropol*. 2011;35(suppl 2):319-323.
46. Raz A, Harris CS, de Jong V, Braude H. Is there a place for (deceptive) placebos within clinical practice? *Am J Bioeth*. 2009;9(12):52-54.
47. Miller FG, Colloca L, Kaptchuk TJ. The placebo effect: illness and interpersonal healing. *Perspect Biol Med*. 2009;52(4):518-539.
48. Lui F, Colloca L, Duzzi D, et al. Neural bases of conditioned placebo analgesia. *Pain*. 2010;151(3):816-824.
49. Benedetti F, Lanotte M, Colloca L, et al. Electrophysiological properties of thalamic, subthalamic and nigral neurons during the anti-parkinsonian placebo response. *J Physiol*. 2009;587(Pt 15):3869-3883.
50. Thomas KB. Time and the consultation in general practice. *Br Med J*. 1978;2(6143):1000.
51. Colloca L, Lopiano L, Lanotte M, Benedetti F. Overt versus covert treatment for pain, anxiety, and Parkinson's disease. *Lancet Neurol*. 2004;3(11):679-684.

52. Kaptchuk TJ, Kelley JM, Conboy LA, et al. Components of placebo effect: randomised controlled trial in patients with irritable bowel syndrome. *BMJ*. 2008;336(7651):999-1003.
53. Brody H, Colloca L, Miller FG. The placebo phenomenon: implications for the ethics of shared decision-making. *J Gen Intern Med*. 2012;27(6):739-742.
54. Colloca L, Miller FG. The nocebo effect and its relevance for clinical practice. *Psychosom Med*. 2011;73(7):598-603.
55. Colloca L, Finniss D. Nocebo effects, patient-clinician communication, and therapeutic outcomes. *JAMA*. 2012;307(6):567-568.
56. Amanzio M, Corazzini LL, Vase L, Benedetti F. A systematic review of adverse events in placebo groups of anti-migraine clinical trials. *Pain*. 2009;146(3):261-269.
57. Reuter U, Sanchez del Rio M, Carpay JA, et al. Placebo adverse events in headache trials: headache as an adverse event of placebo. *Cephalalgia*. 2003;23(7):496-503.
58. Wells RE, Kaptchuk TJ. To tell the truth, the whole truth, may do patients harm: the problem of the nocebo effect for informed consent. *Am J Bioeth*. 2012;12(3):22-29.
59. Miller FG. Clarifying the nocebo effect and its ethical implications. *Am J Bioeth*. 2012;12(3):30-31.
60. Varelmann D, Pancaro C, Cappiello EC, Camann WR. Nocebo-induced hyperalgesia during local anesthetic injection. *Anesth Analg*. 2010;110(3):868-870.
61. Colloca L, Miller FG. Harnessing the placebo effect: the need for translational research. *Philos Trans R Soc Lond B Biol Sci*. 2011;366(1572):1922-1930.
62. Cohen N, Ader R, Bovbjerg D. Conditioned effects of cyclophosphamide. Enhancement of delayed-type hypersensitivity in the mouse. *Ann N Y Acad Sci*. 1987;496:553-560.
63. Ader R, Cohen N. Behaviorally conditioned immunosuppression and murine systemic lupus erythematosus. *Science*. 1982;215(4539):1534-1536.
64. Nolan TA, Price DD, Caudle RM, et al. Placebo-induced analgesia in an operant pain model in rats. *Pain*. 2012;153(10):2009-2016.
65. Guo JY, Wang JY, Luo F. Dissection of placebo analgesia in mice: the conditions for activation of opioid and non-opioid systems. *J Psychopharmacol*. 2010;24(10):1561-1567.
66. Zhang RR, Zhang WC, Wang JY, Guo JY. The opioid placebo analgesia is mediated exclusively through mu-opioid receptor in rat. *Int J Neuropsychopharmacol*. 2013;16(4):849-856.
67. Ader R, Mercurio MG, Walton J, et al. Conditioned pharmacotherapeutic effects: a preliminary study. *Psychosom Med*. 2010;72(2):192-197.
68. Sandler AD, Bodfish JW. Open-label use of placebos in the treatment of ADHD: a pilot study. *Child Care Health Dev*. 2008;34(1):104-110.
69. Sandler AD, Glesne CE, Bodfish JW. Conditioned placebo dose reduction: a new treatment in attention-deficit hyperactivity disorder? *J Dev Behav Pediatr*. 2010;31(5):369-375.
70. Mayer DJ. Acupuncture: an evidence-based review of the clinical literature. *Annu Rev Med*. 2000;51:49-63.
71. Egbert LD, Battit GE, Welch CE, Bartlett MK. Reduction of postoperative pain by encouragement and instruction of patients. a study of doctor–patient rapport. *N Engl J Med*. 1964;270:825-827.
72. Hull SC, Colloca L, Avins A, et al. A survey of patients' attitudes about the use of placebo treatments. *BMJ*. 2013;346:f3757.

Index

Note: Page numbers with "f" denote figures; "t" tables; and "b" boxes.

A

Acceptance commitment therapy (ACT), 253
ACTH. *See* Adrenocorticotropic hormone (ACTH)
Active pain-relieving substance, 270–271
Active placebo, 111
    acupuncture, 164–165
    individual differences in, 111
    physical therapy, 165
    surgery, 164
    technical interventions, 165
'Active placebo action', 284
Acupuncture, 250
    modulate acupuncture effects, 121–122
    neuroimaging, contribution of, 119–123
    placebo/sham studies, challenges and issues in, 118
    placebo treatment, form of, 116–118
    sham acupuncture evoked nocebo effect, 120–121
    sham acupuncture evoked placebo effect, 119–120
    sham procedure in, 164
    stimulation, brain network related to, 122–123
    subjective and objective measurements in, 118–119
    treatment, efficacy of, 116–118
    trials, 186
Acupuncture-naive patients, estimation of, 165f
Acute placebo responses, neurotransmitter system in, 31
Adaptive resonance theory (ART), 84–85
Additive effects, interactive effects *vs.*, 133–134, 134f
Additive model assumption, 161, 161f
ADHD. *See* Attention deficit hyperactivity disorder (ADHD)
Adrenocorticotropic hormone (ACTH), 12, 19–20
Alleviating symptoms, of disease, 3
Allodynic effects, 138
Aminophylline, for angina pectoris, 3–4
Amygdala (AMY), 29
Analgesic treatments
    conditioning role and prior experience in, 129
    placebo and nocebo effects, predictors for, 134
    role of expectation in, 127–129
Analysis of variance (ANOVA), 121
Angina pectoris, aminophylline for, 3–4
Animal magnetism, 2

Animal models, placebo analgesia, 17
    effects in, 15–17
    pros and cons of, 20–22
Animal training procedure, 17, 18f, 20
Antagonistic responses/verbal information, evidence base for, 107
Anterior insular cortex (aINS), 28–29
Antidepressant effect, 17
Anti-hypertensive treatment, 184
Antiulcer drugs, clinical trials of, 184
Anterior prefrontal cortex (aPFC), 38–39
Anxiety levels, reduction in, 73–74
Anxiety relief, 26
Arthroscopic debridement patients, 228–229
Articular rheumatism, management of, 3
Aspirin-induced placebo effect, 19f
Attention deficit hyperactivity disorder (ADHD), 133
Authorized concealment, 283
Authorized deception, 281

B

Balanced cross-over design (BCD), 162–163, 163f
    double-blind cross-over design, 162–163, 163f
    methodologic limitation, 163
Balanced placebo design (BPD), 152, 161–162, 162f
    BCD and, 163
    pitfalls of, 162
    variant of, 162
Baseline resting-state connectivity, of right fronto-parietal network, 93–94
Bayesian model
    of pain perception, 38–40
    theory, 37–38
Beliefs About Medicine Questionnaire (BMQ), 258–259
Best-practice-guideline-oriented therapy, 193
Beta blocker evaluation of survival trial (BEST), 185
Biochemical systems, 107
Biomedicine (BM), 189
Biosense DMR system, 230
Bipolar disorder, 185–186
Blind method, 3–4
Blood oxygenation level-dependent (BOLD), 27
BOLD. *See* Blood oxygenation level-dependent (BOLD)
BPD. *See* Balanced placebo design (BPD)
Brain, 84
    connect intuitive models, function of, 84
    imaging methods, 277–278
        and experimental approaches, 95–96
        experimental procedures and, 95f–96f, 97

    placebo analgesia effects, meta-analyses of, 209
    imaging, placebo designs for, 85f, 86
    predictors
        brain imaging study findings, 99–100
        personality and, 89–97
        placebo responses, individual differences in, 89
    structure of, 87, 94
Bronchoconstrictor carbachol, 138
Buprenorphine, thoracotomized patients treated with, 145–146

C

CAD. *See* Coronary artery disease (CAD)
Cardiovascular diseases, 259
CBP. *See* Chronic back pain (CBP)
Caffeine, 106–110
    in active placebo, 107–110
    compensatory and drug-like conditioned responses, 107
    conditioned stimuli, 107
    placebo responses, 106–107
    as reinforcer, 106
    'stomach activity', 107
Caffeine-associated stimuli, conditioned arousal to, 108–109
Cannabis sativa, 9
Capsules containing lactoses, 74–75
Carisoprodol, 109, 152
    EMG, 109, 110f
    serum levels, 111
Carver's behavioral activation scale, 96
CB1 receptors, 12
Central sensitization, 46
Cerebral cortex, opioid neuronal network in, 17
Cholecystokinin (CCK), 9, 12–13, 46, 79, 132
Chronic back pain (CBP), 227
Chronic pain
    acupuncture treatment on, 116, 117f
    placebo effect on, 150
    RCT in, 237
    and symptom severity, 93
'Classic conditioning', 270
Clinical pain, 150–151
Cochrane meta-analysis, 164
Cognition, balanced placebo design, 152
Cognitive behavioral therapy (CBT), 253
Cognitive dissonance, concept of, 37–38
Communication, operationalized, 245t–247t
Comparative effectiveness research (CER), 167
Compensatory reaction, 111
Compensatory responses
    of hypoarousal, 108
    nocebo effect and, 111–112

Complementary and alternative medicine (CAM), 238–239, 282
   BM model, 190
   in Europe, 198
   of IBS, 239–240
   as medical counterculture, 190
   myth of, 196
   NCCAM and, 190–191
   OAM and, 190
   patient–doctor relationships, 197–198
   psychoneuroimmunologic research, 190–191
   specificity and efficacy paradox, 191–194
Computer-controlled infusion pump, 152
COMT val158met functional single nucleotide variant, 145–146
Concept of expectation, 155
Concomitant activity, in autonomic nervous system, 74–75
Conditioned analgesic responses, 129
Conditioned cues, 139–140, 144f
Conditioned hyperalgesia, 104
Conditioned hypoalgesia, 104
Conditioned nocebo stimuli, 263
Conditioned response (CR), 15–16, 139–140, 268
   in immune system, 139–140
Conditioned stimulus (CS), 15–17, 109, 139–140, 268
   coffee, smell and taste of, 108–109
Conditioning-based approaches, 133
Conditioning processes, 5
Conditioning hypothesis, 26
Conditioning modulated N2–P2 amplitude reductions, verbal suggestions of, 141f
Connectivity analyses, 11
Conundrum, 98–99
'Contextualized informed consent', 283
Control group healing rates, 184
Convergence-facilitation theory, 218
Coronary artery bypass graphting (CABG), 229–230
Coronary artery disease (CAD), 229–230
Correlational approach, 76
Cross-validated regression procedure, 27
CR. See Conditioned response (CR)
CS. See Conditioned stimulus (CS)

D

DBRPC trials. See Double-blinded randomized placebo-controlled (DBRPC) trials
Deception, 281
Declaration of Helsinki (DoH), 170
Delayed response design, 163–164, 163f
   randomized run-in and withdrawal periods, 163–164
   variant of, 163–164
Demand characteristics, 177–178
   defined, 177
   demonstration of, 177
Descending inhibition, and anti-hyperalgesia, 216f
Descending pain control system
   anatomy of, 53–55, 54f
   hypothalamus/PAG and RVM, 59

opioid-dependent system of, 56
Descending pain-modulating system, 54, 57–60
Descending pain-inhibitory system, 78
Design, 151–152
   cross-over designs, 160–161
   single-blind vs. double-blind, 154
   within-subjects vs. between- subjects, 152–153
Desire
   and emotions, 221–222
   for relief, patient experiences of, 219–221
Desire–expectation model, 74
Diffuse noxious inhibitory control (DNIC), 48
Diffusion tensor imaging data, 59–60, 94
Direct-to-consumer advertising (DTCA), 185
Discriminative stimulus complex, 107–108
Dopamine (DA), 94
Dopaminergic mechanisms, formation of, 30–31
DOR receptors, 11
Dorsal anterior cingulate cortex, 110
Dorsal horn circuitry, 216f
Dorsal part of anterior cingulate cortex (dACC), 56
Dorsolateral prefrontal cortex (dlPFC), 28–29, 28f, 56–57
Double-blinded randomized placebo-controlled (DBRPC) trials, 169, 169f
Double-blinding, 228
   brain-imaging clinical trial, 227
   cross-over design, 162–163, 163f
   paradigm, 145–146
   parallel-group design, 162f
D-Phe-Cys-Tyr-D-Trp-Orn-Thr-Pen-Thr-NH2 (CTOP), 20
   rACC and, 20
Drug
   efficacy
      neural mechanisms, effects of, 129–132
      and tolerability, 133
   rehabilitation, 106–107
   unconditioned effect of, 104
DTCA. See Direct-to-consumer advertising (DTCA)
Dual-processing theory, 41

E

EAS. See Electroacupuncture stimulation (EAS)
Efficacy paradox, 191–194, 192f, 235–236, 236f
   indirect evidence of, 193
Emotion
   desire and, 221–222
   individual differences in, 90
   and motivation, 73–74
   pain, cognitive control of, 120f, 122
   placebo analgesia and opioid activity, 78
   regulation, placebo responses and, 221–222
Emotional ratings, pain and, 74–75
Endogenous analgesia, expectation effects on, 49–50
Endogenous opioid system, 5, 9–11, 56, 60
   CCK-1 and CCK-2 receptors, 10
   in conditioned response, 129
   DOR and KOR receptors, 11

   and placebo analgesia, 6–7
   pharmacologic approach with, 11, 60
   specific placebo analgesic responses, 11
   in vivo receptor-binding techniques, 11
Endogenous pain modulation, 48f
Endogenous substances, with health-promoting effects, 145
Enhanced error related positivity, 94–95
Electroacupuncture stimulation (EAS), 123
Electroconvulsive therapy (ECT), 185
Electroencephalography (EEG)
   ERPs in, 151
   pain anticipation, 38
   pain perception, 38–40
   placebo analgesia, 40–41
   recordings, 42
Event-related potentials (ERPs), 94, 151
   N2/P2 components in, 76
   P2 wave in, 75–76
   painful stimulation, elicited by, 76
Expectancy effects
   determinants for, 244f
   manipulation model, 119
Expectations
   anticipation and, 209
   and emotions, 208–209
   patient experiences of, 219–221
Expectation-based model, 61
Experimental chamber affects, 16
Experimental ischemic arm pain model, 10
Experimental pain, 151–152

F

Facilitatory mechanisms
   descending, 46–47
   placebo and nocebo effects on, 47
   spinal sensitization, 46
False-positive error, 149–151
FCP. See Free-choice paradigm (FCP)
FDA. See Food and Drug Administration (FDA)
Fear of pain (FoP), 75–77
Fear-potentiated startles, FoP, 77
Fentanyl, 6–7
Fibromyalgia (FM), 40
Food and Drug Administration (FDA), 170
Forced expiratory volume in 1 second (FEV(1)), 118–119
Forced swim-ming test (FST), 19
Free-choice paradigm (FCP), 167–169, 168f
   DBRPC trails, 169, 169f
   drugs, statistical superiority of, 168
   patients, selection behavior of, 168
   RCT and, 169
FST. See Forced swimming test (FST)
Functional bowel disorders, 218–219
Functional brain imaging data, and individual differences in, 90–93, 95f–96f
Functional magnetic resonance imaging (fMRI), 56, 151
   acquisition and analysis techniques, 61
   MRI scanner to, 121
   and PET, 57, 119
   and placebo analgesia, 57–60
   on placebo analgesic responses, 76–77

INDEX

spinal cord, 64
top-down regulations, 86
T2-weighted protocol, 62

G

Gamma-aminobutyric acid (GABA), 46
Gascoyne's powder, 2
Generic healing responses, 194
German Acupuncture Trial for Chronic Low Back Pain (GERAC), 164, 193
Grounded theories on cognition (GTC), 84
Growth hormone (GH), and inhibiting cortisol secretion, 140
GTC. *See* Grounded theories on cognition (GTC)

H

Hannover Functional Ability Questionnaire, 267–268
Hawthorne effect, 25, 178–179
 existence of, 179
 mechanisms of, 178–179
 pain, placebo effects for, 179
 US Food and Drug Administration, 179
Healed ulcers, in placebo-treated patients, 184
Healing rituals, 3
Heart rate (HR), 74f
Heat-pain sensitivity, 215–216
Hemagglutinating antibody, 16–17
'Hidden condition', 127
Hidden treatment paradigm, 159–162
High-intensity laser stimulations, 40
High titers, 16–17
Holistic (bio-psycho-social) approach, 227–228
Homeostatic theory, of placebo effects, 108
Hot-plate test, 17, 139–140
H2-receptor-antagonists, for peptic ulcer, 184
Hyperalgesic nocebo responses, fMRI analysis of, 121
Hypothalamic–pituitary–adrenal (HPA), 19–20, 121

I

IBS. *See* Irritable bowel syndrome (IBS)
Imaging markers, of brain activation, 119
'Imitation' rods, 2
Immunomodulatory placebo effects, 16–17
Immunosuppressant cyclophosphamide, 139–140
Immunosuppressive drug cyclosporin A, 129
Imprinting control region (ICR), 17
Independent component analysis (ICA), 123
Individuals in pain, painkiller administered to, 105
Individual responses, low consistency of, 98–99
Induced pain, 150
Inhibitory mechanisms
 diffuse noxious inhibitory controls, 48
 higher centers, control of, 48–49
 placebo effects on, 49–50
 spinal mechanism, 47–48
Innocuous stimuli, 62
'Instruction', 270
Integrated model, 38
Integrative medicine (IM), 189

Interactive effects, additive effects *vs.*, 133–134, 134f
Internal mammary artery ligation, 5
Interventional radiologic procedure, 262–263
In vivo receptor-binding techniques, 11
Iontophoretic pain stimulation, 140
Irritable bowel syndrome (IBS), 133, 138, 144, 207–208, 250
 evoked and spontaneous pain, placebo effects on, 219
 fMRI using, 216
 hyperalgesia, animal models of, 218
 hyperalgesia and anti-hyperalgesia, psychologic contributions to, 218–222
 pain, CNS modulation of, 222–223
 visceral and somatic hyperalgesia in, 215–216
Ischemic arm pain, 25–26

K

Knee irrigation, 229
KOR receptors, 11

L

Laser evoked potentials (LEPs), 38, 41, 140
 amplitude of, 94
 SPN, 38–39, 41
Lateral prefrontal cortex, pre-SMA and, 93
Learning mechanisms, 87
 exploiting, 133
Learning processes, 5
Least absolute shrinkage and selection operator (LASSO), 93
Ligation of the internal mammary artery (LIMA), 229–230
Lithium chloride (LiCl), 16–17
Long conditioning paradigm, 142f
Low back pain (LBP), 229
Low-intensity painful stimuli, 138

M

Machine-initiated therapy, doctor-initiated *vs.*, 152
Maladaptive function, 258
McGill Pain Questionnaire (MPQ), 31, 34f
Mean placebo effect, 76
Mean plasma cortisol, 79, 79f
Medial thalamus (mTHA), 29
Mediated expectancy effects, 175–176
Mesmerism, 2
Meta-analyses, 210
 of placebo mechanism, 206–207
 recent developments in, 210–211
Meta regression analysis, 128–129, 278
N-methyl-D-aspartate (NMDA) receptors, 46, 218
Methylphenidate, cocaine-addicted patients in, 132
Mice
 behavioral despair tests in, 18–20
 dissection of placebo analgesia in, 17–18
MID. *See* Monetary incentive delay (MID)
Microinjection, of naloxone, 20
Migraine prevention, 193
Mind–brain injuries and stresses (MBIS), 227–228

Modulation of pain perception, 120f, 121–122
Molecular imaging, 33
 using PET, 94
Molecular mechanisms, of placebo responses
 dopaminergic mechanisms, formation of, 30–31
 regional endogenous opioid neurotransmission
 placebo analgesia theories, 31–33
 placebo-induced activation of, 28–34
Monetary incentive delay (MID), 30–31
Morphine-agonistic conditioned response, 104
Morphine effects, 16
Morphine-induced placebo effect, 19f
Motivational interviewing (MI), 253
Myocardial infarction, 132–133

N

Naloxone, 6–7
 reversibility, 25–26
Naloxone-induced blockade, of opioid system, 56
Naltrexone, 78
National Institutes of Health (NIH), 170
Natural history (NH), 217f
 and PL conditions, 219–221
 on visceral pain intensity ratings, 220f
Naturalistic conditioned stimuli, 109
'Needles', 118
Negative emotions
 opioids, effectiveness of, 78
 individual differences in, 75–78
 reduction in, 74–75
 methodologic issues and empirical studies, 74–75
'Negative placebo effects', 120–121
Nervous system regulatory mechanisms, 86–87
Neural mechanisms, of placebo responses, 83–84
Neurocognitive explanatory model, of placebo responses, 84
Neuroendocrine systems, central opioid and, 94
Neuroimaging method, 119–123, 132
Neuropathic pain trials, 128–129, 185–186
Neutral stimulus, 109, 154
Neurophysiologic processing, of painful stimuli, 94
Nocebo effects, 138
 diagnosis and test results, communicating, 259–260
 illnesses and medications, 258–259
 meta-analyses of, 257–258
 minimization of, 264f
 traditional definition of, 257
 treatment, initiating, 260–262
 treatment experience, role of, 263
 treatment implementation, 262–263
Nocebogenic effect, 262
 of communication strategies, 259–260
 of fear-avoidance beliefs, 259–260
Nocebo hyperalgesia, 12–13
 by cholecystokinin, 12–13

opioid/cannabinoid and cholecystokinin
    systems in, 9, 10f, 12–13
  spinal mechanisms of, 49
Nocebo-induced side effects, consequences
  of, 258
Nocebo responses
  and compensatory responses, 111–112
  negative placebo effects, 78–79
Nociceptive afferents, 46
Nociceptive flexion reflex (NFR), 74f
Nociceptive processing, 55
Nociceptive stimulus, 11
Nociceptive thermal stimulation, 61
Non-additive model, 162, 162f
Nonhuman primates, genomes of, 22
Non-specific antagonist, for CCK-1 and
    CCK-2 receptors, 10
'Nonspecific effects', 166
Nonsteroidal anti-inflammatory drugs
    (NSAIDs), 12, 105, 210, 260
Non-noxious stimulus perception, 40
Nonspecific psychologic processes, 38
Noradrenergic system, 54
Noxious stimulation, 209
  selective inhibitory effect on, 54
Noxious thermal stimulation, 11
N2–P2 amplitude reductions, conditioning
    modulated, 141f
NSAIDs. See Nonsteroidal anti-inflammatory
    drugs (NSAIDs)
Nucleus accumbens (NAC), 28–29, 28f, 94
  BOLD signal, 31
  DA activity in, 94
  dopaminergic mechanisms, 30–31
  individual differences in, 94
  MID task, 30–31
  opioid neurotransmission, 29–30
  synaptic activity, 30–31
Numerical rating scale (NRS), 77f, 143f, 150

O

Objective physiologic measurements,
    149–150
Observational cues, 144f
Office of Alternative Medicine (OAM), 190
Open–hidden paradigm, 5, 127, 128f, 152
  clinical implications based on, 270–271, 271f
'Open condition', 127
Opioid antagonists, pharmacologic approach
    with, 11
Opioidergic activity
  in dorsal ACC and PAG, 94
  OFC/dlPFC/PAG and rostral ACC, 94
Opioidergic descending pain
  control system, 20
  modulatory system, 96–97
Opioidergic underpinnings, of placebo
    analgesic effects, 56
Opioid-induced analgesia, 11, 131f
  expectancy modulation of, 132
Opioid neuronal network, in cerebral cortex, 17
μ-opioid receptors (MOR)
  antagonist naloxone, blocking with, 9–10, 57
  neurotransmission, 11
μ-opioid receptor-mediated
    neurotransmission, 20, 33, 34f

reduction in, 78
μ-opioid-receptor-selective radiotracer, 20
μ-opioid system activation, magnitude of,
    32–33
Oral placebo, 237
Orbitofrontal cortex (OFC), 29
Osteoarthritis
  pain, 228–229
  using PET imaging, 122–123
  sham-controlled surgery trial for, 280
Overt–covert paradigm. See Open-hidden
    paradigm
Over the counter (OTC) medicines, 185
Over-investigation, 259

P

PAG. See Periaqueductal gray (PAG)
Pain
  angina, 229–230
  anticipation and, 38–40
  complexity of, 227–228
  EEG measures of, 38
  emotional ratings, 74–75
  exacerbation, explicit nocebo instructions
    of, 261–262
  headache, 229
  intensity ratings, 28–29
    VAS, 216f–217f, 220f
  LBP, 229
  management
    clinical studies, expectancy in, 244–249
    communication affect, 243–244
    emotional communication, impact of,
      249–251
    multi-disciplinary approaches to, 228
    patient–clinician relationship, contextual
      factor in, 243
    promoting patient involvement,
      251–253
    psychosocial interventions in, 253
    selection of studies, 244
  modulatory control system, 132
  non-pharmacologic therapies for, 127
  osteoarthritis, 228–229
  perception, 38–40
    patient–practitioner interaction effects,
      244, 248f
  placebo and brain stimulation for, 230–231
  placebo-induced relief of, 90
  processing, 90
  predictor of, 38–39
  quantification of, 150–151
  rating, hippocampal activation, 119
  report, basic design of, 151–152, 151f
  responsiveness, 263
    pain modulatory vs., 60
  stimuli responses, 40
  surgical approach to, 228
  threshold, 150
  top-down modulation of, 137
  tolerance, 150
  treatment of
    clinical trial methodology and
      decision-making, 240–241
    direct comparisons, evidence from,
      237–238

  efficacy paradox, 235–236
    indirect comparisons, evidence from,
      238–239
    literature, hypotheses from, 236–237
  unpleasantness, 150
Painful heel lancing, 263
Painful stimulations, 106, 154
  conditioning phase, pre-test to, 154
Pain-induction procedure, 74
'Pain ladder', 260
Pain-processing regions, BOLD responses in,
    56–57
Pain-reducing medication, 138, 272, 272f
Pain-responsive regions, reduced activity
    in, 93
Parenteral placebos, effect of, 237
Partial reinforcement scheme, 133
Patient-centered approach, 251–252
Patient-controlled analgesia (PCA), 78
Pavlovian conditioning, 145, 263
  of drug responses, 133
PCA. See Patient-controlled analgesia (PCA)
Perception–action loop, core concept of, 84
Percutaneous myocardial laser
    revascularization (PMLR), 230
Periaqueductal gray (PAG), 26–27, 29, 54, 54f,
    123, 209
  from amygdala, 55
  and dlPFC, 56–57
  and medulla, 87
  role of, 54
  RVM system, 55
Peripheral-central modulation, 122
Peripheral impulse input
  and central facilitation, 224
  visceral and somatic hyperalgesia, 216–218
Peripheral physiologic mechanisms, CNS
    and, 132
PET. See Positron emission tomography
    (PET)
Pharmacologic conditioning procedure, 109,
    129, 130f
  randomized placebo-controlled trials of,
    128–129
Pharmacologic pain management, 260
Pills, color of, 184
Pious fraud, 3
Placebo-activated endogenous opioid
    systems, 11
Placebo acupuncture
  applying real and, 250
  with low and high expectancy, 121
Placebo agents, and effects, 204–205, 204t
Placebo analgesia
  brain correlates of, 95–96, 95f–96f
  brain imaging studies, advances from,
    90–97
  brain predictors of, 89–97
  in 20th century, 3–4
  conceptualizations and definitions of,
    203–205
  conditioning (learning) processes, 5
  definitions and conceptualization, 1–2
  development of, 25–26
  experimental controls, 4–5
  facilitatory mechanisms, 46–47

fear, causal effect of, 77
history of, 7
identification of, prerequisite for, 98
individual differences in, 89–90, 91t
  limitations of, 97–99
induction of, methodological issues, 154
inhibitory mechanisms, 47–50
investigating, magnitude of, 209–210
magnitude of, 207t
  factors influencing, 207–210
  meta-analyses, current status of, 210–211
mechanisms, 5–7
medical setting, controls in, 2
meta-analyses of, 205–207, 206t
  clinical trials *vs.* placebo mechanism studies, 205–207
  pain, natural history of, 205
neurochemistry of
  cholecystokinin, nocebo hyperalgesia by, 12–13
  endocannabinoids, involving of, 11–12
  endogenous opioids, 9–11
  neuroimaging literature of, 119
  and nocebo hyperalgesia, 149–150
  NSAIDs, 12
  opioid/cannabinoid and cholecystokinin systems in, 9, 10f, 12–13
pain, 45–46
  anticipation and perception, 38–40
  parietal and prefrontal cortex, 95–96
  putative pain processing brain regions, 93
in rodents
  in animal models, 17
  animals, 15–17
  behavioral despair tests, 18–20
  mice, dissection of, 17–18
  opioid receptors involving, 20
  pros and cons of, 20–22
subjective and objective measurements in, 118–119
surgical procedures, 5
theoretical models of, 37–38
as treatment, 2–3
Placebo animal model, 20–21
Placebo anti-hyperalgesic mechanisms
  anti-hyperalgesia, neurochemical basis of, 223–224
  neuroimaging of, 222–223
Placebo capsules, 75
Placebo-controlled trials, 2
  cross-over trial, 162–163
Placebo effects
  applications of, 272, 272f
  clinical application of, 269–270
  concept, 151–152
  doctor-patient interaction with, 4, 4f
  experimental models of, 107
  mechanisms of, 271
  medication and prior experience, attitudes towards, 274
  negative impact on, 90, 91t
  responses and reward processing, interdependence of, 274
  shaping, 272
    conditioning, treatment effects by, 273–274
    instructions, expectancy by, 272

Placebo-induced activation
  blockade of, 10
  personality predictors of, 33–34
  placebo analgesia and, 31–33
  of regional endogenous opioid neurotransmission, 28–30
Placebo-induced midbrain, and brainstem responses, 59f
Placebo-like effect, 209–210
Placebo manipulation, on spontaneous pain, 150
Placebo mechanisms
  conditioning (learning) processes, 5
  endogenous opioids, 5
  experimental validity, 6
  'responders' and 'non-responders' group, 5
Placebo/naloxone and fentanyl (PNF), 6–7, 6f
Placebo nonresponders, 75
Placebo perturbation, 85
Placebo research
  design, 151–152
    single-blind *vs.* double-blind, 154
    within-subjects *vs.* between- subjects, 152–153
  ethics of, 169–170
    NIH and FDA guidelines for, 170
    RCTs and, 169
  expectations, measurement of, 155
  induced pain and clinical pain, 150
  pain, quantification of, 150–151
  methodological issues, 154
  placebo analgesia and nocebo hyperalgesia, 149–150
  pre-test, 153
  response bias, 151
Placebo responses
  associative learning, 139–143
  brain correlates of, 95f–96f
  clinicians attitudes towards, 278–279
  clinician–patient relationship, 282–283
  in clinical practice, 270–274, 278
  deception associated with, 281–282
  declaration of Helsinki, 280
  drug response from, 160t
  emotional regulation and, 221–222
  evolutionary principles, 145–146
  expectations, 144–145
  fMRI study on, 76–77
  healthy controls and patients, comparison of, 269–270
  and homeostasis, 103
    active placebo, 111
    compensatory responses and nocebo effect, 111–112
    theoretical background, 103–104
  individual, brain correlates of, 95f–96f
  instructional learning, 137–139
  laboratory encounters, deception in, 280–281
  magnitude of, 16
  minimize *vs.* maximize, 159–161
  molecular mechanisms of
    dopaminergic mechanisms, formation of, 30–31
    placebo analgesia theories, 31–33

placebo-induced activation of, 28–34
regional endogenous opioid neurotransmission
  neuroimaging, 93
  contribution of, 119–123
patients attitudes, 279–280
patients, nocebo and implications with, 283–284
patient–physician relationship impact on, 98–99
in patients, 267–269
person and situation characteristics, 99, 100f
predictors of, 90
proglumide enhanced, 79
rats, placebo response in, 20
social learning, 143–144
theory, 107
translational research, 284
Placebo treatment
  adherence of, 185
  bias responses to
    demand characteristics, 177–178
    Hawthorne effect, 178–179
    objective outcomes, importance of, 181
    response shift, 179–181
    theoretical model, 175–177, 176f
  color and number, 184
  culture of, 184
  form of, 184
  history of, 185–186
  hype of, 185
  surgery of, 186
Positive control clomipramine, 19–20
Positive and Negative Affectivity Scale (PANAS), 28–29
Positron emission tomography (PET), 26–27, 32, 48–49, 56, 78, 85–86, 94, 151
  raclopride and fMRI, 30–31
Post-operative pain model, 56, 150, 244–248
  reduction in, 248
Post-puncture headache, 128
Post-traumatic stress disorder (PTSD), 228
Potentiation
  analgesics, net-analgesic effect of, 132
  of startle reflex, 77
'Preference design', 167
Prescientific medicine, 2
Profile of Mood States inventory (POMS), 28
Prolonged drug action monitoring, 163
Psychologic comorbidity, 40
Psychoneuroimmunologic research, 190–191
Psychotherapy research, stemming from, 194

Q
Quantitative sensory testing, 222

R
Radiofrequency fields (RF), of mobile phone, 258
Randomized controlled trials (RCTs), 83, 115–116, 127, 159, 205, 236, 249, 259
  nonspecific effects, placebo effect in, 166, 166f
  traditional and novel design features, 160, 160t

Receptor mechanisms, and central sensitization, 218
Rectal lidocaine (RL), 217f
 on visceral pain intensity ratings, 220f
Rectal placebo (RP), 217f
 expectancy/desire and anxiety, contributions of, 219–221, 220t
 on visceral pain intensity ratings, 220f
Red-light-associated painful stimulus, 154, 155f
Reframing continuous drug intake, 133
Repetitive transcranial magnetic stimulation (rTMS), 230
Response bias, 151
Response shift, 179–181
 magnitude of, 180–181
 placebo effect, 180
 research in, 180
 target construction, self-evaluation of, 179
 US Air Force, 179–180
Reward-related activity, 96
Rheumatoid arthritis patients, 179
Right fronto-parietal network, with rostral ACC, 93–94
Robust analgesic responses, 141–143
Rogerian nondirective psychotherapeutic approach, 244
Rostral anterior cingulate cortex (rACC), 26–27, 56, 85
 and orbitofrontal cortex, 11
 and dlPFC, 56–57
Rostral ventromedial medulla (RVM), 54, 54f
 functional properties of, 54–55
 hypothalamus and, 57
 modulatory effects of, 55
Roter's Interaction Analysis System (RIAS), 251
RVM. See Rostral ventromedial medulla (RVM)

S
Saliency, 263
Self-confirming feedback mechanism, 223
Self-healing responses
 anxiety, relationship and alleviation of, 196–198
 background, 189–191
 CAM, 191–194
 common myth, 194–195
 general healing effects, Jerome D Frank's model of, 194
 nonspecific and placebo effects, 199
 patients and mobilizing resources, 198–199
 power, insignia of, 198
 ritual, 195–196
'Self-regulation' system, 115
Sensations hypothesis, 26
Sensitive verbal instructions, 283
Sensory ambiguity, 39f
Sensory stimulus, 85–86
Scopolamine
 effects of, 16

hydrobromide effects, 16
Sham acupuncture, 239
 controlled trials, 5
 needle, 118
 'inertness' and blindness of, 165–166
 irrigation, 229
 Sham surgery, 228–229
Short duration pain stimuli, 206–207
Siegel's theory, 104
Signal detection theory, 178
Signalled morphine injections, 104
Skin conductance responses (SCR), 74f
Social observation, in placebo analgesia, 143
Somatic focusing, 221
Somatic hyperalgesia
 IBS in, 215–216
 tonic peripheral impulse input, 216–218
Somatosensory activity, 93
Somatosensory evoked potentials (SEP), 49–50
Specific placebo analgesic responses, 11
Spinal cord, 53–54, 54f
 BOLD responses, 61–62, 62f, 64
 direct evidence, 61
 dorsal horn of, 55, 216, 216f
 mechanisms, 47–48
 nociceptive processing, 54–55
  descending control of, 54
  endogenous modulation of, 61–62
  modulation of, 63f, 64
 placebo analgesia and, 60–64
 size of, 62f
SPN. See Stimulus-preceding negativity (SPN)
Spontaneous remissions, 149
Spontaneous variation, of symptoms, 166
Startle reflexes, and skin conductance responses, 111
Stepwise regression analysis, 74–75
Stimulus-preceding negativity (SPN), 38–39, 39f, 42f
 LEPs, 38–39, 41
 measurement of, 41
Streitberger needle stimulation, 122–123, 165, 165f
Stress, 73–76, 78–79
Subgenual anterior cingulate cortex (sgACC), 132
Subsequent effective procedure, 141–143, 142f
Supernatural mechanism, 236
Supraspinal mechanism, hot-plate test, 17
Symptomatic relief, 149–150

T
Tail suspension test (TST), 19
TCM system. See Traditional Chinese Medicine (TCM) system
The powerful placebo: From Ancient Priest to Modern Physician (Shapiro and Shapiro), 115
Therapeutic aggravation, 195–196

Thermal grill illusion, 85
Thermal pain, 76
Three-armed design, 152
Top-down regulation, 26–27
Total mood disturbance (TMD), 28
Tract-finding algorithm, priori cortical and subcortical regions, 94
Traditional Chinese Medicine (TCM) system, 115, 189
Traditional medicine (TM), 189
Transcutaneous electrical nerve stimulation (TENS), 49, 49f, 118
Trauma spectrum response (TSR), 228
Treatment-modulating effects, 257
'Treatment as usual' (TAU), 166
Tricyclic antidepressant trials, placebo groups of, 128–129
Trinitrobenzene sulfonic acid (TNBS), 218
TST. See Tail suspension test (TST)

U
Unconditioned response (UR), 268
Unconditioned stimulus (US), 15–16, 37, 139–140
 in central nervous system, 104, 104f
 conditioning with
  increase in pain, 106–110
  individuals in pain, 105
Unsignalled morphine injections, 104
US Food and Drug Administration, 179

V
Verbal cues, 144f
Verbally-induced expectations, 26
Verbal suggestions, for pain relief, 207–208
Verum acupuncture
 with low and high expectancy, 121
 sham acupuncture treatment vs., 122
Visceral-evoked neural activity, 222
Visceral hyperalgesia
 IBS in, 215–216
 tonic peripheral impulse input, 216–218
Visual analog scale (VAS), 27, 57–58, 58f, 150
 intensity ratings, 32–33, 32f
 pain intensity, rectal placebo and rectal lidocaine scores on, 207–208, 208f
Voxel-based morphometry, 94

W
Waiting-list (WL) control
 disadvantages of, 167
 TAU and, 166
Wide dynamic range (WDR), 46
World Medical Association (WMA), 170

Z
Zelen design, 167, 167f

Color Plates

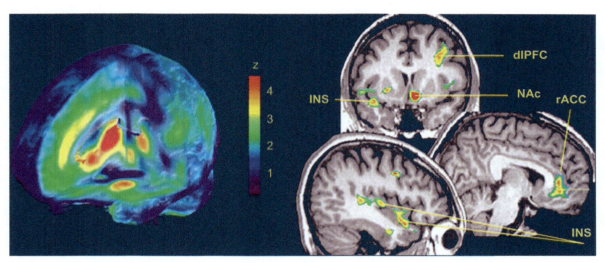

FIGURE 4.1 Placebo-induced activation of regional μ-opioid receptor-mediated neurotransmission. Left: distribution of μ-opioid receptors in the human brain, in a 3D rendering. Right: some of the areas in which significant activation of μ-opioid neurotransmission during sustained pain were observed after the introduction of a placebo with expectation of analgesia. INS: insula; dlPFC: dorsolateral prefrontal cortex; NAC: nucleus accumbens; rACC: rostral anterior cingulate cortex.

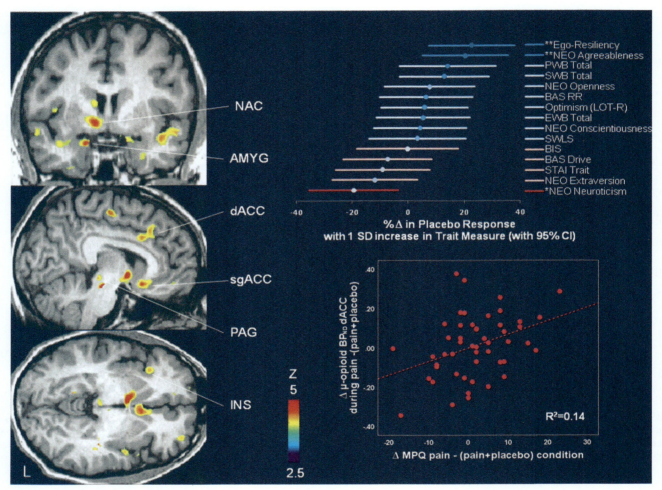

FIGURE 4.3 Personality traits effect on placebo-induced activation of regional μ-opioid receptor mediated neurotransmission. Left: Regions of greater μ-opioid system activation during placebo administration in subjects with high levels of Ego Resilience, Straightforwardness and Altruism and low levels of Angry Hostility. Upper right: Simple linear regression representing percent change in Placebo Response associated with 1 SD increase in Trait Measure (with 95% Confidence Intervals). (** indicates $p < 0.01$; * indicates $p < 0.05$). Lower right: Correlations between Δ μ-opioid BP_{ND} in the dACC during pain compared to (pain + placebo) and Δ in pain ratings (MPQ) during placebo administration. INS: insula; NAC: nucleus accumbens; r/sgACC: rostral and subgenual anterior cingulate cortex; AMYG: amygdala; PAG: periaqueductal gray.

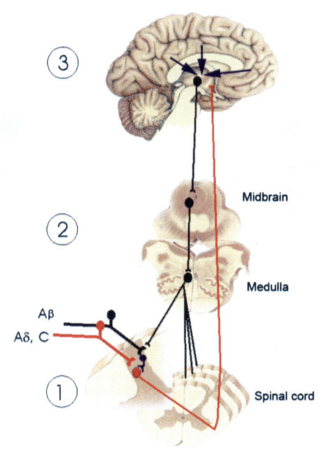

FIGURE 6.1 A schematic representation of the three main levels of endogenous pain modulation: (1) spinal inhibitory mechanisms, (2) inhibitory mechanisms descending from the brainstem, and (3) inhibitory mechanisms descending from higher centers. As described in the text, placebo and nocebo responses act by modulating these mechanisms and changing the spinal cord response to nociceptive activity.

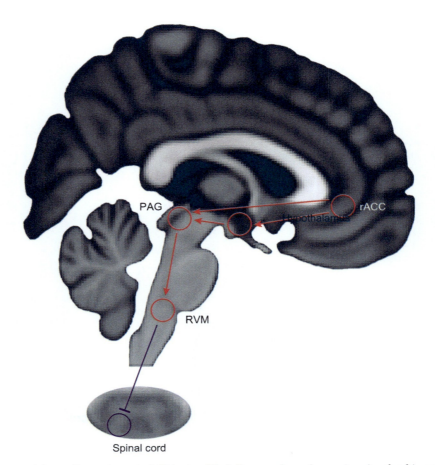

FIGURE 7.1 Neuroanatomy of descending pain control. This simplified diagram shows key regions involved in opioidergic descending pain control, as identified by both animal studies and human imaging studies. The endpoint of this system is the spinal cord, where nociceptive processing is inhibited by projections from the RVM (blue arrow). The RVM in turn receives a substantial input from the PAG, which is innervated by the hypothalamus as well as medial prefrontal regions (red arrows). Note that several connections (such as reciprocal ones) are omitted for the sake of clarity and that several non-midline regions (such as the amygdala) are not depicted. The sagittal T1-weighted brain section stems from the MNI152 brain, whereas the transversal T2*-weighted spinal cord section stems from a recent spinal cord study (Eippert et al, unpublished data). PAG: periaqueductal gray; rACC: rostral anterior cingulate cortex; RVM: rostral ventromedial medulla.

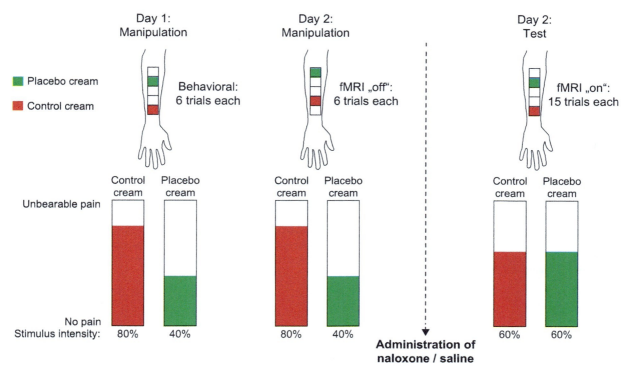

FIGURE 7.2 Experimental paradigm of pharmacologic fMRI study on placebo analgesia. Participants were recruited with the understanding that we were investigating the effects of a peripherally acting analgesic ('lidocaine' cream) on brain responses to noxious heat. The experiment took place on two consecutive days and consisted of three phases: manipulation day 1, manipulation day 2, and test day 2. Before each phase, subjects were treated with two identical creams on their left forearm and were told that one cream was a highly effective pain reliever, whereas the other served as sensory control. During the manipulation phases (which consisted of six trials under placebo cream and control cream, respectively), painful stimulation on the placebo-treated patch was surreptitiously lowered (from 80 [score on a visual analog scale (VAS)] under control to 40 under placebo) to convince the subjects that they had received a potent analgesic cream and to create expectations of future pain relief when treated with this cream. On day 2, the manipulation phase was carried out inside the (resting) MR scanner, to reactivate and strengthen the expectations of pain relief in this context. Before the test phase started, subjects either received an injection of saline or naloxone. fMRI data were collected during the test phase, which consisted of 15 trials under each condition. Importantly, during this phase the strength of painful stimulation was identical on both skin patches (60 on a VAS), in order to test for placebo analgesic effects. Note that in the spinal imaging study (see section 'Spinal fMRI of placebo analgesia'), we omitted the day 1 manipulation session—as subjects had participated in the previous study—and also did not administer any drugs. Reproduced and modified, with permission, from reference 128.

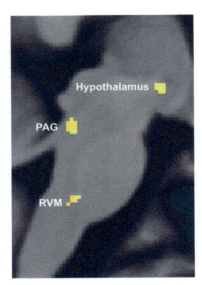

FIGURE 7.3 Placebo-induced midbrain and brainstem responses. This sagittal slice shows hypothalamus, PAG and RVM responses that were significantly stronger under placebo than under control; the response is overlaid on the group-averaged T1 image. Importantly, the responses in these key regions of descending pain control were significantly weaker under naloxone, indicating that these responses are opioid-dependent. All three structures furthermore showed responses that were correlated with the strength of the behavioral placebo effect. Reproduced and modified, with permission, from reference 128.

FIGURE 7.4 Brain and spinal cord size. Transversal slices through the brain (left) and the cervical spinal cord (middle) at the same scale show how minuscule the spine is in relation to the brain. The enlarged section (right) indicates that a standard in-plane voxel size of 3 × 3 mm would be much too coarse to image the spinal cord. Therefore, we used a 1 × 1 mm in-plane voxel size, which is more adequate to disentangle white and gray matter within the spinal cord, as well as to dissociate responses in the anterior–posterior and left–right dimensions. Note that due to the imaging sequence used, cerebrospinal fluid is black in the brain section and white in the spinal cord section, whereas gray matter is dark in the brain section and white in the spinal cord section.

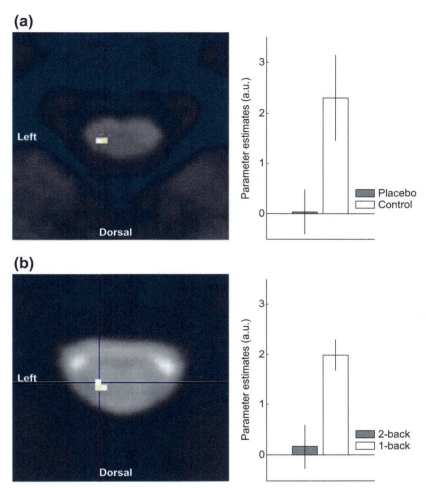

FIGURE 7.5 Modulation of nociceptive processing in the human spinal cord. (a) In the spinal fMRI study on placebo analgesia, we observed significant responses to the painful stimulation in the ipsilateral dorsal spinal cord (where nociceptive afferents terminate), as shown by the transversal section (level C6); the response is overlaid on the group-averaged T1 image. The group-averaged parameter estimates (reflecting the strength of activation) on the right were obtained from the voxel that exhibited the strongest response to pain and clearly show a significant reduction under placebo compared to control. (b) A similar result was observed in the spinal fMRI study on distraction, where pain-related responses in the ipsilateral dorsal spinal cord (at a location nearly identical to the one shown in panel (a)) were significantly reduced when participants where distracted from pain by high working memory load under the 2-back condition (see transversal section and parameter estimates); the response is overlaid on the group-averaged T2* image. Reproduced and modified, with permission, from references 195 and 205.

 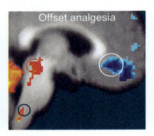 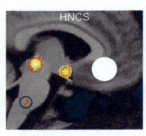 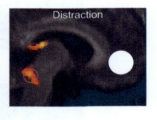

FIGURE 7.6 Brainstem and rACC responses in different forms of pain modulation. Key regions of the descending pain control system show responses in different forms of pain modulation, such as placebo,[128] offset analgesia,[222] heterotopic noxious conditioning stimulation (HNCS)[218] and distraction.[215] There is a general overlap of activated regions—most clearly seen in the PAG—but response locations obviously vary, as do the underlying mechanisms (for example, offset analgesia is mediated by non-opioidergic mechanisms,[223] while the other depicted forms of pain modulation do have an opiodergic component). Black circles indicate the RVM, red circles indicate the PAG, yellow circles indicate the hypothalamus and white circles indicate the rACC (filled circles indicate that these regions were used as seeds in functional connectivity analyses). Reproduced and modified, with permission, from references 128, 215, 218 and 222. Note that this figure has been reproduced with permission of the International Association for the Study of Pain® (IASP). The figure may not be reproduced for any other purpose without permission.

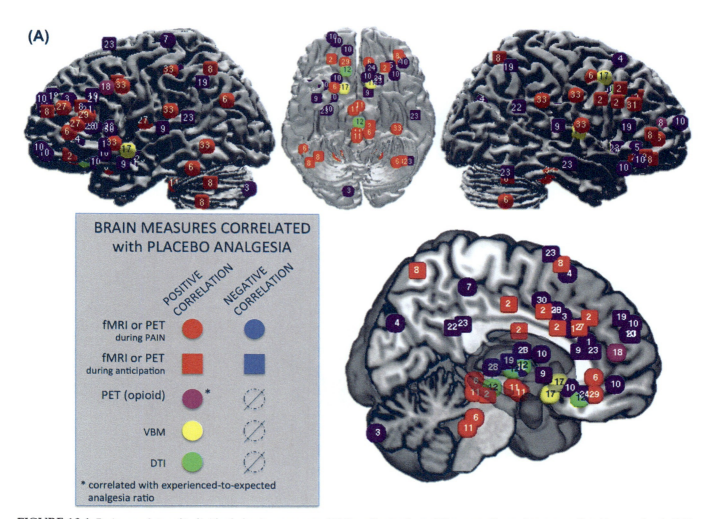

FIGURE 10.1 Brain correlates of individual placebo responses. (A) Coordinates from different studies and contrasts, listed by number in Table 10.2, at which brain measures are reported to correlate with individual differences in PA or to differ between groups of placebo responders and non-responders. Positive and negative correlations from fMRI and PET studies are shown in red and blue, respectively, with spheres and cubes for those points with correlating activity during pain and anticipation, respectively. Magenta spheres denote coordinates at which PET studies found increased opioid activity correlated with the ratio of experienced analgesia to expected analgesia. And locations of gray matter density (VBM) and white matter integrity (DTI) correlates (all positive) with placebo analgesia are marked with yellow and green spheres, respectively.

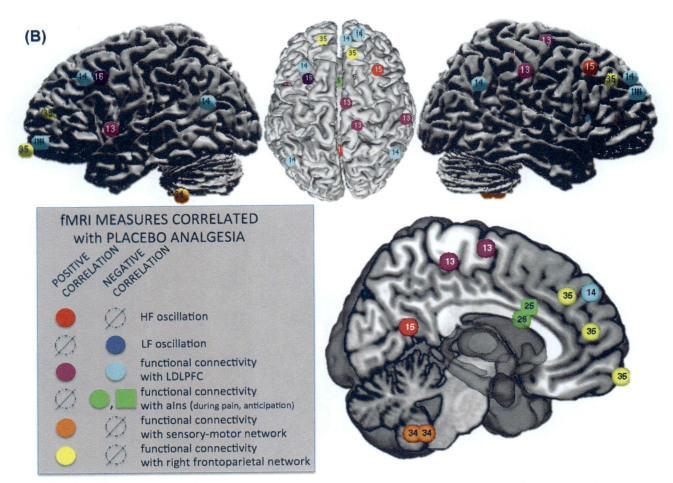

FIGURE 10.1 Cont'd (B) Coordinates at which oscillatory measures and functional connectivity findings from fMRI studies are reported to correlate with placebo analgesia (individual studies and contrasts are listed, by number, in Table 10.2). Included are high frequency band BOLD oscillations (HF) that were positively correlated with placebo analgesia (red); low frequency band BOLD oscillations (LF) that were negatively correlated with placebo analgesia (blue); locations where functional connectivity with left dorsolateral prefrontal cortex (LDLPFC) was positively (magenta) or negatively (cyan) correlated with placebo analgesia; locations where functional connectivity with anterior insula (aIns) during pain (green spheres) or anticipation (green cubes) was negatively correlated with placebo analgesia; locations where functional connectivity with the sensory-motor network was correlated positively with placebo analgesia (orange); and locations where functional connectivity with the right frontoparietal network was correlated positively with placebo analgesia (yellow).

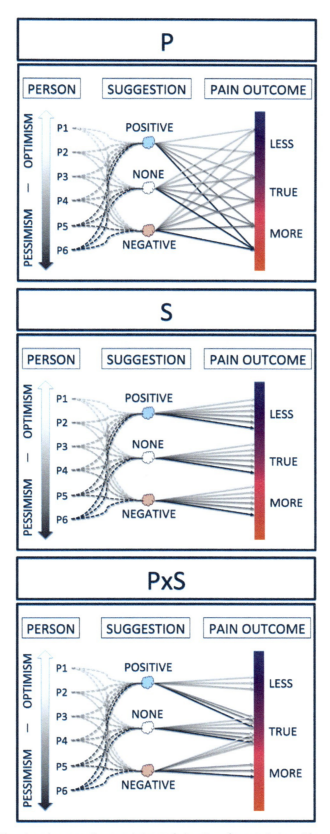

FIGURE 10.2 Illustration of possible roles of person characteristics and situation characteristics with respect to placebo response. Each panel shows diagrammatically, from left to right, a set of individuals with varying trait optimism/pessimism levels (an example of a person-level variable) experiencing a positive, negative, or neutral suggestion about treatment efficacy (a situational variable), and subsequently rating the intensity of a painful stimulus. In the three panels, the set of outcomes are determined by: (P) only the person-level variable; (S) only the situational variable; and (P × S) an interaction between the two variables. In the interaction case, analgesia is shown by the right kind of person in the right situation, and simple effects of each variable need neither exist nor be easily detectable if they do exist.

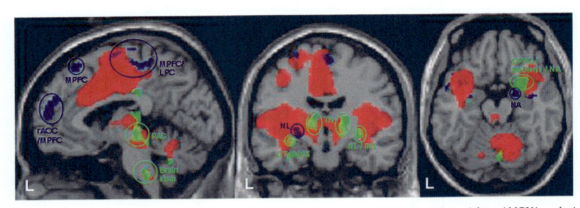

FIGURE 12.2 Representative brain regions involved in expectancy (blue) and acupuncture treatment (green) from ANOVA analysis across four groups.[44] The red color indicates the mask of high pain minus low pain across four groups. L indicates left side of the brain, R indicates right side of the brain. rACC: rostral anterior cingulate cortex; MPFC: medial prefrontal cortex; LPC: paracentral lobule; PAG: periaqueductal gray; NL: lentiform nucleus; INS: insula; OPFC: orbital prefrontal cortex; NA: amygdala.

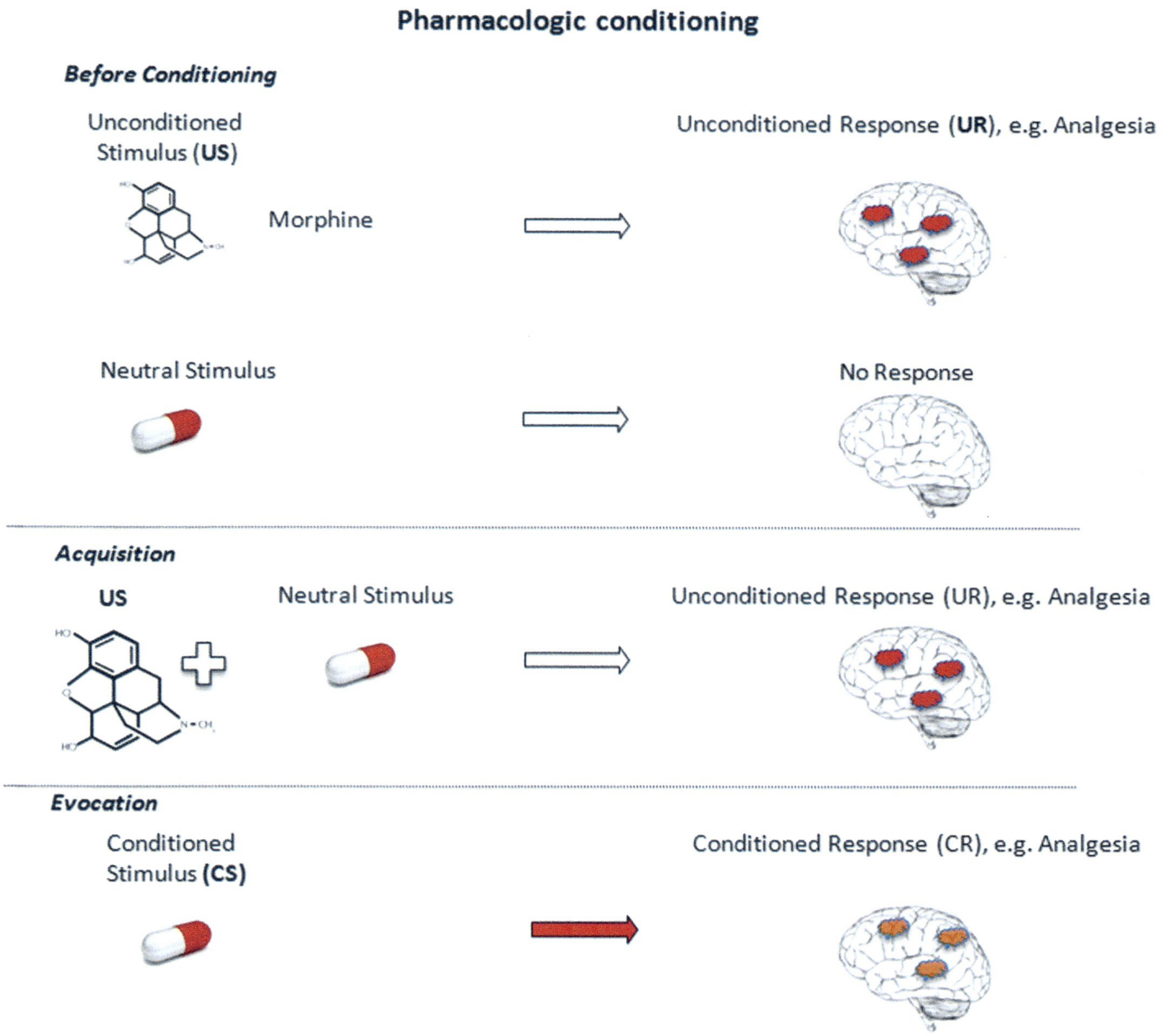

FIGURE 13.2 Pharmacologic conditioning. In the context of behavioral conditioning (B), the unconditioned stimulus (e.g. a pharmacologic agent) is inducing a response in the CNS (unconditioned response/UR); a neutral stimulus (e.g. environmental stimuli, an inert substance) is inducing no such response. During the acquisition phase, the neutral stimulus is paired with the unconditioned stimulus (US). After one or several pairings of the neutral stimulus with the US, the neutral stimulus becomes the conditioned stimulus (CS). During evocation, the CS is now able to mimic the effects formally induced by the US.

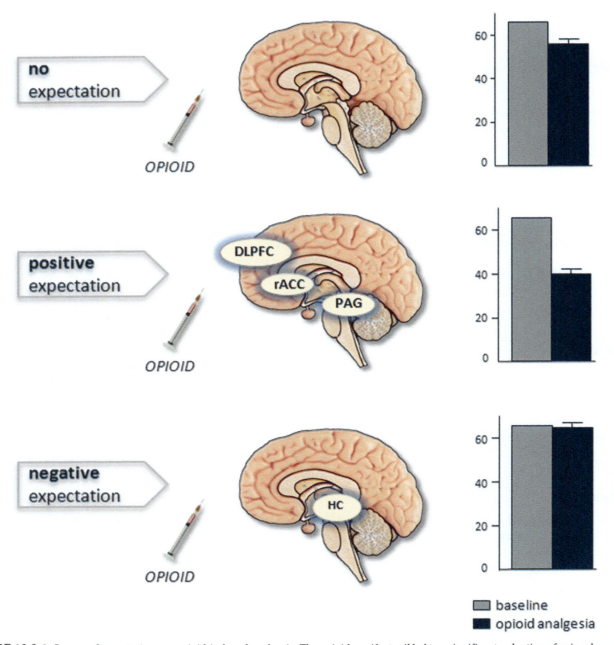

FIGURE 13.3 Influence of expectations on opioid-induced analgesia. The opioid remifentanil led to a significant reduction of pain when participants were not aware of the time-point of drug application, reflecting the pharmacologic effect with no expectations (top row). The analgesic effect was, however, doubled when participants were informed about application onset and expected a reduction of pain (middle row). Conversely, the drug effect was completely abolished in the condition where participants expected the drug to exacerbate pain (bottom row). (See Bingel et al[23] for details.) DLPFC, dorsolateral prefrontal cortex; rACC, rostral anterior cingulate cortex; PAG, periaqueductal gray; HC, hippocampus.

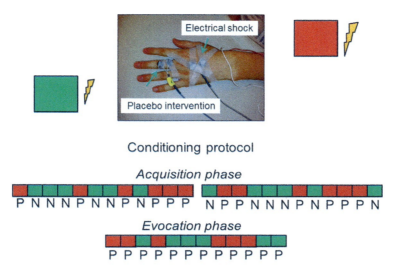

FIGURE 15.3 Experimental paradigm to investigate the role of conditioning and prior experience. Participants were informed that a green light displayed on a computer screen indicates activation of the electrode pasted on their middle finger which, in turn, would induce analgesia by virtue of a sub-threshold stimulation. Conversely, a red light indicates that the electrode is not activated, so that they would experience pain (which serves as control). The intensity of stimulation was manipulated to give the experience of analgesia in association with the presentation of a green light during the acquisition phase of conditioning (P=pain; N=no pain). No intensity manipulation was performed during the evocation phase when red and green lights were set at the control level of pain. Placebo responses were calculated as the differences between pain reports under red and green stimuli in the evocation phase. Adapted from Colloca and Benedetti (2006).[50]

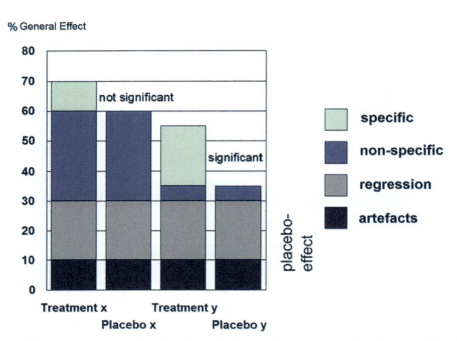

FIGURE 19.1 Efficacy paradox: an intervention with small and nonsignificant effects (treatment x) may be more effective overall than an efficacious one (treatment y); adapted from Walach.[59]

Printed and bound by CPI Group (UK) Ltd, Croydon, CR0 4YY

08/06/2025

01896880-0004